Strong Uniformity and Large Dynamical Systems

Strong Uniformity and Large Dynamical Systems

Jozsef Beck

Rutgers University, USA

World Scientific

NEW JERSEY • LONDON • SINGAPORE • BEIJING • SHANGHAI • HONG KONG • TAIPEI • CHENNAI • TOKYO

Published by

World Scientific Publishing Co. Pte. Ltd.
5 Toh Tuck Link, Singapore 596224
USA office: 27 Warren Street, Suite 401-402, Hackensack, NJ 07601
UK office: 57 Shelton Street, Covent Garden, London WC2H 9HE

Library of Congress Control Number: 2017017762

British Library Cataloguing-in-Publication Data
A catalogue record for this book is available from the British Library.

STRONG UNIFORMITY AND LARGE DYNAMICAL SYSTEMS

ISBN 978-981-4740-74-6

Typeset by Stallion Press
Email: enquiries@stallionpress.com

Printed in Singapore

Preface

Why does the typical time evolution of a "large" mechanical system (i.e., a system with many degrees of freedom, like gas in a container), starting from "off-equilibrium", approach "equilibrium" in a short time, and remain in "equilibrium" for a very, very long time? In what sense does "equilibrium" reflect "randomness"? Basically the same questions were raised in physics in the second half of the 19th century when Maxwell, Boltzmann and Gibbs developed the foundations of statistical mechanics. In this book we study the same general global questions about large (= many-particle) systems, but our approach is completely different from the well-known probabilistic machinery of statistical mechanics.

We also use probability theory, but it is not our primary tool. What we do is at the crossroads of number theory/uniform distribution and dynamical systems/quantitative ergodic theory. It is pure mathematics with rigorous proofs; nevertheless, we borrow some motivations and intuitions from physics.

Consider the following concrete (idealized) mechanical model that we may call "off-equilibriun ideal gas" (or off-equilibrium Bernoulli model of gases). Assume that there are N particles moving around in a cubic container, bouncing back and forth on the walls like billiard balls. Let N be large — e.g., in the range of the Avogadro number, roughly 10^{24} — so the system imitates the motion of gas molecules in a box. Assume that the time evolution of the system starts from an explicit *far-from-equilibrium* initial point configuration; say, "Big Bang" where all particles start from the same point, or something similar to Big Bang. The particles move on straight lines like point billiards until they hit a wall ("elastic collision"). Two typical point-particles in the 3-space do not collide; so, we assume that there is no particle–particle interaction. To determine the time evolution of the system, we have to say something about the initial velocities of the particles. We consider the most important velocity distribution in physics: assume that the particles have 3-dimensional Gaussian (= normal) initial velocity distribution (physicists call it Maxwellian distribution). The initial point configuration is explicitly given (like Big Bang); the intial velocities of different particles are chosen independently — this defines a measure (in fact, a product measure, due to independence) that makes it possible to talk about the "typical" time evolution of this large billiards-in-a-box system. (This is just the simplest model; we also study far more general models that mimick "energy dispersal" on a microscopic level, see Sec. 10; and completely different "closed orbit systems", see Secs. 6–7, etc.)

We focus on the following global questions: In what precise sense does the typical time evolution of this large system (= off-equilibrium ideal gas) approach "equilibrium"? How fast is the approach to "equilibrium"? Does the system really remain in "equilibrium" for a very long time? In what precise sense does "equilibrium" exhibit "randomness"?

Here "equilibrium" means spatial equilibrium, since the Gaussian initial velocity distribution is already the equilibrium velocity distribution, as it was discovered by Maxwell.

Statistical mechanics has a complete theory for the probabilistic model of the *equilibrium* ideal gas, based on the partition function. Unfortunately it is not clear at all — to say the least — how to extend that theory for the non-equilibrium case. Especially that our model is mainly *deterministic*, due to the billiard orbits, and only partly random, due to the independent choice of velocities. This is why we wrote this book.

(Note that there is a large literature of studying the motion of a single particle, or a few particles, as part of the large system — like the Lorentz gas models and other scattering billiards (see e.g. in [Ch-Ma2006]). What we do here is completely different: we focus on the *whole* system, answering global questions.)

Here is a brief summary of Boltzmann's answer to the basic questions. According to Boltzmann, the first step is to switch from the 3-dimensional cubic container — that we like to call the "particle space" — to the $6N$-dimensional phase space (each particle has 3 space coordinates and 3 moments), where a single point represents the *microstate* of the whole N-particle system at a given time instant t. Boltzmann introduced the concept of *macrostate*: macrostates are the "observable states" — a macrostate is basically a "large set of microstates that look the same". Boltzmann's key insight was that the "equilibrium macrostate" must contain *vastly* more microstates than any "off-equilibrium macrostate". Thus, it is a reasonable expectation that a system starting from off-equilibrium — which represents an atypical microstate — evolves through macrostates occupying progressively larger volumes in phase space, and eventually reaching the "equilibrium macrostate". Boltzmann's explanation why the system remains in the "equilibrium macrostate" for a very long time was to combine the so-called *Probability Postulate* with the fact that the "equilibrium macrostate" represents the overwhelming majority of the phase space. Boltzmann's *Probability Postulate* states that "the larger the macrostate, the greater the probability of finding a microstate in it". And it is complemented with Boltzmann's classical definition that "the entropy of the system is the logarithm of the probability of its macrostate", carved on Boltzmann's gravestone. (In fact, "Boltzmann's entropy formula" was formulated by Planck. Note that in physics literature "probability" is often replaced by the "number of microstates that look the same for a macroscopic observer", which is simply the frequency interpretation of probability.)

Well, this is a great insight/intuition. Many physicists find Boltzmann's argument a perfectly convincing explanation that settles the issue. Mathematicians, on the other hand, point out that Boltzmann's argument is nowhere near to a mathematical proof, and call it a framework, a first step toward the solution. This is what the distinguished mathematician A.I. Khinchin wrote in 1943 in his book *Mathematical Foundations of Statistical Mechanics*: "All existing attempts to give a general proof of this postulate (= Probability Postulate, including Boltzmann's definition of entropy) must be considered as an aggregate of logical and mathematical errors superimposed on a general confusion in the definition of the basic quantities"; see p. 142 in the English translation [Kh49]. Unfortunately most physics textbooks and lectures repeat the old intuition, claiming that it is well-established — there is no need for any rigorous mathematical theorem. (This is quite troubling, since many physicists consider Boltzmann's formula *Entropy* $= k \log W$ the second most important formula in physics after Einstein's $E = mc^2$.) The low level of mathematics in general, or the lack of it, was our main motivation to write this book.

The first logical difficulty in Boltzmann's argument is that in physics macrostates are well-defined only in equilibrium. When the system is far from equilibrium, it is not clear at all how to define macrostates. (This is why we do not use the concept of macrostate at all; in this book "state of a system" always means a microstate, which is simply *all positions and all velocities at a time instant.*)

The second difficulty is that it basically ignores the dynamical aspect. To put it in a nutshell: if a system is in an atypical microstate, it does *not* automatically evolve into an equilibrium macrostate *just because* the latter is typical! Indeed, for a simple pure mathematical illustration consider the sequence $T_n(x) = x^{(n+2)/2}$ for $n \geq 0$, i.e., $x, x^{3/2}, x^2, x^{5/2}, x^3, \ldots$, which represents the "time evolution" of a real number $x > 0$. If $x \geq 2$ is a non-square integer, the members of this sequence are alternately integers and irrational numbers. The integers represent a negligible (countable) subset of the real numbers, so it is plausible to call the integers "atypical" and the irrationals "typical". So if $x \geq 2$ is a non-square integer, in the corresponding sequence the "atypical" members have the same density as the "typical" members. In other words, starting from an "atypical microstate" (= non-square integer), the sequence does *not* evolve into the "majority macrostate" of "typical microstates", i.e., the set of irrationals — instead, it oscillates between the typical and atypical microstates.

A second example comes from the "time evolution" of human society, which contradicts Boltzmann's argument. Indeed, a typical country consists of a usually small upper class ("rich people"), a usually not so small middle class, and a large lower class ("poor people"); see for example India. If a system is a "large family" (in a broad sense), and we define "entropy" as the negative of the total family fortune (e.g., the negative of the total bank accounts, using modern financial terms), then Boltzmann's argument would predict entropy increase in the form that an upper class family (a small minority) would quickly turn into a middle class family, and later would quickly become a lower class family (like the overwhelming majority of families), since to be poor is "typical". Nevertheless, as history teaches us, upper class families do not evolve: they usually remain upper class for a long time; there is no entropy increase; there is no fast approach to equilibrium.

These two examples demonstrate the absurdity of Boltzmann's argument in general *without* any extra condition. So, we definitely need some extra condition.

To solidify Boltzmann's argument, we have to identify properties of the dynamics of the system that guarantee the evolution of how atypical (= unlikely) microstates evolve into typical (= very likely) microstates. We have to answer the question "why does a probability argument work for the short-time dynamics of the system?" It means to justify the *Probability Postulate* on a realistic time scale, i.e., to justify the following kind of approximation "phase-space average"$\approx$"short-time average" in a quantitative form. We may call it the "short-time ergodic problem".

We may summarize this long book in one sentence: the typical time evolution of a large off-equilibrium system of non-interacting particles represents a "kind of" uniquely ergodic time-flow in the configuration space, and to justify the Probability Postulate, we solve the "short-time ergodic problem" by proving, and repeatedly applying, "short-time ergodic theorems". (Of course we will clarify what "kind of" means.)

Perhaps this surprises the reader, since at first sight traditional ergodic theory seems to be the perfect tool to handle this kind of problem. So, the reader is probably wondering: why do we need new "short-time ergodic theorems"? Before addressing this question, here we briefly elaborate on ergodicity, including a short historic sketch. The first "ergodic hypothesis" was stated by Boltzmann, who also was the first to use the terminology. The "naive ergodic hypothesis" of Boltzmann stated that each surface of constant energy in the phase space consists of a single trajectory of the system. It was found out soon that the naive hypothesis was false, since no curve with well-defined arclength can fill out a whole many dimensional surface of constant energy (a curve with well-defined arclength has zero 2-dimensional Lebesgue measure).

After this failed attempt, Boltzmann replaced the naive hypothesis with the "quasi-ergodic hypothesis", according to which every trajectory of the system, although not filling completely out the energy surface on which it is situated, constitutes at least a *dense* subset of the surface. That is, it gets arbitrarily close to every point of the surface. However, the quasi-ergodic hypothesis turned out to be weak to derive the coincidence of the time and the phase-space averages, so it was eventually rejected.

A mathematical breakthrough came from the proofs of the first so-called "ergodic theorems" (J. von Neumann and G.D. Birkhoff), which started the subject of "ergodic theory"; see e.g. the books [Wal82] and [C-F-S82].

Let us return now to the "short-time ergodic problem", and answer the question of "how come traditional ergodic theory, and in particular Birkhoff's theorem, does not solve the *short-time ergodic problem*?" Indeed, the message of Birkhoff's well-known individual ergodic theorem is exactly the equality "phase-space average" = "asymptotic time average". Well, the first problem with traditional ergodic theory is that "asymptotic time average" means to take the infinite time limit (i.e., $t \to \infty$), and because Birkhoff's theorem does not give any estimation on the error term, it does not say anything about the realistic time scale.

The second problem is that (traditional measure-theoretic) ergodic theory ignores zero measure sets, and a fixed initial point configuration (Big Bang, say) represents a zero set in the phase space.

To solve the "short-time ergodic problem" we do not use traditional ergodic theory: we develop our own tools (mostly) from scratch. Our key concept is "strong (and robust) uniformity in the configuration space" that we briefly explain here as follows. First note that in our models there is no particle–particle interaction — this is why it suffices to study the $3N$-dimensional configuration space instead of the $6N$-dimensional phase space. Note that "robust uniformity" actually means "dimension-free, complexity-free and start-free strong uniformity". The property of "dimension-free" is absolutely crucial here, since the dimension of the configuration space is *extremely high*. Moreover, "complexity-free strong uniformity" means that our theorems hold for arbitrarily complicated measurable "test sets" in the configuration space. This is equally important, since as it turns out, even the simplest properties of "equilibrium" and "randomness" in the particle space (= gas container) are represented by *extremely* complicated "test sets" in the configuration space (to be explained later). Finally, "start-free" is obviously needed, because we want to prove theorems that hold for *arbitrary* fixed initial point configurations, representing a concrete off-equilibrium configuration; e.g., Big Bang. We give the detailed precise definitions later. (We may say, the starting point of this book was a result in Uniform Distribution: our solution of the continuous version of Khinchin's strong uniformity conjecture; see Sec. 2.)

Another shortcoming of Boltzmann's argument is that it is basically "soft"; so, it needs a "hard" quantitative upgrading. We prove quantitative theorems, which guarantee — under mild condition — that the typical time evolution of the system exhibits robust uniformity in a *short time*. The main part of this book is about the applications of these "short-time ergodic theorems" for large classes of "kind of" uniquely ergodic time-flows in the configuration space. These applications can be summarized in four sentences as follows.

From Non-Equilbrium to Equilibrium and Randomness, and beyond. *The family of time evolutions of a large system with noninteracting particles, starting from off-equilibrium, exhibits robust uniformity in a short time. It means that the typical time evolution quickly reaches configuration space equilibrium, demonstrating "advanced randomness", and the system stays in this state for a very, very long time (long-term stability). By using our "short-time ergodic theorems", we prove both static and dynamic "advanced randomness"; we call them, respectively, snapshot randomness and time-lapse randomness.*

These applications of our "short-time ergodic theorems" for large classes of "kind of" uniquely ergodic systems explain why the typical time evolution of a large system, starting from off-equilibrium, quickly reaches equilibrium in such a way that the laws of equilibrium statistical mechanics "set in" instantly, and hold on for a very, very long time.

We complete our nutshell summary of this book by the following diagram:

Robust Uniformity

= Dimension-Free, Start-Free and Complexity-

Free Strong Uniformity ⟶ Short-Time Ergodic Theorems

⟶ Short Time Average = Configuration Space Average

± negligible error, which holds independently of the initial

configuration (= unique ergodicity)

⟶ fast approach to Configuration Space Equilibrium

⟶ long term stability of different forms of equilibria in the

particle space;

Square Root Fluctuation Equilibrium, CLT and Poisson Snapshot Equilibrium, where CLT stands for the central limit theorem.

The subject of this book is on the borderline of number theory (in fact, uniform distribution), dynamical systems (dynamical systems of many non-interacting particles, mimicking off-equilibrium ideal gas and "energy spreading" in liquids and solids), and ergodic theory (unique ergodicity on the high-dimensional torus). Also, the extremely high dimension of the configuration space requires the application of nontrivial combinatorics. We do not study hyperbolic systems; nevertheless, our models still exhibit advanced randomness — snapshot randomness and time-lapse randomness. And everything happens on a realistic time scale! The message is that we do not necessarily need hyperbolicity to demonstrate quick approach to advanced randomness.

Our main tools are Fourier analysis and combinatorics; we also apply some classical results from probability theory.

This book is written, above all, for the mathematician, but because the motivation comes from physics, hopefully some physicists — especially those who are working in non-equilibrium physics — will also find it useful.

The notation is rather standard. We use log for the natural (base e) logarithm, use $\log_2$ for the binary (base 2) logarithm, and for a (usually finite) set A we use $|A|$ to denote the number of elements of A. For simplicity the d-dimensional Lebesgue measure in the familiar low dimensions $d = 1$, 2, and 3 is called, respectively, *length*, *area*, and *volume*. The d-dimensional Lebesgue measure with $d > 3$ is usually denoted by vol_d. We use $\lfloor x \rfloor$, $\lceil x \rceil$, and $\{x\}$ for the lower integral part, the upper integral part, and the fractional part of a real number x in this order (so $x = \lfloor x \rfloor + \{x\}$), and we frequently use $\|x\|$ to denote the distance of x from the nearest integer. We often say "x modulo one", which (usually) means the fractional part $\{x\}$.

Contents

Chapter 1

From Uniform Distribution to the Time-Evolution of Large Off-Equilibrium Systems

1. Traditional Uniform Distribution and Weyl's Criterion

We can briefly summarize the book by the following diagram

$$\begin{array}{ccc} \text{Uniform Distribution} & \Longrightarrow & \text{Short-Time evolution of large systems} \\ \downarrow & & \uparrow \\ \text{Strong (and Robust) Uniformity} & \longrightarrow & \text{Short-Time Ergodic Theorems} \end{array}$$

To explain what the diagram means, we have to begin with Uniform Distribution (see Secs. 1 and 2) and the high-dimensional configuration space of a large system (see Sec. 3).

The single most important theorem in Uniform Distribution is (arguably) the Kronecker–Weyl equidistribution theorem, which is an upgrading of Kronecker's density theorem. Kronecker's density theorem answers a basic question raised by Dirichlet's simultaneous diophantine approximation theorem, and Dirichlet's theorem is historically the first *recurrence* type result (i.e., the first weak ergodic type result). So, the starting point is in fact Dirichlet's classical theorem, which goes as follows. Let $d \geq 1$ be an arbitrary integer, let $\alpha_1, \ldots, \alpha_d$ be an arbitrary sequence of d real numbers, let N_0 be an arbitrarily large but fixed integer, and let $\varepsilon > 0$ be an arbitrarily small but fixed real. Then there are infinitely many positive integers $n_0 = n_0(\alpha_1, \ldots, \alpha_d; \varepsilon)$ such that

$$\|n_0\alpha_j\| < \varepsilon \quad \text{for every} \quad 1 \leq j \leq d, \tag{1.1}$$

where $\|x\|$ denotes the distance of a real x from the nearest integer. The well-known pigeonhole principle proof gives the quantitative upper bound

$$n_0 \leq (\lceil 1/\varepsilon \rceil)^d \tag{1.2}$$

for the smallest positive integer n_0 satisfying (1.1) (here, as usual, $\lceil x \rceil$ denotes the upper integral part of a real). The qualitative form of Dirichlet's theorem is equivalent to the following recurrence type statement: the infinite arithmetic progression $n\mathbf{a} = (n\alpha_1, \ldots, n\alpha_d)$, $n \geq 0$, starting from the origin $\mathbf{0} \in \mathbb{R}^d$ of the d-space, and taken modulo one, returns arbitrarily close to the origin.

A far-reaching generalization of the recurrence phenomenon was formulated by Poincaré. Let U be a measure-preserving transformation of a probability space $(\Omega, \mathcal{A}, \mu)$ (i.e., $\mu(\Omega) = 1$). If $A \in \mathcal{A}$ is an arbitrary measurable subset of Ω with positive measure, $\mu(A) > 0$, then there is some point $x \in A$ and an integer $n \geq 1$ with $U^n x \in A$. In fact, "almost every" $x \in A$ is good in the sense that the set of bad x's has μ-measure zero. By choosing Ω to be the d-dimensional unit torus $[0,1)^d$, $U = U(\mathbf{a})$ to be the translation by the vector $\mathbf{a} = (\alpha_1, \ldots, \alpha_d)$, and A to be a small ball centered at the origin, Poincaré's Recurrence Theorem — a forerunner of the ergodic theorem — clearly implies the qualitative form of Dirichlet's theorem.

The next big step is to move from Recurrence to Density. Let $\mathbf{b} = (\beta_1, \ldots, \beta_d) \in [0,1)^d$ be an arbitrary point in the d-dimensional unit cube. Under what condition can we guarantee that the infinite arithmetic progression $n\mathbf{a} = (n\alpha_1, \ldots, n\alpha_d)$, $n \geq 0$, taken modulo one (i.e., we take the fractional parts of the coordinates), gets arbitrarily close to $\mathbf{b}$? A simple sufficient and necessary condition was discovered by Kronecker in 1884. Kronecker's famous theorem says that, the infinite arithmetic progression $n\mathbf{a} = (n\alpha_1, \ldots, n\alpha_d)$, $n \geq 0$, taken modulo one, is dense in the unit torus $[0,1)^d$ if and only if the $d+1$ real numbers $1, \alpha_1, \ldots, \alpha_d$ are linearly independent over the rationals. The latter means that

$$k_1\alpha_1 + \cdots + k_d\alpha_d = k_0, \quad k_j \in \mathbb{Z} \text{ implies } k_j = 0,\ 0 \leq j \leq d.$$

The continuous version of Kronecker's theorem is about the straight line $t\mathbf{a}$, $t \geq 0$ modulo one. The continuous version says that $t\mathbf{a}$, $t \geq 0$ modulo one (i.e., the torus line starting from the origin) is dense in the unit torus $[0,1)^d$ if and only if $\alpha_1, \ldots, \alpha_d$ are independent over the rationals. (Note that $[0,1)^d$ can be interpreted in two slightly different ways: (1) the half-open unit cube, and (2) the unit torus. We often mix up the two interpretations; we hope this minor ambiguity does not confuse the reader.)

Since we apply the continuous Kronecker theorem in Sec. 3 (see Theorem 3.1), we include the following beautiful proof, due to Bohr. The basic idea is shockingly simple: It suffices to show that, under the condition of linear independence of $\alpha_1, \ldots, \alpha_d$, the complex exponential sum (where of course $\mathrm{i} = \sqrt{-1}$)

$$1 + \sum_{j=1}^{d} e^{2\pi \mathrm{i}(t\alpha_j - \beta_j)} \quad \text{gets arbitrarily close to } d+1. \tag{1.3}$$

To prove (1.3), we just take a very high power of the left-hand side sum in (1.3), evaluate a related integral defined over a very long interval $0 \le t \le T$, and (1.3) follows via routine calculations.

The first step in the proof of (1.3) is to apply the multinomial theorem

$$\begin{aligned}
&\left(1 + \sum_{j=1}^{d} \exp\left(2\pi \mathrm{i}(t\alpha_j - \beta_j)\right)\right)^N \\
&= \sum_{\substack{(k_0,k_1,\ldots,k_d)\in\mathbb{Z}^{d+1}:\\ k_j\ge 0, \sum_{j=0}^{d} k_j = N}} \frac{N!}{k_0! k_1! \cdots k_d!} 1^{k_0} \exp\left(2\pi \mathrm{i} \sum_{j=1}^{d} k_j (t\alpha_j - \beta_j)\right) \\
&= \sum_{\substack{(k_0,k_1,\ldots,k_d)\in\mathbb{Z}^{d+1}:\\ k_j\ge 0, \sum_{j=0}^{d} k_j = N}} \frac{N!}{k_0! k_1! \cdots k_d!} \exp\left(-2\pi \mathrm{i} \sum_{j=1}^{d} k_j \beta_j\right) \\
&\quad \times \exp\left(2\pi \mathrm{i} t \sum_{j=1}^{d} k_j \alpha_j\right).
\end{aligned} \tag{1.4}$$

Notice that the condition of linear independence of $\alpha_1, \ldots, \alpha_d$ implies the following "irreducibility" property of the sum in the last line of (1.4): if $(k_0^{(\ell)}, k_1^{(\ell)}, \ldots, k_d^{(\ell)}) \in \mathbb{Z}^{d+1}$, $\ell = 1, 2$ are two distinct $(d+1)$-dimensional integral vectors such that

$$k_j^{(\ell)} \ge 0 \quad \text{and} \quad \sum_{j=0}^{d} k_j^{(\ell)} = N \text{ for both } \ell = 1, 2,$$

then

$$\sum_{j=1}^{d} k_j^{(1)} \alpha_j \ne \sum_{j=1}^{d} k_j^{(2)} \alpha_j. \tag{1.5}$$

It is well known from elementary combinatorics that the multinomial theorem in (1.4) has $\binom{N+d}{d}$ different multinomial coefficients, and by the "irreducibility" property (1.5) we cannot reduce the last line in (1.4) to a sum of less than $\binom{N+d}{d}$ terms.

Let

$$C^* = \max_{\substack{(k_0,k_1,\ldots,k_d)\in\mathbb{Z}^{d+1}:\\ k_j\geq 0,\sum_{j=0}^{d} k_j=N}} \frac{N!}{k_0!k_1!\cdots k_d!} \tag{1.6}$$

denote the largest multinomial coefficient in the last line of (1.4).

Clearly

$$C^* = \text{Largest Coefficient} \geq \text{Average Coefficient} = \frac{(d+1)^N}{\binom{N+d}{d}}. \tag{1.7}$$

We rewrite (1.4) in the short form

$$\left(1+\sum_{j=1}^{d} e^{2\pi \mathrm{i}(t\alpha_j-\beta_j)}\right)^N = \sum_{r=1}^{\binom{N+d}{d}} C_r e^{2\pi \mathrm{i}(t\gamma_r-\theta_r)}, \tag{1.8}$$

where the γ_r's are all different and (see (1.6) and (1.7))

$$\max_{1\leq r\leq \binom{N+d}{d}} C_r = C_{r_0} = C^* \geq \frac{(d+1)^N}{\binom{N+d}{d}}. \tag{1.9}$$

Next we consider the following integral related to (1.8)

$$\begin{aligned}
&\int_0^T \left(1+\sum_{j=1}^{d} e^{2\pi \mathrm{i}(t\alpha_j-\beta_j)}\right)^N e^{-2\pi \mathrm{i}(t\gamma_{r_0}-\theta_{r_0})}\,dt \\
&= \int_0^T \left(\sum_{r=1}^{\binom{N+d}{d}} C_r e^{2\pi \mathrm{i}(t\gamma_r-\theta_r)}\right) e^{-2\pi \mathrm{i}(t\gamma_{r_0}-\theta_{r_0})}\,dt \\
&= TC_{r_0} + \sum_{\substack{1\leq r\leq \binom{N+d}{d}:\\ r\neq r_0}} C_r \int_0^T e^{2\pi \mathrm{i}\left(t(\gamma_r-\gamma_{r_0})-(\theta_r-\theta_{r_0})\right)}\,dt.
\end{aligned} \tag{1.10}$$

We use the elementary fact from calculus that for an arbitrary real $c \neq 0$,

$$\frac{1}{T}\int_0^T e^{2\pi i ct}\,dt = \frac{e^{2\pi i cT}-1}{2\pi i cT},$$

which implies

$$\left|\frac{1}{T}\int_0^T e^{2\pi i ct}\,dt\right| \leq \frac{1}{\pi|c|T}. \tag{1.11}$$

Combining (1.10) and (1.11) we obtain that

$$\left|\frac{1}{T}\int_0^T \left(1+\sum_{j=1}^{d} e^{2\pi i(t\alpha_j-\beta_j)}\right)^N e^{-2\pi i(t\gamma_{r_0}-\theta_{r_0})}\,dt - C_{r_0}\right| \leq \frac{1}{2}C_{r_0} \tag{1.12}$$

for all $T \geq T^*$, where T^* is a sufficiently large threshold.

By taking the limit $T \to \infty$ in (1.12), and using (1.9),

$$\begin{aligned}
&\limsup_{t\to\infty}\left|1+\sum_{j=1}^{d} e^{2\pi i(t\alpha_j-\beta_j)}\right| \\
&\geq \limsup_{T\to\infty}\left(\left|\frac{1}{T}\int_0^T \left(1+\sum_{j=1}^{d} e^{2\pi i(t\alpha_j-\beta_j)}\right)^N e^{-2\pi i(t\gamma_{r_0}-\theta_{r_0})}\,dt\right|\right)^{1/N} \\
&\geq \left(\frac{1}{2}C_{r_0}\right)^{1/N} = \left(\frac{1}{2}C^*\right)^{1/N} \geq \left(\frac{(d+1)^N}{2\binom{N+d}{d}}\right)^{1/N}.
\end{aligned} \tag{1.13}$$

Next we take the limit $N \to \infty$; then (1.13) implies

$$\begin{aligned}
&\limsup_{t\to\infty}\left|1+\sum_{j=1}^{d} e^{2\pi i(t\alpha_j-\beta_j)}\right| \geq \lim_{N\to\infty}\left(\frac{(d+1)^N}{2\binom{N+d}{d}}\right)^{1/N} \\
&\geq \lim_{N\to\infty}\left(\frac{(d+1)^N}{2(N+d)^d}\right)^{1/N} = \frac{d+1}{\lim_{N\to\infty} 2^{1/N}(N+d)^{d/N}} = d+1.
\end{aligned} \tag{1.14}$$

It follows from (1.14) that, given any $T_0 > 0$ and $\varepsilon > 0$, there exists a $t > T_0$ such that

$$\left|1+\sum_{j=1}^{d} e^{2\pi \mathrm{i}(t\alpha_j-\beta_j)}-(d+1)\right| \le \varepsilon,$$

which completes the proof of the continuous version of Kronecker's theorem. □

Note that the continuous form of Kronecker's theorem easily implies the discrete form. Indeed, we can assume, without loss of generality, that in the discrete form $\alpha_1, \dots, \alpha_d$ are all between 0 and 1. Let $N_0 > 0$ be arbitrarily large but fixed. Applying the continuous form for the $d+1$ linearly independent numbers $1, \alpha_1, \dots, \alpha_d$, there exists a real $t_0 > N_0 + 1$ such that

$$\|t_0\| < \frac{1}{2}\varepsilon,\ \|t_0\alpha_j - \beta_j\| < \frac{1}{2}\varepsilon,\ 1 \le j \le d. \tag{1.15}$$

Let n_0 denote the nearest integer to t_0. Then by (1.15), $n_0 > N_0$, and by the triangle inequality

$$\|n_0\alpha_j - \beta_j\| \le \|t_0\alpha_j - \beta_j\| + |t_0 - n_0| < \frac{1}{2}\varepsilon + \frac{1}{2}\varepsilon = \varepsilon,\ 1 \le j \le d,$$

completing the proof of the discrete version of Kronecker's theorem. □

In the original form of Kronecker's theorem the starting point was the origin. Since the torus is translation invariant, Kronecker's theorem remains true if we replace the origin with any other starting point in the torus.

The next big step is to upgrade Kronecker's Density Theorem to Uniformity. This was done by H. Weyl, who in the process created — almost single-handedly — the theory of Uniform Distribution (see [We16], and also the books [Ku-Ni74] and [Dr-Ti97]). Weyl introduced the following basic definition: an infinite sequence of points $\mathbf{x}_1, \mathbf{x}_2, \mathbf{x}_3, \dots$ in the d-space $\mathbb{R}^d$ is said to be *uniformly distributed modulo one* if for every axis-parallel box (Cartesian product of intervals) in the unit cube $B = I_1 \times \cdots \times I_d \subset [0,1)^d$

$$\lim_{N\to\infty} \frac{1}{N} \sum_{\substack{1\le j\le N: \\ \mathbf{x}_j\in B \text{ modulo one}}} 1 = \mathrm{vol}_d(B) = \prod_{k=1}^{d} \mathrm{length}(I_k), \tag{1.16}$$

where vol_d denotes the d-dimensional volume.

The proof of Kronecker's theorem, using complex exponential sums, already demonstrated the power of Fourier analysis. Weyl formulated 3 criteria for sequences uniformly distributed modulo one. The most important is criterion (c), which emphasizes the close relation between Uniform Distribution and Fourier Analysis. (Fourier Analysis remains a key tool throughout the book.)

Weyl's criterion. *An infinite sequence of points* $\mathbf{x}_1, \mathbf{x}_2, \mathbf{x}_3, \ldots$ *in the d-space* $\mathbb{R}^d$ *is uniformly distributed modulo one if and only if any of the following three criteria holds*:

(a) *for all real-valued Riemann-integrable functions* $f : I^d \to \mathbb{R}$ *with* $I^d = [0,1]^d$, *the corresponding Riemann sums converge to the integral*

$$\lim_{N\to\infty} \frac{1}{N} \sum_{1\le j\le N} f(\{\mathbf{x}_j\}) = \int_{I^d} f\, d\mathrm{V}; \tag{1.17}$$

(b) *for all real-valued continuous functions* $f : I^d \to \mathbb{R}$ (1.17) *holds*;
(c) *for all complex exponential functions*

$$f(\mathbf{y}) = f_{\mathbf{n}}(\mathbf{y}) = e^{2\pi \mathrm{i} \mathbf{n}\cdot\mathbf{y}}, \quad \mathbf{n} \in \mathbb{Z}^d \setminus \mathbf{0}$$

(1.17) *holds.*

Here $\{\mathbf{z}\} = (\{z_1\}, \ldots, \{z_d\})$ *denotes the vector in* $[0,1)^d$ *formed by the fractional parts of the coordinates of a vector* $\mathbf{z} \in \mathbb{R}^d$, $d\mathrm{V}$ *stands for integration with respect to the d-dimensional Lebesgue measure,* $\mathrm{i} = \sqrt{-1}$, *and finally* $\mathbf{n} \cdot \mathbf{y} = n_1 y_1 + \cdots + n_d y_d$ *stands for the usual dot product of vectors.*

The message of Weyl's criterion is that there are 4 equivalent ways to characterize the infinite sequences of points that are uniformly distributed modulo one (the 4 equivalent ways are (1.16), (a), (b) and (c)).

The proof of Weyl's criterion is based on a subtle approximation argument, where the key ingredient is Weierstrass's well-known approximation theorem with trigonometric sums (for any $\varepsilon > 0$, every continuous 1-periodic function can be uniformly ε-approximated by a finite sum of complex exponential functions $f_{\mathbf{n}}(\mathbf{y}) = e^{2\pi \mathrm{i}\mathbf{n}\cdot\mathbf{y}}$, $\mathbf{n} \in \mathbb{Z}^d$). For the details of the proof; see e.g. Drmota–Tichy [Dr-Ti97].

It is easy to reformulate Weyl's criterion (c) in the following more familiar form: an infinite sequence of points $\mathbf{x}_1, \mathbf{x}_2, \mathbf{x}_3, \ldots$ in the d-space $\mathbb{R}^d$ is uniformly distributed modulo one if and only if

$$\lim_{N\to\infty}\frac{1}{N}\sum_{1\le j\le N} e^{2\pi i \mathbf{n}\cdot\mathbf{x}_j} = 0 \tag{1.18}$$

for all nonzero lattice points $\mathbf{n} \in \mathbb{Z}^d \setminus \mathbf{0}$. (Notice that (1.3) was also a complex exponential sum.)

It is the form (1.18) that most authors call *Weyl's criterion.*

The Weyl's criterion can be easily extended to the continuous case; it was done by Weyl himself. Let $\mathbf{r}(t) \in \mathbb{R}^d$ be an arbitrary continuous parametrized curve defined for all $0 \le t < \infty$. Choosing the analog of (1.17) as our definition, we say that the curve $\mathbf{r}(t) \in \mathbb{R}^d$ is *uniformly distributed modulo one* if for every Riemann integrable function $f = f(x_1, \ldots, x_d)$ defined on the unit torus $I^d = [0,1)^d$ we have

$$\lim_{T\to\infty}\frac{1}{T}\int_0^T f(\mathbf{r}(t))\,dt = \int_{I^d} f\,d\mathrm{V}. \tag{1.19}$$

In the special case of the characteristic function $f = \chi_S$, where $S \subset I^d$ is a Jordan measurable subset, the left-hand side of (1.19) is the frequency of the time t the parametrized curve $\mathbf{r}(t)$ — representing the motion of a particle — spends in the given subset S, and the right-hand side is the volume of S.

(Note that for the class of bounded sets in $\mathbb{R}^d$ the following 3 properties are equivalent: (1) S is Jordan measurable, (2) $f = \chi_S$ is Riemann integrable, (3) the boundary of S has d-dimensional Lebesgue measure zero.)

The curve version of Weyl's criterion is the perfect analog of the discrete case (1.18) : (1.19) is equivalent to

$$\lim_{T\to\infty}\frac{1}{T}\int_{t=0}^T e^{2\pi i \mathbf{n}\cdot\mathbf{r}(t)}\,dt = 0 \tag{1.20}$$

for all $\mathbf{n} \in \mathbb{Z}^d \setminus \mathbf{0}$.

Note that (1.19) is also equivalent to the alternative weaker requirement that (1.19) holds for every $f = \chi_B$, where B runs over the axis-parallel boxes in the unit cube $[0,1]^d$ (analog of (1.16)), and of course f is defined on the unit torus $I^d = [0,1)^d$.

A striking illustration of the power of Weyl's criterion is Weyl's upgrading of Kronecker's density theorem to uniform distribution modulo one; see (1.22) and (1.23) below. To give due credit to both mathematicians, we refer to it as the Kronecker–Weyl equidistribution theorem. The Kronecker–Weyl theorem has two parts: the continuous part and the discrete part, and they both work in every dimension. For simplicity we start with the 2-dimensional continuous case. Then the problem is to describe the distribution of a straight line in the plane modulo one, i.e., the distribution of a torus line in the unit torus, $[0,1)^2$. If the slope of the line is rational, then of course the torus line in the unit torus is periodic. On the other hand, if the slope is irrational (and this is the hard part), then the torus line is *dense* in the unit torus $[0,1)^2$ (proved by Kronecker in 1884), and this was upgraded in 1916 by H. Weyl to the more subtle property of *uniform distribution*.

One may argue that the Kronecker–Weyl equidistribution theorem is an "ergodic" theorem: it makes a precise statement about the equality of the time-average and the space-average. (There is, however, a crucial technical difference: the traditional measure-theoretic ergodic theorems are expressed in terms of the Lebesgue integral; the Kronecker–Weyl theorem, on the other hand, is restricted to the Riemann integral. It is true, however, that the Kronecker–Weyl theorem is the first, and arguably the most important, example of what we call now *unique ergodicity*. We will return to unique ergodicity in Sec. 10.)

Unlike Birkhoff's individual ergodic theorem, which does not have a quantitative form, the Kronecker–Weyl theorem has quantitative versions (see e.g. the Erdős–Turán–Koksma inequality in (3.21)).

The general d-dimensional continuous form of the Kronecker–Weyl equidistribution theorem goes as follows. Let $\alpha_1, \ldots, \alpha_d$ be arbitrary reals, and write $\mathbf{a} = (\alpha_1, \ldots, \alpha_d)$. Consider the straight line $t\mathbf{a}$, $-\infty < t < \infty$ in the d-space $\mathbb{R}^d$ passing through the origin $\mathbf{0}$; this straight line modulo one is a torus line in the d-dimensional unit torus $[0,1)^d$, $d \geq 2$. The continuous Kronecker–Weyl theorem says that, if $\alpha_1, \ldots, \alpha_d$ are linearly independent over the rationals, then the straight line $t\mathbf{a}$, $0 < t < \infty$ is uniformly distributed modulo one in the d-dimensional unit torus $[0,1)^d$. That is, for every Riemann integrable function f defined on the unit torus $I^d = [0,1)^d$,

$$\lim_{T\to\infty} \frac{1}{T} \int_0^T f(t\mathbf{a})\, dt = \int_{I^d} f\, d\mathrm{V}. \tag{1.21}$$

If $f = \chi_S$, where χ_S is the 0,1 valued characteristic function of a "not too ugly" subset $S \subset [0,1)^d$, then the integral on the right-hand side of (1.21) becomes the (d-dimensional) volume of S. Here the word "not too ugly" is a warning: S *cannot* be an arbitrary Lebesgue measurable subset of the d-dimensional unit cube, and in general, (1.21) is trivially false for Lebesgue integrable f (instead of Riemann integrable). The reason is very simple: the straight line $t\mathbf{a}$, $0 < t < \infty$ modulo one has d-dimensional Lebesgue measure zero (if $d \geq 2$), so by removing the torus line from an arbitrary test set $S \subset [0,1)^d$ of positive Lebesgue measure, we obtain a subset S_0 of S, which has the same positive Lebesgue measure as S, and the torus line does not intersect S_0 at all — this contradicts (1.21) with the choice $f = \chi_{S_0}$.

The precise meaning of "not too ugly" above is Jordan measurable.

It is important to point out that (1.21) remains true if the torus line $t\mathbf{a}$, $0 < t < \infty$ is replaced by any translated copy $\mathbf{w} + t\mathbf{a}$, $0 < t < \infty$. Indeed, we just use the function $g(\mathbf{x}) = f(\mathbf{x} - \mathbf{w})$ instead of f, and note that g and f have the same integral.

The beauty of the Weyl's criterion is that it implies the d-dimensional Kronecker–Weyl theorem in *one line*. Indeed, in the continuous case

$$\left| \int_0^T e^{2\pi i t \mathbf{n}\cdot\mathbf{a}}\, dt \right| = \left| \frac{e^{2\pi i T \mathbf{n}\cdot\mathbf{a}} - 1}{2\pi i \mathbf{n}\cdot\mathbf{a}} \right| \leq \frac{1}{\pi|\mathbf{n}\cdot\mathbf{a}|}, \tag{1.22}$$

and because $\mathbf{n}\cdot\mathbf{a} \neq 0$ (due to the linear independence of the coordinates of $\mathbf{a} = (\alpha_1, \ldots, \alpha_d)$ over the rationals), (1.20) applies, and the proof is complete.

In the discrete case the integral of the exponential function is replaced by a geometric series, so its evaluation is just as simple: again with $\mathbf{a} = (\alpha_1, \ldots, \alpha_d)$ we have

$$\left| \sum_{j=1}^N e^{2\pi i j \mathbf{n}\cdot\mathbf{a}} \right| = \left| \frac{e^{2\pi i N \mathbf{n}\cdot\mathbf{a}} - 1}{1 - e^{2\pi i \mathbf{n}\cdot\mathbf{a}}} \right| \leq \left| \frac{2}{1 - e^{2\pi i \mathbf{n}\cdot\mathbf{a}}} \right|, \tag{1.23}$$

and because $1 \neq e^{2\pi i \mathbf{n}\cdot\mathbf{a}}$ (due to the linear independence of $1, \alpha_1, \ldots, \alpha_d$ over the rationals), (1.18) applies, and the proof is complete. □

Similarly to Kronecker's theorem, in the original form of the Kronecker–Weyl theorem the starting point was the origin. Since the torus is translation invariant, the Kronecker–Weyl theorem remains true if we replace the origin with any other starting point in the torus.

Weyl's criterion is a qualitative result, and due to its central role in uniform distribution, we often need a quantitative (or finite) version. A nearly optimal quantitative version for the class of axis-parallel boxes $B \subset [0,1)^d$ is the Erdős–Turán–Koksma Inequality — it covers both the discrete and continuous cases. We will formulate the general continuous version in Sec. 3; see (3.21).

2. Strong Uniformity

Traditional Uniform Distribution — which is built around Weyl's criterion and "nice" test sets such as axis-parallel rectangles and boxes — does not go beyond Riemann integral. Strong uniformity (in a broad sense) refers to the extension from Riemann integral to Lebesgue measure/integral. It seems like a minor change, but it has surprisingly far-reaching consequences. First of all, we have the issue of "nice vs. ugly" test sets. An axis-parallel box is certainly "nice", and a Lebesgue measurable sets can be arbitrarily complicated ("ugly") — the latter motivates the concept of *complexity-free strong uniformity*. In fact, we discuss three different aspects of strong uniformity: *start-free strong uniformity*, *complexity-free strong uniformity*, and *dimension-free strong uniformity*. The three different aspects are all crucial to achieve our goal: to describe the fast approach to equilibrium in large off-equilibrium systems (see e.g. the Remarks after Theorem 5.1). The combination of the three concepts is what we are going to call later in Sec. 4 as *robust uniformity*. Sections 2 and 3 are preparations for Sec. 4, where we formulate a main result, Theorem 4.2, our first "short-time ergodic theorem". Sections 5–8 are the first applications of Theorem 4.2 (and many more come later).

As far as we know the subject of strong uniformity started with the following old conjecture of Khinchin [Kh23] from 1923, involving Lebesgue measure. Prove that, given a Lebesgue measurable set $S \subset [0,1]$, the sequence $\alpha, 2\alpha, 3\alpha, \ldots$ modulo one is uniformly distributed with respect to S for almost every α. Formally, the conjecture states that

$$\lim_{n\to\infty} \frac{1}{n} \sum_{\substack{1\le k\le n:\\ \{k\alpha\}\in S}} 1 = \text{length}(S) \quad \text{for almost every } \alpha. \tag{2.1}$$

Here, as usual, $0 \le \{x\} < 1$ denotes the fractional part of a real number x, and length stands for the one-dimensional Lebesgue measure.

Khinchin's conjecture remained among the most famous open problems in the subject of Uniform Distribution for several decades. The likely reason why the conjecture resisted every attack is that researchers were convinced about its truth, and wanted to prove a positive result.

Researchers were convinced about the positive solution, because there were (at least) three somewhat related positive results, suggesting that Khinchin's conjecture was "quite plausable".

The first one is the well-known one-dimensional equidistribution theorem, which states that (2.1) holds for every irrational α if $S = [a, b)$ is an arbitrary subinterval of $[0, 1)$. Formally,

$$\lim_{n\to\infty} \frac{1}{n} \sum_{\substack{1\le k\le n:\\ a\le\{k\alpha\}<b}} 1 = b - a \tag{2.2}$$

holds for every $0 < a < b < 1$ and every irrational α.

A second related positive result comes from replacing the sequence $\alpha, 2\alpha, 3\alpha, \ldots$ by the translated copy $\beta + \alpha, \beta + 2\alpha, \beta + 3\alpha, \ldots$, translated by a typical real number β, that is, if we start from a typical β instead of 0. More precisely, for every Lebesgue measurable set $S \subset [0, 1]$ and for every irrational α, we have

$$\lim_{n\to\infty} \frac{1}{n} \sum_{\substack{1\le k\le n:\\ \{\beta+k\alpha\}\in S}} 1 = \text{length}(S) \quad \text{for almost every } \beta. \tag{2.3}$$

Notice that (2.3) is a special case of Birkhoff's well-known individual ergodic theorem.

A third positive result was proved by Raikov [Ra36]: (2.1) holds if the linear sequence $k\alpha$ is replaced by the exponential sequence $2^k\alpha$, or in general, by any sequence $q^k\alpha$, where q is a fixed integer greater than one ($k = 1, 2, 3, \ldots$). (Later F. Riesz noticed that Raikov's theorem is another special case of Birkhoff's theorem; see [Ri45].)

However, despite the three related positive results mentioned above (see (2.2), (2.3) and Raikov's theorem), conjecture (2.1) turned out to be false. In 1970 Marstrand [Mar70] proved that there exists an open set $S \subset [0, 1]$

with length$(S) < 1$ such that

$$\limsup_{n\to\infty} \frac{1}{n} \sum_{\substack{1\le k\le n:\\ \{k\alpha\}\in S}} 1 = 1 \quad \text{for every } \alpha.$$

The fact that open sets are the simplest in the Borel hierarchy makes Marstrand's negative result even more suprising. Marstrand's result demonstrates that Khinchin was too optimistic.

The good news is that recently we succeeded to "save" Khinchin's conjecture in the *continuous* case, i.e., we could prove a positive result by switching from a sequence (arithmetic progression) to the continuous torus line. Of course, we also have to increase the dimension of the underlying set. That is, we replace the unit interval $[0, 1)$ modulo one with the 2-dimensional unit torus $[0, 1)^2 = I^2$, and replace the arithmetic progression $\alpha, 2\alpha, 3\alpha, \ldots$ starting from 0 with the straight line $(t\cos\theta, t\sin\theta)$, $t \geq 0$ starting from the origin $(0, 0)$ with angle θ. The positive solution of the continuous Khinchin's conjecture represents the starting point of this book.

But before formulating the positive solution in the continuous case, it is natural to recall the 2-dimensional continuous version of the Kronecker–Weyl theorem. It says that the torus line $(t\cos\theta, t\sin\theta)$, $t \geq 0$ modulo 1 is uniformly distributed with respect to every axis-parallel rectangle S in the unit square $[0, 1]^2 = I^2$ if and only if the slope $\tan\theta$ is irrational. This raises the question: What happens if one replaces the rectangle S with an arbitrary Lebesgue measurable test set $S \subset [0, 1]^2$? Can one prove uniformity with respect to S for almost every angle θ? This is what it means to prove "strong uniformity" in the continuous case.

The adjective "strong" in "strong uniformity" is justified by the fact that Lebesgue measurable sets form the largest class of sets for which we can define measure, and uniformity is meaningless without a measure.

Strong uniformity in the *narrow sense* means that it must work for *every* starting point, not just for almost every (which is the main difference between Khinchin's conjecture and Birkhoff's ergodic theorem). We refer to the extra property of "it works for every starting point" later as *start-free*.

As we said above, we proved strong uniformity in the continuous case. Actually we proved much more: the error term turned out to be polylogarithmically small! To emphasize the surprisingly small error term we refer to this as "superuniformity".

Here is the precise statement. Let $S \subset [0,1]^2 = I^2$ be an arbitrary Lebesgue measurable set in the unit square, and assume that $0 < \text{area}(S) < 1$, where area denotes the 2-dimensional Lebesgue measure. Let $\theta \in [0, 2\pi)$ be an arbitrary angle, and consider the straight line $(t\cos\theta, t\sin\theta)$, $t \geq 0$ starting from the origin $(0,0)$ with angle θ. Let $T_S(\theta)$ denote the time the line $(t\cos\theta, t\sin\theta)$ modulo one spends in the given set S as $0 \leq t \leq T$ (straight line modulo one = torus line). Formally,

$$T_S(\theta) = \text{length}\{t \in [0,T] : (\{t\cos\theta\}, \{t\sin\theta\}) \in S\}.$$

Note that it may happen that the set $\{t \in [0,T] : (\{t\cos\theta\}, \{t\sin\theta\}) \in S\}$ is not measurable for some particular angle $\theta \in [0, 2\pi)$, and so $T_S(\theta)$ is not defined. But this technical nuisance happens only for a negligible set of angles: $T_S(\theta)$ is certainly well-defined for almost every $\theta \in [0, 2\pi)$. This follows from some general results in Lebesgue Integral. We are referring to Fubini's theorem and what we call the *change of variables* in multiple integrals, applied in the special case of polar coordinates. (The relevance of polar coordinates is clear from the fact that the lines $(t\cos\theta, t\sin\theta)$ are passing through the origin.)

Uniformity of the torus line $(t\cos\theta, t\sin\theta)$ modulo one relative to S means that

$$\lim_{T\to\infty} \frac{T_S(\theta) - \text{area}(S)T}{T} = 0. \tag{2.4}$$

We could replace the factor of T in the denominator of (2.4) with the much smaller $(\log T)^{3+\varepsilon}$ for almost every θ. (We recall that $\log x$ and $\log_2 x$ denote, respectively, the natural and the binary logarithm of x.)

Theorem A1. (see Beck [Be2015]) *Let $S \subset [0,1)^2$ be an arbitrary Lebesgue measurable set in the unit square with $0 < \text{area}(S) < 1$. Then for every $\varepsilon > 0$,*

$$\lim_{T\to\infty} \frac{T_S(\theta) - \text{area}(S)T}{(\log T)^{3+\varepsilon}} = 0$$

for almost every angle θ.

(We mention Theorems A, B, C without proofs; they are not applied in the rest of the book. We use them to illustrate the concepts of *complexity-free* and *start-free.*)

We can of course rewrite Theorem A1 in the equivalent form

$$T_S(\theta) = \text{area}(S)T + o((\log T)^{3+\varepsilon}) \tag{2.5}$$

for almost every angle θ. Notice that the polylogarithmic error term is shockingly small compared to the linear main term area$(S)T$. This is why we call Theorem A1 a *superuniformity* result.

It is well possible that the already small error term in (2.5) can be further improved.

What makes the *continuous* superuniformity result Theorem A1 particularly interesting is the sharp contrast with the *discrete* Khinchin's conjecture, where there is no uniformity at all!

Here is a simple intuitive argument which may help to explain the striking difference between the discrete and continuous versions of Khinchin's conjecture. The well-known Weyl's criterion applied to a torus line ("continuous case") leads to the integral

$$\left|\int_0^T e^{2\pi \mathrm{i} t\mathbf{n}\cdot\mathbf{a}}\,dt\right| = \left|\frac{e^{2\pi \mathrm{i}T\mathbf{n}\cdot\mathbf{a}}-1}{2\pi \mathrm{i}\mathbf{n}\cdot\mathbf{a}}\right| \le \frac{1}{\pi|\mathbf{n}\cdot\mathbf{a}|},$$

where the coordinates of $\mathbf{a} = (\alpha_1, \alpha_2)$ are linearly independent over the rationals.

The discrete analog is the sum

$$\left|\sum_{j=1}^{N} e^{2\pi \mathrm{i} j\mathbf{n}\cdot\mathbf{a}}\right| = \left|\frac{e^{2\pi \mathrm{i}N\mathbf{n}\cdot\mathbf{a}}-1}{1-e^{2\pi \mathrm{i}\mathbf{n}\cdot\mathbf{a}}}\right| \le \left|\frac{2}{1-e^{2\pi \mathrm{i}\mathbf{n}\cdot\mathbf{a}}}\right|,$$

where $1, \alpha_1, \alpha_2$ are linearly independent over the rationals. Note that

$$\frac{4}{\pi} \le \frac{|1-e^{2\pi \mathrm{i}\mathbf{n}\cdot\mathbf{a}}|}{\pi\|\mathbf{n}\cdot\mathbf{a}\|} \le 2$$

(where $\|u\|$ is the distance from the nearest integer), that is,

$$\left|1-e^{2\pi \mathrm{i}\mathbf{n}\cdot\mathbf{a}}\right| \approx \pi\|\mathbf{n}\cdot\mathbf{a}\|.$$

The problem of "small denominator" in the discrete case means that $\|\mathbf{n}\cdot\mathbf{a}\|$ is "small," and the analog problem of "small denominator" in the continuous case means that $|\mathbf{n}\cdot\mathbf{a}|$ is "small". We have the trivial inequality $\|\mathbf{n}\cdot\mathbf{a}\| \le |\mathbf{n}\cdot\mathbf{a}|$, but the converse is obviously false: here the right-hand side can be much larger than the left-hand side. This quantitative difference between $\|\mathbf{n}\cdot\mathbf{a}\|$ and $|\mathbf{n}\cdot\mathbf{a}|$ gives an intuitive explanation for the fundamental difference between the discrete and continuous cases of Khinchin's conjecture.

In [Be2015] we derived Theorem A1 from a more general quantitative result.

Theorem A2. *Let $f \in L_2$ be an arbitrary real-valued Lebesgue square integrable function on the unit square $[0,1)^2 = I^2$. Then for every $\varepsilon > 0$ there is a subset $\mathcal{A} = \mathcal{A}(f;\varepsilon)$ of the interval $[0,2\pi)$ such that*

$$\frac{1}{2\pi}\text{length}(\mathcal{A}) \geq 1-\varepsilon,$$

and for every $\theta \in \mathcal{A}$ and $T \geq 8$,

$$\left|\int_0^T f(t\cos\theta, t\sin\theta)\,dt - T\int_{I^2} f(\mathbf{y})\,d\mathbf{y}\right| \leq \int_0^8 |f(t\cos\theta, t\sin\theta)|\,dt$$

$$+\frac{12\cdot 10^4\sigma_0(f)}{\varepsilon}(\log_2 T+1)^3(\log(\log_2 T+1))^2 + 12\sigma_0(f), \qquad (2.6)$$

where

$$\sigma_0^2(f) = \int_{I^2}\left(f(\mathbf{y}) - \int_{I^2} f(\mathbf{z})\,d\mathbf{z}\right)^2 d\mathbf{y}$$

is the "variance" of f.

To derive Theorem A1 from Theorem A2, choose $f = \chi_s$, $\varepsilon = 2^{-n}$, $n = 1,2,3,\ldots$, and consider the union set

$$\mathcal{A} = \bigcup_{n=1}^{\infty}\mathcal{A}(f;2^{-n}).$$

Then length $(\mathcal{A}) = 2\pi$; and by Theorem A2, for every $\varepsilon > 0$ and every $\theta \in \mathcal{A}$,

$$\lim_{T\to\infty}\frac{\int_0^T f(t\cos\theta, t\sin\theta)\,dt - T\int_{I^2} f(\mathbf{y})\,d\mathbf{y}}{(\log T)^{3+\varepsilon}} = 0,$$

proving Theorem A1.

Note that the upper bound in Theorem A2 must contain a function of $1/\varepsilon$. Indeed, if the slope of the torus line is close to a rational with small denominator, then a not too long line segment of such a torus line is far from being uniform.

In Theorems A1 and A2 the starting point of the torus line can be any specific point (since the torus is translation-invariant). We may say that these theorems are *start-free.*

Here is an interesting remark about the concept of *start-free*. Theorem A2 is start-free in the following precise sense: given an arbitrary $\varepsilon > 0$, an arbitrary starting point $\mathbf{s} = (s_1, s_2) \in [0,1)^2$, and an arbitrary Lebesgue measurable test set $S \subset [0,1)^2$, there is a subset $\mathcal{A} = \mathcal{A}(S; \mathbf{s}; \varepsilon)$ of the interval $[0, 2\pi)$ such that

$$\frac{1}{2\pi}\text{length}(\mathcal{A}) \geq 1 - \varepsilon,$$

and for every $\theta \in \mathcal{A}$ and $T \geq 8$,

$$\begin{aligned}|\text{length}\{t \in [0,T] :\ (\{t\cos\theta\} + s_1, \{t\sin\theta\} + s_2) \in S\}\\ - \text{area}(S)T| \leq c_0(\varepsilon)(\log T)^{3+\varepsilon},\end{aligned}$$

where the constant factor $c_0(\varepsilon)$ depends only on $\varepsilon > 0$. It raises the following natural question. Can we upgrade this result as follows: given an arbitrary $\varepsilon > 0$, and an arbitrary Lebesgue measurable test set $S \subset [0,1)^2$, there is a subset $\mathcal{A} = \mathcal{A}(S; \varepsilon)$ of the interval $[0, 2\pi)$ such that

$$\frac{1}{2\pi}\text{length}(\mathcal{A}) \geq 1 - \varepsilon,$$

and for every $\theta \in \mathcal{A}$ and $T \geq 8$,

$$\begin{aligned}|\text{length}\{t \in [0,T] :\ (\{t\cos\theta\} + s_1, \{t\sin\theta\} + s_2) \in S\}\\ - \text{area}(S)T| \leq c_0(\varepsilon)(\log T)^{3+\varepsilon},\end{aligned}$$

holds simultaneously for all starting points $\mathbf{s} = (s_1, s_2) \in [0,1)^2$? In other words, can we upgrade *start-free* to *simultaneously start-free*? Well, the answer is no. Any similar conjecture about *simultaneously start-free* is completely destroyed by the existence of the following "Besicovitch type set" $S^* \subset [0,1)^2$ of the 2-dimensional unit torus: (1) S^* has 2-dimensional Lebesgue measure zero, and (2) for every integer $n \geq 1$, S^* contains a torus line segment of length n in every direction $\theta \in [0, 2\pi)$. To construct such a set S^*, we recall a well-known theorem of Besicovitch, which states that there exists a plane set $S_0 \subset \mathbb{R}^2$ of 2-dimensional Lebesgue measure zero with the property that S_0 contains a unit line segment in every direction $\theta \in [0, 2\pi)$. Replacing "unit line segment" with "line segment of length k" for every positive integer $k = 1, 2, 3, \ldots$, and taking the union, we obtain a plane set $S_1 \subset \mathbb{R}^2$ of 2-dimensional Lebesgue measure zero with the property that for every integer $n \geq 1$, S_1 contains a line segment of length n in every direction $\theta \in [0, 2\pi)$. Next we take S_1 modulo one with respect to both coordinates; thus we obtain the desired subset $S^* \subset [0,1)^2$ of the 2-dimensional unit torus.

Probably the reader is wondering what happens in higher dimensions, i.e., replacing the unit square with the unit cube $[0,1]^d$, $d \geq 3$. Again we study the "strong" uniformity of a typical torus lines starting from the origin ("strong" means that the test set is an arbitrary measurable set).

Let $S \subset [0,1]^d = I^d$ be an arbitrary Lebesgue measurable set in the unit cube of dimension $d \geq 3$, and assume that $0 < \text{vol}_d(S) < 1$, where vol_d denotes the d-dimensional Lebesgue measure. Let $\mathbf{e} \in \mathbf{S}^{d-1}$ be an arbitrary unit vector in the d-dimensional Euclidean space $\mathbb{R}^d$; $\mathbf{S}^{d-1}$ denotes the unit sphere in $\mathbb{R}^d$. Consider the straight line $t\mathbf{e}$, $t \geq 0$ starting from the origin $\mathbf{0} \in \mathbb{R}^d$. Let $T_S(\mathbf{e})$ denote the time the line $t\mathbf{e}$ modulo one spends in the given set S as $0 \leq t \leq T$ (line modulo one = torus line).

Uniformity of the torus line $t\mathbf{e}$ (mod 1) relative to S means that

$$\lim_{T\to\infty} \frac{T_S(\mathbf{e}) - \text{vol}_d(S)T}{T} = 0. \tag{2.7}$$

In the 3-dimensional case we can replace the factor of T in the denominator of (2.7) with the substantially smaller $T^{1/4}(\log T)^{3+\varepsilon}$ for almost every direction $\mathbf{e} \in \mathbf{S}^2$ in the 3-space. In the d-dimensional case with $d \geq 4$ we can replace the factor of T in the denominator of (2.7) with $T^{\frac{1}{2}-\frac{1}{2(d-1)}}(\log T)^{3+\varepsilon}$ for almost every direction $\mathbf{e} \in \mathbf{S}^{d-1}$ in the d-space $\mathbb{R}^d$.

Theorem B1. (see [Be2015]) (a) *Let $S \subset [0,1)^3$ be an arbitrary Lebesgue measurable set in the unit cube with $0 < \text{vol}(S) < 1$. Then for every $\varepsilon > 0$,*

$$\lim_{T\to\infty} \frac{T_S(\mathbf{e}) - \text{vol}(S)T}{T^{1/4}(\log T)^{3+\varepsilon}} = 0 \tag{2.8}$$

for almost every direction $\mathbf{e} \in \mathbf{S}^2$ in the 3-space.

(b) *In the d-dimensional case $S \subset [0,1)^d$ with $d \geq 4$, we have the perfect analog of (2.8) where the factor of $T^{1/4}$ in (2.8) is replaced by $T^{\frac{1}{2}-\frac{1}{2(d-1)}}$ for almost every direction $\mathbf{e} \in \mathbf{S}^{d-1}$ in the d-space $\mathbb{R}^d$.*

Theorem B1 can be derived from the following quantitative result exactly the same way as we derived Theorem A1 from Theorem A2.

Theorem B2. (a) *Let $f \in L_2$ be an arbitrary real-valued Lebesgue square integrable function on the unit cube $[0,1)^3 = I^3$. Then for every $\varepsilon > 0$ there is a subset $\mathcal{A} = \mathcal{A}(f;\varepsilon)$ of the unit sphere $\mathbf{S}^2$ such that*

$$\frac{1}{4\pi}\text{SurfaceArea}(\mathcal{A}) \geq 1 - \varepsilon,$$

and for every direction $\mathbf{e} \in \mathcal{A}$ and $T \geq 8$,

$$\left| \int_0^T f(t\mathbf{e})\,dt - T\int_{I^3} f(\mathbf{y})\,d\mathbf{y} \right| \leq \int_0^8 |f(t\mathbf{e})|\,dt + \frac{12 \cdot 10^5 \sigma_0(f)}{\varepsilon} T^{1/4} (\log_2 T + 1)^3 (\log(\log_2 T + 1))^2 + 12\sigma_0(f), \tag{2.9}$$

where again

$$\sigma_0^2(f) = \int_{I^3} \left(f(\mathbf{y}) - \int_{I^3} f(\mathbf{z})\,d\mathbf{z} \right)^2 d\mathbf{y}$$

is the "variance" of f.

(b) *In the d-dimensional case with $d \geq 4$, we have the perfect analog of (2.9) where the factor of $T^{1/4}$ in (2.9) is replaced by $c_0(d)T^{\frac{1}{2}-\frac{1}{2(d-1)}}$, and $c_0(d)$ is a positive absolute constant that depends only on the dimension.*

If we make the choice $d = 2$ in Theorem B1 (or B2) then we obtain Theorem A1 (or A2). Perhaps the reader is wondering, why did we formulate two separate theorems if the first one is the special case of the second one in the 2-dimensional case. Well, the answer is that we wanted to emphasize superuniformity — meaning strong uniformity with polylogarithmic error term — which is *not* the case in dimensions $d \geq 3$. This follows from the next result, which tells that Theorem B1 (and B2) is best possible apart from polylogarithmic factor: the exponent $\frac{1}{2} - \frac{1}{2(d-1)}$ of T in the error term of Theorem B1 (and B2) is best possible.

We point out the crucial fact that in the results mentioned so far in this section, the upper bound on the error does not depend on the complexity (= ugliness) of the test set S or test function f. We may call them *complexity-free and start-free strong uniformity* results.

Note that both Theorems A1 and B1 are about the torus line, which is the simplest curve on the torus. We can define a simple motion on the torus by assuming that a particle moves on the torus line with unit speed.

Theorem C below is a general result about the limitations of the time discrepancy of a motion of a particle in the unit torus $[0,1)^d$, $d \geq 3$.

Let

$$\Gamma = \{\mathbf{x}(t) = (\{x_1(t)\}, \ldots, \{x_d(t)\}) : \ 0 \leq t \leq T\} \tag{2.10}$$

be an arbitrary (continuous) parametrized curve on the d-dimensional unit torus $[0,1)^d$, $d \geq 2$ with total time T; here each coordinate $x_j(t)$, $1 \leq j \leq d$ is a continuous function of t, and $\{x\}$ denotes, as usual, the fractional part of a real number x. Note that the parametrized curve Γ in (2.10) represents the motion of a particle on the torus, and we constantly use this interpretation below; we refer to T as the "total traveling time". Let $S \subset [0,1)^d$ be an arbitrary measurable subset, and let $\text{vol}(S)$ denote the volume, i.e., the d-dimensional Lebesgue measure. Let $T_S(\Gamma)$ denote the time the particle spends in the given set S; formally,

$$T_S(\Gamma) = \{0 \leq t \leq T : \ \mathbf{x}(t) \in S\}. \tag{2.11}$$

We call $T_S(\Gamma)$ the *actual time*, and we compare it to the *expected time*, which — assuming perfect uniformity — is proportional to the volume expected time $= T \cdot \text{vol}(S)$. The difference of the actual time and the expected time is called *time discrepancy*; formally,

$$\text{time discrepancy} = D_S(\Gamma) = T_S(\Gamma) - T \cdot \text{vol}(S). \tag{2.12}$$

To prove a non-trivial result, we need an extra assumption. We assume that "T = total traveling time = arclength," which is equivalent to the requirement that the average speed is one.

Theorem C. *For every integer $d \geq 3$ and real $T > 1$, there exists an integer $m = m(d,T) \geq 2$ such that we can construct m measurable subsets $S_1, \ldots, S_m$ of the unit torus $[0,1)^d$ with the following property: given any parametrized curve*

$$\Gamma = \{\mathbf{x}(t) = (\{x_1(t)\}, \ldots, \{x_d(t)\}) : \ 0 \leq t \leq T\}$$

of arclength T on the torus $[0,1)^d$ (i.e., the average speed is one)

$$|D_{S_j}(\Gamma)| > c_1(d) T^{\frac{1}{2} - \frac{1}{2(d-1)}} \tag{2.13}$$

holds for at least two-thirds of the m subsets $S_1, \ldots, S_m$. Here $c_1(d) > 0$ is a constant depending only on the dimension $d \geq 3$. In particular, $c_1(3) = 1/500$ is a good choice for $d = 3$.

Theorem C implies, via a standard averaging argument, that Theorem B2 is best possible apart from polylogarithmic factor of T. The explanation goes as follows. First note that every torus line in Theorem B2 is determined by its direction $\mathbf{e} \in \mathbf{S}^{d-1}$, and using the ($(d-1)$-dimensional) surface area on the unit sphere $\mathbf{S}^{d-1}$, it is meaningful to talk about the "majority of torus lines," or more precisely, about "$1-\varepsilon$ part of all torus lines passing through the origin". Now assume that, for some $d \geq 3$ and $T > 1$, there exists a continuous family of parametrized curves

$$\{\Gamma_\omega : \omega \in \Omega\} \tag{2.14}$$

on the torus $[0,1)^d$ such that there is a probability measure μ on the index-set Ω (i.e., $\mu(\Omega) = 1$, so it is meaningful to talk about $1-\varepsilon$ part), and the family of curves in (2.14) beats Theorem B2 in the following quantitative sense: Given any measurable subset $S \subset [0,1)^d$,

$$|D_S(\Gamma_\omega)| < c_1(d) T^{\frac{1}{2} - \frac{1}{2(d-1)}} \tag{2.15}$$

holds for at least two-thirds of the curves Γ_ω, $\omega \in \Omega$, in the sense of the μ-measure.

We show that this contradicts Theorem C. Indeed, we apply (2.15) for the $m = m(d,T) \geq 2$ sets S_j, $1 \leq j \leq m$ whose existence is guaranteed by Theorem C. Thus for every $1 \leq j \leq m$ there exists a (measurable) subset Ω_j of the index-set such that $\mu(\Omega_j) \geq 2/3$, and

$$|D_{S_j}(\Gamma_\omega)| < c_1(d) T^{\frac{1}{2} - \frac{1}{2(d-1)}}$$

holds for all $\omega \in \Omega_j$. The fact

$$\frac{1}{m} \sum_{j=1}^{m} \mu(\Omega_j) \geq 2/3$$

immediately implies that there must exist an index $\omega_0 \in \Omega$ which is contained by at least $2m/3$ of the m sets Ω_j, $1 \leq j \leq m$. In other words, there is a curve Γ_{ω_0} such that (see (2.15))

$$|D_{S_j}(\Gamma_{\omega_0})| < c_1(d) T^{\frac{1}{2} - \frac{1}{2(d-1)}} \tag{2.16}$$

holds for at least two-thirds of the m sets $S_1, \ldots, S_m$. But (2.16) clearly contradicts Theorem C, and this contradiction proves that Theorem B2 is nearly best possible in every dimension $d \geq 3$: we cannot replace the error term

$$T^{\frac{1}{2} - \frac{1}{2(d-1)}} (\log T)^{3+\varepsilon} \quad \text{with } o(T^{\frac{1}{2} - \frac{1}{2(d-1)}}).$$

For the proofs of the results mentioned in this section, we refer the reader to [Be2015].

In the book we sharply distinguish between two concepts of "space": the low-dimensional particle space (= intuitively the "gas container"), which is usually 3-dimensional, and the high-dimensional configuration space, where the *whole* system is represented by a single point. To describe the time evolution of large off-equilibrium systems, it is natural to work in the high-dimensional configuration space. Then the time evolution of the system means a motion of this single point, i.e., a parametrized curve.

(Note in advance that we will also sharply distinguish between two concepts of "equilibrium": equilibrium in the particle space; see *square-root equilibrium*, and equilibrium in the configuration space; see *conf-space equilibrium* — see Secs. 4 and 5.)

In an off-equilibrium real life gas model the number of particles N is typically in the range of the Avogadro number (close to 10^{24}), and the time scale is just a few seconds. Therefore, it is natural to ask what happens if the dimension d is *much* larger than the time parameter T (in seconds, say). We answer this question in Sec. 4. But first in Sec. 3 we elaborate on the concept of configuration space.

3. High-Dimensional Configuration Space of Large Systems and Unrealistic Time Scale

In the "off-equilibrium ideal gas" model N point particles are moving around in a cubic container — say, the unit cube $[0, 1]^3$ — bouncing back and force on the walls like billiard balls. To study the time evolution of such a large billiard system, we use the geometric trick of *unfolding* that converts a billiard orbit (= zig-zag) into a torus line (unfolding was introduced in [Kö-Sz13]).

The figure shows the 2-dimensional case. It illustrates the geometric trick of unfolding the billiard orbit (= "reflecting ray") inside the unit square to a straight line in the entire plane. The transformation of *unfolding* simply means that we keep reflecting the square itself in the respective side and unfold the piecewise linear billiard path to a straight line.

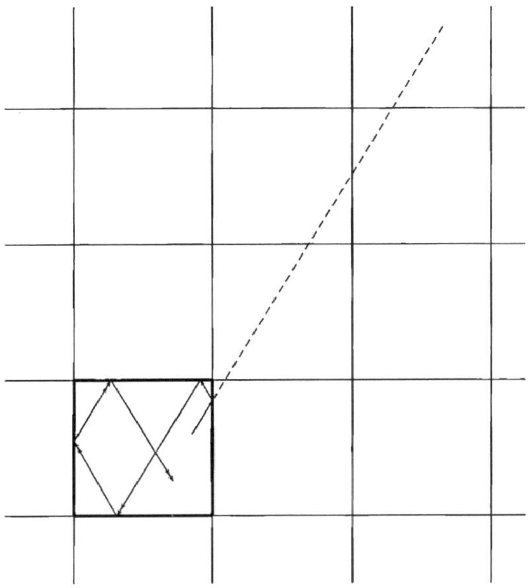

Fig. 3.1

Two (straight) lines in the plane correspond to the same billiard path if and only if they differ by a translation through an integral vector where both coordinates are even, i.e., where the vector is from the "double" square lattice $2\mathbb{Z} \times 2\mathbb{Z}$. In other words, the problem of the long-term distribution of a billiard path in the unit square is equivalent to the distribution of the corresponding torus-line in the 2×2 square.

The law of reflection implies that there are at most four different directions along the billiard path (the initial direction is preserved modulo $\pi/2$, which is one-fourth of the whole angle 2π; the same holds for any rectangle). If we use unit speed, then of course *arclength* and *time* are the same.

Formally, a billiard path in the unit square $[0,1]^2$ has the form

$$\mathbf{x}(t) = (x_1(t), x_2(t)), \quad 0 < t < \infty \quad \text{with}$$

$$x_j(t) = 2\|(y_j + t\alpha_j)/2\|, \quad j = 1, 2, \tag{3.1}$$

where $\mathbf{a} = (\alpha_1, \alpha_2)$ is a nonzero vector, and $\|z\|$ denotes the distance of a real number z from the nearest integer. Here $\mathbf{y} = (y_1, y_2)$ is the starting point, and $\mathbf{a}$ gives the initial direction; so $\theta = \arctan \frac{\alpha_2}{\alpha_1}$ ("inverse tangent") is the initial angle.

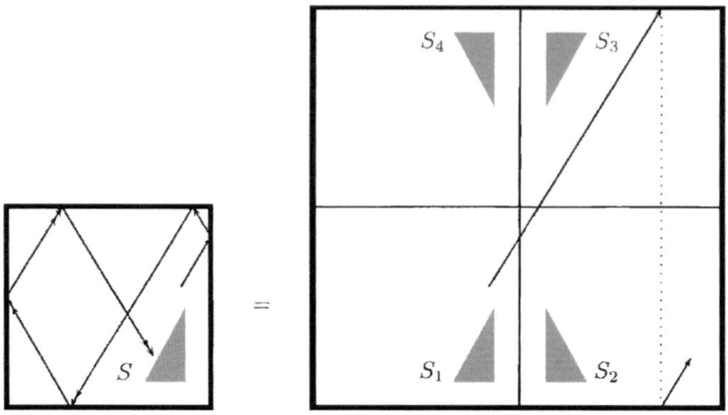

Fig. 3.2

It follows from unfolding that a billiard path in the unit square intersects a given test set S at time t exactly when the corresponding torus-line in the 2×2 square intersects the union set $S_1 \cup S_2 \cup S_3 \cup S_4$ (see Fig. 3.2).

Indeed, unfolding defines four reflected copies S_1, S_2, S_3, S_4 of S (each of the four unit squares on the figure above contains a reflected copy of the given test set S). In the last step we shrink the underlying 2×2 square to the unit square $I^2 = [0,1)^2$.

Of course, the test set S can be upgraded to any "not too ugly" periodic function f, i.e., $f(x,y)$ is a Riemann integrable periodic complex-valued function with period one in both variables. In the case of the figure, f means the $0,1$ valued characteristic function of the union $S_1 \cup S_2 \cup S_3 \cup S_4$ of the four half-size reflected copies of the given subset $S \subset I^2 = [0,1)^2$.

Note that the unfolding of the billiard path in the h-dimensional unit cube $[0,1]^h$ with $h \neq 2$ can be defined in the analog way. Formally, unfolding means the map

$$2\|x/2\| \to \{x/2\} \text{ applied to each coordinate,} \tag{3.2}$$

where $0 \leq \{x\} < 1$ denotes the fractional part of a real number x (i.e., $x = \lfloor x \rfloor + \{x\}$ where $\lfloor x \rfloor$ is the largest integer $\leq x$).

We use the fact that for every set $S \subset [0,1)$,

$$2\|x/2\| \in S \iff \{x/2\} \in \frac{1}{2}S \cup \left(1 - \frac{1}{2}S\right), \tag{3.3}$$

where

$$\frac{1}{2}S = \{y/2: \; y \in S\} \quad \text{and} \quad 1 - \frac{1}{2}S = \left\{1 - \frac{y}{2}: \; y \in S\right\}. \tag{3.4}$$

It follows from unfolding that a billiard path in the h-dimensional unit cube $[0,1]^h$ intersects a given test set $S \subset [0,1]^h$ at time t exactly when the corresponding torus-line in the h-dimensional $2 \times \cdots \times 2$ (hyper)cube intersects the union of the 2^h reflected copies of S. (Note that each of the 2^h unit (hyper)cubes contains a reflected copy of the given test set S.) In the last step we shrink the h-dimensional $2 \times \cdots \times 2$ (hyper)cube to the unit (hyper)cube $I^h = [0,1)^h$.

A constant speed piecewise linear point billiard motion in I^h is defined by the equation

$$\begin{aligned} \mathbf{x}(t) &= (x_1(t), \ldots, x_h(t)), \; 0 < t < \infty \text{ with} \\ x_i(t) &= 2\|(y_i + t\alpha_i)/2\|, \; 1 \le i \le h, \end{aligned} \tag{3.5}$$

where $\mathbf{y} = (y_1, \ldots, y_h)$ is the starting point and the nonzero vector $\mathbf{a} = (\alpha_1, \ldots, \alpha_k)$ is the initial direction.

Next we study the time evolution of a *large off-equilibrium system* of N point-billiards in a cube $[0,1]^h$, and assume that the hN velocity coordinates are linearly independent over the rationals.

To demonstrate the power of the concept of *configuration space*, we prove a surprising "superdiscrepancy" result, which is a simple application of Kronecker's well-known density theorem in the configuration space. We start with a special case: let S be the right half $[\frac{1}{2}, 1] \times [0,1] \cdots \times [0,1]$ of the unit cube $[0,1]^h$ (let, say, $h = 3$). Whatever initial point configuration we had at the start, if we wait long enough, it will happen that all N particles will be in S at the same time. And what is more, the system will return to this "absurd" state infinitely many times.

We call it "absurd," because in real life such a 3-dimensional cubic container would collapse, due to the outside pressure.

It is easy to upgrade this to an even more absurd result. Under the same condition ("linear independence") it will happen that, given any *rational* box (i.e., Cartesian product of closed intervals that have rational endpoints, and the volume is positive), all N particles will be in this box at the same time, and the system will return to each rational box infinitely many times.

(The set of rational boxes is countable, and every box of positive volume contains a rational box.)

The most general result of this type is the following.

Theorem 3.1. (Superdiscrepancy) *Assume that the hN initial velocity coordinates of a point-billiard system in the h-dimensional unit cube $I^h = [0,1]^h$ are linearly independent over the rationals. Then the time evolution of the system exhibits the following property that we call superdiscrepancy. Given any sequence $B_1, B_2, \ldots, B_N$ of N rational boxes in I^h and any ordering of the N particles, starting from an arbitrary initial point configuration, it will happen that the first particle will be in the first rational box B_1, the second particle will be in the second rational box B_2, and so on, the Nth particle will be in the Nth rational box B_N, all at the same time, and the system will return to each such $(B_1, \ldots, B_N)$-configuration infinitely many times.*

The superdiscrepancy result is somewhat similar to the classical Zermelo–Poincaré paradox in statistical mechanics. Both say, very roughly speaking, that if one waits long enough, "crazy things will happen".

Theorem 3.1 explains why we cannot expect convergence to "snapshot equilibrium" as $t \to \infty$ for the individual time evolution of large off-equilibrium systems. We have to downgrade it to *approach to snapshot equilibrium*, emphasizing the temporary nature of the static aspects of (snapshot) equilibrium on the individual level.

We show that Theorem 3.1 is a straightforward corollary of Kronecker's density theorem. Indeed, if the motion of the kth point billiard in $I^h = [0,1]^h$ is described by $\mathbf{x}_k(t)$, $0 \le t < \infty$ in (3.5), then the time-evolution of the whole system is represented by the following parametrized curve in the hN-dimensional cube $[0,1]^d = [0,1]^{hN}$ that we call the trajectory of the system

$$\begin{aligned}\Gamma(t) &= (\mathbf{x}_1(t), \mathbf{x}_2(t), \ldots, \mathbf{x}_N(t)) \\ &= (2\|(y_{1,1} + v_{1,1}t)/2\|,\ 2\|(y_{1,2} + v_{1,2}t)/2\|, \ldots, 2\|(y_{1,h} + v_{1,h}t)/2\|, \ldots, \\ &\qquad 2\|(y_{N,1} + v_{N,1}t)/2\|,\ 2\|(y_{N,2} + v_{N,2}t)/2\|, \ldots, 2\|(y_{N,h} + v_{1,h}t)/2\|).\end{aligned} \tag{3.6}$$

Here $[0,1]^d = [0,1]^{hN}$ is the *configuration space*, $\Gamma(0) = \mathcal{Y} = \{\mathbf{y}_1, \ldots, \mathbf{y}_N\}$ is the initial configuration, and $\Gamma'(0) = \mathcal{V} = (\mathbf{v}_1, \ldots, \mathbf{v}_N)$ describes the initial velocities.

Let

$$S_{\mathcal{C}}(B) = B_1 \times B_2 \times \cdots \times B_N \tag{3.7}$$

be the Cartesian product of N rational boxes in this order; the index $\mathcal{C}$ indicates that $S_{\mathcal{C}}(B) \subset I^d$ is a subset of the configuration space.

Applying d-dimensional *unfolding*, we see that the distribution of the (d-dimensional) billiard path $\Gamma(t)$, $0 \le t \le T$ (see (3.6)) with respect to the test set $S_{\mathcal{C}}(\Delta)$ (see (3.7)) is equivalent to the distribution of the corresponding d-dimensional torus line (with $d = hN$; see (3.2))

$$\begin{aligned}\Gamma^*(t) &= (\{y_{1,1} + v_{1,1}t\}, \{y_{1,2} + v_{1,2}t\}, \ldots, \{y_{1,h} + v_{1,h}t\}, \ldots, \\ &\qquad \{y_{N,1} + v_{N,1}t\}, \{y_{N,2} + v_{N,2}t\}, \ldots, \{y_{N,h} + v_{1,h}t\}) \\ &= \Gamma(0) + \Gamma'(0)t \ \text{ modulo one}\end{aligned} \tag{3.8}$$

with starting point $\Gamma(0)$ and direction vector $\Gamma'(0)$ ("derivative") with respect to the union

$$\bigcup_{i=1}^{2^{hN}} S_{\mathcal{C},i}(B) \tag{3.9}$$

of the 2^{hN} "reflected halves" of $S_{\mathcal{C}}(B)$.

By hypothesis the hN coordinates $v_{i,j}$, $1 \le i \le N$, $1 \le j \le h$ of the direction vector $\Gamma'(0)$ are linearly independent over the rationals, so we can apply the Kronecker's density theorem to any of the 2^{hN} "reflected halves" of $S_{\mathcal{C}}(B)$, and Theorem 3.1 follows. □

Next we repeat the proof of the Superdiscrepancy result (Theorem 3.1) by replacing Kronecker's density theorem with the Kronecker–Weyl equidistribution theorem (continuous version), and obtain an "asymptotic CLT" (central limit theorem).

For simplicity assume $h = 3$, i.e., we study a system of N point billiards in the unit cube $[0,1]^3$, where the motion of the kth point billiard in $[0,1]^3$ is described by $\mathbf{x}_k(t)$, $0 \le t < \infty$. As in (1.6), the time-evolution of the system is represented by the parametrized curve in the $3N$-dimensional cube $[0,1]^d = [0,1]^{3N}$ (=configuration space) that we call the trajectory of the system

$$\begin{aligned}\Gamma(t) &= (\mathbf{x}_1(t), \mathbf{x}_2(t), \ldots, \mathbf{x}_N(t)) \\ &= (2\|(y_{1,1}+v_{1,1}t)/2\|, 2\|(y_{1,2}+v_{1,2}t)/2\|, 2\|(y_{1,3}+v_{1,3}t)/2\|, \ldots, \\ &\qquad 2\|(y_{N,1}+v_{N,1}t)/2\|, 2\|(y_{N,2}+v_{N,2}t)/2\|, 2\|(y_{N,3}+v_{1,3}t)/2\|),\end{aligned} \tag{3.10}$$

where $\Gamma(0) = \mathcal{Y} = \{\mathbf{y}_1, \ldots, \mathbf{y}_N\}$ is the initial configuration, and $\Gamma'(0) = \mathcal{V} = (\mathbf{v}_1, \ldots, \mathbf{v}_N)$ describes the initial velocities.

Let $S \subset [0,1]^3$ be an arbitrary but fixed Jordan measurable subset of the unit cube with volume $0 < \text{vol}(S) < 1$. Let $N_S(t) = N_S(\mathcal{Y}; \mathcal{V}; t)$ denote the point-counting function

$$N_S(t) = N_S(\mathcal{Y}; \mathcal{V}; t) = \sum_{\substack{1 \le k \le N: \\ \mathbf{x}_k(t) \in S}} 1. \tag{3.11}$$

First we prove the following asymptotic binomial distribution type result.

Theorem 3.2. (asymptotic binomial distribution) *Consider the N-particle billiard model in the 3-dimensional unit cube $[0,1]^3$ defined in (3.10). Assume that the $3N$ initial velocity coordinates*

$$\{v_{k,1}, v_{k,2}, v_{k,3} :\ 1 \le k \le N\}$$

in (3.10) are linearly independent over the rationals. Then

$$\lim_{T\to\infty} \frac{1}{T} \text{length}\{t \in [0,T] :\ N_S(\mathcal{Y}; \mathcal{V}; t) = n\} = \binom{N}{n} p^n q^{N-n} \tag{3.12}$$

for all Jordan measurable subsets $S \subset [0,1]^3$ of the unit cube, $p = \text{vol}(S)$ denotes the volume, $q = 1 - p$, the point-counting function $N_S(t)$ is defined in (3.11), and length simply means the one-dimensional Lebesgue measure.

Proof of Theorem 3.2. We show that, applying the Kronecker–Weyl theorem in the configuration space, and combining it with unfolding (as we did in the proof of Theorem 3.1), (3.12) follows.

Following the proof of Theorem 3.1, let $S^{(1)} = S \subset [0,1]^3$, and let $S^{(0)} = [0,1]^3 \setminus S$ be the complement of S, and for every N-dimensional vector $\Delta = (\delta_1, \delta_2, \ldots, \delta_N)$ with coordinates $\delta_i = 0$ or 1, let

$$S_{\mathcal{C}}(\Delta) = S^{(\delta_1)} \times S^{(\delta_2)} \times \cdots \times S^{(\delta_N)} \tag{3.13}$$

be the Cartesian product of the N sets $S^{(\delta_i)}$, $1 \le i \le N$ in this order (the index $\mathcal{C}$ indicates the configuration space). If the vector Δ has n coordinates with 1 and $N - n$ coordinates with 0, then

$$\mathrm{vol}_d(S_{\mathcal{C}}(\Delta)) = p^n q^{N-n}, \tag{3.14}$$

where vol_d stands for the $d = 3N$-dimensional Lebesgue measure.

We repeat the argument of (3.8) and (3.9): the distribution of the billiard path $\Gamma(t)$, $0 \le t \le T$ with respect to a test set $S_{\mathcal{C}}(\Delta)$ (see (3.13)) is equivalent — via *unfolding* — to the distribution of the corresponding torus line

$$\begin{aligned}\Gamma^*(t) &= (\{y_{1,1} + v_{1,1}t\}, \{y_{1,2} + v_{1,2}t\}, \{y_{1,3} + v_{1,3}t\}, \ldots,\\ &\qquad \{y_{N,1} + v_{N,1}t\}, \{y_{N,2} + v_{N,2}t\}, \{y_{N,3} + v_{1,3}t\})\\ &= \Gamma(0) + \Gamma'(0)t \ \text{ modulo one}\end{aligned} \tag{3.15}$$

with starting point $\Gamma(0)$ and direction vector $\Gamma'(0)$ with respect to the union

$$\bigcup_{i=1}^{2^{3N}} S_{\mathcal{C},i}(\Delta)$$

of the 2^{3N} "reflected halves" of $S_{\mathcal{C}}(\Delta)$. In other words, by using the notation $S^*(\Delta) \subset [0,1]^{3N}$ for this union set, we have

$$\mathrm{length}\{t \in [0,T] : \ \Gamma(t) \in S_{\mathcal{C}}(\Delta)\} = \mathrm{length}\{t \in [0,T] : \ \Gamma^*(t) \in S^*(\Delta)\}.$$

Clearly (see (3.14))

$$\mathrm{vol}_d(S^*(\Delta)) = \mathrm{vol}_d(S_{\mathcal{C}}(\Delta)) = p^n q^{N-n}. \tag{3.16}$$

By hypothesis the coordinates $v_{i,j}$, $1 \le i \le N$, $1 \le j \le 3$ of the direction vector $\Gamma'(0)$ ("derivative") are linearly independent over the rationals, so we can apply the Kronecker–Weyl equidistribution theorem (continuous form) to the 0, 1 valued characteristic function of a set $S^*(\Delta) \subset [0,1]^{3N}$, and obtain

$$\lim_{T\to\infty} \frac{1}{T}\mathrm{length}\{t \in [0,T] : \ \Gamma^*(t) \in S^*(\Delta)\} = \mathrm{vol}_d(S^*(\Delta)) = p^n q^{N-n}, \tag{3.17}$$

where in the last step we used (3.16), and $\Gamma^*(t)$ is defined in (3.15). By (3.17),

$$\lim_{T\to\infty}\frac{1}{T}\text{length}\{t\in[0,T]:\ N_S(t)=n\}$$

$$=\sum_{\substack{\Delta \text{ has } n\\ \text{coordinates } 1}}\lim_{T\to\infty}\frac{1}{T}\text{length}\{t\in[0,T]:\ \Gamma(t)\in S_{\mathcal{C}}(\Delta)\}$$

$$=\sum_{\substack{\Delta \text{ has } n\\ \text{coordinates } 1}}\lim_{T\to\infty}\frac{1}{T}\text{length}\{t\in[0,T]:\ \Gamma^*(t)\in S^*(\Delta)\}$$

$$=\binom{N}{n}p^n q^{N-n},$$

which proves Theorem 3.2. □

Remark. Under the condition of Theorem 3.2 the flow $\Gamma'(0)t$ in the torus $[0,1)^{3N}$ (see (3.15)) is ergodic, so we can apply Birkhoff's individual ergodic theorem. It says that in Theorem 3.2 we can replace the class of Jordan measurable sets $S\subset[0,1]^3$ with an arbitrary but fixed Lebesgue measurable test set $S\subset[0,1]^3$. The disadvantage is that it works only for *almost every* initial configuration $\Gamma(0)=\mathcal{Y}=\{\mathbf{y}_1,\ldots,\mathbf{y}_N\}$, and a typical initial configuration represents equilibrium at the start. Since our main goal is to study the approach to equilibrium starting from an arbitrary far-from-equilibrium initial configuration, this corollary of Birkhoff's theorem is not particularly useful for us.

Combining Theorem 3.2 with the classical DeMoivre–Laplace theorem with error term, we obtain the following result (CLT stand for the central limit theorem).

Theorem 3.3. (asymptotic CLT) *Under the conditions of Theorem* 3.2, *with* $N_S(t)=N_S(\mathcal{Y};\mathcal{V};t)$,

$$\sup_{\infty<y<\infty}\left|\lim_{T\to\infty}\frac{1}{T}\text{length}\left\{t\in[0,T]:\ \frac{N_S(t)-pN}{\sqrt{pqN}}\le y\right\}\right.$$
$$\left.-\frac{1}{\sqrt{2\pi}}\int_{-\infty}^{y}e^{-u^2/2}\,du\right|\le\frac{1}{\sqrt{pqN}},\tag{3.18}$$

where $p=\text{vol}(S)$ *and* $q=1-p$. □

Note that the error term $O(1/\sqrt{pqN})$ in (3.18) is the inevitable discrepancy between the (continuous) standard normal distribution and the (discrete) normalized binomial distribution with parameters N and $0 < p < 1$.

Theorems 3.1–3.3 are asymptotic results, describing the (very!) long-term time evolution of a large N-particle system, starting from an arbitrary initial point configuration (e.g., "Big Bang," where all particles start from the same point). In particular, Theorem 3.3 says that a *large* billiard system exhibits *asymptotic time-lapse randomness* as $t \to \infty$.

If the volume $p = \text{vol}(S)$ of the test set S is very small — say, it is around $1/N$ — then of course (3.18) becomes useless (since the error term $O(1/\sqrt{Npq})$ is in the range of 1). However, Theorem 3.2 still works, and implies the following result.

Theorem 3.4. (asymptotic Poisson law) *Write* $\lambda = pN = \text{vol}(S)N$. *Under the conditions of Theorem* 3.2, *for every integer* $\ell \geq 0$,

$$\lim_{T\to\infty} \frac{1}{T}\text{length}\{t \in [0,T] : \ N_S(t) = \ell\} = \frac{\lambda^\ell e^{-\lambda}}{\ell!} + O(\lambda^2/N). \qquad (3.19)$$

□

Notice that the probability distribution

$$\frac{\lambda^\ell e^{-\lambda}}{\ell!}, \quad \ell = 0, 1, 2, 3, \ldots$$

on the right-hand side of (3.19) is the familiar Poisson distribution with parameter $\lambda > 0$.

Note that (3.19) immediately follows from Theorem 3.2 combined with the simple approximation

$$\binom{N}{\ell}(\lambda/N)^\ell(1-\lambda/N)^{N-\ell} = \frac{\lambda^\ell e^{-\lambda}}{\ell!} + O(\lambda^2/N).$$

If $\lambda = pN = \text{vol}(S)N$ is in the "constant range" (i.e., the volume of the test set S is very small), then by (3.19), the point-counting function $N_S(t)$ has Poisson distribution with negligible error (since the error term $O(\lambda^2/N)$ is very small).

The requirement in Theorem 3.2 — linear independence of the $3N$ velocity coordinates — is easily satisfied by choosing a *typical* velocity distribution, i.e., almost all choices satisfy linear independence.

Perhaps the simplest *explicit* example of arbitrarily large finite sets of real numbers that are linearly independent over the rationals is the following:

Explicit Example: $\sqrt{2}, \sqrt{3}, \sqrt{5}, \sqrt{7}, \sqrt{11}, \ldots, \sqrt{p}$, where p denotes primes.

The linear independence of this example easily follows from the well-known fact from algebra and number theory that the exponentially long symmetric product of 2^d factors

$$\prod_{\delta_j=\pm 1, 1\le j\le d} \left(\sum_{j=1}^{d} \delta_j h_j \sqrt{p_j}\right) \quad \text{is a nonzero integer,} \tag{3.20}$$

assuming $(h_1, \ldots, h_d) \in \mathbb{Z}^d \setminus \mathbf{0}$.

This corollary of the Kronecker–Weyl theorem is very elegant and interesting, but, unfortunately, it has a fundamental shortcoming: it do not say anything about the crucial question of *how long does it take to reach CLT-equilibrium*? In most real life experiments, the transition from nonequilibrium to equilibrium is very fast, and this is not reflected in the statement.

The underlying problem is that the Weyl's criterion is a qualitative result, since in the limit process, where we approximate a "nice" function with a finite Fourier series, we lose control over the quantitative aspects of the error term. This is why in Theorem 3.3 we cannot provide any quantitative answer to question of *how long does it take to reach CLT-equilibrium.*

In fact, we cannot even expect any explicit error term for the trivial reason that the class of Jordan measurable test sets is far too large. Indeed, let T be large, and let $L_0(\mathbf{e}, T)$ denote the line segment $\{t\mathbf{e} : 0 \le t \le T\}$ modulo one. Given any Jordan measurable test set $A \subset I^d = [0,1)^d$ in the configuration space I^d with positive measure, the difference $A_1 = A \setminus L_0(\mathbf{e}, T)$ is still Jordan measurable, and has the same measure (d-dimensional volume) as A. Let $f = \chi_{A_1}$, then $\frac{1}{T}\int_0^T f(t\mathbf{e})\, dt = 0$, and

$$\int_{I^d} f \, d\mathrm{V} = \mathrm{vol}_d(A) = \text{nonzero constant.}$$

Since T can be arbitrarily large, this explains why we cannot expect any explicit error term $o(1)$ tending to zero as $T \to \infty$.

Since the class of Jordan measurable test sets $A \subset [0,1)^d$ (meaning $f = \chi_S$) is far too large, we may consider a much smaller class. The simplest class — the most natural choice used since the beginning of Uniform Distribution — is the class of axis-parallel boxes B in $[0,1)^d$, i.e., the Cartesian products $B = I_1 \times I_2 \times \cdots \times I_d$ of subintervals $I_i \subset [0,1)$. For this class there is a "quantitative continuous Weyl's criterion"; see e.g., in [Dr-Ti97].

Continuous Erdős–Turán–Koksma Inequality: Let $\mathbf{r}(t) : [0,T] \to [0,1)^d = I^d$ be an arbitrary continuous curve and $H \geq 1$ be an arbitrary integer; then with $f = \chi_B$, where B denotes axis-parallel boxes and of course f is defined on the unit torus I^d, we have

$$\max_{B=I_1\times\cdots\times I_d\subset I^d} \left| \frac{1}{T}\int_0^T f(\mathbf{r}(t))\,dt - \int_{I^d} f\,d\mathrm{V} \right|$$
$$\leq \left(\frac{3}{2}\right)^d \left(\frac{2}{H+1} + \sum_{\substack{\mathbf{h}\in\mathbb{Z}^d:\\ 1\leq\|\mathbf{h}\|_\infty\leq H}} \frac{1}{\Lambda(\mathbf{h})} \left| \frac{1}{T}\int_0^T e^{2\pi \mathrm{i}\mathbf{h}\cdot\mathbf{r}(t)}\,dt \right| \right), \quad (3.21)$$

where of course $\mathrm{i} = \sqrt{-1}$,

$$\|\mathbf{h}\|_\infty = \max_{1\leq j\leq d} |h_j|$$

is the maximum norm of the vector $\mathbf{h} = (h_1, \ldots, h_d) \in \mathbb{Z}^d$, and

$$\Lambda(\mathbf{h}) = \prod_{1\leq j\leq d} (1 + |h_j|).$$

An obvious weakness of the Erdős–Turán–Koksma inequality is the factor $(3/2)^d$, which makes it totally useless in large dimensions (e.g., when d is around the Avogadro number). But even if we assume that someone could get rid of the exponentially large factor $(3/2)^d$, there remains another, less obvious, but just as dangerous problem with (3.21). The Kronecker–Weyl theorem is the special case $\mathbf{r}(t) = t\mathbf{e}$ with some $\mathbf{e} = (\alpha_1, \ldots, \alpha_d)$, and we have

$$\frac{1}{T}\int_0^T e^{2\pi \mathrm{i}\mathbf{h}\cdot t\mathbf{e}}\,dt = \frac{e^{2\pi \mathrm{i}T\mathbf{h}\cdot\mathbf{e}} - 1}{2\pi T\mathrm{i}\mathbf{h}\cdot\mathbf{e}}. \quad (3.22)$$

Of course we have the trivial upper bound

$$\left|\frac{1}{T}\int_0^T e^{2\pi i \mathbf{h}\cdot t\mathbf{e}}\,dt\right| \le 1,$$

but, because the infinite sum

$$\sum_{1\le\|\mathbf{h}\|_\infty<\infty} \frac{1}{\Lambda(\mathbf{h})} = \prod_{j=1}^{d}\left(\sum_{h_j\in\mathbb{Z}} \frac{1}{1+|h_j|}\right) - 1 = \infty$$

is clearly divergent, we need a nontrivial upper bound in (3.22), which requires a nontrivial lower bound for the absolute value of the linear combination

$$\mathbf{h}\cdot\mathbf{e} = \sum_{j=1}^{d} h_j\alpha_j$$

in the denominator on the right-hand side of (3.22), where $\alpha_1,\dots,\alpha_d$ are linearly independent over the rationals.

Unfortunately, the quantitative aspects of linear independence of concrete sets of real numbers, even for such a simple case as the Explicit Example (see before (3.20)), lead to notoriously difficult number-theoretic problems that are considered hopeless for contemporary research. To give a nutshell illustration, note that the linear independence of the Explicit Example says that (p_j stands for primes, and h_j are integers)

$$\sum_{j=1}^{d} h_j\sqrt{p_j} \ne 0 \text{ unless } h_j = 0,\ 1\le j\le d.$$

Even for the simplest family of axis-parallel d-dimensional boxes, we need a quantitative lower bound on the distance between a linear combination $\sum_j h_j\sqrt{p_j}$ and 0. (3.20) implies the lower bound

$$\left|\sum_{j=1}^{d} h_j\sqrt{p_j}\right| \ge \left(\sum_{j=1}^{d} |h_j|\sqrt{p_j}\right)^{-2^d+1}, \tag{3.23}$$

but, unfortunately, lower bound (3.23) is *extremely* weak for large d — due to the double exponential nature of the right-hand side. This shows that, even for such a simple set of linearly independent real numbers as the Explicit Example, and even for such a simple test set as a box in the configuration space, we do not have a satisfying quantitative result. (Note that W.M. Schmidt proved a lower bound that is asymptotically much better than (3.23), but, unfortunately, his deep result is *ineffective* — a notorious, and very annoying, technical problem. This prevents us from using Schmidt's theorem: to prove results on a "realistic time scale" we need effective estimations.)

Another illustration of how little we know about linear independence of explicit sets of reals is the more than 100 years old open problem that whether or not the pair π, e of the two most famous special numbers is linearly independent over the rationals.

This forces us to switch from studying the time evolution of a large off-equilibrium system with *explicit* initial velocities — where we *cannot* prove anything about realistic time scale — to *typical* initial velocities (and arbitrary Lebesgue measurable test sets in the configuration space, i.e., strong uniformity in high dimensions), where we *can* prove many interesting results about realistic time scale — this is exactly the subject of this book. To define *typical*, we need a probability distribution. In this book *typical* means that the initial velocities are taken independently (for each particle) from the 3-dimensional normal (= Gaussian) distribution.

We conclude this section with a short discussion about the (multi-dimensional) Gaussian (velocity) distribution. The 3-dimensional Gaussian velocity distribution is isotropic, and it is determined by the following density function for the speed:

$$\frac{1}{\sigma^3}\sqrt{\frac{2}{\pi}}y^2 e^{-y^2/(2\sigma^2)} \tag{3.24}$$

for all $y \geq 0$, where the physical meaning of parameter $\sigma > 0$ is $\sigma = \sqrt{k_0 T/m}$; here m is the mass, T is the absolute temperature, and k_0 is the Boltzmann's constant. (Physicists call (3.24) the Maxwellian speed distribution; it reflects physical reality for gas particles in thermal equilibrium — first discovered by Maxwell.)

Using the substitution $x = y/\sigma$ in (3.24) we have

$$\frac{1}{\sigma^3}\int_0^u y^2 e^{-y^2/(2\sigma^2)}\,dy = \int_0^{u/\sigma} x^2 e^{-x^2/2}\,dx,$$

which implies that the Gaussian speed distributions in (3.24) are basically the same. So, in the rest we assume that $\sigma = 1$:

$$\sqrt{\frac{2}{\pi}} y^2 e^{-y^2/2} \quad \text{for all } y \geq 0. \tag{3.25}$$

To define a large system of N point billiards with Gaussian initial velocites, we assume that ρ_j (representing "radius"), $1 \leq j \leq N$ are N independent, identically distributed positive random variables, with common distribution

$$\Pr[\rho_j \leq u] = \sqrt{\frac{2}{\pi}} \int_0^u y^2 e^{-y^2/2}\, dy \tag{3.26}$$

for all positive $u > 0$ (here the function $e^{-y^2/2}$ comes from the standard normal distribution, and y^2 is motivated by the surface area $4\pi y^2$ of a sphere of radius $y > 0$). We also assume that $\mathbf{u}_j$ (= direction), $1 \leq j \leq N$ are N identically distributed random variables, where each $\mathbf{u}_j$ is uniformly distributed on the unit sphere $\mathbf{S}^2$, and the $2N$ random variables $\mathbf{u}_j$, ρ_j, $1 \leq j \leq N$ are mutually independent. Formally, we consider the product measure on the Cartesian product set

$$\begin{aligned}(\rho_1, \mathbf{u}_1, \ldots, \rho_N, \mathbf{u}_N) &\in [0, \infty) \times \mathbf{S}^2 \times \cdots \times [0, \infty) \times \mathbf{S}^2 \\ &= ([0, \infty) \times \mathbf{S}^2)^N = \Omega_{\text{Gauss}},\end{aligned}$$

where the factor measure is a probability measure (we divide the surface area on the unit sphere by 4π). So, the product measure, denoted by $\text{ProdMeas}_{\text{Gauss}}$, is also a probability measure on $([0, \infty) \times \mathbf{S}^2)^N = \Omega_{\text{Gauss}}$.

The key fact is that $\text{ProdMeas}_{\text{Gauss}}$ is a d-dimensional Gaussian distribution with $d = 3N$.

The actual speed of the jth particle is a constant (depending on the particle) $|\mathbf{x}_j'(t)| = v_j(t) = \rho_j$, $0 \leq t \leq T$, and the whole billiard-path (piecewise linear zig-zag) is determined by the initial velocity $\mathbf{x}_j'(0) = \mathbf{v}_j(0) = \rho_j \mathbf{u}_j$. (In fact, the piecewise linear path is determined by the unit vector $\mathbf{u}_j$ alone.)

Let us return to (3.26); using integration by parts, we have

$$\sqrt{\frac{2}{\pi}} \int_0^\infty y \cdot y^2 e^{-y^2/2}\, dy = 2\sqrt{\frac{2}{\pi}} \int_0^\infty y \cdot e^{-y^2/2}\, dy = \sqrt{\frac{8}{\pi}},$$

that is,

$$\sqrt{8/\pi} \text{ is the mean speed.} \tag{3.27}$$

Typical Gaussian distribution precisely means $(1-\varepsilon)$-part with some small but fixed $\varepsilon > 0$ (say, $\varepsilon = 10^{-2}$ means 99%) of the product set $\Omega_{\text{Gauss}} = ([0,\infty)\times\mathbf{S}^2)^N$, equiped with the product measure $\text{ProdMeas}_{\text{Gauss}}$. Formally,

$$\begin{aligned}(\rho_1, \mathbf{u}_1, \ldots, \rho_N, \mathbf{u}_N) \in \mathcal{A}_1 \quad &\text{where} \quad \mathcal{A}_1 \subset ([0,\infty)\times\mathbf{S}^2)^N \\ = \Omega_{\text{Gauss}} \quad &\text{with } \text{ProdMeas}_{\text{Gauss}}(\mathcal{A}_1) = 1-\varepsilon.\end{aligned}$$

We also need the Gaussian speed distribution for arbitrary dimension $d \geq 1$. In the one-dimensional case we have the density function

$$\sqrt{\frac{2}{\pi}} e^{-y^2/2} \quad \text{for } y > 0,$$

and 0 otherwise; and in general, for every odd dimension $d \geq 3$ we have the density function

$$\sqrt{\frac{2}{\pi}} \frac{y^{d-1} e^{-y^2/2}}{(d-2)!!} \quad \text{for } y > 0, \tag{3.28}$$

and 0 otherwise (here we use the well-known notation $k!! = k(k-2)(k-4)\cdots$, where the last factor is 1 or 2 depending on the parity of k). Finally, for every even $d \geq 2$, we have the density function

$$\frac{y^{d-1} e^{-y^2/2}}{(d-2)!!} = \frac{y^{d-1} e^{-y^2/2}}{2^{\frac{d}{2}-1}(\frac{d}{2}-1)!} \quad \text{for } y > 0, \tag{3.29}$$

and 0 otherwise.

The disadvantage of definition (3.28)–(3.29) (describing the speed distribution) is that it comes without any *a priori* motivation for the "weird" constant factors in (3.28)–(3.29). An equivalent alternative definition — which is certainly more instructive — is to describe the Gaussian velocity distribution $\mathbf{v} = (v_1, \ldots, v_d)$ with its multivariate density function

$$f(x_1, \ldots, x_d) = \prod_{j=1}^{d} \frac{1}{\sqrt{2\pi}} e^{-x_j^2/2}. \tag{3.30}$$

Since (3.30) is the product of the factors $\frac{1}{\sqrt{2\pi}}e^{-x_j^2/2}$, the coordinates $v_1, \ldots, v_d$ of the velocity $\mathbf{v}$ are independent random variables with standard normal distribution each. This is why (3.30) is called the density function of the d-dimensional standard normal distribution.

We can rewrite (3.30) in the form

$$f(\mathbf{x}) = (2\pi)^{-d/2}e^{-|\mathbf{x}|^2/2}, \tag{3.31}$$

where $|\mathbf{x}| = \sqrt{x_1^2 + \ldots + x_d^2}$ is the usual Euclidean distance. So (3.31) implies that the distribution of the velocity $\mathbf{v}$ — i.e., the distribution of the d-dimensional standard normal distribution — is isotropic. Since $|\mathbf{v}|$ is the speed, by (3.31) the density function of the speed distribution is

$$g(y) = (2\pi)^{-d/2}\text{SurfArea}(\mathbf{S}^{d-1}(y))e^{-y^2/2}, \tag{3.32}$$

where $\mathbf{S}^{d-1}(y) = \{\mathbf{x} \in \mathbb{R}^d : \ |\mathbf{x}| = y\}$ is the sphere in the d-space of radius y, and SurfArea($\mathbf{S}^{d-1}(y)$) stands for its surface area.

The next step is to show that (3.32) gives back (3.28)–(3.29), and thus it explains, in retrospect, the geometric meaning of the *ad hoc* constant factors in (3.28)–(3.29). Since a solid sphere (= ball) is the union of concentric spheres, we can express SurfArea($\mathbf{S}^{d-1}(y)$) in terms of the derivative of the volume (warning: $\frac{\mathrm{d}}{\mathrm{d}y}$ indicates derivative and d is the dimension)

$$\begin{aligned}\text{SurfArea}(\mathbf{S}^{d-1}(y)) &= \frac{\mathrm{d}}{\mathrm{d}y}\ \text{Volume(solid sphere } \mathbf{S}^{d-1}(y))\\ &= dy^{d-1}C^*(d),\end{aligned} \tag{3.33}$$

where $C^*(d)$ is the (hyper)volume of the d-dimensional unit ball. A well-known recurrence formula on trigonometric integrals gives that $C^*(d)$ is equal to the ratio of $\pi^{d/2}$ and the Gamma function at $\frac{d}{2}+1$, i.e.,

$$\begin{aligned}C^*(d) &= \text{volume of the } d\text{-dimensional unit ball}\\ &= \frac{\pi^{d/2}}{(d/2)!} \quad \text{or} \quad \frac{2^{\lceil d/2\rceil}\pi^{\lfloor d/2\rfloor}}{1\cdot 3\cdot 5\cdot 7\cdots d}\end{aligned} \tag{3.34}$$

if d is even or odd (in the odd case $\lceil x\rceil$ and $\lfloor x\rfloor$ denote the upper and lower integral parts of a real number x). Combining (3.32), (3.33) and (3.34), we obtain, respectively, (3.28) and (3.29). This proves the equivalence of the two definitions of the multi-dimensional Gaussian distribution.

4. Dimension-Free Strong Uniformity on a Realistic Time Scale

We study strong uniformity in very high dimensions; e.g., when the dimension is in the range of the Avogadro number (around 10^{24}). Note that Theorem B2(b) is about arbitrary dimension d, but, unfortunately, it does not help, because of the unspecified constant factor $c_0(d)$ in the upper bound for the discrepancy. Our proof of Theorem B2(b) in [Be2015] gives a very weak exponential upper bound on $c_0(d)$, which makes it totally useless in high dimensional applications. It would be most useful, therefore, to have an upper bound on the discrepancy that does *not* depend on the dimension. And indeed, in this section we are able to formulate a result that is basically *dimension-free*.

Note that the diameter of the d-dimensional unit cube $[0,1]^d$ is $\sqrt{d}$. Moreover, it is an easy exercise in probability theory to prove that the distance between two randomly chosen points in $[0,1]^d$ is $\sqrt{d/6}+o(\sqrt{d})$ with probability close to one if d is large. These two facts explain why it is natural to modify the time-discrepancy

$$\int_0^T f(t\mathbf{e})\,dt - T\int_{I^d} f\,d\mathrm{V} \tag{4.1}$$

in Theorem B2(b) by replacing t with $t\sqrt{d}$, and to study

$$\int_0^T f(t\sqrt{d}\mathbf{e})\,dt - T\int_{I^d} f\,d\mathrm{V} \tag{4.2}$$

instead of (4.1), where $\mathbf{e}\in\mathbf{S}^{d-1}$ is a d-dimensional unit vector. The effect of the switch from (4.1) to (4.2) is modest in small dimensions, but it becomes substantial in very large dimensions.

In fact, we need the following slightly more general notation: for $0\le T_1<T_2$ and $\mathbf{v}\in\mathbb{R}^d\setminus\mathbf{0}$ write (vol_d is the d-dimensional Lebesgue measure)

$$\begin{aligned}\mathrm{TimeDiscrep}_f(\mathbf{v};T_1,T_2) &= D_f(\mathbf{v};T_1,T_2)\\ &= \int_{T_1}^{T_2} f(t\mathbf{v})\,dt-(T_2-T_1)\int_{I^d} f\,d\mathrm{V}.\end{aligned} \tag{4.3}$$

So (4.3) is the special case $D_f(\sqrt{d}\mathbf{e};0,T)$.

For simplicity, in Theorem 4.1 below we just consider the case of test sets $f = \chi_S$, where $S \subset [0,1)^d$. Write

$$\begin{aligned}\text{TimeDiscrep}_S(\mathbf{v}; T_1, T_2) &= D_{\chi_S}(\mathbf{v}; T_1, T_2) \\ &= \int_{T_1}^{T_2} \chi_S(t\mathbf{v})\, dt - (T_2 - T_1)\text{vol}_d(S). \quad (4.4)\end{aligned}$$

Notice that Theorem B1(b) immediately implies the following soft qualitative result. *Let $S \subset [0,1)^d$ be an arbitrary Lebesgue measurable set with $0 < \text{vol}_d(S) < 1$; then for almost every direction $\mathbf{e} \in \mathbf{S}^{d-1}$ in the d-space*

$$\text{TimeDiscrep}_S(\mathbf{v}; 0, T) = O(T^{1/2}).$$

The following is a "dimension-free" (to be explained below) quantitative version of this qualitative result.

Theorem 4.1. *Let $S \subset [0,1)^d$ be an arbitrary measurable test set in the d-dimensional unit torus with $d \geq 10^3$. Let $p = \text{vol}_d(S)$ be the d-dimensional Lebesgue measure of S. Let $T = T_0 = T_0(d) > 0$ be the solution of the equation*

$$100dTe^{-\pi^2 T^2/2} = 1.$$

Note that

$$T_0 = T_0(d) = \frac{\sqrt{2}}{\pi}\sqrt{\log d} + o(1),$$

where $o(1) = o_d(1) \to 0$ as $d \to \infty$.

Given any $0 < \varepsilon < 1$, there exists a measurable subset $\mathcal{A} = \mathcal{A}(d;\varepsilon) \subset \mathbf{S}^{d-1}$ of the (hyper)sphere such that the normalized surface area of $\mathcal{A}$ is $> 1 - \varepsilon$, and

$$\begin{aligned}&|\text{TimeDiscrep}_S(\sqrt{d}\mathbf{e}; T_0, T_1)| \\ &\leq \sqrt{p(1-p)}\left(\frac{50}{\sqrt{\varepsilon}}\sqrt{T_1}(2 + \log_2 T_1)^3 + 5\right) \quad (4.5)\end{aligned}$$

holds for every $\mathbf{e} \in \mathcal{A}$ and every $T_1 > \max\{3T_0, 10\}$.

Remarks. It is easy to extend Theorem 4.1 for square-integrable test functions $f \in L_2$; we leave it to the reader.

The requirement $d \geq 10^3$ is purely technical. Theorem 4.1 should be true for all dimensions less than 10^3.

The crucial fact here is that the upper bound on the error in (4.5) does not depend on the dimension d. This is why we call Theorem 4.1 a dimension-free result (despite the fact that the threshold $T_0 = T_0(d)$ does depend on the dimension).

Also the upper bound on the error in (4.5) does not depend on the complexity (= ugliness) of the test set S. The common starting point of the torus lines $t\sqrt{d}\mathbf{e}$, $t \geq 0$ is the origin, but of course we could choose any other common starting point (since the torus is translation invariant). The order of the error term

$$\sqrt{T_1}(2 + \log_2 T_1)^3 = T_1^{1/2+o(1)}$$

is nearly square-root size, which is basically best possible. Indeed, as we said above, the error term in Theorem B2

$$T^{\frac{1}{2}-\frac{1}{2(d-1)}}$$

is best possible (apart from a polylogarithmic factor), and $\frac{1}{2} - \frac{1}{2(d-1)}$ converges to $\frac{1}{2}$ as $d \to \infty$. Square-root size upper bound for the error term is very good, since uniformity requires much less: any sublinear upper bound suffices. These facts justify the claim that Theorem 4.1 is a *dimension-free, start-free and complexity-free strong uniformity* result. For brevity we combine these properties into a single concept that we call *robust uniformity*. Thus we say that Theorem 4.1 exhibits *robust uniformity*, where

$$\text{Robust Uniformity} = \text{Dimension-Free, Start-Free and Complexity-Free Strong Uniformity.} \tag{4.6}$$

The only dependence on the dimension d in Theorem 4.1 is in the threshold $T_0 = T_0(d)$, which is an extremely weak dependence. Indeed, $T_0(d)$ is shockingly small: it is a square-root logarithmic function of d. For example, if $d = 10^{1000}$ then $T_0 \leq 25$.

Perhaps the reader is wondering whether or not we need the strange threshold $T_0 = T_0(d)$ in Theorem 4.1 (in Theorem B2 we do not have such a threshold). It is natural to ask the question: why do we ignore the initial part

$$0 \leq t \leq T_0 = \frac{\sqrt{2}}{\pi}\sqrt{\log d} + o(1) \tag{4.7}$$

of the torus lines? Is it really necessary? The answer is yes: we have to ignore an initial part. In Sec. 23 we prove why it is impossible to expect any kind of uniformity in the slightly shorter interval

$$0 \le t < \frac{\sqrt{\log d}}{2\pi} - o(1). \tag{4.8}$$

Note that interval (4.8) is more than one-third of (4.7); they have the same order $\sqrt{\log d}$.

We may call $T_0 = T_0(d)$ in Theorem 4.1 the threshold for *configuration space equilibrium*, or *conf-space equilibrium* in short. It is the threshold when the typical time evolution of a system with $N = d/3$ particles and Gaussian initial velocity distribution reaches equilibrium in the configuration space. (Note that $\sqrt{T_1}$ in (4.5) represents the best possible square-root fluctuation.)

We derive Theorem 4.1 from a more general result Theorem 4.2; see below (we carry out the deduction in Sec. 26). Theorem 4.2 is about the "Gaussian" square-integral of (4.3)

$$\begin{aligned}\Delta_f^2(\text{Gauss}; T_1, T_2) &= (2\pi)^{-d/2} \int_{\mathbf{v}\in\mathbb{R}^d} |\text{TimeDiscrep}_f(\mathbf{v}; T_1, T_2)|^2 e^{-|\mathbf{v}|^2/2}\, d\mathbf{v} \\ &= (2\pi)^{-d/2} \int_{\mathbf{v}\in\mathbb{R}^d} |D_f(\mathbf{v}; T_1, T_2)|^2 e^{-|\mathbf{v}|^2/2}\, d\mathbf{v} \\ &= \int_{\rho=0}^{\infty} \int_{\mathbf{e}\in\mathbf{S}^{d-1}} |D_f(\rho\mathbf{e}; T_1, T_2)|^2 \frac{\rho^{d-1}e^{-\rho^2/2}}{C_d}\, d\rho\, d\text{SA}^\star(\mathbf{e}),\end{aligned} \tag{4.9}$$

where $d\text{SA}^\star(\mathbf{e})$ denotes the integration with respect to the normalized surface area on the sphere $\mathbf{e} \in \mathbf{S}^{d-1}$, i.e., $\text{SA}^\star(\mathbf{S}^{d-1}) = 1$, and

$$C_d = \left\{\sqrt{\frac{\pi}{2}} \text{ or } 1\right\} \cdot (d-2)!! \tag{4.10}$$

with

$$\left\{\sqrt{\frac{\pi}{2}} \text{ or } 1\right\} \text{ is 1 for even } d \text{ and } \sqrt{\frac{\pi}{2}} \text{ for odd } d, \tag{4.11}$$

where in (4.10) we use the well-known notation $k!! = k(k-2)(k-4)\cdots$, and the last factor is 1 or 2 depending on the parity of k. In (4.9) the vector $\mathbf{v} = \rho\mathbf{e}$ has d-dimensional standard normal (= Gaussian) distribution. This explains the use of "Gauss" in $\Delta_f^2(\text{Gauss}; T_1, T_2)$ (see (4.29)). Note that

(4.9)–(4.11) immediately follow from (3.32)–(3.34) at the end of Sec. 3; in particular, ρ^{d-1}/C_d is the surface area of the sphere of radius ρ in $\mathbb{R}^d$.

Theorem 4.2. *Let $1 \le U < W$ be real numbers and $d \ge 2$ be an integer such that*

$$e^{\pi^2 U^2/2} \ge 3dU. \tag{4.12}$$

Then for every test function $f \in L_2(I^d)$ with $I^d = [0,1)^d$,

$$\begin{aligned}\Delta_f^2(\text{Gauss}; U, W) &= (2\pi)^{-d/2} \int_{\mathbf{v}\in\mathbb{R}^d} (\text{TimeDiscrep}_f(\mathbf{v}; U, W))^2 e^{-|\mathbf{v}|^2/2}\, d\mathbf{v} \\ &= (2\pi)^{-d/2} \int_{\mathbf{v}\in\mathbb{R}^d} \left(\int_U^W f(t\mathbf{v})\, dt - (W-U) \int_{I^d} f\, dV \right)^2 \\ &\quad \times e^{-|\mathbf{v}|^2/2}\, d\mathbf{v} \le 10\sigma_0^2(f) \lceil \log_2(W/U) \rceil (W-U+1),\end{aligned} \tag{4.13}$$

where

$$\sigma_0^2(f) = \int_{I^d} |f|^2\, dV - \left| \int_{I^d} f\, dV \right|^2 = \int_{I^d} \left| f(\mathbf{y}) - \int_{I^d} f\, dV \right|^2 d\mathbf{y},$$

noting that in the special case of a characteristic function $f = \chi_S$, $S \subset I^d$,

$$\sigma_0^2(f) = \text{vol}_d(S)(1 - \text{vol}_d(S)) \le \text{vol}_d(S).$$

Note that Theorem 4.2 is a main result that we apply a lot — Theorem 4.1 is just an interesting corollary. Theorem 4.2 goes far beyond being a "strong uniformity" result: it is a *dimension-free, complexity-free and start-free strong uniformity* result; or using (4.6), we may say that Theorem 4.2 exhibits *robust uniformity.* Sections 5–8 are all about the applications of Theorem 4.2 in the very high dimensional configuration space (and many more come later).

The value of the constant 10 is of course accidental, and it is basically irrelevant in the applications.

Note that $\Delta_f^2(\text{Gauss}; U, W)$ is the average square-error, and, intuitively speaking, we may refer to

$$\left(\sigma_0^2(f)(W-U)\right)^{1/2} = \sigma_0(f)\sqrt{W-U}$$

as the "inevitable random error".

Condition (4.12) is equivalent to

$$U \geq \frac{\sqrt{2}}{\pi}\sqrt{\log d} + o(1). \tag{4.14}$$

The square-root-logarithmic (4.14) is the (shockingly small!) threshold for *conf-space equilibrium*. It is the threshold when the typical time evolution of a system with $N = d/3$ particles and Gaussian initial velocity distribution reaches equilibrium in the configuration space. (Note that Theorem 4.2 implies the best possible square-root fluctuation; see Theorem 4.3 below.)

We apply Theorem 4.2 as a "short-time ergodic theorem": it justifies the approximation "configuration space average" ≈ "short-time average" in a quantitative form. These applications can be summarized in the following vague statement.

Metatheorem. *The typical time evolution of a large system with non-interacting particles, starting from an arbitrary (off-equilibrium) initial configuration, reaches conf-space equilibrium on a realistic time scale, exhibiting robust uniformity in the configuration space. In this equilibrium state the system demonstrates "advanced randomness," and stays in this state for a very, very long time (long-term stability).*

This book is an attempt to turn the vague Metatheorem into precise theorems.

What happens if the Gaussian initial velocity distribution is replaced by other initial velocity distribution? Can we still prove an analog of Theorem 4.2? We return to this important question in Secs. 13 and 14.

Theorem 4.2 is complemented with the following result.

Theorem 4.3. *Let* $\mathbf{n}_0$ *be one of the six lattice points* $(\pm 1, 0, 0), (0, \pm 1, 0), (0, 0, \pm 1)$ *in the neighborhood of the origin, and let* $f_0(\mathbf{u}) = e^{2\pi i \mathbf{n}_0 \cdot \mathbf{u}}$. *Then for every* $0 \leq U < W < \infty$,

$$\Delta_f^2(\text{Gauss}; U, W) \geq \sigma_0^2(f_0)\frac{W-U}{\sqrt{2\pi}} - \frac{1}{5}.$$

In view of Theorem 4.3 the upper bound in (4.13) is sharp apart from a constant factor. What makes it particularly interesting is that the test function f_0 is analytic, i.e., it is as nice as it gets.

We postpone the proofs of the theorems in this section to Secs. 9 and 26.

5. Rapid Approach and Long-Term Stability of Square-Root Equilibrium

We recall the key diagram from the Preface:

Robust Uniformity

= Dimension-Free, Start-Free and Complexity-Free Strong Uniformity

$\longrightarrow$ Short Time Ergodic Theorems

$\longrightarrow$ Short Term Time Average = Configuration Space Average plus negligible error,

which holds independently of the initial configuration (= unique ergodicity)

$\longrightarrow$ fast approach to Configuration Space Equilibrium

$\longrightarrow$ long term stability of different forms of equilibria in the particle space :

Square Root Fluctuation Equilibrium, CLT and Poisson Snapshot Equilibrium,

where CLT stands for the central limit theorem. Our goal is to work out the missing details. In Sec. 4 we already formulated our first theorem about *robust uniformity in the configuration space*, see Theorems 4.2, and as the diagram shows, we are going to apply it, and its extensions in Secs. 13–14, as a "short-time ergodic theorem". According to the diagram, the term "short-time ergodic theorem" refers to the approximation property "conf-space average" $\approx$ "short-time average". The key point here is that we replace the "long-time average" in ergodic theorem (which usually means unrealistic time scale) with realistic "short-time average". We demonstrate how this approximation property implies fast approach to a state of "randomness" and its long-term stability.

Motivated by Theorem 3.3 (asymptotic CLT), it is a natural intuition to visualize "snapshot equilibrium" in the particle space (="gas container") as a state where the system exhibits square-root fluctuation. More precisely, given a test set, the point-counting function (see (3.11)) should differ from the *expected value* (i.e., N times the volume) by $O(\sqrt{N})$. In other words, it is a good intuition to visualize "snapshot equilibrium" as a *square-root fluctuation equilibrium in the particle space*, or simply *square-root equilibrium.*

Warning: Using *square-root equilibrium* as the definition of "snapshot equilibrium," the statement "once the system reaches (snapshot) equilibrium (in the particle space), it stays in (snapshot) equilibrium forever" is certainly *untrue* for the unlimited time evolution of a typical trajectory of the system (i.e., $t \to \infty$).

Perhaps the reader's first reaction to justify this claim would be to recall the well-known Zermelo–Poincaré paradox. Well, it has the right message, but, strictly speaking, the Poincaré recurrence theorem does not apply here. Indeed, the Poincaré recurrence theorem is about the time evolution starting from *almost every* initial condition, and what we study here is the time evolution starting from an arbitrary explicit initial point configuration, which represents a zero set in the space of all initial conditions. What we can use instead is Theorem 3.1 ("superdiscrepancy") in the point billiard model, which gives a rigorous proof. Indeed, it tells us that, given an *arbitrary* initial configuration of starting points, if the initial velocity coordinates are linearly independent over the rationals (representing "typical" directions), then the time evolution of this individual trajectory of the system eventually violates square-root equilibrium in the worst possible way. In fact, it violates square-root equilibrium (in the particle space) infinitely many times.

In this section we focus on the following natural question: after reaching "square-root equilibrium in the particle space" (= "snapshot equilibrium"), *how long* does the time evolution of a typical individual trajectory (of the point billiard system) stay in "square-root equilibrium?"

Proving long-term stability of square-root equilibrium in the particle space with respect to an arbitrary but fixed measurable test set. The N-point-billiard model in the cube can be reduced to the *torus-billiard* model with N point-billiards moving on torus lines via *unfolding*. We study the simplest model where the particles independently have Gaussian initial velocity distribution in the 3-space. That is, the *set* of the N particles in the particle space $I^3 = [0,1)^3$ at time t is

$$\begin{aligned}\mathcal{Y}(\text{Gauss};\omega;t) &= \mathcal{Y}(\text{Gauss};\rho_1,\mathbf{e}_1,\ldots,\rho_N,\mathbf{e}_N;t)\\ &= \{\mathbf{y}_1+\rho_1 t\mathbf{e}_1,\ldots,\mathbf{y}_N+\rho_N t\mathbf{e}_N\} \text{ modulo one}\\ &= \{\{\mathbf{y}_1+\rho_1 t\mathbf{e}_1\},\ldots,\{\mathbf{y}_N+\rho_N t\mathbf{e}_N\}\}, \qquad (5.1)\end{aligned}$$

where

$$\mathcal{Y} = \{\mathbf{y}_1, \mathbf{y}_2, \ldots, \mathbf{y}_N\} \subset [0,1]^3$$

is the N-element set of initial point configuration, and the initial velocities of the particles are independent random variables having the same speed distribution with density $g(u) = \frac{\sqrt{2}}{\pi}u^2 e^{-u^2/2}$, $0 \le u < \infty$, which is the density of the speed distribution of the 3-dimensional Gaussian velocity distribution (see the end of Sec. 3). So, the trajectory of the kth particle is $\mathbf{y}_k + \rho_k t\mathbf{e}_k \in \mathbb{R}^3$ modulo one, $1 \le k \le N$, where $\Pr[\rho_k \le u] = \frac{\sqrt{2}}{\pi}\int_0^u z^2 e^{-z^2/2}\,dz$.

Warning: In the last line of (5.1) we used the slightly ambiguous notation where the outside $\{\ldots\}$ represents a set, and the inside $\{\ldots\}$ represents fractional part applied for the three coordinates, clarifying the meaning of modulo one; we hope this minor ambiguity does not confuse the reader.

Now we switch to the configuration space: the curve in the configuration space $I^d = [0,1)^d$ with $d = 3N$, representing the time evolution of the *whole* system (5.1), is the following *straight line* in $\mathbb{R}^d$ modulo one:

$$\vec{\mathcal{Y}}(\text{Gauss}; \omega; t) = \vec{\mathcal{Y}} + t\mathbf{v}(\omega) \quad \text{modulo one,} \tag{5.2}$$

where

$$\omega = (\rho_1, \mathbf{e}_1, \rho_2, \mathbf{e}_2, \ldots, \rho_N, \mathbf{e}_N) \in \Omega_{\text{Gauss}} = \left([0,\infty) \times \mathbf{S}^2\right)^N \tag{5.3}$$

and

$$\mathbf{v}(\omega) = (\rho_1\mathbf{e}_1, \rho_2\mathbf{e}_2, \ldots, \rho_N\mathbf{e}_N). \tag{5.4}$$

As we explained at the end of Sec. 3, the product space Ω_{Gauss} is equipped with the product measure $\text{ProdMeas}_{\text{Gauss}}$, where the half-line $[0,\infty)$ has the probability density function $g(u) = \frac{\sqrt{2}}{\pi}u^2 e^{-u^2/2}$. Here the arrow on top of $\vec{\mathcal{Y}}$ indicates the $3N$-dimensional *vector*

$$\vec{\mathcal{Y}} = (\mathbf{y}_1, \mathbf{y}_2, \ldots, \mathbf{y}_N)$$

formed from the N-element initial point configuration $\mathcal{Y} \subset [0,1]^3$.

Key Fact: since the particles independently have Gaussian initial velocity distribution in the 3-space, $\text{ProdMeas}_{\text{Gauss}}$ is a d-dimensional Gaussian distribution with $d = 3N$.

Let $B \subset I^3 = [0,1)^3$ be an arbitrary but fixed measurable test set in the particle space (="gas container"), where vol(B) denotes the 3-dimensional Lebesgue measure. Assume that N is large; we study the following question. Is it true that, once a typical time evolution of the Gaussian torus-billiard system reaches square-root equilibrium in the particle space, then it stays in that state in the quantitative sense of factor 30 (say)

$$||\mathcal{Y}(\text{Gauss};\omega;t) \cap B| - \text{vol}(B)N| \leq 30\sqrt{N} \tag{5.5}$$

for an extremely long time (with the possible exception of a totally negligible set of t's)? Of course, the choice of 30 in (5.5) is accidental.

Note that square-root equilibrium is clearly the best that we can hope for. Indeed, it immediately follows from the time-lapse central limit theorem Theorem 3.3, which gives a precise description of the (typical) square-root size fluctuations.

(Note that the square-root equilibrium in the particle space

$$|\mathcal{Y}(\text{Gauss};\omega;t) \cap B| = \text{vol}(B)N + O\left(\sqrt{N}\right),$$

see (5.5), can be upgraded to the sharper inequality

$$|\mathcal{Y}(\text{Gauss};\omega;t) \cap B| = \text{vol}(B)N + O\left(\sqrt{\text{vol}(B)N}\right).$$

For the sake of simplicity we work with the former concept (5.5). We return to the latter at the end of this section.)

By using Theorem 4.2 (a "robust uniformity" result; see (4.13)) we give a positive answer to this question. This is the first application of Theorem 4.2 where it acts like a *short-time ergodic theorem* in the configuration space (more applications come in the next sections). It means that "the configuration space average nearly equals the short-term time average". The good news is that the configuration space average can be easily computed with direct application of probability theory (since the configuration space is a product space with product measure; see the application of Bernstein's Large Deviation Inequality in (5.9) below). Moreover, Theorem 4.2 has the advantage that it works for arbitrary off-equilibrium initial configuration (unlike Birkhoff's ergodic theorem that works only for typical initial condition, and a typical initial condition represents equilibrium — which is the trivial case, since we study off-equilibrium dynamics). Theorem 4.2 has an explicit error term (unlike the ergodic theorem that does not have an explicit error term), and we can use it to describe the time evolution in realistic short-term time scale. The details go as follows.

As usual, we employ the notation

$$\vec{\mathcal{Y}} = (\mathbf{y}_1, \ldots, \mathbf{y}_N) \in I^d$$

(where $\mathcal{Y} = \{\mathbf{y}_1, \ldots, \mathbf{y}_N\}$ is the given N-set of initial point configuration) to denote the corresponding point in the configuration space $I^d = [0,1)^{3N}$.

The family of time evolutions $\mathcal{Y}(\text{Gauss}; \omega; t)$, $\omega \in \Omega_{\text{Gauss}}$ of the 3-dimensional Gaussian torus-billiard model (in the particle space I^3) is represented by the family of torus lines (see (5.2))

$$\vec{\mathcal{Y}}(\text{Gauss}; \omega; t) = \vec{\mathcal{Y}} + t\mathbf{v}(\omega) \quad \text{modulo one} \tag{5.6}$$

in the configuration space I^d, all starting from the *same point* $\vec{\mathcal{Y}} \in I^d$.

For an arbitrary $\gamma > 0$ define the following — very complicated and "ugly"! — test set in the configuration space

$$S(B;\gamma) = \left\{ \vec{\mathcal{Z}} \in I^d : \; ||\mathcal{Z} \cap B| - \text{vol}(B)N| > \gamma\sqrt{N} \right\}, \tag{5.7}$$

where

$$\vec{\mathcal{Z}} = (z_1, \ldots, z_{3N}) \quad \text{and} \quad \mathcal{Z} = \{\mathbf{z}_1, \ldots, \mathbf{z}_N\}$$

with $\mathbf{z}_k = (z_{3k-2}, z_{3k-1}, z_{3k})$, $1 \le k \le N$.

We recall Bernstein's inequality (see e.g. in [F71], a well-known large deviation type inequality in probability theory, which will be applied several times in this book.

Bernstein inequality. *Let $X_1, \ldots, X_n$ be real-valued independent random variables with zero expectation $\mathbf{E}X_i = 0$ and $|X_i| \le M$, $1 \le i \le n$. Then, for all positive τ,*

$$\Pr\left[\left|\sum_{i=1}^n X_i\right| \ge \tau\right] \le 2\exp\left(-\frac{\tau^2/2}{(\sum_{i=1}^n \mathbf{E}X_i^2) + (\tau M/3)}\right).$$

In the important case of $\Pr[X_i = 1$ or $0] = p_i$ or $q_i = 1 - p_i$, the general result implies for every positive γ,

$$\Pr\left[\left|\sum_{i=1}^n (X_i - p_i)\right| \ge \gamma\left(\sum_{i=1}^n p_i q_i\right)^{1/2}\right]$$

$$\le 2\exp\left(-\frac{\gamma^2/2}{1 + \gamma\left(\sum_{i=1}^n p_i q_i\right)^{-1/2}/3}\right). \tag{5.8}$$

In the special case of the binomial distribution $\Pr[X_i = 1 \text{ or } 0] = p$ *or* $q = 1 - p$, *we have for every positive* γ,

$$\Pr\left[\left|\sum_{i=1}^{n}(X_i - p)\right| \geq \gamma\sqrt{npq}\right]$$

$$= \sum_{\substack{0\leq k\leq n: \\ |k-pn|\geq\gamma\sqrt{pqn}}} \binom{n}{k} p^k q^{n-k} \leq 2\exp\left(-\frac{\gamma^2/2}{1+\frac{\gamma}{3\sqrt{npq}}}\right), \tag{5.9}$$

where $\sqrt{npq}$ *is the standard deviation of the binomial distribution.*

In the symmetric case $p = q = 1/2$ *we have the particularly simple upper bound for every positive* $\gamma > 0$:

$$\Pr\left[\sum_{i=1}^{n}\left(X_i - \frac{1}{2}\right) \geq \gamma\sqrt{n}/2\right] = 2^{-n} \sum_{\substack{0\leq k\leq n: \\ k-n/2\geq\gamma\sqrt{n}/2}} \binom{n}{k} \leq e^{-\gamma^2/2}. \tag{5.10}$$

By using (5.9) with $p = \text{vol}(B)$ we have

$$\text{vol}_d(S(B;\gamma)) \leq 2\exp\left(-\frac{\gamma^2(2p(1-p))^{-1}}{1+\frac{\gamma}{3p(1-p)\sqrt{N}}}\right) \leq 2\exp\left(-\frac{2\gamma^2}{1+\frac{2\gamma}{\sqrt{N}}}\right), \tag{5.11}$$

where the last inequality comes from the fact $p(1-p) \leq 1/4$. The reason why we could apply Bernstein's inequality is that vol_d is a product measure, and so the $d = 3N$-dimensional volume $\text{vol}_d(S(B;\gamma))$ represents a large deviation probability for N independent random variables.

For example,

$$\text{if } \gamma = 30 \quad \text{and} \quad N \geq 10^6$$
$$\text{then (5.11) gives } 2\exp\left(-\frac{2\gamma^2}{1+\frac{2\gamma}{\sqrt{N}}}\right) < 10^{-700}, \tag{5.12}$$

which is *extremely small.* The long-term stability of (say) 30-square-root equilibrium (in the particle space) is based on this numerical fact. More precisely, we make use of (5.1)–(5.12) via Theorem 4.2 as follows.

Since the torus I^d is translation invariant, we apply Theorem 4.2 with $f = \chi_S$ where S is the translated copy $S = S(B;\gamma) - \vec{\mathcal{Y}}$ of $S(B;\gamma)$ in the torus I^d.

By (5.11),

$$\mathrm{vol}_d(S) = \mathrm{vol}_d(S(B;\gamma)) < 2\exp\left(-\frac{2\gamma^2}{1+\frac{2\gamma}{\sqrt{N}}}\right),$$

and using it in Theorem 4.2 with $W = 2^k U$, we obtain

$$\begin{aligned}
&(2\pi)^{-d/2}\int_{\mathbf{v}\in\mathbb{R}^d}\left(\int_U^{2^kU}\chi_S(t\mathbf{v})\,dt - \mathrm{vol}_d(S)(2^k-1)U\right)^2 e^{-|\mathbf{v}|^2/2}\,d\mathbf{v}\\
&= (2\pi)^{-d/2}\int_{\mathbf{v}\in\mathbb{R}^d}\left(\mathrm{TimeDiscrep}_S(\mathbf{v};U,2^kU)\right)^2 e^{-|\mathbf{v}|^2/2}\,d\mathbf{v}\\
&\le 2\exp\left(-\frac{2\gamma^2}{1+\frac{2\gamma}{\sqrt{N}}}\right)10k\left((2^k-1)U+1\right), \qquad (5.13)
\end{aligned}$$

assuming of course $U \ge 1$ and $e^{\pi^2U^2/2} \ge 3dU$. By (5.6)–(5.7) and using $S = S(B;\gamma) - \vec{\mathcal{Y}}$, we have

$$\begin{aligned}
&(2\pi)^{-d/2}\int_{\mathbf{v}\in\mathbb{R}^d}\left(\int_U^{2^kU}\chi_S(t\mathbf{v})\,dt - \mathrm{vol}_d(S)(2^k-1)U\right)^2 e^{-|\mathbf{v}|^2/2}\,d\mathbf{v}\\
&= \int_{\omega\in\Omega_{\mathrm{Gauss}}}\left(\int_U^{2^kU}\chi_S(t\mathbf{v}(\omega))dt - \mathrm{vol}_d(S)(2^k-1)U\right)^2\\
&\quad\times d\mathrm{ProdMeas}_{\mathrm{Gauss}}(\omega) = \int_{\omega\in\Omega_{\mathrm{Gauss}}}\Big(\mathrm{length}\big\{U\le t\le 2^kU :\\
&\qquad ||\mathcal{Y}(\mathrm{Gauss};\omega;t)\cap B| - \mathrm{vol}(B)N| > \gamma\sqrt{N}\big\}\\
&\quad - \mathrm{vol}_d(S(B;\gamma))(2^k-1)U\Big)^2\, d\mathrm{ProdMeas}_{\mathrm{Gauss}}(\omega), \qquad (5.14)
\end{aligned}$$

where length stands for the one-dimensional Lebesgue measure.

Combining (5.13) and (5.14), we obtain the following result.

Theorem 5.1. *Let* $\mathcal{Y}(\mathrm{Gauss};\omega;t)$, $\omega\in\Omega_{\mathrm{Gauss}}$ *be the* 3*-dimensional Gaussian torus-billiard model, and let* $B\subset[0,1)^3$ *be a measurable test with* 3*-dimensional Lebesgue measure* $\mathrm{vol}(B)$. *Assume that*

$$U \ge 1 \quad \textit{and} \quad e^{\pi^2U^2/2} \ge 3dU.$$

Then for every $\gamma > 0$ *and every integer* $k \geq 1$,

$$\int_{\omega\in\Omega_{\text{Gauss}}} \Bigg(\text{length}\left\{U \leq t \leq 2^k U : \ ||\mathcal{Y}(\text{Gauss};\omega;t) \cap B| \right.$$

$$\left. - \ \text{vol}(B)N| > \gamma\sqrt{N}\right\}$$

$$- \text{vol}_d(S(B;\gamma))(2^k - 1)U\Bigg)^2 \ d\text{ProdMeas}_{\text{Gauss}}(\omega)$$

$$\leq 2\exp\left(-\frac{2\gamma^2}{1+\frac{2\gamma}{\sqrt{N}}}\right) 10k\left((2^k-1)U+1\right), \tag{5.15}$$

where

$$\text{vol}_d(S(B;\gamma)) < 2\exp\left(-\frac{2\gamma^2}{1+\frac{2\gamma}{\sqrt{N}}}\right). \qquad \square$$

Remarks. In the proof we used all three aspects — start-free, complexity-free, dimension-free — of Robust Uniformity in Theorem 4.2. Indeed, we needed *start-free*, since the starting point

$$\vec{\mathcal{Y}} = (\mathbf{y}_1, \ldots, \mathbf{y}_N) \in I^d$$

of the torus line (5.2) in the configuration space $I^d = [0,1)^{3N}$ was arbitrary. We also needed, *complexity-free*, since the test set (5.7) was extremely complicated. Finally, we needed *dimension-free*, since for a large system the configuration space has high dimension.

Probably the reader does not find Theorem 5.1 very pretty, but it is an extremely powerful result. To illustrate the power of Theorem 5.1, let $\gamma = 30$, $U = 4$, $k = 100$, and $N = 10^{27}$; so $d = 3N = 3\cdot 10^{27}$. Then by (5.12) and (5.15),

$$\int_{\omega\in\Omega_{\text{Gauss}}} \Bigg(\text{length}\{4 \leq t \leq 4\cdot 2^{100} : \ ||\mathcal{Y}(\text{Gauss};\omega;t)\cap B|$$

$$- \text{vol}(B)N| > 30\sqrt{N}\} - \text{vol}_d(S(B;30))(2^{100}-1)4\Bigg)^2$$

$$\times d\text{ProdMeas}_{\text{Gauss}}(\omega) \leq 10^{-700}\cdot 400\cdot 2^{100}\cdot 20 < 10^{-661}. \tag{5.16}$$

Let $\Omega_{\text{Gauss}}(\text{bad})$ be the set of those $\omega \in \Omega_{\text{Gauss}}$ for which

$$\text{length}\left\{4 \leq t \leq 4\cdot 2^{100} : \ ||\mathcal{Y}(\text{Gauss};\omega;t)\cap B|\right.$$

$$\left. - \text{vol}(B)N| > 30\sqrt{N}\right\} \geq 10^{-220}. \tag{5.17}$$

We claim that (5.16) implies

$$\mathrm{ProdMeas}_{\mathrm{Gauss}}(\Omega_{\mathrm{Gauss}}(\mathrm{bad})) \leq 10^{-220}. \tag{5.18}$$

Indeed, otherwise

$$\begin{aligned}
&\int_{\omega\in\Omega_{\mathrm{Gauss}}} \Big(\mathrm{length}\Big\{4 \leq t \leq 4\cdot 2^{100} : ||\mathcal{Y}(\mathrm{Gauss};\omega;t)\cap B| \\
&\quad - \mathrm{vol}(B)N| > 30\sqrt{N}\Big\} \\
&\quad - \mathrm{vol}_d(S(B;30))(2^{100}-1)4\Big)^2 \, d\mathrm{ProdMeas}_{\mathrm{Gauss}}(\omega) \\
&\geq \int_{\omega\in\Omega_{\mathrm{Gauss}}(\mathrm{bad})} \Big(\mathrm{length}\Big\{4 \leq t \leq 4\cdot 2^{100} : ||\mathcal{Y}(\mathrm{Gauss};\omega;t)\cap B| \\
&\quad - \mathrm{vol}(B)N| > 30\sqrt{N}\Big\} \\
&\quad - \mathrm{vol}_d(S(B;30))(2^{100}-1)4\Big)^2 \, d\mathrm{ProdMeas}_{\mathrm{Gauss}}(\omega) \\
&\geq 10^{-220}\left(10^{-220} - 10^{-600}\right)^2 > 10^{-661},
\end{aligned} \tag{5.19}$$

which contradicts (5.16). In the last step we used the fact

$$\mathrm{vol}_d(S(B;30))(2^{100}-1)4 \leq 10^{-700}\cdot 10^{31} < 10^{-600}.$$

Note that the choice of $N = 10^{27}$ was "realistic" in the sense that there are roughly 10^{27} gas molecules in a cubic box of volume $1\,\mathrm{m}^3$. In the classical Bernoulli gas model the gas molecules are represented by point billiards. Using *unfolding* (see the beginning of Sec. 3) we can reduce the billiards-in-a-box model to a torus-billiard model. The threshold $U = 4$ represents — roughly speaking — the relaxation distance, i.e., the necessary number of "jumps" per particle in the torus-billiard model (which is half of the number of bounces in the billiards-in-a-box model) to reach square-root equilibrium (in the particle space) for the typical time evolution of the Gaussian case. Assume that the gas molecules have average speed 10^3 meter per second. For this Gaussian system it takes only a few milliseconds to reach square-root equilibrium. Now (5.17) and (5.18) have the following interpretation. Choosing an arbitrary (measurable) test set $B \subset [0,1)^3$ in the "gas container" (= particle space) and an arbitrary N-element initial point configuration $\mathcal{Y}$, for the totally overwhelming majority of the initial velocities (Gaussian distribution), the number of particles in B remains very close to the expected value $\text{vol}(B)N$ for an extremely long time, with the possible exception of a totally negligible set of times t.

Indeed, for every (measurable) test set $B \subset [0,1)^3$ and every $N = 10^{27}$-element initial point configuration $\mathcal{Y} \subset [0,1)^3$, there exists a subset $\Omega_{\text{Gauss}}(\text{good})$ where

$$\Omega_{\text{Gauss}}(\text{good}) = \Omega_{\text{Gauss}} \setminus \Omega_{\text{Gauss}}(\text{bad})$$

with

$$\text{ProdMeas}_{\text{Gauss}}(\Omega_{\text{Gauss}}(\text{good})) \geq 1 - 10^{-220}$$

(see (5.18)), representing a totally overwhelming majority, such that for every $\omega \in \Omega_{\text{Gauss}}(\text{good})$,

$$\left| |\mathcal{Y}(\text{Gauss};\omega;t) \cap B| - \text{vol}(B)10^{27} \right| \leq 30 \cdot 10^{13.5} \tag{5.20}$$

holds for every $4 \leq t \leq 4 \cdot 2^{100}$ with the possible exception of a set of times t of total length $< 10^{-220}$, see (5.17). The latter actually represents less than 10^{-223} seconds, which is a ridiculously short time.

Note that $4 \leq t \leq 4 \cdot 2^{100}$ represents a time interval of length about 10^{27} seconds, which is an incredibly long time: it is roughly billion times the age of the universe.

Finally, by (5.20)

$$\frac{1}{N} \left| |\mathcal{Y}(\text{Gauss};\omega;t) \cap B| - \text{vol}(B) \right| \leq 3 \cdot 10^{-12.5} < 10^{-12}, \tag{5.21}$$

which can be interpreted as almost constant density for an incredibly long time.

What happens if we increase the number N of particles? To answer the question, we replace $N = 10^{27}$ with $N = 10^{80}$ (say). We keep $\gamma = 30$ and $W = 2^k U$ with $k = 100$, and switch to $U = 7$, $N = 10^{80}$; it follows that $d = 3 \cdot 10^{80}$. Note that 10^{80} is around the estimated number of particles in the observable universe. Then we have the perfect analogue of (5.16)

$$\begin{aligned} \int_{\omega \in \Omega_{\text{Gauss}}} & \Big(\text{length}\left\{7 \leq t \leq 7 \cdot 2^{100} : \ \left||\mathcal{Y}(\text{Gauss};\omega;t) \cap B| \right.\right. \\ & \quad \left.\left. - \text{vol}(B)10^{80}\right| > 30 \cdot 10^{40}\right\} \\ & - \text{vol}_d(S(B;30))(2^{100}-1)4\Big)^2 \, d\text{ProdMeas}_{\text{Gauss}}(\omega) < 10^{-661}. \end{aligned} \tag{5.22}$$

Similarly, let $\Omega_{\text{Gauss}}(\text{bad})$ be the set of those $\omega \in \Omega_{\text{Gauss}}$ for which

$$\begin{aligned} & \text{length}\left\{7 \leq t \leq 7 \cdot 2^{100} : \ \left||\mathcal{Y}(\text{Gauss};\omega;t) \cap B| \right.\right. \\ & \quad \left.\left. - \text{vol}(B)10^{80}\right| > 30 \cdot 10^{40}\right\} \geq 10^{-220}. \end{aligned} \tag{5.23}$$

Then we have the analogue of (5.18)

$$\text{ProdMeas}_{\text{Gauss}}(\Omega_{\text{Gauss}}(\text{bad})) \leq 10^{-220}. \tag{5.24}$$

So again, for every (measurable) test set $B \subset [0,1)^3$ and every $N = 10^{80}$-element initial point configuration $\mathcal{Y} \subset [0,1)^3$, there exists a subset $\Omega_{\text{Gauss}}(\text{good}) \subset \Omega_{\text{Gauss}}$ with

$$\text{ProdMeas}_{\text{Gauss}}(\Omega_1(\text{good})) \geq 1 - 10^{-220}$$

(see (5.24)), representing a totally overwhelming majority, such that for every $\omega \in \Omega_{\text{Gauss}}(\text{good})$

$$\Big| |\mathcal{Y}(\text{Gauss};\omega;t) \cap B| - \text{vol}(B)10^{80} \Big| \leq 30 \cdot 10^{40} \tag{5.25}$$

holds for every $7 \leq t \leq 7 \cdot 2^{100}$ with the possible exception of a set of times t of total length $<10^{-220}$, see (5.23). The only difference is that by (5.22)

$$\frac{1}{N} \left| |\mathcal{Y}(\text{Gauss};\omega;t) \cap B| - \text{vol}(B) \right| \leq 3 \cdot 10^{-39},$$

which is much sharper than (5.21). That is, the larger the N, the closer we get to constant density.

Definition (5.7)

$$S(B;\gamma) = \left\{ \vec{\mathcal{Z}} \in I^d : \ \left| |\mathcal{Z} \cap B| - \text{vol}(B)N \right| > \gamma\sqrt{N} \right\} \tag{5.26}$$

becomes useless for "small" test sets B with

$$\text{vol}(B) = O\left(N^{-1/2}\right). \tag{5.27}$$

A natural way to overcome this technical restriction is to replace the square-root equilibrium concept (5.5)

$$|\mathcal{Y}(\text{Gauss};\omega;t) \cap B| = \text{vol}(B)N + O\left(\sqrt{N}\right)$$

with the sharper *relative* square-root equilibrium concept

$$|\mathcal{Y}(\text{Gauss};\omega;t) \cap B| = \text{vol}(B)N + O\left(\sqrt{\text{vol}(B)N}\right).$$

It means, we replace (5.26) with the following modified set in the configuration space

$$S'(B;\gamma) = \left\{ \vec{\mathcal{Z}} \in I^d : \ \left| |\mathcal{Z} \cap B| - \text{vol}(B)N \right| > \gamma\sqrt{pqN} \right\}, \tag{5.28}$$

where $p = \text{vol}(B) \leq 1/2$ and $q = 1 - p \geq 1/2$. Simply repeating the argument above — using an appropriate form of the Large Deviation Theorem in probability theory (usually a tail probability estimation of the binomial distribution) — we can easily prove useful *relative* square-root discrepancy even for test sets as small as

$$p = \text{vol}(B) \geq C \cdot N^{-1}, \tag{5.29}$$

where C is a "large" absolute constant. Note that (5.29) describes a much larger family of test sets than

$$\text{vol}(B) \geq C \cdot N^{-1/2},$$

and (5.29) is best possible in the sense that it is the largest class for which the *relative* square-root discrepancy estimation is still useful.

What happens if we want to prove long-term stability of square-root equilibrium with respect to a whole family of nice sets (instead of a fixed measurable test set)? Of course we cannot expect that a system stays in square-root equilibrium with respect to *all measurable* test sets simultaneously. Indeed, this follows from a well-known measure-theoretic argument: the union of the N trajectories of the N particles has volume zero, so the system does not visit the *complement* of the union at all, despite the fact that the complement has volume one, so the majority of the particles should be in the complement in the overwhelming majority of the time.

Of course long-term stability does not mean that the system stays in equilibrium forever. Indeed, the "superdiscrepancy" result Theorem 3.1 implies that, given any kind of equilibrium state, the system will leave this equilibrium for an infinite sequence of times tending to infinity.

6. Non-ergodic Time-flow: Closed Orbit Spherical Systems

Closed orbit systems represent, in some sense, the complete opposite of the box/torus billiard models that we have been studying so far. Note that closed orbits already show up in the 2-dimensional billiard: it is the case of rational slope at the start. But what we study is the typical time evolution of a large off-equilibrium system, and a typical point billiard path in the unit square $[0, 1]^2$ has irrational slope. Irrational slope implies (via the Kronecker–Weyl theorem) that the billiard path is uniformly distributed in the unit square. The two cases — irrational vs. rational slope — lead to diametrically opposite behaviors: uniformity vs. periodicity.

In this section we switch from the box/torus billiard models to spherical billiard models, where *every* orbit is closed. Indeed, the geodesics in the (algebraic) torus are the torus lines, and the geodesics on the sphere $\mathbf{S}^2 \subset \mathbb{R}^3$ are the great circles, which are closed orbits. The obvious difference is that a great circle is certainly *not* uniformly distributed on the sphere. We may say, somewhat loosely speaking, that the typical time-flow of the torus-billiard system is *ergodic* (and what is more, uniquely ergodic); on the other hand, the time-flow of the spherical billiard system is *not ergodic*. (Of course a closed orbit is still a torus — a one-dimensional torus — but its dimension is less than the dimension of the particle space.)

Despite this fundamental difference between the (at least two-dimensional) box/torus billiard systems and the closed orbit spherical billiard systems, we can still prove similar results about the approach to equilibrium, and its stability. First we define the *Gaussian GreatCircle model*, which is an analog of the 2-dimensional Gaussian torus-billiard model. Let $\mathcal{Y} \subset \mathbf{S}^2$ be an arbitrary N-element point set on the unit sphere — it represents the initial configuration of a spherical system of N particles. Let $\mathbf{y}_k \in \mathcal{Y}$ be the starting point of the kth particle. The plane passing through the origin and being perpendicular to the $(\mathbf{y}_k, -\mathbf{y}_k)$-diameter (where $-\mathbf{y}_k$ is the antipodal point of $\mathbf{y}_k$) intersects the unit sphere $\mathbf{S}^2$ in a great circle that we call the $\mathbf{y}_k$-equator. (If $\mathbf{y}_k$ is the North or South Pole $(0, 0, \pm 1)$, then we get back the usual Equator.) Each half great circle in $\mathbf{S}^2$ going from $\mathbf{y}_k$ to $-\mathbf{y}_k$ intersects the $\mathbf{y}_k$-equator in one point; this gives a natural parametrization of the family of great circles passing through $\mathbf{y}_k$. The parameter, denoted by α_k, runs between 0 and 2π. We may refer to parameter α_k as the *angle*. We choose a great circle in the family by choosing an angle α_k with uniform distribution in the interval $[0, 2\pi)$. The kth particle moves with constant speed on this great circle. *Gaussian* means that we choose the speed v_k (here the speed can be positive and negative) of the kth particle by the one-dimensional normal distribution, that is, the density function of the distribution of the speed $v = v_k$ is $(2\pi)^{-1/2} e^{-v^2/2}$ (negative speed means that the particle moves in the opposite direction). The pair (α_k, v_k) of the angle and the speed determine the initial velocity. Finally, we choose the N initial velocites (α_k, v_k), $1 \leq k \leq N$ *independently* of each other. This means that

$$\omega = (\alpha_1, \ldots, \alpha_N, v_1, \ldots, v_N) \in \Omega(\text{Gauss}; GCsphere) = [0, 2\pi)^N \times \mathbb{R}^N$$

is the initial velocity space. Let

$$\mathrm{ProdMeas}_{\mathrm{Gauss},GC} = (2\pi)^{-N}\mathrm{vol}_N \times \mathrm{Gauss}_N$$

denote the corresponding product measure. Here of course vol_N stands for the N-dimensional Lebesgue measure and Gauss_N is the N-dimensional Gauss measure in the N-space $\mathbb{R}^N$ with density function

$$(2\pi)^{-N/2}e^{-|\mathbf{v}|^2/2},$$

representing the N-dimensional standard normal distribution.

Next we define a second spherical model, which is motivated by Archimedes's famous discovery that the surface area of a spherical belt depends only on the "height" of the belt. The "height" is exactly the length of the projection on the perpendicular diameter. This is why in the second spherical model the speed of the motion of the kth particle on the corresponding great circle is not constant. Instead it is the motion of the *projection* on the $(\mathbf{y}_k, -\mathbf{y}_k)$-diameter that has constant speed (where $-\mathbf{y}_k$ is the antipodal point of $\mathbf{y}_k$). Constant projection speed on the diameter with endpoints $\mathbf{y}_k$ and $-\mathbf{y}_k$ means that the actual speed of the kth particle on the great circle (= orbit) depends on the arclength distance from $\mathbf{y}_k$. Indeed, if $0 \le \beta \le \pi$ denotes the arclength, then the actual speed of the kth particle on its orbit is $u_k/\sin\beta$, where u_k is the constant projection speed. Notice that the actual speed is infinite at $\mathbf{y}_k$ and $-\mathbf{y}_k$.

More precisely, we define the *Gaussian GreatCircleDiameter model* as follows. As usual, let $\mathcal{Y} \subset \mathbf{S}^2$ be an arbitrary N-element point set on the unit sphere — it represents the initial configuration of a spherical system of N particles. Let $\mathbf{y}_k \in \mathcal{Y}$ be the starting point of the kth particle. Again we choose a great circle in $\mathbf{S}^2$ passing through $\mathbf{y}_k$ the same way by choosing an angle α_k with uniform distribution in the interval $[0, 2\pi)$. The kth particle moves on this great circle in such a way that its projection on the $(\mathbf{y}_k, -\mathbf{y}_k)$-diameter has constant speed u_k (positive or negative; $-\mathbf{y}_k$ is the antipodal point of $\mathbf{y}_k$). We refer to this constant speed as the projection-speed. We choose the projection-speed u_k of the kth particle by the one-dimensional normal distribution, that is, the density function of the distribution of the projection-speed $u = u_k$ is $(2\pi)^{-1/2}e^{-u^2/2}$. The pair (α_k, u_k) of the angle and the projection-speed determine the initial velocity. Finally, we choose the N initial velocites (α_k, u_k), $1 \le k \le N$ *independently* of each other. This means that the Gaussian GreatCircleDiameter system has exactly the same initial velocity space and product measure as the Gaussian GreatCircle system. Nevertheless, to emphasize the difference in the interpretation (constant speed v_k versus constant projection-speed u_k) we write an extra D (= Diameter) as follows:

$$\omega = (\alpha_1, \ldots, \alpha_N, u_1, \ldots, u_N) \in \Omega(\text{Gauss}; GCDsphere) = [0, 2\pi)^N \times \mathbb{R}^N$$

where

$$\text{ProdMeas}_{\text{Gauss},GCD} = (2\pi)^{-N}\text{vol}_N \times \text{Gauss}_N$$

denotes the corresponding product measure.

We study how the two spherical models, the (Gaussian) GreatCircle model and the GreatCircleDiameter model, approach equilibrium and stay in equilibrium. Again we use Theorem 4.2 to describe the long-term stability of equilibrium (in the particle space = sphere).

Despite the fact that the GreatCircle model is simpler and more natural, the GreatCircleDiameter model has the simpler and more elegant result. So we start with the latter.

Again let $B \subset \mathbf{S}^2$ be an arbitrary but fixed measurable test set; here $\text{SA}^*(B)$ denotes the normalized surface area (so $\text{SA}^*(\mathbf{S}^2) = 1$), and let $\mathcal{Y}_{\omega;t}$ denote the (Gaussian) GreatCircleDiameter model with initial configuration $\mathcal{Y} \subset \mathbf{S}^2$ and

$$\omega = (\omega^*, \mathbf{u}) \in \Omega(\text{Gauss}; GCDsphere)$$

where

$$\omega^* = (\alpha_1, \ldots, \alpha_N) \in [0, 2\pi)^N \quad \text{and} \quad \mathbf{u} = (u_1, \ldots, u_N) \in \mathbb{R}^N.$$

We ask the same question as for the (at least 2-dimensional) torus-billiard model. Assume that N is large; is it true that, once a typical time evolution of the system reaches square-root equilibrium (in the particle space), then it stays in that state for an extremely long time (with the possible exception of a totally negligible set of t's)? We give a positive answer to this question by applying Theorem 4.2 as a *short-time ergodic theorem* in the configuration space (in fact, it works as a "large deviation theorem").

For every $\omega^* \in [0, 2\pi)^N$, let $GC_k = GC_k(\omega^*)$ denote the orbit of the kth particle: it is a great circle containing the antipodal points $\mathbf{y}_k \in \mathcal{Y}$ and $-\mathbf{y}_k$ with angle α_k. Since the projection of the motion on the $(\mathbf{y}_k, -\mathbf{y}_k)$-diameter has constant speed, we replace the motion of the kth particle on the great circle $GC_k = GC_k(\omega)$ with the constant speed motion in the interval $[0, 4)$ interpreted as a 1-dimensional torus ("projection"). Of course $[0, 4)$ comes from the fact that we double-count the diameter: first we go from $\mathbf{y}_k$ to $-\mathbf{y}_k$, and the second part is the reverse trip going back from $-\mathbf{y}_k$ to $\mathbf{y}_k$. For notational convenience we divide by 4 in the sense that we replace the constant speed motion in the 1-dimensional torus $[0,4)$ with the constant speed motion in the 1-dimensional unit torus $[0,1)$. This means that, for every *fixed* $\omega^* \in [0,2\pi)^N$, the family of time evolutions $\mathcal{Y}_{\omega;t}$, $\omega = (\omega^*, \mathbf{u}) \in \Omega(\text{Gauss}; GCDsphere)$ of the (Gaussian) GreatCircleDiameter model with initial configuration $\mathcal{Y} \subset \mathbf{S}^2$ is represented by the family of torus lines in the configuration space $I^N = [0,1)^N$, all starting from the origin $\mathbf{0} \in I^N$.

The great circle $GC_k = GC_k(\omega^*)$ (i.e., the orbit of the kth particle) intersects the given 2-dimensional measurable test set $B \subset \mathbf{S}^2$ in a one-dimensional *measurable* set for almost every $\omega^* \in [0,2\pi)^N$ (set of initial angles). This is Fubini's theorem in the theory of Lebesgue measure. Consider the projection of the intersection $GC_k(\omega^*) \cap B$ on the double copy of the $(\mathbf{y}_k, -\mathbf{y}_k)$-diameter, where the latter is represented as the interval $[0,4)$. Let $B_k = B_k(\omega^*) \subset [0,4)$ denote this projection, and again we apply the natural division by 4:

$$\frac{1}{4}B_k(\omega^*) \subset (0,1), \text{ where we use the standard notation}$$

$$\frac{1}{4}A = \left\{\frac{1}{4}x : x \in A\right\}. \tag{6.1}$$

Write $b_k(\omega^*) = \frac{1}{4}\text{length}(B_k(\omega^*))$, where, as usual, length stands for the 1-dimensional Lebesgue measure. The mean value of $b_k(\omega^*)$, $\omega^* \in [0,2\pi)^N$ is clearly equal to the normalized surface area of the test set $B \subset \mathbf{S}^2$; formally,

$$(2\pi)^{-N} \int_{[0,2\pi)^N} b_k(\omega^*)\, d\omega^* = \text{SA}^*(B) = \frac{1}{4\pi}\text{SurfaceArea}(B) \tag{6.2}$$

(where $d\omega^*$ represents the N-dimensional Lebesgue measure).

Write

$$\text{BadSet} = \left\{\omega^* \in [0,2\pi)^N : \left|\sum_{k=1}^{N} b_k(\omega^*) - \text{SA}^*(B)N\right| > 30\sqrt{N}\right\}. \tag{6.3}$$

Of course, we could choose here any other "large deviation factor" γ (like we did in Theorem 5.1). Note that $\gamma = 30$ is a good choice for the illustration below.

Since the initial velocities are chosen independently, we can apply Bernstein's large deviation inequality (5.8)

$$(2\pi)^{-N}\text{vol}_d\,(\text{BadSet}) \le 2\exp\left(-\frac{450}{1+10N^{-1/2}}\right). \tag{6.4}$$

Let $\omega^* \in [0, 2\pi)^N$ be *fixed*, and write

$$S(B;\omega^*) = \Bigg\{\vec{\mathcal{Z}} = (z_1, \ldots, z_N) \in I^N :$$

$$\left|\left|\left\{1 \le k \le N : \; z_k \in \frac{1}{4}B_k(\omega^*)\right\}\right| - \sum_{k=1}^{N} b_k(\omega^*)\right| > 60\sqrt{N}\Bigg\}. \tag{6.5}$$

Again using Bernstein's large deviation inequality (5.8), we obtain

$$\text{vol}_N(S(B;\omega^*)) \le 2\exp\left(-\frac{1800}{1+20N^{-1/2}}\right), \tag{6.6}$$

which corresponds to (5.11)–(5.12), where we used (5.9). (Here in (6.6) we use (5.8) in a trivial way, and to compensate for that loss, we replace the factor 30 with 60.)

We apply Theorem 4.2 with $\gamma = 60$ and $f = \chi_S$ where $S = S(B;\omega^*) \subset I^N$. Thus, for every $\omega^* \in [0, 2\pi)^N$, integer $k \ge 1$ and $U \ge 1$ satisfying $e^{\pi^2 U^2/2} > 3UN$,

$$(2\pi)^{-N/2}\int_{\mathbf{u}\in\mathbb{R}^N}\Bigg(\text{length}\Bigg\{U \le t \le 2^k U : \Bigg||\mathcal{Y}_{\omega;t} \cap B|$$

$$-\sum_{k=1}^{N} b_k(\omega^*)\Bigg| > 60\sqrt{N}\Bigg\}$$

$$-\,\text{vol}_N(S(B;\omega^*))(2^k-1)U\Bigg)^2 e^{-|\mathbf{u}|^2/2}\,d\mathbf{u}$$

$$\le 2\exp\left(-\frac{1800}{1+20N^{-1/2}}\right)10k\left((2^k-1)U+1\right), \tag{6.7}$$

where

$$\omega = (\omega^*, \mathbf{u}) \in \Omega(\text{Gauss}; \mathit{GCDsphere}). \tag{6.8}$$

To illustrate the power of (6.7) (combined with (6.4)), let

$$U = 4, \;\; k = 100 \quad \text{and} \quad N = 10^{27}. \tag{6.9}$$

Then by (6.7), for every $\omega^* \in [0, 2\pi)^N$,

$$(2\pi)^{-N/2} \int_{\mathbf{u}\in\mathbb{R}^N} \left(\text{length}\left\{4 \le t \le 4\cdot 2^{100} : \left| |\mathcal{Y}_{\omega;t} \cap B| - \sum_{k=1}^{N} b_k(\omega^*)\right| > 60\cdot 10^{13}\right\} - \text{vol}_N(S(B;\omega^*))(2^{100}-1)4\right)^2 e^{-|\mathbf{u}|^2/2}\, d\mathbf{u}$$
$$\le 2\exp\left(-\frac{1800}{1+20N^{-1/2}}\right)\cdot 400\cdot 2^{100}\cdot 20$$
$$< 10^{-700}\cdot 400\cdot 2^{100}\cdot 20 < 10^{-661}. \tag{6.10}$$

Let $\mathbb{R}^N(\omega^*;\text{bad})$ be the set of those $\mathbf{u}\in\mathbb{R}^N$ for which

$$\text{length}\left\{4 \le t \le 4\cdot 2^{100} : \left| |\mathcal{Y}_{\omega;t} \cap B| - \sum_{k=1}^{N} b_k(\omega^*)\right| > 60\cdot 10^{13}\right\} \ge 10^{-220}, \tag{6.11}$$

where

$$\omega = (\omega^*, \mathbf{u}) \in \Omega(\text{Gauss}; GCDsphere).$$

We claim that (6.11) implies

$$\text{Gauss}_N(\mathbb{R}^N(\omega^*;\text{bad})) = (2\pi)^{-N/2}\int_{\mathbf{u}\in\mathbb{R}^N(\omega^*;\text{bad})} e^{-|\mathbf{u}|^2/2}\, d\mathbf{u} \le 10^{-220}. \tag{6.12}$$

Indeed, otherwise

$$(2\pi)^{-N/2} \int_{\mathbf{u}\in\mathbb{R}^N} \left(\text{length}\left\{4 \le t \le 4\cdot 2^{100} : \left| |\mathcal{Y}_{\omega;t} \cap B| - \sum_{k=1}^{N} b_k(\omega^*)\right| > 60\cdot 10^{13}\right\} - \text{vol}_N(S(B;\omega^*))(2^{100}-1)4\right)^2 e^{-|\mathbf{u}|^2/2}\, d\mathbf{u}$$

$$\geq (2\pi)^{-N/2} \int_{\mathbf{u}\in\mathbb{R}^N(\omega^*;bad)} \Bigg(\text{length} \Bigg\{ 4 \leq t \leq 4\cdot 2^{100} :$$

$$\left| |\mathcal{Y}_{\omega;t} \cap B| - \sum_{k=1}^{N} b_k(\omega^*) \right| > 60 \cdot 10^{13} \Bigg\}$$

$$- \text{vol}_N(S(B;\omega^*))(2^{100}-1)4 \Bigg)^2 e^{-|\mathbf{u}|^2/2}\, d\mathbf{u}$$

$$\geq 10^{-220}\left(10^{-220} - 10^{-600}\right)^2 > 10^{-661},$$

which contradicts (6.10). Here we used

$$\text{vol}_d(S(B;\omega^*))(2^{100}-1)4 \leq 10^{-700}\cdot 10^{31} < 10^{-600}.$$

Again the threshold $U = 4$ represents — roughly speaking — the necessary number of "rounds" per particle in the (Gaussian) GreatCircleDiameter model to reach square-root equilibrium (in the particle space for the typical time evolution). Assume that the average projection-speed is 10^3 meter per second. For this system it takes only a few milliseconds to reach square-root equilibrium (in the particle space). Now (6.3)–(6.4) and (6.11)–(6.12) have the following interpretation. Choosing an arbitrary (measurable) test set $B \subset \mathbf{S}^2$ on the unit surface with normalized surface area $\text{SA}^*(B)$ and an arbitrary N-element initial point configuration $\mathcal{Y}$, for the totally overwhelming majority of the initial velocities (Gaussian distribution), the number of particles in B remains very close to the expected value $\text{SA}^*(B)N$ for an extremely long time, with the possible exception of a totally negligible set of times t.

Indeed, for every (measurable) test set $B \subset \mathbf{S}^2$ and every $N = 10^{27}$-element initial point configuration $\mathcal{Y} \subset \mathbf{S}^2$, there exists a subset

$$\Omega(\text{Gauss}; GCDsphere; \text{good}) = \Big\{\omega = (\omega^*, \mathbf{u}) \in \Omega(\text{Gauss}; GCDsphere) :$$

$$\omega^* \notin \text{BadSet and } \mathbf{u} \notin \mathbb{R}^N(\omega^*; \text{bad})\Big\}$$

with

$$\text{ProdMeas}_{\text{Gauss},GCD}(\Omega(\text{Gauss}; GCDsphere; \text{good}))$$

$$\geq 1 - 10^{-220} - \frac{1}{2}10^{-190} > 1 - 10^{-190}$$

(see (6.3)–(6.4) and (6.12); in particular, we used the numerical fact that

$$2\exp\left(-\frac{450}{1+10N^{-1/2}}\right) < \frac{1}{2}10^{-190}$$

holds for $N = 10^{27}$) representing a totally overwhelming majority, such that for every $\omega \in \Omega(\text{Gauss}; GCDsphere; \text{good})$,

$$\big||\mathcal{Y}_{\omega;t} \cap B| - \text{SA}^*(B)10^{27}\big| \le 90 \cdot 10^{13.5} \tag{6.13}$$

(where 90 comes from $90 = 30 + 60$; see (6.3) and the definition of $\mathbb{R}^N(\omega^*; \text{bad})$) holds for every $4 \le t \le 4 \cdot 2^{100}$ with the possible exception of a set of times t of total length $< 10^{-220}$, see (6.11). The latter represents less than 10^{-223} seconds — a ridiculously short time.

Note that $4 \le t \le 4 \cdot 2^{100}$ represents an incredibly long time interval: it is roughly billion times the age of the universe.

Finally, by (6.13)

$$\frac{1}{N}\,||\mathcal{Y}_{\omega;t} \cap B| - \text{vol}(B)| \le 9 \cdot 10^{-12}, \tag{6.14}$$

which can be interpreted as almost constant density for an incredibly long time.

Summarizing, once a typical time evolution of this (Gaussian) GreatCircleDiameter system reaches square-root equilibrium, then it stays in that state in the quantitative sense of 90-square-root equilibrium

$$||\mathcal{Y}_{\omega;t} \cap B| - \text{SA}^*(B)N| \le 90\sqrt{N}$$

for an incredibly long time (with the possible exception of a totally negligible set of t's).

Here is a nutshell summary of what we did in Secs. 5–6 so far. Despite the fundamental difference between the (at least 2-dimensional) torus-billiard model and the GreatCircleDiameter model on the sphere (a typical torus line is uniformly distributed, but a great circle — a closed orbit — is certainly not uniform on the sphere) the two systems show striking similarities. Both Gaussian billiard systems reach square-root equilibrium (in the particle space) superexponentially fast, and stay in square-root equilibrium for an incredibly long time.

What about the (simpler) GreatCircle model on the sphere? Well, we show that it is similar to the GreatCircleDiameter model with one substantial difference: the normalized surface area is replaced by another measure which *depends* on the initial configuration $\mathcal{Y} \subset \mathbf{S}^2$.

The details go as follows. We repeat the argument of the case of the (Gaussian) GreatCircleDiameter model with some necessary changes. Let $\mathbf{e} \in \mathbf{S}^2$ be an arbitrary point on the unit sphere; the first novelty is that we need to define an inhomogeneous probability measure $\text{SA}^*_{\mathbf{e}}(\dots)$ on the unit sphere that depends on the given point $\mathbf{e}$ (and also $\text{SA}^*_{\mathbf{e}}(\mathbf{S}^2) = 1$). By symmetry, we can assume that $\mathbf{e} = \mathbf{e}_0 = (0, 0, 1)$ is the North Pole, and we use the spherical coordinates

$$x = x(\vartheta, \phi) = \sin\vartheta\cos\phi, \quad y = y(\vartheta, \phi) = \sin\vartheta\sin\phi,$$

$$z = z(\vartheta, \phi) = \cos\phi, \quad 0 \le \vartheta < 2\pi, 0 \le \phi < \pi$$

to parametrize the unit sphere. For an arbitrary measurable set $A \subset \mathbf{S}^2$ let

$$\text{SA}^*_{\mathbf{e}_0}(A) = \frac{1}{2\pi^2}\text{area}\,(\{(\vartheta, \phi) \in [0, 2\pi) \times [0, \pi) : (x(\vartheta, \phi), y(\vartheta, \phi), z(\vartheta, \phi)) \in A\}), \tag{6.15}$$

where area stands for the 2-dimensional Lebesgue measure. For a general $\mathbf{e} \in \mathbf{S}^2$ different from the North Pole, let $\text{Rot}_{\mathbf{e}} \in \text{SO}(3)$ be a rotation moving $\mathbf{e}$ to the North Pole $\mathbf{e}_0 = (0, 0, 1)$, and define the measure $\text{SA}^{**}_{\mathbf{e}}(\dots)$ as

$$\text{SA}^{**}_{\mathbf{e}}(A) = \text{SA}^{**}_{\mathbf{e}_0}(\text{Rot}_{\mathbf{e}}A). \tag{6.16}$$

Let $B \subset \mathbf{S}^2$ be an arbitrary but fixed measurable test set, and let $\mathcal{Y}_{\omega;t}$ denote the (Gaussian) GreatCircle model with initial configuration $\mathcal{Y} \subset \mathbf{S}^2$ and

$$\omega = (\omega^*, \mathbf{v}) \in \Omega(\text{Gauss}; GCsphere)$$

where

$$\omega^* = (\alpha_1, \dots, \alpha_N) \in [0, 2\pi)^N \quad \text{and} \quad \mathbf{v} = (v_1, \dots, v_N) \in \mathbb{R}^N.$$

We ask the usual question. Assume that N is large; is it true that, once a typical time evolution of the system reaches square-root equilibrium, then it stays in that state for an extremely long time (with the possible exception of a totally negligible set of t's)? Again we use Theorem 4.2 to give a positive answer.

For every $\omega^* \in [0, 2\pi)^N$, let $GC_k = GC_k(\omega^*)$ denote the orbit of the kth particle: it is a great circle containing the antipodal points $\mathbf{y}_k \in \mathcal{Y}$ and $-\mathbf{y}_k$ with angle α_k. The motion of the kth particle on the great circle $GC_k = GC_k(\omega)$ is equivalent to a motion in the interval $[0, 2\pi)$ interpreted as a 1-dimensional torus. For notational convenience we divide by 2π in the sense that we replace the constant speed motion in the 1-dimensional torus $[0, 2\pi)$

with the constant speed motion in the 1-dimensional unit torus $[0, 1)$. This means that, for every *fixed* $\omega^* \in [0, 2\pi)^N$, the family of time evolutions $\mathcal{Y}_{\omega;t}$, $\omega = (\omega^*, \mathbf{v}) \in \Omega(\text{Gauss}; GCsphere)$ of the Gaussian GreatCircle model with initial configuration $\mathcal{Y} \subset \mathbf{S}^2$ is represented by the family of torus lines in the configuration space $I^N = [0, 1)^N$, all starting from the origin $\mathbf{0} \in I^N$.

The great circle $GC_k = GC_k(\omega^*)$ (i.e., the orbit of the kth particle) intersects the given 2-dimensional measurable test set $B \subset \mathbf{S}^2$ in a 1-dimensional *measurable* set for almost every $\omega^* \in [0, 2\pi)^N$ (Fubini's theorem). Write

$$b_k(\omega^*) = \frac{1}{2\pi} \text{ length } (GC_k(\omega^*) \cap B) \tag{6.17}$$

(length stands for the 1-dimensional Lebesgue measure). The mean value of $b_k(\omega^*)$, $\omega^* \in [0, 2\pi)^N$ is clearly equal to the $\text{SA}^{**}_{\mathbf{y}_k}$-measure (see (6.15)–(6.16)) of the test set $B \subset \mathbf{S}^2$; formally,

$$(2\pi)^{-N} \int_{[0,2\pi)^N} b_k(\omega^*)\, d\omega^* = \text{SA}^{**}_{\mathbf{y}_k}(B). \tag{6.18}$$

Again we choose $\gamma = 30$ as the "large deviation factor," and write

$$\text{BadSet} = \left\{ \omega^* \in [0, 2\pi)^N : \left| \sum_{k=1}^{N} b_k(\omega^*) - \sum_{k=1}^{N} \text{SA}^{**}_{\mathbf{y}_k}(B) \right| > 30\sqrt{N} \right\}. \tag{6.19}$$

Since the initial velocities are chosen independently, by Bernstein's large deviation inequality (5.8),

$$(2\pi)^{-N} \text{vol}_d\, (\text{BadSet}) \le 2 \exp\left(-\frac{450}{1 + 10N^{-1/2}} \right), \tag{6.20}$$

a perfect analog of (6.4).

Let $\omega^* \in [0, 2\pi)^N$ be *fixed*, and write

$$S(B; \omega^*) = \left\{ \vec{\mathcal{Z}} = (z_1, \ldots, z_N) \in I^N : \right.$$
$$\left| \left| \left\{ 1 \le k \le N : \; z_k \in \frac{1}{2\pi}(GC_k(\omega^*) \cap B) \right\} \right| \right.$$
$$\left. \left. - \sum_{k=1}^{N} b_k(\omega^*) \right| > 60\sqrt{N} \right\}. \tag{6.21}$$

Again using Bernstein's large deviation inequality (5.8), we obtain

$$\mathrm{vol}_N(S(B;\omega^*)) \le 2\exp\left(-\frac{1800}{1+20N^{-1/2}}\right), \tag{6.22}$$

which is the perfect analogue of (6.6). Again we apply Theorem 4.2 with $f = \chi_S$ where $S = S(B;\omega^*) \subset I^N$, and obtain the perfect analogue of (5.14): let $k \ge 1$ be an integer, for every $\omega^* \in [0,2\pi)^N$,

$$(2\pi)^{-N/2}\int_{\mathbf{v}\in\mathbb{R}^N}\left(\text{length}\left\{U \le t \le 2^k U : \left| |\mathcal{Y}_{\omega;t}\cap B| - \sum_{k=1}^{N} b_k(\omega^*)\right| > 60\sqrt{N}\right\} - \mathrm{vol}_N(S(B;\omega^*))(2^k-1)U\right)^2 e^{-|\mathbf{v}|^2/2}\,d\mathbf{v}$$

$$\le 2\exp\left(-\frac{1800}{1+20N^{-1/2}}\right)10k\left((2^k-1)U+1\right), \tag{6.23}$$

where

$$\omega = (\omega^*, \mathbf{v}) \in \Omega(\text{Gauss}; GCsphere).$$

Again for illustration we choose $U = 4$, $k = 100$ and $N = 10^{27}$. Then by (6.23), for every $\omega^* \in [0,2\pi)^N$,

$$(2\pi)^{-N/2}\int_{\mathbf{v}\in\mathbb{R}^N}\left(\text{length}\left\{4 \le t \le 4\cdot 2^{100} : \left| |\mathcal{Y}_{\omega;t}\cap B| - \sum_{k=1}^{N} b_k(\omega^*)\right| > 60\cdot 10^{13}\right\} - \mathrm{vol}_N(S(B;\omega^*))(2^{100}-1)4\right)^2 e^{-|\mathbf{v}|^2/2}\,d\mathbf{v}$$

$$\le 2\exp\left(-\frac{1800}{1+20N^{-1/2}}\right)\cdot 400\cdot 2^{100}\cdot 20 < 10^{-661}. \tag{6.24}$$

Again let $\mathbb{R}^N(\omega^*;\text{bad})$ be the set of those $\mathbf{v}\in\mathbb{R}^N$ for which

$$\text{length}\left\{4 \le t \le 4\cdot 2^{100} : \left| |\mathcal{Y}_{\omega;t}\cap B| - \sum_{k=1}^{N} b_k(\omega^*)\right| > 60\cdot 10^{13}\right\} \ge 10^{-220}, \tag{6.25}$$

where

$$\omega = (\omega^*, \mathbf{v}) \in \Omega(\text{Gauss}; GCsphere).$$

Again we have

$$\text{Gauss}_N(\mathbb{R}^N(\omega^*; \text{bad})) = (2\pi)^{-N/2} \int_{\mathbf{v}\in\mathbb{R}^N(\omega^*;\text{bad})} e^{-|\mathbf{v}|^2/2}\, d\mathbf{v} \leq 10^{-220}, \tag{6.26}$$

which is the perfect analog of (6.12).

Again the threshold $U = 4$ represents — roughly speaking — the necessary number of "rounds" per particle in the Gaussian GreatCircle model to reach square-root equilibrium (in the particle space for the typical time evolution). The only difference is that in the definition of square-root equilibrium we replace the homogeneous measure (= surface area) with an inhomogeneous measure that depends on the initial configuration $\mathcal{Y} = \{\mathbf{y}_1, \ldots, \mathbf{y}_N\}$:

$$\text{SA}^{**}_{\mathcal{Y}}(\ldots) = \frac{1}{N}\sum_{k=1}^{N} \text{SA}^{**}_{\mathbf{y}_k}(\ldots), \tag{6.27}$$

see (6.15) and (6.16).

Again assume that the average speed is 10^3 meter per second. Again it takes only a few milliseconds to reach square-root equilibrium with respect to the inhomogeneous measure $\text{SA}^{**}_{\mathcal{Y}}$ (see (6.27)). Similarly to (6.3)–(6.4) and (6.11)–(6.12), we have the following interpretation of (6.5)–(6.6) and (6.11)–(6.12). Choosing an arbitrary (measurable) test set $B \subset \mathbf{S}^2$ on the unit surface and an arbitrary N-element initial point configuration $\mathcal{Y}$, for the totally overwhelming majority of the initial velocities (Gaussian distribution), the number of particles in B remains very close to $\text{SA}^{**}_{\mathcal{Y}}(B)N$ for an extremely long time, with the possible exception of a totally negligible set of times t.

This proves long-term stability of square-root equilibrium with respect to an arbitary (possibly "ugly") but fixed measurable test set $B \subset \mathbf{S}^2$.

Summarizing, we showed that the (Gaussian) GreatCircle model and the GreatCircleDiameter model are very similar. The concrete numerical values remain exactly the same; the only change is the appearance of the inhomogeneous measure $\text{SA}^{**}_{\mathcal{Y}}$ in the first model, replacing the normalized surface area SA^* in the second model.

In the next section we study models where the closed orbits are line segments.

7. Closed Orbit Polar Systems

We know a lot about the asymptotic behavior of the billiard path in a box, but, unfortunately, we know *much, much* less about the billiard motion in a general underlying set. It is quite humiliating how little we know about the simplest and most natural problems about billiards. For example, let Δ be an arbitrary triangle: is it true that the billiard in triangle Δ is uniform for almost every initial condition (= starting point and direction)? We just have a very modest goal here: all what we want is the weakest quantitative form of uniformity: is it true or not that, as $T \to \infty$, for almost every initial condition, the actual time the (constant speed) billiard path spends in a given nice subset of Δ as t runs in $0 < t < T$ equals the expected time ($= T \cdot$ relative area) plus $o(T)$? Shockingly, this innocent-looking question is a famous long-standing open problem. All what we know is that the answer to this question is affirmative in the narrow special case where the triangle is "rational," meaning that the angles are rational multiples of π. The proof of this special case is already difficult; see [Ke-Ma-Sm86] (unfortunately this proof technique does not seem to work in higher dimension). Also, this special case is a "soft" qualitative result, and does not give any reasonable quantitative estimation on the speed of convergence to uniformity.

For comparison, note that replacing the triangle with a rectangle, we have *superuniformity(!)*; see Theorems A1–A2 in Sec. 2. On the other hand, for many other natural shapes the billiard is very bad, uniformity clearly fails. For example, if the billiard table is a circular disk (= solid circle), then *no* infinite billiard path is uniformly distributed inside the circular disk. (Here of course we talk about uniformity in the traditional sense of Weyl, i.e., with respect to all Riemann integrable test functions, or equivalently, all Jordan measurable test sets.)

As far as we know, the only other cases beyond the rectangle billiard, where we have satisfying quantitative results, are the equilateral triangle billiard, $\pi/3$-right-triangle billiard, $\pi/4$-right-triangle billiard, and finally, the geodesic flow on an equifacial tetrahedron. We briefly elaborate on this. At the beginning of Sec. 3 we explained how *unfolding* converts the square billiard orbit (= zig-zag) into a straight line on the plane. This raises the following question: For what other shapes can the unfolding used in Sec. 3 (= iterated reflection across a line) convert the billiard orbit (= zig-zag) into a straight line on the plane? Well, it is clearly necessary that the process of *unfolding* applied to the given shape (= billiard table) gives a perfect tiling of the plane. Let us start with triangles; let $\mathcal{T}$ be a triangle. In Sec. 3 unfolding of $\mathcal{T}$ means the infinite process where we select a side of $\mathcal{T}$, and reflect $\mathcal{T}$ across this side; we obtain a new polygon $\mathcal{T}_1$; next we select a side of $\mathcal{T}_1$, and reflect $\mathcal{T}_1$ in this side; we obtain another polygon $\mathcal{T}_2$; and so on — we keep doing this in all possible ways. When does the complete infinite sequence $\mathcal{T}, \mathcal{T}_1, \mathcal{T}_2, \ldots$ of congruent triangles tile the whole plane? We clearly need that the angles of $\mathcal{T}$ have the form $2\pi/k_1$, $2\pi/k_2$, $2\pi/k_3$, where k_i are positive integers. Since reflection across a line changes the orientation of a triangle, we also need that every k_i is even, i.e. $k_i = 2\ell_i$, $1 \leq i \leq 3$. Since the sum of degrees in a triangle is 180 (or π), we need to find the positive integer solutions of the equation

$$\frac{1}{\ell_1} + \frac{1}{\ell_2} + \frac{1}{\ell_3} = 1 \tag{7.1}$$

with $\ell_i \geq 2$. We can assume $2 \leq \ell_1 \leq \ell_2 \leq \ell_3$. Clearly $\ell_1 \leq 3$, so either $\ell_1 = 3$ or $\ell_1 = 2$. If $\ell_1 = 3$ then $(3,3,3)$ is the only solution of (7.1). If $\ell_1 = 2$ then $(2,3,6)$ and $(2,4,4)$ are the only solutions of (7.1). This gives a very short list of special triangles: (1) equilateral triangle, (2) right triangle with angle $\pi/3$ (= half of an equilateral triangle). (3) isosceles right triangle (= half of a square).

If we switch from triangles to quadrilaterals, then Eq. (7.1) changes to

$$\frac{1}{\ell_1} + \frac{1}{\ell_2} + \frac{1}{\ell_3} + \frac{1}{\ell_4} = 2, \tag{7.2}$$

which has only one integer solution with $\ell_i \geq 2$; namely, $\ell_1 = \ell_2 = \ell_3 = \ell_4 = 2$, representing the rectangle.

If we switch from quadrilaterals to polygons of $r \geq 5$ sides, then Eq. (7.2) changes to

$$\frac{1}{\ell_1} + \frac{1}{\ell_2} + \cdots + \frac{1}{\ell_r} = r - 2, \tag{7.3}$$

which has no integer solution with $\ell_i \geq 2$. Indeed,

$$\frac{1}{2} \geq \max_i \frac{1}{\ell_i} \geq \frac{r-2}{r} > \frac{1}{2} \text{ if } r \geq 5.$$

So far everything was based on reflection across a line; but in geometry there is also a different kind of reflection called "reflection in a point". Reflection across a line led to the disappointingly short list of the three special triangles mentioned above. In sharp contrast, using reflection in a point we have far more freedom: we can tile the plane with *any* triangle. Indeed, we simply start with an arbitrary parallelogram lattice, and split every parallelogram into two congruent triangles by means of the same diagonal (i.e., the diagonals are all parallel). Neighboring triangles that share a side are congruent via reflection in the midpoint of the common side.

Next we explain how this triangle tiling of the plane with an arbitrary (acute) triangle helps to describe the uniformity of the *geodesics on an equifacial tetrahedron surface*. Equifacial means that the four faces of the tetrahedron are congruent triangles; the simplest special case is the regular tetrahedron.

It is easy to see that polygon billiards and geodesics on a flat surface are basically the same. Indeed, we may look at a polygon billiard as a geodesic on a *two-face* polygon: with top face and bottom face. Indeed, when a point billiard on the top face hits a side, it goes under to the bottom face, keeping the linear orbit. (And of course the other way around for the bottom face.)

Consider a triangle tiling of the plane with an acute triangle Δ. We can easily fold Δ into an equifacial tetrahedron surface in the 3-space. First we obtain four congruent half triangles by considering the midpoints; formally, if A, B, C are the vertices of Δ, then let C_1, A_1, B_1 be the midpoints of the sides AB, BC, CA of Δ in this order. Next we simply fold along the three line segments in Δ which join the midpoints A_1, B_1, C_1 of two of the three sides. Then the three vertices A, B, C come together in a point D in the 3-space, and we obtain an equifacial tetrahedron surface $\mathrm{TetS} = \mathrm{TetS}(\Delta)$ with vertices A_1, B_1, C_1, D. (This is where we use the fact that Δ is acute: starting with a right triangle Δ, A_1, B_1, C_1, D turn out to be coplanar, providing a pathological tetrahedron. The case of obtuse Δ is even "more pathological".) Note that this way the boundary points of the triangle Δ glue together in pairs; for example, B_1A becomes identical to B_1C via the reflection in the midpoint B_1.

The simplest equifacial tetrahedron is of course the regular tetrahedron.

Let L be a straight line in the plane that contains triangle Δ. For simplicity, assume that L intersects the inside of Δ; let $Q \in L \cap \Delta$ be such a common point. It is easy to see that the whole straight line L in the plane is wrapped up as a *geodesic* on the equifacial tetrahedron surface

$$\mathrm{TetS} = \mathrm{TetS}(\Delta) = \mathrm{TetS}(A_1 B_1 C_1 D).$$

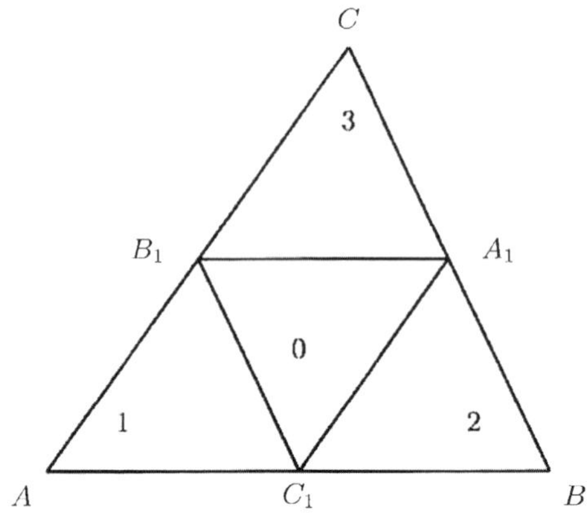

Fig. 7.1

Note that in general, a geodesic on a polyhedron surface consists of straight line segments that fit together to a straight line at the boundary, where the line segment hits an edge: the consecutive line segments become collinear when we turn the two neighboring faces (that share the edge) into coplanar position.

The geometric explanation why a straight line L in the plane is wrapped up as a *geodesic* on the equifacial tetrahedron surface

$$\text{TetS} = \text{TetS}(\Delta) = \text{TetS}(A_1B_1C_1D)$$

is based on the effect of reflection in a point. Indeed, moving from point Q along the line L in one direction, we reach a congruent neighbor Δ_1 of Δ in the triangle lattice, and the fact that Δ_1 is congruent to Δ via reflection in the midpoint of the common side means that the part of L in Δ_1 corresponds exactly to the next part of the geodesic on the tetrahedron surface TetS. Similarly, when moving further along L we reach a neighbor Δ_2 of Δ_1 in the triangle lattice, then the part of L in Δ_2 corresponds exactly to the next part of the geodesic on the tetrahedron surface TetS, and so on.

An alternative way to say the same thing in terms of the smaller half triangles goes as follows. We label the midpoint triangle $A_1B_1C_1$ with 0, label AC_1B_1 with 1, label BA_1C_1 with 2, and label CB_1A_1 with 3; this way the four faces of tetrahedron surface TetS are labeled with 0,1,2,3. Next we divide every congruent copy of Δ in the triangle lattice into four congruent half triangles ("halves"), and extend the labeling of the four halves of Δ to the whole plane by using the following simple rule: the midpoint triangles all have the same label 0, and if two half triangles have exactly one common vertex, then they have the same label (1 or 2 or 3). Now wrapping-up-a-line means that, if a motion on a straight line L in the plane is in a midpoint triangle labeled i, then the corresponding motion on the geodesic (= wrapped up line) is on the face of the tetrahedron surface

$$\mathrm{TetS} = \mathrm{TetS}(\Delta) = \mathrm{TetS}(A_1 B_1 C_1 D)$$

with the same label i ($i = 0, 1, 2, 3$).

The next figure illustrates reflection of $\Delta = ABC$ in point A_1, where the latter is rrepresented by the dot in the middle.

Figure 7.2 illustrates the extension of "reflection in a point" over the whole plane.

Here $g_1 g_2$ is a geodesic of the tetrahedron surface, where g_2 is the directed continuation of g_1, obtained by reflection of the dashed line segment with arrow (= linear extension of g_1) in A_1 (= dot in the middle). This illustrates how geodesics are unfolded into straight lines.

Let S be an arbitrary measurable subset of triangle $\Delta = ABC$. Again let A_1 denote the midpoint of BC, and let Δ_1 denote the reflection of Δ in the midpoint A_1. Similarly, let S^* denote the reflection of S in the midpoint A_1, and let A_2 denote the reflection of A in the midpoint A_1. Let $\mathcal{P}$ denote the parallelogram with vertices A, B, A_2, C, and let $\mathcal{L} = \mathcal{L}(\Delta)$ denote the parallelogram lattice on the plane with fundamental parallelogram $\mathcal{P}$.

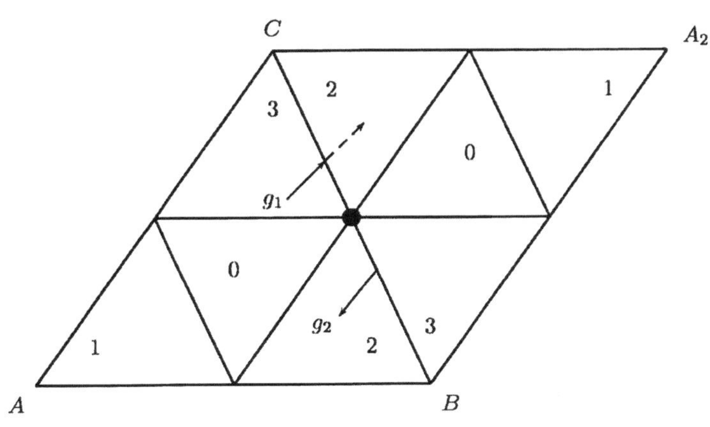

Fig. 7.2

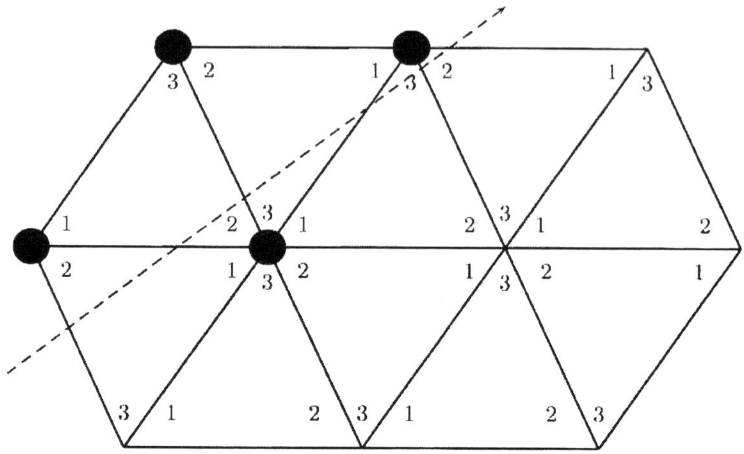

Fig. 7.3

In Fig. 7.3 the four big dots are the vertices of fundamental parallelogram $\mathcal{P}$, and the dashed straight line with arrow is an unfolded geodesic of the equifacial tetrahedron. (Since every midpoint triangle has label 0, in the last figure we omitted — for notational simplicity — the midpoint triangles.)

The geometric method of "wrapping up a straight line to a geodesic on an equifacial tetrahedron surface" explains why the uniformity of a straight line L (see the dashed straight line with arrow) modulo $\mathcal{P}$ in the plane with respect to test set $\widetilde{S} = S \cup S^*$ is equivalent to the uniformity of the corresponding geodesic (= wrapped up copy of line L) on the equifacial tetrahedron surface TetS = TetS(Δ) with respect to test set S.

The uniformity of a straight line L modulo $\mathcal{P}$ is analogous to the problem of the uniformity of a torus line modulo $[0, 1)^2$. Indeed, every parallelogram lattice $\mathcal{L}$ in the plane can be mapped to the integer square lattice $\mathbb{Z}^2$ via some linear transformation of the plane. This means, whatever result is available on the uniformity of a 2-dimensional torus line, it also applies (with straightforward modification) for the uniformity of the geodesic flow on any equifacial tetrahedron surface.

This "transference principle" comes from unfolding via reflection in a point. It is an analog of the "transference principle" that comes from unfolding via reflection across a line. Applying the latter guarantees that, whatever result is available on the uniformity of a 2-dimensional torus line, it also applies for the uniformity of a square/rectangle billiard orbit. Similarly, whatever result is available on the uniformity of a 2-dimensional torus line, it also applies for the uniformity of the three special triangle billiard orbits: (1) equilateral triangle, (2) right triangle with angle $\pi/3$, (3) isosceles right triangle.

It is very surprising that this geometric argument — wrapping up a straight line via reflection in a point — which works so well for the (say) regular tetrahedron surface, does not seem to work for the seemingly equally simple cube surface. In fact, we do not know any other polyhedron, or any other "reasonable" surface, for which this method works.

Physicists call a dynamical system "integrable" if it is (in some natural way) equivalent to the torus line in a square. The short list — (1) rectangle billiard, (2) equilateral triangle billiard, (3) $\pi/3$-right-triangle billiard, (4) $\pi/4$-right-triangle billiard, (5) equifacial tetrahedron geodesic — represents the known cases of "integrable" billiard/geodesic dynamical systems. Is the list complete? This is a very interesting open problem. These are the only cases when we have good quantitative results.

The disappointing (or even humiliating!) lack of quantitative results for general billiards/geodesics motivates the introduction of the following drastic simplification of the billiard model that we call the PolarBilliard model. Let $A \subset \mathbb{R}^h$ be an arbitrary bounded closed convex set in the h-space (noting that our favorite dimensions are $h = 2$ and 3). Let $P \in A$ be an arbitrary point in A, and let $\vec{L}$ be a half-line starting from P. By definition the intersection $\vec{L} \cap A$ is a line segment; we denote it by $I_A(P; \vec{L})$, or $I_A(P; \mathbf{e})$, where $\mathbf{e} \in \mathbf{S}^{h-1}$ is the direction of the half-line $\vec{L}$. The PolarBilliard motion simply means that a particle moves with constant speed (to be specified later), starting from P, on the line segment $I_A(P; \vec{L})$ back and forth like a 1-dimensional billiard, i.e., it bounces back when it hits the boundary of A and also when it hits P. The advantage of this simplified billiard model is that we can describe the approach to equilibrium, and the stability of equilibrium.

Let $\mathcal{Y} = \{\mathbf{y}_1, \ldots, \mathbf{y}_N\}$ be an arbitrary N-element set in the convex set A; it represents the initial configuration of the N-particle PolarBilliard model. The kth particle starts from $\mathbf{y}_k$, and moves with constant speed v_k on a line segment $I_A(\mathbf{y}_k; \mathbf{e}_k)$ back and forth like a 1-dimensional billiard. Here $\mathbf{e}_k \in \mathbf{S}^{h-1}$ is a unit vector on the unit sphere, and the speed

$v_k = v_k(\mathbf{y}_k; \mathbf{e}_k)$ depends on both $\mathbf{y}_k$ and $\mathbf{e}_k$ (see below). We assume that every $\mathbf{e}_k$ is uniformly distributed on the sphere $\mathbf{S}^{h-1}$, and every v_k has the form

$$v_k = u_k \text{length}(I_A(\mathbf{y}_k; \mathbf{e}_k)), \tag{7.4}$$

where u_k has standard normal distribution. (So the speed can be negative, meaning that the particle goes backward — notice that it is exactly the same motion as the forward motion!) Moreover, we assume that $\mathbf{e}_1, \ldots, \mathbf{e}_N$, $u_1, \ldots, u_N$ are $2N$ independent random variables. In other words,

$$\omega = (\mathbf{e}_1, \ldots, \mathbf{e}_N, u_1, \ldots, u_N) \in \Omega(\text{Gauss}; A; PolB) = \left(\mathbf{S}^{h-1}\right)^N \times \mathbb{R}^N, \tag{7.5}$$

is the initial velocity space (of the whole system) with the product measure

$$\text{ProdMeas}_{\text{Gauss};A;PolB} = (\text{SA}^*)^N \times \text{Gauss}_N, \tag{7.6}$$

where SA^* denotes the normalized surface area on $\mathbf{S}^{h-1}$ (i.e., SA^* $(\mathbf{S}^{h-1}) = 1$), and, as usual, Gauss_N denotes the N-dimensional Gauss measure (= N-dimensional standard normal distribution). The initial configuration $\mathcal{Y}$, the vector

$$\omega = (\mathbf{e}_1, \ldots, \mathbf{e}_N, u_1, \ldots, u_N),$$

and Eq. (7.4) determine the time evolution of the N-particle PolarBilliard system; as usual, we denote it by $\mathcal{Y}_{\omega,t}$. We refer to it as the Gaussian PolarBilliard model.

There is some similarity between the (Gaussian) GreatCircle model and the PolarBilliard model. In both models each particle moves on its own closed orbit (great circle or line segment) with its own constant speed (depending on the particle).

We study the usual problem. Assume that N is large; is it true that, once a typical time evolution of the system (= Gaussian PolarBilliard model) reaches square-root equilibrium (in the particle space), then it stays in that state for an extremely long time (with the possible exception of a totally negligible set of t's)? We recall that in the GreatCircle model the equilibrium was defined in terms of an inhomogeneous measure that depends on the inital configuration $\mathcal{Y}$. We show — by using Theorem 4.2 as a *short-time ergodic theorem* in the configuration space (in fact, it works as a "large deviation theorem") — that, similarly to the GreatCircle model, the PolarBilliard model exhibits long-term stability in equilibrium, and the equilibrium is defined in terms of an inhomogeneous measure that depends on the initial configuration $\mathcal{Y}$.

In Sec. 6 we also introduced a second — at first sight less natural — spherical model, the (Gaussian) GreatCircleDiameter model, which turned out to exhibit a simpler time evolution. Indeed, in the second spherical model the equilibrium is defined in terms of a homogeneous measure independent of the inital configuration $\mathcal{Y}$. This raises the natural question: Is there a Polar model for which the equilibrium is defined in terms of a measure independent of the initial configuration $\mathcal{Y}$?

The answer is yes: we show that the PolarSpeedUp model (to be defined below) has the desired property. Note that the PolarSpeedUp model is similar to the PolarBilliard model, but there are two crucial changes: (1) the speed of the kth particle on the line segment $I_A(\mathbf{y}_k;\mathbf{e}_k)$ is not constant; (2) the direction vector $\mathbf{e}_k$ is not necessarily uniformly distributed on the unit sphere $\mathbf{S}^{h-1}$. The details go as follows.

Again we need a bounded closed convex set in the h-space $A \subset \mathbb{R}^h$— it is the underlying set. Again let $\mathcal{Y} = \{\mathbf{y}_1, \dots, \mathbf{y}_N\}$ be an arbitrary N-element set in the convex set A; it represents the initial configuration of the N-particle PolarSpeedUp model. Again the kth particle starts from $\mathbf{y}_k$, and moves on a line segment $I_A(\mathbf{y}_k;\mathbf{e}_k)$ until it hits the boundary of A, back and forth like a 1-dimensional billiard. Again $\mathbf{e}_k \in \mathbf{S}^{h-1}$ is a unit vector on the unit sphere. The (absolute value of the) speed of the kth particle is

$$|u_k| r^{-(h-1)} a_k^h / h, \tag{7.7}$$

where r is the distance of the particle from the starting point $\mathbf{y}_k$, $a_k = \text{length}(I_A(\mathbf{y}_k;\mathbf{e}_k))$, and u_k is a real number to be specified later.

As we said above, $\mathbf{e}_k$ is *not* necessarily uniformly distributed on the sphere $\mathbf{S}^{h-1}$. To describe the distribution, we define a measure $\text{meas}_{A;\mathbf{y}_k}$ on $\mathbf{S}^{h-1}$, which depends on both A and $\mathbf{y}_k$. For an arbitrary measurable subset $W \subset \mathbf{S}^{h-1}$, let

$$\text{meas}_{A;\mathbf{y}_k}(W) = \frac{\int_{\mathbf{e}\in W} \left(\text{length}(I_A(\mathbf{y}_k;\mathbf{e}))\right)^h \, d\text{SA}^*(\mathbf{e})}{\int_{\mathbf{e}\in \mathbf{S}^{h-1}} \left(\text{length}(I_A(\mathbf{y}_k;\mathbf{e}))\right)^h \, d\text{SA}^*(\mathbf{e})}, \tag{7.8}$$

where SA^* is the normalized surface area.

We assume that the direction of the kth particle $\mathbf{e}_k$ is $\text{meas}_{A;\mathbf{y}_k}$-uniformly distributed on the sphere $\mathbf{S}^{h-1}$ (see (7.8)), and every u_k has standard normal distribution. Moreover, we assume that $\mathbf{e}_1, \dots, \mathbf{e}_N, u_1, \dots, u_N$

are $2N$ independent random variables. In other words,

$$\begin{aligned}\omega &= (\mathbf{e}_1, \dots, \mathbf{e}_N, u_1, \dots, u_N) \in \Omega(\text{Gauss}; A; PolSU) \\ &= \left(\mathbf{S}^{h-1}\right)^N \times \mathbb{R}^N, \end{aligned} \tag{7.9}$$

is the initial velocity space (of the whole system) with the product measure

$$\text{ProdMeas}_{\text{Gauss};A;PolSU} = (\text{SA}^*)^N \times \text{Gauss}_N, \tag{7.10}$$

similarly to (7.5)–(7.6). The initial configuration $\mathcal{Y}$, the vector

$$\omega = (\mathbf{e}_1, \dots, \mathbf{e}_N, u_1, \dots, u_N),$$

and Eq. (7.7) determine the time evolution of the N-particle PolarSpeedUp system; as usual, we denote it by $\mathcal{Y}_{\omega,t}$. We refer to it as the Gaussian PolarSpeedUp model.

It is not an accident that in (7.7) we used the letter r familiar from polar coordinates. The motivation for the mysterious equation (7.7) comes from the volume formula

$$\begin{aligned}\text{vol}_h(A) &= \int_{\mathbf{e}\in\mathbf{S}^{h-1}} \left(\int_0^{a_k(\mathbf{e})} x^{h-1}\, dx \right) d\text{SA}^*(\mathbf{e}) \\ &= \text{SuArea}(\mathbf{S}^{h-1}) \int_{\mathbf{e}\in\mathbf{S}^{h-1}} \left(\int_0 a_k(\mathbf{e}) x^{h-1}\, dx \right) d\text{SA}^*(\mathbf{e}) \\ &= \text{SuArea}(\mathbf{S}^{h-1}) \int_{\mathbf{e}\in\mathbf{S}^{h-1}} \frac{1}{h} a_k^h(\mathbf{e})\, d\text{SA}^*(\mathbf{e}), \end{aligned} \tag{7.11}$$

where $d\text{SA}(\mathbf{e})$, $\mathbf{e} \in \mathbf{S}^{h-1}$ represents the surface area, $d\text{SA}^*(\mathbf{e})$ represents the normalized surface area (i.e., $\text{SA}^*(\mathbf{S}^{h-1}) = 1$), and

$$a_k(\mathbf{e}) = \text{length}(I_A(\mathbf{y}_k; \mathbf{e})).$$

Note that (7.11) means to compute the volume of the translated copy $A - \mathbf{y}_k$ of A (i.e., $\mathbf{y}_k$ moves to the origin) by switching to the polar coordinates.

Similarly to Sec. 6, first we study the typical time evolution of the (Gaussian) PolarSpeedUp model, which is somewhat similar to the (Gaussian) GreatCircleDiameter model. Indeed, in both models the kth particle has infinite speed whenever it returns to the starting point $\mathbf{y}_k$. We closely follow the argument in Sec. 6.

Again let $B \subset A$ be an arbitrary but fixed measurable test set. Let $\mathcal{Y}_{\omega;t}$ denote the (Gaussian) ChordSpeedUp model with initial configuration $\mathcal{Y} \subset A$ and

$$\omega = (\omega^*, \mathbf{u}) \in \Omega(\text{Gauss}; A; PolSU),$$

where

$$\omega^* = (\mathbf{e}_1, \ldots, \mathbf{e}_N) \in \left(\mathbf{S}^{h-1}\right)^N \quad \text{and} \quad \mathbf{u} = (u_1, \ldots, u_N) \in \mathbb{R}^N.$$

For every $\omega^* \in \left(\mathbf{S}^{h-1}\right)^N$, let

$$I_k = I_k(\mathbf{e}_k) = I_k(\omega^*) = I_k(A; \mathcal{Y}; \omega^*)$$

denote the orbit of the kth particle: it is a line segment starting from $\mathbf{y}_k$ in the direction $\mathbf{e}_k \in \mathbf{S}^{h-1}$ until it hits the boundary of A.

For notational convenience write $I_k = I_k(\omega^*) = [0, a_k]$, where

$$a_k = a_k(\mathbf{e}_k) = a_k(\omega^*) = \text{length}(I_k)$$

and $0 \in I_k$ represents the starting point $\mathbf{y}_k$ of the motion of the kth particle. By (7.7) the motion of the kth particle has speed

$$|u_k| \cdot x^{-(h-1)} a_k^h / h \text{ at } x \in I_k, \tag{7.12}$$

meaning the absolute value of the speed. So, the kth particle starts from 0 with infinite speed, goes (by slowing down) to the right endpoint a_k, where it bounces back by reversing the velocity vector. It keeps speeding up to infinite speed at 0, where it bounces back reversing the velocity vector, and so on. This is a (varying speed) billiard motion. To obtain a (verying speed) torus motion, we apply the trick of *unfolding* in its simplest 1-dimensional form. Let

$$\widetilde{I}_k = \widetilde{I}_k(\mathbf{e}_k) = \widetilde{I}_k(\omega^*) = [0, 2a_k)$$

denote the interval twice as long as I_k, and we extend the motion on I_k to $\widetilde{I}_k \setminus I_k$ by taking the *reflection*. That is, just like the motion on I_k, the kth particle starts from 0 with infinite speed, and goes (by slowing down) to the right endpoint a_k; then it keeps going to the right speeding up to infinite speed at $2a_k$. Then it jumps back to 0, where it keeps going to the right slowing down; and so on, extending periodically, always going to the right. In the torus motion on $\widetilde{I}_k$ the speed of the kth particle is

$$|u_k| \cdot x^{-(h-1)} a_k^h / h \text{ at } x \in I_k, \quad \text{and}$$
$$|u_k| \cdot (2a_k - x)^{-(h-1)} a_k^h / h \text{ at } x \in \widetilde{I}_k \setminus I_k. \tag{7.13}$$

Consider the function

$$f_k(x) = a_k^{-h} x^h. \tag{7.14}$$

The derivative of f_k is

$$f_k'(x) = h a_k^{-h} x^{h-1}. \tag{7.15}$$

The function f_k maps the interval $[0, a_k]$ to the unit interval $J_k = [0, 1]$. Consider a particle moving with constant speed u_k from the left endpoint of J_k to its right endpoint. The motion induced by the inverse mapping f_k^{-1} represents a varying speed motion on the interval $[0, a_k]$ from left to right: at $x \in [0, a_k]$ the induced speed is (the reciprocal in (7.15))

$$\frac{u_k}{h a_k^{-h} x^{h-1}} = u_k x^{-(h-1)} a_k^h / h, \tag{7.16}$$

which is exactly (7.12).

Next we extend the mapping $f_k : I_k = [0, a_k] \to J_k = [0, 1]$ by the usual trick of reflection to convert the billiard motion to a torus motion ("unfolding"). Thus we obtain

$$f_k : \widetilde{I_k} = [0, 2a_k) \to \widetilde{J_k} = [0, 2).$$

Consider now a torus motion of a particle moving with constant speed u_k from the left endpoint 0 of $\widetilde{J_k}$ to its right endpoint 2. Then the motion induced by the inverse mapping f_k^{-1} represents a varying speed torus motion from left to right on the interval $\widetilde{I_k}$, where the speed is exactly (7.13).

Since $\widetilde{J_k}$ has length 2, and for the application of Theorem 4.2 it is convenient to work with the unit torus, we "divide by 2" to obtain a torus motion with constant speed u_k in the unit interval $[0, 1)$.

This means that, for every *fixed* $\omega^* \in \left(\mathbf{S}^{h-1}\right)^N$, the family of time evolutions $\mathcal{Y}_{\omega;t}$, $\omega = (\omega^*, \mathbf{u}) \in \Omega(\text{Gauss}; A; \mathit{PolSU})$ of the Gaussian Polar-SpeedUp model with initial configuration $\mathcal{Y} \subset A$ is represented by the family of torus lines in the configuration space $[0, 1)^N$, all starting from the origin $\mathbf{0} \in [0, 1)^N$.

The line sement $I_k(\mathbf{y}_k; \mathbf{e}_k)$, i.e., the orbit of the kth particle that we prefer to represent in the form $[0, a_k(\mathbf{e}_k)]$, intersects the given (measurable) h-dimensional test set $B \subset A$ in a 1-dimensional *measurable* set for almost every $\omega^* \in \left(\mathbf{S}^{h-1}\right)^N$ (N-tuple of directions). This is Fubini's theorem. Let

$$B_k(\omega^*) = B_k(\mathbf{e}_k) \subset [0, a_k(\mathbf{e}_k)]$$

represent the intersection $I_k(\mathbf{y}_k; \mathbf{e}_k) \cap B$ as a subset of $[0, a_k(\mathbf{e}_k)]$; let

$$b_k(\omega^*) = b_k(\mathbf{e}_k) = \frac{\int_{B_k(\omega^*)} x^{h-1}\, dx}{\int_0^{a_k} x^{h-1}\, dx} = \frac{\int_{B_k(\omega^*)} x^{h-1}\, dx}{a_k^h(\mathbf{e}_k)/h} \tag{7.17}$$

be the induced normalized one-dimensional measure of $B_k(\omega^*)$, induced by the mapping f_k.

The $\text{meas}_{A;\mathbf{y}_k}$-mean value (i.e., a weighted average) of $b_k(\omega^*)$, $\omega^* \in \left(\mathbf{S}^{h-1}\right)^N$ is

$$\begin{aligned}
&\int_{\mathbf{e}\in\mathbf{S}^{h-1}} b_k(\mathbf{e})\, d\text{meas}_{A;\mathbf{y}_k}(\mathbf{e}) \\
&\quad = \int_{\mathbf{e}\in\mathbf{S}^{h-1}} \frac{\int_0^{a_k(bfe)} \chi_B(\mathbf{y}_k + x\mathbf{e})x^{h-1}\, dx}{a_k^h(\mathbf{e})/h} a_k^h(\mathbf{e})\, d\text{SA}^*(\mathbf{e}) \\
&\quad \cdot \left(\int_{\mathbf{e}\in\mathbf{S}^{h-1}} (\text{length}(I_A(\mathbf{y}_k;\mathbf{e})))^h\, d\text{SA}^*(\mathbf{e})\right)^{-1} \\
&\quad = h\text{vol}_h(B) \cdot (h\text{vol}_h(A))^{-1} = \frac{\text{vol}_h(B)}{\text{vol}_h(A)},
\end{aligned} \tag{7.18}$$

where in the last step we used the polar coordinate volume formula (7.11).

By similarity we can assume that $\text{vol}_h(A) = 1$. Then (7.18) implies that the $\text{meas}_{A;\mathbf{y}_k}$-mean value of $b_k(\omega^*)$, $\omega^* \in (\mathbf{S}^{h-1})^N$ is $\text{vol}_h(B)$.

Write

$$\text{BadSet} = \left\{\omega^* \in \left(\mathbf{S}^{h-1}\right)^N : \left|\sum_{k=1}^N b_k(\omega^*) - \text{vol}_h(B)N\right| > 30\sqrt{N}\right\}. \tag{7.19}$$

Of course, we could choose here a general "large deviation factor" γ (like we did in Theorem 5.1). Note that $\gamma = 30$ is a good choice for our usual illustration (see below).

Let $\text{meas}_{A;\mathbf{y}_k}^N$ denote the product measure on the product space $(\mathbf{S}^{h-1})^N$ (i.e., each factor $\mathbf{S}^{h-1}$ has the same measure $\text{meas}_{A;\mathbf{y}_k}$).

Since the initial velocities are chosen independently, by Bernstein's large deviation inequality (5.8),

$$\text{meas}^N_{\mathbf{A};\mathbf{y}_k}(\text{BadSet}) \le 2\exp\left(-\frac{450}{1+10N^{-1/2}}\right). \tag{7.20}$$

Let $\omega^* \in \left(\mathbf{S}^{h-1}\right)^N$ be *fixed*, and write

$$S(B;\omega^*) = \left\{ \vec{\mathcal{Z}} = (z_1,\ldots,z_N) \in [0,1)^N : \right.$$

$$\left. \left| |\{1 \le k \le N :\ z_k \in B_k(\omega^*)\}| - \sum_{k=1}^{N} b_k(\omega^*) \right| > 60\sqrt{N} \right\}. \tag{7.21}$$

Using Bernstein's large deviation inequality (5.8), we obtain

$$\text{vol}_N(S(B;\omega^*)) \le 2\exp\left(-\frac{1800}{1+20N^{-1/2}}\right). \tag{7.22}$$

Like in Secs. 5–6, we apply Theorem 4.2 with $\gamma = 60$ and $f = \chi_S$ where $S = S(B;\omega^*) \subset [0,1)^N$. Thus, we obtain the following perfect analog of (6.7): for every $\omega^* \in [0,2\pi)^N$, integer $k \ge 1$ and $U \ge 1$ satisfying $e^{\pi^2U^2/2} > 3UN$,

$$(2\pi)^{-N/2} \int_{\mathbf{u}\in\mathbb{R}^N} \left(\text{length} \left\{ U \le t \le 2^k U : \left| |\mathcal{Y}_{\omega;t} \cap B| \right. \right. \right.$$

$$\left. \left. - \sum_{k=1}^{N} b_k(\omega^*) \right| > 60\sqrt{N} \right\}$$

$$\left. - \text{vol}_N(S(B;\omega^*))(2^k-1)U \right)^2 e^{-|\mathbf{u}|^2/2}\, d\mathbf{u}$$

$$\le 2\exp\left(-\frac{1800}{1+20N^{-1/2}}\right) 10k((2^k-1)U+1), \tag{7.23}$$

where

$$\omega = (\omega^*, \mathbf{u}) \in \Omega(\text{Gauss}; A; PolSU).$$

As usual, to illustrate (7.23), we choose $U = 4$, $k = 100$ and $N = 10^{27}$. Then we obtain the perfect analogue of (6.10)–(6.12).

Summarizing, the threshold $U = 4$ represents — roughly speaking — the necessary number of zig-zags per particle in the (Gaussian) PolarSpeedUp model to reach square-root equilibrium (in the particle space for the typical time evolution). As usual, assume that the average speed is 10^3 meter per second. For this system it takes only a few milliseconds to reach square-root equilibrium. Choosing an arbitrary (measurable) test set $B \subset A$ — where A is an h-dimensional convex set of volume $\text{vol}_h(A) = 1$ with $h \geq 2$ — and an arbitrary N-element initial point configuration $\mathcal{Y}$, for the totally overwhelming majority of the initial velocities (Gaussian distribution), the number of particles in B remains very close to the expected value $\text{vol}_h(B)N$ for an extremely long time, with the possible exception of a totally negligible set of times t.

Indeed, for every (measurable) test set $B \subset A$ and every $N = 10^{27}$-element initial point configuration $\mathcal{Y} \subset A$, there exists a subset $\Omega(\text{Gauss}; A; PolSU; \text{good}) \subset \Omega(\text{Gauss}; A; PolSU)$ with

$$\text{ProdMeas}_{\text{Gauss};A;PolSU}(\Omega(\text{Gauss}; A; PolSU; \text{good}))$$
$$\geq 1 - 10^{-220} - \frac{1}{2}10^{-190} > 1 - 10^{-190},$$

representing a totally overwhelming majority, such that for every initial condition $\omega \in \Omega(\text{Gauss}; A; PolSU; \text{good})$,

$$\left||\mathcal{Y}_{\omega;t} \cap B| - \text{vol}_d(B)10^{27}\right| \leq 90 \cdot 10^{13.5}$$

(where 90 comes from $90 = 30 + 60$) holds for every $4 \leq t \leq 4 \cdot 2^{100}$ with the possible exception of a set of times t of total length $< 10^{-220}$. The possible exception represents less than 10^{-223} seconds — a ridiculously short time.

Note that $4 \leq t \leq 4 \cdot 2^{100}$ represents an incredibly long time interval: it is roughly billion times the age of the universe.

That is, once a typical time evolution of this (Gaussian) PolarSpeedUp system reaches square-root equilibrium (in the particle space), then it stays in that state in the quantitative sense of 90-square-root equilibrium

$$||\mathcal{Y}_{\omega;t} \cap B| - \text{vol}_h(B)N| \leq 90\sqrt{N}$$

for an incredibly long time (with the possible exception of a totally negligible set of t's).

What about the seemingly simpler (Gaussian) PolarBilliard model? Exactly like in Sec. 6, it is similar to the PolarSpeedUp model with the usual difference: the (normalized) volume is replaced by another probability measure $\text{vol}_{A;\mathcal{Y}}$ which *depends* on the initial configuration $\mathcal{Y} \subset A$.

To define the measure $\mathrm{vol}_{A;\mathcal{Y}}$, let $P \in A$ be an arbitrary point. For every direction $\mathbf{e} \in \mathbf{S}^{h-1}$ let $I(P;\mathbf{e})$ denote the line segment starting from $P \in A$, and going in the direction $\mathbf{e}$ until it hits the boundary of A. Let $W \subset A$ be an arbitrary measurable set, and write

$$\mathrm{vol}_{A;P}(W) = \int_{\mathbf{e}\in\mathbf{S}^{h-1}} \frac{\mathrm{length}(W \cap I(P;\mathbf{e}))}{\mathrm{length}(I(P;\mathbf{e}))}\, d\mathbf{SA}^*(\mathbf{e}).$$

Given an initial configuration $\mathcal{Y} = \{\mathbf{y}_1, \ldots, \mathbf{y}_N\}$, let

$$\mathrm{vol}_{A;\mathcal{Y}}(\ldots) = \frac{1}{N}\sum_{k=1}^{N} \mathrm{vol}_{A;\mathbf{y}_k}(\ldots). \tag{7.24}$$

It defines a measure that depends on the initial configuration $\mathcal{Y}$.

We can simply repeat the argument of the case of the (Gaussian) PolarSpeedUp model. The only difference is that in the PolarBilliard model the definition of square-root equilibrium (in the particle space) requires the inhomogeneous measure $\mathrm{vol}_{A;\mathcal{Y}}$ (see (7.24)) instead of the normalized volume.

This proves long-term stability of square-root equilibrium with respect to an arbitary (possibly "ugly") but fixed measurable test set $B \subset A$ (where A is an h-dimensional bounded convex set).

Finally note that, similarly to Sec. 5, we can easily extend the results in Sec. 6–7 from test set size $\geq C \cdot N^{-1/2}$ to the much smaller (and best possible!) size $\geq C \cdot N^{-1}$; see the argument of (5.26)–(5.29) at the end of Sec. 5.

8. Snapshot Randomness (I): Poisson

Time-lapse randomness versus snapshot randomness. Theorems 3.3 and 3.4 have the following common message. The (extremely!) long-term time evolution of a large system with noninteracting particles exhibits "asymptotic randomness" (central limit theorem, Poisson law). These "soft" results express the *dynamic* aspects of randomness on an unlimited time scale, i.e., where the time t runs in a very, very long interval. We call it *time-lapse randomness* on an infinite time scale. This raises the natural question: What can we say about the short-term time-evolution? Do we have any kind of *time-lapse randomness* on a realistic time scale? The answer is yes; we will give a detailed discussion starting in Sec. 17.

What we study here is the completely different *static* aspect of randomness, meaning that t is fixed — we call it *snapshot randomness.* We prove that the overwhelming majority of the time evolutions of a Gaussian system achieves nearly perfect snapshot randomness superexponentially fast, and stays in that state for an incredibly long time.

The best way to explain *snapshot randomness* is to introduce a game, an Alice–Bob type game. In fact, it is an "Alice–Wicky–Bob game": in this version Alice has a sister called Wicky. Bob is a superman, and the two sisters are superwomen. Alice is a nice girl, and Wicky is a wicked girl (to be explained below). Superhumans have no "complexity limitation"; e.g., either one can carry out a case-study of size $10^{10^{1000}}$ (say) in a split second. Alice, the nice girl, is honest, and Wicky, the wicked girl, is an unpredictable liar. Wicky and Bob are adversaries; Alice is neutral to Bob. They are in three different rooms, separated by doors that are locked. Each sister has an N-element point set of her own in the unit cube $[0,1]^3$ ($N = 10^{100}$, say). Alice's N-set is "truly random": Alice created her N-set in $[0,1]^3$ in the hard way by honestly carrying out N independent trials with uniform distribution each (as a superwoman, she can do it in a split second). Wicky claims that her N-set is also "truly random," but, because she is an unpredictable liar, nobody knows what Wicky actually did; nobody knows whether Wicky's N-set is "truly random" or not. Now we open the door between the two sisters: Alice and Wicky put their N-sets into a black box. Next we open Bob's door, and give the black box to Bob, who pulls out one of the two N-sets (despite being superman, he does not know which one is due to Alice and which one is due to Wicky). Bob has to decide whether the N-set he picked is "truly random" or not. He can perform one test (say). Note that Bob's test can be arbitrarily complex (since he is a superman). The sisters do not know in advance Bob's test, and Bob does not know what Wicky actually did. All what Bob knows is that Alice created an honest random N-set (with uniform distribution), and the other sister — Wicky, Bob's adversary — tries to fool him. What test should Bob perform? What test would help him to make a reliable decision on the status of his N-set ("truly random" or not)? On the other hand, what is the optimal strategy for Wicky to fool Bob?

We think these are very interesting questions of fundamental theoretical importance in both complexity theory and probability theory. Unfortunately, we know very little about this. All what we can say are some suggestions for Bob. The two most important limit theorems in probability theory are the central limit theorem (CLT) and the Poisson limit theorem. These limit theorems motivate two tests: the *CLT-partition test* and the *Poisson partition test*.

Intuitively speaking, we say that the time evolution of the system exhibits "complete CLT or Poisson randomness" if the bell-shaped curve (standard normal distribution) or the Poisson distribution shows up with striking precision (we want optimal, or nearly optimal, error term).

The Poisson partition test is somewhat simpler, so, first we discuss the Poisson test. We will return to the CLT test in Sec. 15.

We employ the standard notation: given an N-element set in the 3-dimensional unit cube

$$\mathcal{Z} = \{\mathbf{z}_1, \ldots, \mathbf{z}_N\} \subset [0,1)^3,$$

let

$$\vec{\mathcal{Z}} = (\mathbf{z}_1, \ldots, \mathbf{z}_N) \in I^d$$

denote the corresponding point in the configuration space $I^d = [0,1)^d$ of dimension $d = 3N$.

Poisson partition test. Bob makes a partition $[0,1]^3 = B_1 \cup B_2 \cup \ldots \cup B_N$ of the unit cube where each B_i is measurable and $\text{vol}(B_i) = 1/N$ (3-dimensional Lebesgue measure). Let

$$\mathcal{Z} = \{\mathbf{z}_1, \ldots, \mathbf{z}_N\} \subset [0,1)^3$$

be an arbitrary N-element set in the 3-dimensional unit cube. Let $b_i = b_i(\mathcal{Z})$ denote the number of points of the given N-set $\mathcal{Z}$ that are contained in B_i, and consider the distribution of the N nonnegative integers $b_i = b_i(\mathcal{Z})$, $1 \le i \le N$. Let $X_k = X_k(\vec{\mathcal{Z}})$ denote the number of b_i's that are equal to k:

$$X_k = X_k(\vec{\mathcal{Z}}) = |\{1 \le i \le N : \; b_i(\mathcal{Z}) = k\}|, k = 0, 1, 2, \ldots. \tag{8.1}$$

We compare the empirical distribution in (8.1) to the Poisson distribution with parameter one:

$$\text{error} = \text{error}(\vec{\mathcal{Z}}) = \sup_{k \ge 0} \left| \frac{X_k(\vec{\mathcal{Z}})}{N} - \frac{e^{-k}}{k!} \right|. \tag{8.2}$$

Since $N^{-1/2}$ is the "random error," if the error in (8.2) satisfies the upper bound

$$\text{error} < N^{-\frac{1}{2}+\varepsilon}, \tag{8.3}$$

then we say that the N-set $\vec{\mathcal{Z}}$ is ε-close to *maximum snapshot randomness* with respect to the Poisson-partition test.

Note that Alice's honest random N-set certainly satisfies (8.3) with probability extremely close to one (see Lemma 8.2 below). Since the two sisters (Alice and Wicky) do not know Bob's partition, and the error term (8.3) is nearly optimal ($N^{-1/2}$ is the optimum), if Bob's N-set happens to satisfy (8.3), then Bob has a good reason to believe that his N-set is "truly random". For example, if $N = 10^{100}$ and $\varepsilon = 2 \cdot 10^{-2}$ (say), then

$$N^{-\frac{1}{2}+\varepsilon} = 10^{-48}$$

is an extremely small error. Such an extremely small (and nearly optimal) error is very convincing: it suggests "complete randomness".

For the usual application of Theorem 4.2, we need to know the volume of the subset

$$S(\text{error} \geq \gamma N^{-1/2}) = \left\{ \vec{\mathcal{Z}} \in I^d : \sup_{k \geq 0} \left| \frac{X_k(\vec{\mathcal{Z}})}{N} - \frac{e^{-k}}{k!} \right| \geq \gamma N^{-1/2} \right\} \tag{8.4}$$

in the configuration space $I^d = [0,1)^d$, $d = 3N$, where $\gamma > 0$ is arbitrary.

The d-dimensional volume

$$\text{vol}_d \left(S(\text{error} \geq \gamma N^{-1/2}) \right) \tag{8.5}$$

is exactly a large deviation type probability related to Alice's random N-set. It actually leads us to a classical *occupancy* problem. There is, however, a (luckily minor) technical problem: (8.5) is a large deviation type probability for *dependent* random variables (see (8.6) below).

A simple combinatorial/probabilistic argument about the occupation numbers. The empirical distribution of Alice's honestly constructed random N-element point set (with uniform distribution) in Bob's N-partition of the unit cube is equivalent to the following N-balls-to-N-bins problem. Suppose that we have thrown N balls independently and uniformly into N bins. That is, the probability that the ith ball goes to the jth bin is $1/N$, and for different balls we make *independent* choices. Let Y_j denote the number of balls in the jth bin at the end of the distribution process.

Let X_0 denote the number of empty bins, and in general, for any integer $k \geq 0$, let X_k be the number of j's such that $Y_j = k$. We know that the expectation of X_0 is N/e, and in general, the expectation of X_k is $N/(k!e)$.

In view of (8.4) and (8.5),

$$\mathrm{vol}_d\left(S(\text{error} \geq \gamma N^{-1/2})\right) = \Pr\left[\sup_{k\geq 0}\left|\frac{X_k}{N} - \frac{e^{-k}}{k!}\right| \geq \gamma N^{-1/2}\right]. \tag{8.6}$$

Thus we need an upper bound on the large deviation probability

$$\Pr\left[\left|X_k - \frac{N}{k!e}\right| > \gamma\sqrt{N}\right]. \tag{8.7}$$

X_k is expressed in terms of the underlying random variables Y_j, $1 \leq j \leq N$, but, unfortunately, the random variables Y_j are not independent (for example, their sum is N, which is deterministic), so we cannot directly apply Bernstein's large deviation type inequality. One way to overcome this technical problem is to introduce some auxiliary random variables W_j, $1 \leq j \leq N$ that are *independent.*

Let W_j, $1 \leq j \leq N$ be independent random variables, all having Poisson distribution with parameter one, i.e., $\Pr[W_j = k] = (k!e)^{-1}$, $k = 0, 1, 2, 3, \ldots$. We study the connection between the system of the old random variables Y_j, $1 \leq j \leq N$ and the system of the new independent random variables W_j, $1 \leq j \leq N$.

We call a sequence of N integers $\ell_1, \ell_2, \ldots, \ell_N$ satisfying $\sum_{j=1}^N \ell_j = N$ with $\ell_j \geq 0$ a *relevant sequence.* Given a relevant sequence $\ell_1, \ell_2, \ldots, \ell_N$, we compare the two probabilities $\Pr[Y_j = \ell_j, 1 \leq j \leq N]$ and $\Pr[W_j = \ell_j, 1 \leq j \leq N]$.

Lemma 8.1. *For every relevant sequence* $\ell_1, \ell_2, \ldots, \ell_N$,

$$\frac{\Pr[Y_j = \ell_j, 1 \leq j \leq N]}{\Pr[W_j = \ell_j, 1 \leq j \leq N]} = \frac{N!}{(N/e)^N}.$$

Remark. We emphasize the crucial fact that the ratio is independent of the sequence $\ell_1, \ell_2, \ldots, \ell_N$.

Proof of Lemma 8.1. We apply elementary combinatorics

$$\begin{aligned}\Pr[Y_j = \ell_j, 1 \le j \le N] \\ &= \binom{N}{\ell_1}\binom{N-\ell_1}{\ell_2}\binom{N-\ell_1-\ell_2}{\ell_3}\cdots\binom{N-\ell_1-\cdots-\ell_{N-1}}{\ell_N} N^{-N} \\ &= \frac{N!}{\ell_1!\ell_2!\cdots\ell_N!}N^{-N}.\end{aligned}$$

On the other hand,

$$\Pr[W_j = \ell_j, 1 \le j \le N] = \prod_{j=1}^{N}(e\ell_j!)^{-1} = \frac{e^{-N}}{\ell_1!\ell_2!\cdots\ell_N!}.$$

Taking the ratio of the two probabilities, Lemma 8.1 follows. □

Now we are ready to estimate the large deviation probability (8.7). We prove

Lemma 8.2. *For every real* $\gamma > 0$

$$\Pr\left[\sup_{k\ge 0}\left|X_k - \frac{N}{k!e}\right| > \gamma\sqrt{N}\right] \le \frac{12\sqrt{N}e^{-2\gamma^2}}{1-e^{-2\gamma^2}}.$$

Remark. Probably the reader is wondering why we have the extra factor of $\sqrt{N}$ here. This factor turns out to be harmless in the application below. Nevertheless, it is an interesting question whether or not there is a version of Lemma 8.2 without the extra factor $\sqrt{N}$. We return to this question at the end of the section in the Concluding Remark.

Proof of Lemma 8.2. We need the well-known Stirling's formula; in fact, we use the following refinement

$$e^{1/(12N+1)}\sqrt{2\pi N} < \frac{N!}{(N/e)^N} < e^{1/(12N)}\sqrt{2\pi N}. \tag{8.8}$$

By Lemma 8.1,

$$\begin{aligned}\Pr\left[X_k > \frac{N}{k!e} + \gamma\sqrt{N}\right] &= \sum_{\substack{(\ell_1,\ldots,\ell_N):\text{relevant sequence}\\ \text{containing}>\frac{N}{k!e}+\gamma\sqrt{N}\ k\text{'s}}} \Pr[Y_j = \ell_j, 1 \le j \le N] \\ &= \frac{N!}{(N/e)^N}\sum_{\substack{(\ell_1,\ldots,\ell_N):\text{relevant sequence}\\ \text{containing}>\frac{N}{k!e}+\gamma\sqrt{N}\ k\text{'s}}} \Pr[W_j = \ell_j, 1 \le j \le N]. \qquad (8.9)\end{aligned}$$

Let $W_j^* = 1$ if $W_j = k$ and $W_j^* = 0$ if $W_j \neq k$, and let $Z_k = \sum_{j=1}^N W_j^*$. Since W_j, $1 \le j \le N$ are independent, W_j^*, $1 \le j \le N$ are also independent.

By (8.9),

$$\Pr\left[X_k > \frac{N}{k!e} + \gamma\sqrt{N}\right] = \frac{N!}{(N/e)^N}$$

$$\sum_{\substack{(\ell_1,\dots,\ell_N)\text{:relevant sequence} \\ \text{containing}> \frac{N}{e}+\gamma\sqrt{N}\ k's}} \Pr[W_j = \ell_j, 1 \le j \le N]$$

$$\le \frac{N!}{(N/e)^N} \Pr\left[Z_k > \frac{N}{k!e} + \gamma\sqrt{N}\right]. \tag{8.10}$$

Similarly,

$$\Pr\left[X_k < \frac{N}{k!e} - \gamma\sqrt{N}\right] \le \frac{N!}{(N/e)^N} \Pr\left[Z_k < \frac{N}{k!e} - \gamma\sqrt{N}\right]. \tag{8.11}$$

Combining (8.10) and (8.11),

$$\Pr\left[\left|X_k - \frac{N}{k!e}\right| > t\sqrt{N}\right] \le \frac{N!}{(N/e)^N} \Pr\left[\left|Z_k - \frac{N}{e}\right| > \gamma\sqrt{N}\right]. \tag{8.12}$$

Since $Z_k = \sum_{j=1}^N W_j^*$ is a sum of N independent Bernoulli variables with probability $p = p_k = (k!e)^{-1}$, we can apply Bernstein's inequality (5.9)

$$\Pr\left[\left|Z_k - \frac{N}{k!e}\right| > \gamma\sqrt{N}\right] \le 2\exp\left(-\frac{\gamma^2(2p_k(1-p_k))^{-1}}{1+\frac{2\gamma}{\sqrt{N}}}\right). \tag{8.13}$$

Combining (8.8), (8.12) and (8.13), we conclude

$$\Pr\left[\sup_{k\ge 0}\left|X_k - \frac{N}{k!e}\right| > \gamma\sqrt{N}\right] \le \sum_{k=0}^{\infty} \Pr\left[\left|X_k - \frac{N}{k!e}\right| > \gamma\sqrt{N}\right]$$

$$\le \frac{N!}{(N/e)^N} \sum_{k=0}^{\infty} \Pr\left[\left|Z_k - \frac{N}{e}\right| > \gamma\sqrt{N}\right]$$

$$\le 3\sqrt{N} \sum_{k=0}^{\infty} 2\exp\left(-\frac{\gamma^2(2p_k(1-p_k))^{-1}}{1+\frac{\gamma}{3p_k(1-p_k)\sqrt{N}}}\right)$$

$$\le 6\sqrt{N}\left(e^{-2\gamma^2}+\sum_{k=1}^{\infty}e^{-2k\gamma^2}\right)$$

$$= 6\sqrt{N}\left(e^{-2\gamma^2}+\frac{e^{-2\gamma^2}}{1-e^{-2\gamma^2}}\right)\le\frac{12\sqrt{N}e^{-2\gamma^2}}{1-e^{-2\gamma^2}},$$

where we used the fact that $p_k = (k!e)^{-1}$ is extremely small if k is large. This completes the proof of Lemma 8.2. □

Applying Theorem 4.2 as a short-time ergodic theorem in the configuration space. Let $\mathcal{B} = \{B_1, B_2, \ldots, B_N\}$ be an arbitrary but fixed measurable partition of the unit cube $[0,1]^3 = B_1 \cup B_2 \cup \ldots \cup B_N$ such that $\mathrm{vol}(B_i) = 1/N$, $1 \le i \le N$ (Bob's partition). Given an N-element set in the 3-dimensional unit cube

$$\mathcal{Z} = \{\mathbf{z}_1, \ldots, \mathbf{z}_N\} \subset [0,1)^3,$$

let $b_i = b_i(\mathcal{Z};\mathcal{B})$ denote the number of points of $\mathcal{Z}$ that are contained in B_i, and consider the distribution of the N nonnegative integers $b_i = b_i(\mathcal{Z};\mathcal{B})$, $1 \le i \le N$. Like in (8.1), let $X_k = X_k(\vec{\mathcal{Z}};\mathcal{B})$ denote the number of b_i's that are equal to k:

$$X_k = X_k(\vec{\mathcal{Z}};\mathcal{B}) = |\{1 \le i \le N : \ b_i(\mathcal{Z};\mathcal{B}) = k\}|, \quad k = 0,1,2,\ldots.$$

Like in (8.2), we compare this empirical distribution to the Poisson distribution with parameter one

$$\sup_{k\ge 0}\left|\frac{X_k(\vec{\mathcal{Z}};\mathcal{B})}{N}-\frac{e^{-k}}{k!}\right|. \tag{8.14}$$

If (8.14) is "close" to $N^{-1/2}$ ("random error") then we say the N-element set $\mathcal{Z}$ exhibits "complete Poisson snapshot randomness" (with respect to the given partition $\mathcal{B}$).

We recall (8.4): for every $\gamma > 0$ let

$$S(\mathrm{error} \ge \gamma N^{-1/2}) = \left\{\vec{\mathcal{Z}} \in I^d : \sup_{k\ge 0}\left|\frac{X_k(\vec{\mathcal{Z}};\mathcal{B})}{N}-\frac{e^{-k}}{k!}\right| \ge \gamma N^{-1/2}\right\}. \tag{8.15}$$

Consider the usual 3-dimensional Gaussian torus-billiard model $\mathcal{Y}_{\omega;t} = \mathcal{Y}(\mathrm{Gauss};\omega;t)$, $\omega \in \Omega_{\mathrm{Gauss}}$.

We study the following question. Assume that N is large; is it true that the overwhelming majority of the time evolutions of this Gaussian billiard system reaches "complete Poisson snapshot randomness" superexponentially fast, and stays in this state for an incredibly long time (with the possible exception of a totally negligible set of t's)? By using Theorem 4.2 we give a positive answer. We basically repeat the argument in Sec. 5.

The family of time evolutions $\mathcal{Y}_{\omega;t}$, $\omega \in \Omega_{\text{Gauss}}$ of the 3-dimensional Gaussian torus-billiard model is represented by the family of torus lines

$$\vec{\mathcal{Y}}_{\omega;t} = \vec{\mathcal{Y}}(\text{Gauss}; \omega; t)$$

in the configuration space $I^d = [0,1)^{3N}$ (= d-dimensional unit torus), all starting from the same point $\vec{\mathcal{Y}} \in I^d$.

Since the torus I^d is translation invariant, we can apply Theorem 4.2 with $f = \chi_S$ where S is the translated copy

$$S = S(\mathcal{B};\gamma) = S(\text{error} \geq \gamma N^{-1/2}) - \vec{\mathcal{Y}} \tag{8.16}$$

of $S(\text{error} \geq \gamma N^{-1/2})$ (see (8.15)) in the torus I^d. Thus, if $U \geq 1$ and $e^{\pi^2 U^2/2} > 3dU$, then for every integer $j \geq 1$,

$$\begin{aligned}
&\int_{\omega\in\Omega_{\text{Gauss}}} \left(\text{length}\left\{ U \leq t \leq 2^j U : \sup_{k\geq 0} \left| \frac{X_k(\vec{\mathcal{Y}}_{\omega;t};\mathcal{B})}{N} - \frac{e^{-k}}{k!} \right| \geq \gamma N^{-1/2} \right\} \right.\\
&\qquad \left. - \text{vol}_d(S)(2^j - 1)U \right)^2 d\text{ProdMeas}_{\text{Gauss}}(\omega)\\
&\quad \leq \text{vol}_d(S) 10j \left((2^j - 1)U + 1 \right),
\end{aligned} \tag{8.17}$$

where $S = S(\mathcal{B};\gamma)$ (see (8.16)).

By (8.6) and Lemma 8.2,

$$\begin{aligned}
\text{vol}_d(S) &= \text{vol}_d\left(S(\text{error} \geq \gamma N^{-1/2}) \right)\\
&= \Pr\left[\sup_{k\geq 0} \left| \frac{X_k}{N} - \frac{e^{-k}}{k!} \right| \geq \gamma N^{-1/2} \right] \leq \frac{12\sqrt{N} e^{-2\gamma^2}}{1 - e^{-2\gamma^2}}.
\end{aligned} \tag{8.18}$$

To illustrate the power of (8.17)–(8.18), we choose the usual numerical values $\gamma = 30$, $U = 4$, $j = 340$ and $N = 10^{27}$; it follows that $d = 3 \cdot 10^{27}$.

Then by (8.17)–(8.18),

$$\int_{\omega\in\Omega_{\text{Gauss}}}\left(\text{length}\left\{4\le t\le 4\cdot 2^{340}:\ \sup_{k\ge 0}\left|\frac{X_k(\vec{\mathcal{Y}}_{\omega;t};\mathcal{B})}{N}-\frac{e^{-k}}{k!}\right|\right.\right.$$

$$\left.\ge 30\cdot 10^{-13.5}\right\}$$

$$\left.-\,\text{vol}_d(S(B))(2^{340}-1)4\right)^2 d\text{ProdMeas}_{\text{Gauss}}(\omega)$$

$$\le\frac{12\cdot 10^{13.5}e^{-2\cdot 30^2}}{1-e^{-2\cdot 30^2}}\cdot 4\cdot 340\cdot 2^{340}\cdot 20<10^{-701}. \tag{8.19}$$

Let $\Omega_{\text{Gauss}}(\text{bad})$ be the set of those $\omega\in\Omega_{\text{Gauss}}$ for which

$$\text{length}\left\{4\le t\le 4\cdot 2^{340}:\ \sup_{k\ge 0}\left|\frac{X_k(\vec{\mathcal{Y}}_{\omega;t};\mathcal{B})}{N}-\frac{e^{-k}}{k!}\right|\ge 30\cdot 10^{-13.5}\right\}$$

$$\ge 10^{-200}. \tag{8.20}$$

Repeating the proof in (5.18)–(5.19), we obtain

$$\text{ProdMeas}_{\text{Gauss}}(\Omega_{\text{Gauss}}(\text{bad}))\le 10^{-300}. \tag{8.21}$$

Again we study the classical Bernoulli gas model where the gas molecules are represented by point billiards ($N=10^{27}$ is a realistic number). Using the trick of *unfolding* we reduce the billiards-in-a-box model to the torus-billiard model. The threshold $U=4$ represents — roughly speaking — the relaxation distance, i.e., the necessary number of "jumps" per particle in the Gaussian torus-billiard model (which is the half of the number of bounces in the analog billiards-in-a-box model) to reach "complete Poisson snapshot randomness". As usual, assume that the gas molecules have average speed 10^3 meter per second. For this system it takes only a few milliseconds to reach "complete Poisson snapshot randomness". Now (8.20) and (8.21) have the following interpretation. Choosing an arbitrary (measurable) Poisson partition $\mathcal{B}=\{B_1,B_2,\ldots,B_N\}$ of the unit cube $[0,1]^3=B_1\cup B_2\cup\ldots\cup B_N$ such that $\text{vol}(B_i)=1/N$, $1\le i\le N$ and an arbitrary N-element initial point configuration $\mathcal{Y}$, for the totally overwhelming majority of the initial velocities (= Gaussian distribution), the distribution of the particles in the partition $\mathcal{B}$ remains very close to the Poisson distribution with parameter one for an extremely long time, with the possible exception of a totally negligible set of times t.

Indeed, for every Poisson partition $\mathcal{B} = \{B_1, B_2, \ldots, B_N\}$ and every $N = 10^{27}$-element initial point configuration $\mathcal{Y} \subset [0,1)^3$, there exists a subset $\Omega_{\text{Gauss}}(\text{good})$ where

$$\Omega_{\text{Gauss}}(\text{good}) = \Omega_{\text{Gauss}} \setminus \Omega_{\text{Gauss}}(\text{bad})$$

with

$$\text{ProdMeas}_{\text{Gauss}}(\Omega_{\text{Gauss}}(\text{good})) \geq 1 - 10^{-300}$$

(see (8.21)), representing a totally overwhelming majority, such that for every $\omega \in \Omega_{\text{Gauss}}(\text{good})$,

$$\sup_{k\geq 0} \left| \frac{X_k(\vec{\mathcal{Y}}_{\omega;t};\mathcal{B})}{N} - \frac{e^{-k}}{k!} \right| \leq 3 \cdot 10^{-12.5} \tag{8.22}$$

holds for every $4 \leq t \leq 4 \cdot 2^{340}$ with the possible exception of a set of times t of total length $< 10^{-200}$, see (8.20). The latter actually represents less than 10^{-203} seconds, which is a ridiculously short time.

Note that $4 \leq t \leq 4 \cdot 2^{340}$ represents a time interval of length about 10^{97} seconds, which is a ridiculously long time (the estimated age of the universe, starting from Big Bang, is less than 10^{20} seconds).

Summarizing, the typical time evolution of this system reaches "complete Poisson snapshot randomness" (see (8.22)) in a few milliseconds (even starting from Big Bang!), and then it remains in the state of "complete Poisson snapshot randomness" for an incredibly long time (with the possible exception of a totally negligible set of t's).

What happens if we replace $N = 10^{27}$ with $N = 10^{100}$ (say)? Everything remains the same, the only change is that (8.22) is replaced by

$$\sup_{k\geq 0} \left| \frac{X_k(\vec{\mathcal{Y}}_{\omega;t};\mathcal{B})}{N} - \frac{e^{-k}}{k!} \right| \leq 30N^{-1/2} = 3 \cdot 10^{-49}. \tag{8.23}$$

That is, independently of the initial configuration, after a few milliseconds the system exhibits the Poisson distribution with striking, and nearly optimal, precision (see (8.23)). This is why we may say that the system is *completely random*.

In the next section we define and study the CLT-partition test (where CLT stands for the central limit theorem).

Concluding Remark. The proof of Lemma 8.2 was based on Lemma 8.1 and Bernstein's large deviation inequality. We show that, replacing Lemma 8.1 with a martingale approach (and, consequently, replacing Bernstein's inequality with the Azuma–Hoeffding inequality), we can get rid of the extra factor $\sqrt{N}$.

A martingale is a sequence $U_0, U_1, \ldots, U_n$ of random variables so that for $0 \leq i < n$, $\mathbf{E}(U_{i+1}|U_i) = X_i$. (Here $\mathbf{E}(U|V)$ denotes the conditional expectation of U given V; we could also write $\mathbf{E}(U|V) = \mathbf{E}(U|\mathcal{V})$, where $\mathcal{V}$ is the σ-algebra generated by the random variable V.) In the application we use the so-called Doob-martingale. Let U be a random variable on the probability space $(\Omega, \mathcal{F}, \mu)$ ($\mathcal{F}$ is a σ-algebra and μ is a positive measure such that $\mu(\Omega) = 1$). Let

$$\mathcal{F}_0 \subset \mathcal{F}_1 \subset \ldots \subset \mathcal{F}_n \subset \mathcal{F}$$

be an increasing sequence of sub-σ-algebras of $\mathcal{F}$ ("filter"), then $U_i = \mathbf{E}(U|\mathcal{F}_i)$, $0 \leq i \leq n$ is a (Doob-)martingale (generated by the filter $\mathcal{F}_0 \subset \ldots \subset \mathcal{F}_n$).

Azuma–Hoeffding inequality (see e.g. [Gr-St92]) Let $U_0, U_1, \ldots, U_n$ be a martingale with $|U_{i+1} - U_i| \leq M$ for $0 \leq i < n$. Then for every $\tau > 0$,

$$\Pr\left[|U_n - U_0| \geq \tau\sqrt{n}\right] \leq 2e^{-\tau^2/(2M)}.$$

We apply this martingale tail inequality for the N-balls-to-N-bins occupancy problem (we place N balls independently and uniformly into N bins). Let $Z_k = j$ if the kth ball goes to the jth bin. Let $X^{(0)}$ denote the number of empty bins at the end (for simplicity we just study this special case). We can view $X^{(0)}$ as a function $f(Z_1, \ldots, Z_N)$. We prove the following result.

Lemma 8.3. *We have*

$$\Pr\left[\left|X^{(0)} - N\left(1 - \frac{1}{N}\right)^N\right| \geq \gamma\sqrt{N}\right] \leq 2e^{-\gamma^2/2}.$$

Note that

$$N\left(1 - \frac{1}{N}\right)^N = \frac{N}{e} + O(1). \tag{8.24}$$

Before deriving Lemma 8.3 from the Azuma–Hoeffding inequality, we want to compare Lemma 8.3 with (8.24) to Lemma 8.2 (the latter in the special case of empty bins, i.e., $k = 0$). An advantage of Lemma 8.3 is that the extra factor $\sqrt{N}$ in the upper bound is gone. An advantage of Lemma 8.2 is that in the exponent we have $2\gamma^2$ instead of $\gamma^2/2$, which makes a big difference. For example, in the applications above with $N = 10^{27}$ or $N = 10^{100}$ and $\gamma = 30$, Lemma 8.2 gives a *much* better upper bound than Lemma 8.3. On the other hand, choosing (say) $N = 10^{100}$ and $\gamma = 6$, Lemma 8.2 becomes useless (the upper bound is greater than 1); on the other hand, Lemma 8.3 still gives a fairly good bound. Summarizing, we may say that, Lemma 8.2 is better if γ is relatively large, and Lemma 8.3 is better if γ is relatively small.

It remains to derive Lemma 8.3 from the Azuma–Hoeffding inequality. Let time t refer to the point at which the first t balls have been thrown. Let $\mathcal{F}_t$ be the σ-algebra generated by the random choice of bins for the first t balls, i.e., the σ-algebra generated by the independent random variables $Z_1, \ldots, Z_t$. Let $U_t = \mathbf{E}(X^{(0)}|\mathcal{F}_t)$, that is, the expected number of empty bins given $Z_1, \ldots, Z_t$. The random variables $U_0, U_1, \ldots, U_N$ form a martingale with

$$U_0 = \mathbf{E}(X^{(0)}) = N\left(1 - \frac{1}{N}\right)^N \quad \text{and} \quad U_N = X^{(0)}.$$

Notice that moving the kth ball from one bin to another can change the number of empty bins by at most one. This implies that $|U_{i+1} - U_i| \leq 1$ for $0 \leq i < N$. Applying the Azuma–Hoeffding inequality with $M = 1$, $n = N$ and $\tau = \gamma$, Lemma 8.3 follows. □

9. Proofs of Theorems 4.2 and 4.3

For technical reasons it is convenient to prove first a special case with an upper bound on the ratio W/U.

Theorem 9.1. *Let $f \in L_2(I^d)$ be a test function, where $I^d = [0,1)^d$. If $1 \leq U < V \leq 2U$ and $e^{\pi^2 U^2/2} \geq 3dU$, then*

$$\Delta_f^2(\text{Gauss}; U, V) \leq \sigma_0^2(f)\,(9(V-U)+1)\,.$$

Note in advance that the technical restriction $U < V \leq 2U$ in Theorem 9.1 can be easily eliminated by a routine application of the Cauchy–Schwarz inequality; for the details see the end of the section. (Needless to say the factor 9 is an accidental constant.)

Proof of Theorem 9.1. We use Fourier analysis in the configuration space $I^d = [0,1)^d$, which has very high dimension in our applications for large off-equilibrium systems. The possibility of high dimension leads to technical difficulties that are combinatorial in nature.

Let $f \in L_2(I^d)$ be a Lebesgue square-integrable function in the d-dimensional unit torus (i.e., we extend f over the whole d-space $\mathbb{R}^d$ periodically), and consider the Fourier expansion of f:

$$f(\mathbf{u}) = \sum_{\mathbf{n}\in\mathbb{Z}^d} a_{\mathbf{n}} e^{2\pi i \mathbf{n}\cdot\mathbf{u}}, \tag{9.1}$$

where

$$a_{\mathbf{n}} = \int_{I^d} f(\mathbf{y}) e^{-2\pi i \mathbf{n}\cdot\mathbf{y}}\, d\mathbf{y}, \quad \mathbf{n} \in \mathbb{Z}^d$$

are the Fourier coefficients, and of course $\mathbf{v}\cdot\mathbf{w} = v^{(1)}w^{(1)} + \cdots + v^{(d)}w^{(d)}$ denotes the dot product of $\mathbf{v} = (v^{(1)}, \ldots, v^{(d)})$ and $\mathbf{w} = (w^{(1)}, \ldots, w^{(d)})$.

Clearly

$$a_{\mathbf{0}} = \int_{I^d} f(\mathbf{y})\, d\mathbf{y}, \text{ and } so \sum_{\mathbf{n}\in\mathbb{Z}^d\setminus\mathbf{0}} |a_{\mathbf{n}}|^2 = \sigma_0^2(f), \tag{9.2}$$

where we used Parseval's formula.

By (9.1) we have

$$f(t\sqrt{d}\mathbf{e}) - \int_{I^d} f\, d\mathrm{V} = \sum_{\mathbf{n}\in\mathbb{Z}^d\setminus\mathbf{0}} a_{\mathbf{n}} e^{2\pi i t\sqrt{d}\mathbf{n}\cdot\mathbf{e}}, \tag{9.3}$$

where $\mathbf{e} \in \mathbf{S}^{d-1}$ is a unit vector in the d-space.

Here we briefly interrupt the proof, and insert a technical remark. Notice that (9.3) is an informal equality: the infinite sum on the right-hand side may be divergent for some unit vector $\mathbf{e} \in \mathbf{S}^{d-1}$ in the d-space. One possible way — the very hard way! — to make (9.3) precise is to prove pointwise convergence by using deep results from Fourier analysis (Carleson, C. Fefferman, and others).

But we do not really need any deep result from Fourier analysis; for us it would be equally good to replace pointwise convergence with Cesaro summability (Fejér kernel), which has a much simpler classical proof, and works better under more general conditions.

What we *actually* do to avoid this kind of technical nuisance is a yet different third way, which is based on the fact that the trigonometric polynomials are dense in the L_2-space. We proceed in two steps. The first step is to prove the theorem in the special case where f is a trigonometric polynomial (in d variables). Then it is trivial to carry out the usual manipulations, e.g., changing the order of finite summation and integration. The second step is the routine limit process: the class of trigonometric polynomials forms a dense subset of the Hilbert space $L_2(I^d)$, and we can complete the proof in the general case with a routine application of Lebesgue's Dominated Convergence Theorem.

By (4.29)–(4.31) and (9.3) we have

$$\Delta_f^2(\text{Gauss}; T_1, T_2) = \int_{\rho=0}^{\infty} \int_{\mathbf{e}\in\mathbf{S}^{d-1}} |D_f(\rho, \mathbf{e}; T_1, T_2)|^2 \times \frac{1}{C_d}\rho^{d-1}e^{-\rho^2/2}\, d\rho\, d\text{SA}^*(\mathbf{e}), \tag{9.4}$$

where

$$D_f(\rho, \mathbf{e}; T_1, T_2) = \int_{T_1}^{T_2} f(\tau\rho\mathbf{e})\, d\tau - (T_2 - T_1)\int_{I^d} f\, d\text{V} = \int_{T_1}^{T_2} \sum_{\mathbf{n}\in\mathbb{Z}^d\setminus\mathbf{0}} a_{\mathbf{n}} e^{2\pi i\tau\mathbf{n}\cdot\rho\mathbf{e}}\, d\tau. \tag{9.5}$$

We need

Lemma 9.1. *For every d-dimensional vector $\mathbf{w} = (w_1, w_2, \ldots, w_d)$ we have*

$$\int_{\rho=0}^{\infty} \int_{\mathbf{e}\in\mathbf{S}^{d-1}} e^{i\mathbf{w}\cdot\rho\mathbf{e}} \frac{1}{C_d}\rho^{d-1}e^{-\rho^2/2}\, d\rho\, d\text{SA}^*(\mathbf{e}) = e^{-|\mathbf{w}|^2/2},$$

where as usual $|\mathbf{w}| = \sqrt{w_1^2 + \cdots + w_d^2}$ denotes the Euclidean norm.

Proof. In the integral

$$\int_{\rho=0}^{\infty} \int_{\mathbf{e}\in\mathbf{S}^{d-1}} e^{i\mathbf{w}\cdot\rho\mathbf{e}} \frac{1}{C_d}\rho^{d-1}e^{-\rho^2/2}\, d\rho\, d\text{SA}^*(\mathbf{e})$$

the vector $\rho\mathbf{e} = \mathbf{v} = (v_1, \ldots, v_d)$ has d-dimensional standard normal (= Gaussian) distribution. Thus we have

$$\int_{\rho=0}^{\infty} \int_{\mathbf{e}\in\mathbf{S}^{d-1}} e^{\mathrm{i}\mathbf{w}\cdot\rho\mathbf{e}} \frac{1}{C_d} \rho^{d-1} e^{-\rho^2/2}\, d\rho\, d\mathrm{SA}^*(\mathbf{e})$$

$$= (2\pi)^{-d/2} \int_{\mathbf{v}\in\mathbb{R}^d} e^{\mathrm{i}\mathbf{w}\cdot\mathbf{v}} e^{-|\mathbf{v}|^2/2}\, d\mathbf{v} = \prod_{j=1}^{d} \frac{1}{\sqrt{2\pi}} \int_{-\infty}^{\infty} e^{\mathrm{i}w_j v_j} e^{-v_j^2/2}\, dv_j$$

$$= \prod_{j=1}^{d} (\text{Fourier transform of } e^{-v^2/2} \text{ at } w_j)$$

$$= \prod_{j=1}^{d} (e^{-w_j^2/2}) = e^{-(w_1^2+\ldots+w_d^2)} = e^{-|\mathbf{w}|^2/2},$$

where in the argument we used the well-known facts that the coordinates $v_1, \ldots, v_d$ of $\mathbf{v}$ are independent random variables having standard normal distribution each, and the Fourier transform of $e^{-x^2/2}$ is itself. This completes the proof of Lemma 9.1. □

Let us return to (9.4) and (9.5). Applying Lemma 9.1 it is easy to prove the following lemma.

Lemma 9.2. *For every $-\infty \le W' < W'' \le \infty$, we have*

$$\Delta_f^2(\text{Gauss}; W', W'')$$

$$= \sum_{\mathbf{n}_1\in\mathbb{Z}^d\setminus\mathbf{0}} \sum_{\mathbf{n}_2\in\mathbb{Z}^d\setminus\mathbf{0}} a_{\mathbf{n}_1}\overline{a_{\mathbf{n}_2}} \int_{W'}^{W''} \int_{W'}^{W''} e^{-2\pi^2|t_1\mathbf{n}_1 - t_2\mathbf{n}_2|^2}\, dt_1\, dt_2.$$

Proof. Using the trivial fact

$$\int_{a_1}^{b_1} h(x)\, dx \int_{a_2}^{b_2} g(y)\, dy = \int_{a_1}^{b_1} \int_{a_2}^{b_2} h(x)g(y)\, dx\, dy$$

in (9.4), we have

$$|D_f(\rho, \mathbf{e}; W', W'')|^2 = \int_{W'}^{W''} \int_{W'}^{W''} \sum_{\mathbf{n}_1\in\mathbb{Z}^d\setminus\mathbf{0}} \sum_{\mathbf{n}_2\in\mathbb{Z}^d\setminus\mathbf{0}} a_{\mathbf{n}_1} \cdot \overline{a_{\mathbf{n}_2}} e^{2\pi\mathrm{i}(t_1\mathbf{n}_1 - t_2\mathbf{n}_2)\cdot\rho\mathbf{e}}\, dt_1\, dt_2. \tag{9.6}$$

Applying (9.6) in (9.5), we obtain

$$\begin{aligned}
\Delta_f^2(\text{Gauss}; W', W'') &= \int_0^\infty \int_{\mathbf{e}\in\mathbf{S}^{d-1}} \left|D_f^2(\rho, \mathbf{e}; W', W'')\right|^2 \\
&\quad \times \frac{\rho^{d-1}e^{-\rho^2/2}}{C_d}\, d\rho\, d\mathrm{SA}^*(\mathbf{e}) \\
&= \int_0^\infty \int_{\mathbf{e}\in\mathbf{S}^{d-1}} \left(\int_{W'}^{W''} \int_{W'}^{W''} \sum_{\mathbf{n}_1\in\mathbb{Z}^d\setminus\mathbf{0}} \sum_{\mathbf{n}_2\in\mathbb{Z}^d\setminus\mathbf{0}} a_{\mathbf{n}_1} \right. \\
&\quad \left. \cdot \overline{a_{\mathbf{n}_2}} e^{2\pi \mathrm{i}(t_1\mathbf{n}_1 - t_2\mathbf{n}_2)\cdot\rho\mathbf{e}}\, dt_1\, dt_2 \right) \frac{\rho^{d-1}e^{-\rho^2/2}}{C_d}\, d\rho\, d\mathrm{SA}^*(\mathbf{e}) \\
&= \sum_{\mathbf{n}_1\in\mathbb{Z}^d\setminus\mathbf{0}} \sum_{\mathbf{n}_2\in\mathbb{Z}^d\setminus\mathbf{0}} a_{\mathbf{n}_1}\overline{a_{\mathbf{n}_2}} \\
&\quad \cdot \int_{W'}^{W''} \int_{W'}^{W''} \left(\int_0^\infty \int_{\mathbf{e}\in\mathbf{S}^{d-1}} e^{2\pi \mathrm{i}(t_1\mathbf{n}_1 - t_2\mathbf{n}_2)} \cdot \rho\mathbf{e} \right. \\
&\quad \left. \cdot \frac{\rho^{d-1}e^{-\rho^2/2}}{C_d}\, d\rho\, d\mathrm{SA}^*(\mathbf{e}) \right) dt_1\, dt_2. \qquad (9.7)
\end{aligned}$$

By Lemma 9.1 we can evaluate the inner integral at the end of (9.7):

$$\begin{aligned}
&\int_0^\infty \int_{\mathbf{e}\in\mathbf{S}^{d-1}} e^{2\pi \mathrm{i}(t_1\mathbf{n}_1 - t_2\mathbf{n}_2)\cdot\rho\mathbf{e}} \cdot \frac{\rho^{d-1}e^{-\rho^2/2}}{C_d}\, d\rho\, d\mathrm{SA}^*(\mathbf{e}) \\
&\quad = e^{-2\pi^2|t_1\mathbf{n}_1 - t_2\mathbf{n}_2|^2},
\end{aligned}$$

and using this in (9.7), Lemma 9.2 follows.

Now we are ready to prove Theorem 9.1. The proof is an elementary brute force combinatorial argument. For every $\mathbf{n} = (n_1, \ldots, n_d) \in \mathbb{Z}^d \setminus \mathbf{0}$ write $L(\mathbf{n}) = \{1 \le i \le d : \ n_i \ne 0\}$. Applying the simple inequality $|a_{\mathbf{n}_1}\overline{a_{\mathbf{n}_2}}| \le (|a_{\mathbf{n}_1}|^2 + |a_{\mathbf{n}_2}|^2)/2$ in Lemma 9.2, we have

$$\begin{aligned}
&\left|\Delta_f^2(\text{Gauss}; U, V)\right| \\
&= \left| \sum_{\mathbf{n}_1\in\mathbb{Z}^d\setminus\mathbf{0}} \sum_{\mathbf{n}_2\in\mathbb{Z}^d\setminus\mathbf{0}} a_{\mathbf{n}_1}\overline{a_{\mathbf{n}_2}} \cdot \int_U^V \int_U^V e^{-2\pi^2|t_1\mathbf{n}_1 - t_2\mathbf{n}_2|^2}\, dt_1\, dt_2 \right| \\
&\le \sum_{\mathbf{n}_1\in\mathbb{Z}^d\setminus\mathbf{0}} \sum_{\mathbf{n}_2\in\mathbb{Z}^d\setminus\mathbf{0}} \frac{1}{2}\left(|a_{\mathbf{n}_1}|^2 + |a_{\mathbf{n}_2}|^2\right) \int_U^V \int_U^V e^{-2\pi^2|t_1\mathbf{n}_1 - t_2\mathbf{n}_2|^2}\, dt_1\, dt_2
\end{aligned}$$

$$= \sum_{\mathbf{n}_1 \in \mathbb{Z}^d \setminus \mathbf{0}} |a_{\mathbf{n}_1}|^2 \sum_{\mathbf{n}_2 \in \mathbb{Z}^d \setminus \mathbf{0}} \int_U^V \int_U^V e^{-2\pi^2 |t_1 \mathbf{n}_1 - t_2 \mathbf{n}_2|^2} \, dt_1 \, dt_2$$

$$= \int_{t_1=U}^V \sum_{\mathbf{n}_1 \in \mathbb{Z}^d \setminus \mathbf{0}} |a_{\mathbf{n}_1}|^2 \left(\int_{t_2=U}^V \sum_{\mathbf{n}_2 \in \mathbb{Z}^d \setminus \mathbf{0}} e^{-2\pi^2 |t_1 \mathbf{n}_1 - t_2 \mathbf{n}_2|^2} \, dt_2 \right) dt_1$$

$$= \sum_{\lambda_1=1}^d \int_{t_1=U}^V \sum_{\substack{\mathbf{n}_1 \in \mathbb{Z}^d \setminus \mathbf{0}: \\ |L(\mathbf{n}_1)|=\lambda_1}} |a_{\mathbf{n}_1}|^2 \left(\sum_{L_{1,2} \subseteq L(\mathbf{n}_1)} \sum_{\lambda_2 = \max\{|L_{1,2}|,1\}}^d \right.$$

$$\left. \cdot \int_{t_2=U}^V \sum_{\substack{\mathbf{n}_2 \in \mathbb{Z}^d \setminus \mathbf{0}: \\ |L(\mathbf{n}_2)|=\lambda_2 \quad \text{and} \quad L(\mathbf{n}_2) \cap L(\mathbf{n}_1) = L_{1,2}}} e^{-2\pi^2 |t_1 \mathbf{n}_1 - t_2 \mathbf{n}_2|^2} \, dt_2 \right) dt_1. \tag{9.8}$$

We fix $t_1 \in [U, V]$, $\mathbf{n}_1 \in \mathbb{Z}^d \setminus \mathbf{0}$, $L_{1,2} \subseteq L(\mathbf{n}_1)$ and λ_2, and focus on the inner integral at the end of (9.8).

Write

$$\lambda_{1,2} = |L_{1,2}| = |\{1 \leq i \leq d : \ n_{1,i} \neq 0 \, \text{and} \, n_{2,i} \neq 0\}|, \tag{9.9}$$

(where, as usual, $|A|$ denotes the number of elements of a finite set A).

Let $k_1(\mathbf{n}_2)$ denote the number of coordinates $n_{2,i} = \pm 1$ of $\mathbf{n}_2$ which also satisfy $n_{1,i} = 0$; let $k_2(\mathbf{n}_2)$ denote the number of coordinates $n_{2,i} = \pm 2$ of $\mathbf{n}_2$ which also satisfy $n_{1,i} = 0$; let $k_3(\mathbf{n}_2)$ denote the number of coordinates $n_{2,i} = \pm 3$ of $\mathbf{n}_2$ which also satisfy $n_{1,i} = 0$; and so on. Note that

$$k_1(\mathbf{n}_2) + k_2(\mathbf{n}_2) + k_3(\mathbf{n}_2) + \cdots = \lambda_2 - \lambda_{1,2}. \tag{9.10}$$

Let $h_0(t_2; \mathbf{n}_2)$ denote the number of coordinates $j \in L_{1,2}$ such that $|t_1 n_{1,j} - t_2 n_{2,j}| < U/2$; let $h_1(t_2; \mathbf{n}_2)$ denote the number of coordinates $j \in L_{1,2}$ such that $U/2 \leq |t_1 n_{1,j} - t_2 n_{2,j}| < 3U/2$; let $h_2(t_2; \mathbf{n}_2)$ denote the number of coordinates $j \in L_{1,2}$ such that $3U/2 \leq |t_1 n_{1,j} - t_2 n_{2,j}| < 5U/2$; let $h_3(t_2; \mathbf{n}_2)$ denote the number of coordinates $j \in L_{1,2}$ such that $5U/2 \leq |t_1 n_{1,j} - t_2 n_{2,j}| < 7U/2$; and so on. Note that

$$h_0(t_2; \mathbf{n}_2) + h_1(t_2; \mathbf{n}_2) + h_2(t_2; \mathbf{n}_2) + h_3(t_2; \mathbf{n}_2) + \cdots = \lambda_{1,2}. \tag{9.11}$$

By definition

$$|t_1\mathbf{n}_1 - t_2\mathbf{n}_2|^2 = \sum_{j\in L_{1,2}} (t_1 n_{1,j} - t_2 n_{2,j})^2 + \sum_{j\in L(\mathbf{n}_1)\setminus L_{1,2}} (t_1 n_{1,j})^2 + \sum_{j\in L(\mathbf{n}_2)\setminus L_{1,2}} (t_2 n_{2,j})^2. \tag{9.12}$$

By using (9.12) and the definitions of $k_i(\mathbf{n}_2)$, $h_i(t_2;\mathbf{n}_2)$, we have

$$\begin{aligned}
&\int_{t_2=U}^{V} \sum_{\substack{\mathbf{n}_2\in\mathbb{Z}^d\setminus\mathbf{0}:\\ |L(\mathbf{n}_2)|=\lambda_2 \quad \text{and} \quad L(\mathbf{n}_2)\cap L(\mathbf{n}_1)=L_{1,2}}} e^{-2\pi^2|t_1\mathbf{n}_1-t_2\mathbf{n}_2|^2}\,dt_2 \\
&\le \exp\left(-2\pi^2 \sum_{j\in L(\mathbf{n}_1)\setminus L_{1,2}} |n_{1,j}|^2U^2\right) \\
&\quad\times \int_{t_2=U}^{V} \sum_{\substack{\mathbf{n}_2\in\mathbb{Z}^d\setminus\mathbf{0}\}:\\ |L(\mathbf{n}_2)|=\lambda_2 \quad \text{and} \quad L(\mathbf{n}_2)\cap L(\mathbf{n}_1)=L_{1,2}}} \\
&\quad\cdot e^{-2\pi^2k_1(\mathbf{n}_2)t_2^2} e^{-2\pi^2k_2(\mathbf{n}_2)2^2t_2^2} e^{-2\pi^2k_3(\mathbf{n}_2)3^2t_2^2}\cdots \\
&\quad\cdot e^{-2\pi^2h_1(t_2;\mathbf{n}_2)(U/2)^2} e^{-2\pi^2h_2(t_2;\mathbf{n}_2)(3U/2)^2} e^{-2\pi^2h_3(t_2;\mathbf{n}_2)(5U/2)^2}\cdots dt_2.
\end{aligned} \tag{9.13}$$

We estimate the long sum at the end of (9.13). By using the definitions of $k_i(\mathbf{n}_2)$, $h_i(t_2;\mathbf{n}_2)$, (9.10) and (9.11), we obtain the upper bound

$$\begin{aligned}
\text{EndSum of (9.13)} &= \sum_{\substack{\mathbf{n}_2\in\mathbb{Z}^d\setminus\mathbf{0}:\\ |L(\mathbf{n}_2)|=\lambda_2 \quad \text{and} \quad L(\mathbf{n}_2)\cap L(\mathbf{n}_1)=L_{1,2}}} e^{-2\pi^2k_1(\mathbf{n}_2)t_2^2} \\
&\quad\cdot e^{-2\pi^2k_2(\mathbf{n}_2)2^2t_2^2} e^{-2\pi^2k_3(\mathbf{n}_2)3^2t_2^2} e^{-2\pi^2k_4(\mathbf{n}_2)4^2t_2^2}\cdots \\
&\quad\cdot e^{-2\pi^2h_1(t_2;\mathbf{n}_2)(U/2)^2} e^{-2\pi^2h_2(t_2;\mathbf{n}_2)(3U/2)^2} e^{-2\pi^2h_3(t_2;\mathbf{n}_2)(5U/2)^2}\cdots \\
&\le \sum_{\substack{(k_1,\ldots,k_r) \text{ with } r\ge1:\\ k_1+\cdots+k_r=\lambda_2-\lambda_{1,2},\, k_j\ge0,\, 1\le j\le r-1 \text{ and } k_r\ge1}} \binom{d-\lambda_1}{k_1} 2^{k_1} e^{-2\pi^2k_1U^2}
\end{aligned}$$

$$\cdot \binom{d-\lambda_1-k_1}{k_2} 2^{k_2} e^{-2\pi^2 k_2 2^2 U^2} \binom{d-\lambda_1-k_1-k_2}{k_3} 2^{k_3} e^{-2\pi^2 k_3 3^2 U^2} \cdots$$

$$\cdot \sum_{\substack{(h_0,h_1,\ldots,h_r) \text{ with } r\geq 0:\\ h_0+h_1+\cdots+h_r=\lambda_{1,2},h_j\geq 0,0\leq j\leq r-1 \text{ and } h_r\geq 1}} \binom{\lambda_{1,2}}{h_0}\binom{\lambda_{1,2}-h_0}{h_1}$$

$$\times 2^{h_1} e^{-2\pi^2 h_1 (U/2)^2}$$

$$\cdot \binom{\lambda_{1,2}-h_0-h_1}{h_2} 2^{h_2} e^{-2\pi^2 h_2 (3U/2)^2} \binom{\lambda_{1,2}-h_0-h_1-h_2}{h_3}$$

$$\times 2^{h_3} e^{-2\pi^2 h_3 (5U/2)^2} \cdots . \tag{9.14}$$

Note that (9.14) includes the pathological case $\lambda_2 - \lambda_{1,2} = 0$ with the natural convention that the summation means the single term $(k_1, \ldots, k_r) = (0)$, and similarly, if $\lambda_{1,2} = 0$ then $(h_0, h_1, \ldots, h_r)$ is just the single term (0).

By using the trivial upper bound

$$\sum_{\substack{(k_1,\ldots,k_r) \text{ with} r\geq 1:\\ k_1+\cdots+k_r=2-\lambda_{1,2},k_j\geq 0,1\leq j\leq r-1 \text{ and } k_r\geq 1}} \binom{d-\lambda_1}{k_1} 2^{k_1} e^{-2\pi^2 k_1 U^2}$$

$$\cdot \binom{d-\lambda_1-k_1}{k_2} 2^{k_2} e^{-2\pi^2 k_2 2^2 U^2} \binom{d-\lambda_1-k_1-k_2}{k_3} 2^{k_3} e^{-2\pi^2 k_3 3^2 U^2} \cdots$$

$$\leq \sum_{\substack{(k_1,\ldots,k_r) \text{ with } r\geq 1:\\ k_1+\cdots+k_r=\lambda_2-\lambda_{1,2},k_j\geq 0,1\leq j\leq r-1 \text{ and } k_r\geq 1}} d^{k_1} 2^{k_1} e^{-2\pi^2 k_1 U^2}$$

$$\cdot d^{k_2} 2^{k_2} e^{-2\pi^2 k_2 2^2 U^2} d^{k_3} 2^{k_3} e^{-2\pi^2 k_3 3^2 U^2} \cdots$$

$$\leq \left(2de^{-2\pi^2 U^2} + 2de^{-2\pi^2 2^2 U^2} + 2de^{-2\pi^2 3^2 U^2} + \cdots\right)^{\lambda_2-\lambda_{1,2}}, \tag{9.15}$$

and applying the multinomial theorem to the end-sum in (9.14)

$$\sum_{\substack{(h_0,h_1,\ldots,h_r) \text{ with } r\geq 0:\\ h_0+h_1+\cdots+h_r=\lambda_{1,2},h_j\geq 0,0\leq j\leq r-1 \text{ and } h_r\geq 1}} \binom{\lambda_{1,2}}{h_0}\binom{\lambda_{1,2}-h_0}{h_1}$$

$$\times 2^{h_1} e^{-2\pi^2 h_1 (U/2)^2}$$

$$\cdot \binom{\lambda_{1,2}-h_0-h_1}{h_2} 2^{h_2} e^{-2\pi^2 h_2 (3U/2)^2} \binom{\lambda_{1,2}-h_0-h_1-h_2}{h_3}$$

$$\times\, 2^{h_3} e^{-2\pi^2 h_3 (5U/2)^2} \cdots$$

$$= \left(1 + 2e^{-\pi^2 U^2/2} + 2e^{-\pi^2 3^2 U^2/2} + 2e^{-\pi^2 5^2 U^2/2} + \cdots\right)^{\lambda_{1,2}}, \quad (9.16)$$

we have

$$\text{EndSum of (9.13)} \le \Big(2de^{-2\pi^2 U^2} + 2de^{-2\pi^2 2^2 U^2}$$
$$+2de^{-2\pi^2 3^2 U^2} + \cdots\Big)^{\lambda_2 - \lambda_{1,2}}$$
$$\cdot \left(1 + 2e^{-\pi^2 U^2/2} + 2e^{-\pi^2 3^2 U^2/2} + 2e^{-\pi^2 5^2 U^2/2} + \cdots\right)^{\lambda_{1,2}}$$
$$\le \left(3de^{-2\pi^2 U^2}\right)^{\lambda_2 - \lambda_{1,2}} \left(1 + 3e^{-\pi^2 U^2/2}\right)^{\lambda_{1,2}}. \quad (9.17)$$

Next we use (9.14)–(9.17) in (9.13):

$$\int_{t_2=U}^{V} \sum_{\substack{\mathbf{n}_2 \in \mathbb{Z}^d \setminus \mathbf{0}: \\ |L(\mathbf{n}_2)| = \lambda_2 \text{ and } L(\mathbf{n}_2) \cap L(\mathbf{n}_1) = L_{1,2}}} e^{-2\pi^2 |t_1 \mathbf{n}_1 - t_2 \mathbf{n}_2|^2}\, dt_2$$
$$\le \exp\left(-2\pi^2(\lambda_1 - \lambda_{1,2})U^2\right)(V - U)\left(3de^{-2\pi^2 U^2}\right)^{\lambda_2 - \lambda_{1,2}}$$
$$\times \left(1 + 3e^{-\pi^2 U^2/2}\right)^{\lambda_{1,2}}. \quad (9.18)$$

Let us return now to (9.8); we have the decomposition

$$\left|\Delta_f^2(\text{Gauss}; U, V)\right| \le \text{Part One} + \text{Part Two} + \text{Part Three} + \text{Part Four}, \quad (9.19)$$

where

$$\text{Part One} = \sum_{\lambda_1=1}^{d-1} \int_{t_1=U}^{V} \sum_{\substack{\mathbf{n}_1 \in \mathbb{Z}^d \setminus \mathbf{0}: \\ |L(\mathbf{n}_1)| = \lambda_1}} |a_{\mathbf{n}_1}|^2 \Bigg(\sum_{L_{1,2} \subseteq L(\mathbf{n}_1)} \sum_{\lambda_2 = \lambda_1 + 1}^{d}$$
$$\cdot \int_{t_2=U}^{V} \sum_{\substack{\mathbf{n}_2 \in \mathbb{Z}^d \setminus \mathbf{0}: \\ |L(\mathbf{n}_2)| = \lambda_2 \text{ and } L(\mathbf{n}_2) \cap L(\mathbf{n}_1) = L_{1,2}}} e^{-2\pi^2 |t_1 \mathbf{n}_1 - t_2 \mathbf{n}_2|^2}\, dt_2 \Bigg)\, dt_1 \quad (9.20)$$

is characterized by the property $\lambda_1 < \lambda_2$,

$$\text{Part Two} = \sum_{\lambda_1=1}^{d} \int_{t_1=U}^{V} \sum_{\substack{\mathbf{n}_1 \in \mathbb{Z}^d \setminus \mathbf{0}: \\ |L(\mathbf{n}_1)|=\lambda_1}} |a_{\mathbf{n}_1}|^2 \left(\sum_{L_{1,2} \subset L(\mathbf{n}_1)} \sum_{\lambda_2=\max\{|L_{1,2}|,1\}}^{\lambda_1 - 1} \right.$$

$$\left. \cdot \int_{t_2=U}^{V} \sum_{\substack{\mathbf{n}_2 \in \mathbb{Z}^d \setminus \mathbf{0}: \\ |L(\mathbf{n}_2)|=\lambda_2 \quad \text{and} \quad L(\mathbf{n}_2) \cap L(\mathbf{n}_1) = L_{1,2}}} e^{-2\pi^2 |t_1 \mathbf{n}_1 - t_2 \mathbf{n}_2|^2} \, dt_2 \right) dt_1 \tag{9.21}$$

is characterized by the property $\lambda_1 > \lambda_2$, and finally, we split the case $\lambda_1 = \lambda_2$ into two subcases according as $L(\mathbf{n}_1) \neq L(\mathbf{n}_2)$ or $L(\mathbf{n}_1) = L(\mathbf{n}_2)$:

$$\text{Part Three} = \sum_{\lambda_1=1}^{d} \int_{t_1=U}^{V} \sum_{\substack{\mathbf{n}_1 \in \mathbb{Z}^d \setminus \mathbf{0}: \\ |L(\mathbf{n}_1)|=\lambda_1}} |a_{\mathbf{n}_1}|^2 \left(\sum_{\substack{L_{1,2} \subset L(\mathbf{n}_1): \\ L_{1,2} \neq L(\mathbf{n}_1)}} \right.$$

$$\left. \cdot \int_{t_2=U}^{V} \sum_{\substack{\mathbf{n}_2 \in \mathbb{Z}^d \setminus \mathbf{0}: \\ |L(\mathbf{n}_2)|=\lambda_1 \text{ and } L(\mathbf{n}_2) \cap L(\mathbf{n}_1) = L_{1,2}}} e^{-2\pi^2 |t_1 \mathbf{n}_1 - t_2 \mathbf{n}_2|^2} \, dt_2 \right) dt_1 \tag{9.22}$$

and

$$\text{Part Four} = \int_{t_1=U}^{V} \sum_{\mathbf{n}_1 \in \mathbb{Z}^d \setminus \mathbf{0}} |a_{\mathbf{n}_1}|^2$$

$$\cdot \left(\int_{t_2=U}^{V} \sum_{\substack{\mathbf{n}_2 \in \mathbb{Z}^d \setminus \mathbf{0}: \\ L(\mathbf{n}_2) = L(\mathbf{n}_1)}} e^{-2\pi^2 |t_1 \mathbf{n}_1 - t_2 \mathbf{n}_2|^2} \, dt_2 \right) dt_1. \tag{9.23}$$

To estimate the last part (9.23), we are going to use a simple but important lemma. First a definition: given real numbers C and C', consider the set

$$\mathcal{B}_U(C; C') = \{t \in [U, 2U] : \text{ there exists } n \in \mathbb{Z} \setminus \{0\} \text{ such that } |C - tn| \leq C'\}.$$

We give an upper bound on the 1-dimensional Lebesgue measure (i.e., the length) of the set $\mathcal{B}_U(C;C')$.

Lemma 9.3. *For arbitrary real numbers C, C' with $|C| \geq U \geq 1$ and $0 < C' < U/2$,*

$$\text{length}(\mathcal{B}_U(C;C')) < 6C'.$$

Proof. We can assume without loss of generality that $C > 0$. Clearly

$$|C - tn| \leq C' \Longleftrightarrow \frac{C - C'}{n} \leq t \leq \frac{C + C'}{n},$$

so

$$\text{length}(\mathcal{B}_U(C;C')) = \sum\nolimits_n^* \frac{2C'}{n},$$

where the summation $\sum_n^*$ is extended over all n's such that

$$\frac{C}{t} - \frac{C'}{t} \leq n \leq \frac{C}{t} + \frac{C'}{t}.$$

Note that

$$\frac{C + C'}{t} \leq \frac{\frac{3}{2}C}{t} \leq \frac{\frac{3}{2}C}{U}$$

and

$$\frac{C - C'}{t} \geq \frac{\frac{1}{2}C}{t} \geq \frac{\frac{1}{2}C}{2U}.$$

Thus we have

$$\begin{aligned}\text{length}(\mathcal{B}_U(C;C')) &\leq \sum_{C/(4U) \leq n \leq 3C/(2U)} \frac{2C'}{n} \\ &\leq 2C'\left(1 + \log \frac{3C/(2U)}{C/(4U)}\right) = 2C'(1 + \log 6),\end{aligned}$$

where we used the well-known fact that

$$\sum_{A \leq n \leq B} \leq 1 + \log(B/A)$$

for all $0 < A < B$. Since $\log 6 < 2$, the proof of Lemma 9.3 is complete. □

Applying (9.18) in (9.20), we have

$$\text{Part One} \le \sum_{\lambda_1=1}^{d-1} \int_{t_1=U}^{V} \sum_{\substack{\mathbf{n}_1\in\mathbb{Z}^d\setminus\mathbf{0}:\\ |L(\mathbf{n}_1)|=\lambda_1}} |a_{\mathbf{n}_1}|^2$$

$$\times \left(\sum_{L_{1,2}\subseteq L(\mathbf{n}_1)} \sum_{\lambda_2=\lambda_1+1}^{d} \cdot e^{-2\pi^2(\lambda_1-\lambda_{1,2})U^2}(V-U) \right.$$

$$\left. \times \left(3de^{-2\pi^2U^2}\right)^{\lambda_2-\lambda_{1,2}} \left(1+3e^{-\pi^2U^2/2}\right)^{\lambda_{1,2}} \right) dt_1$$

$$\le (V-U)^2 \max_{\lambda_{1,2}\le d} \left(1+3e^{-\pi^2U^2/2}\right)^{\lambda_{1,2}}$$

$$\times \sum_{\lambda_2=1}^{d} \left(3de^{-2\pi^2U^2}\right)^{\lambda_2} \cdot \sum_{\lambda_1=1}^{\lambda_2-1} \sum_{\substack{\mathbf{n}_1\in\mathbb{Z}^d\setminus\mathbf{0}:\\ |L(\mathbf{n}_1)|=\lambda_1}} |a_{\mathbf{n}_1}|^2$$

$$\cdot \left(\sum_{\lambda_{1,2}=0}^{\lambda_1} \binom{\lambda_1}{\lambda_{1,2}} \left(3de^{-2\pi^2U^2}\right)^{-\lambda_{1,2}} e^{-2\pi^2(\lambda_1-\lambda_{1,2})U^2} \right)$$

$$= (V-U)^2 \left(1+3e^{-\pi^2U^2/2}\right)^d \sum_{\lambda_2=1}^{d} \left(3de^{-2\pi^2U^2}\right)^{\lambda_2}$$

$$\cdot \sum_{\lambda_1=1}^{\lambda_2-1} \sum_{\substack{\mathbf{n}_1\in\mathbb{Z}^d\setminus\mathbf{0}:\\ |L(\mathbf{n}_1)|=\lambda_1}} |a_{\mathbf{n}_1}|^2 \left(\left(3de^{-2\pi^2U^2}\right)^{-1} + e^{-2\pi^2U^2} \right)^{\lambda_1}. \tag{9.24}$$

By hypothesis

$$dUe^{-\pi^2U^2/2} \le \frac{1}{3}, \tag{9.25}$$

and using it we have

$$\left(\left(3de^{-2\pi^2U^2}\right)^{-1} + e^{-2\pi^2U^2} \right)^{\lambda_1}$$

$$= \left(3de^{-2\pi^2U^2}\right)^{-\lambda_1} \left(1+3de^{-2\pi^2U^2/2}e^{-2\pi^2U^2}\right)^{\lambda_1}$$

$$\le \left(3de^{-2\pi^2U^2}\right)^{-\lambda_1} \left(1+\frac{1}{2d}\right)^d < 2\left(3de^{-2\pi^2U^2}\right)^{-\lambda_1}. \tag{9.26}$$

By using (9.25)–(9.26) in (9.24), we obtain

$$\begin{aligned}
\text{Part One} &\le (V-U)^2\left(1+3(dU)^{-1}\right)^d \\
&\quad\times \sum_{\lambda_2=1}^{d}\sum_{\lambda_1=1}^{\lambda_2-1}\left(3de^{-2\pi^2U^2}\right)^{\lambda_2}\cdot 2\left(3de^{-2\pi^2U^2}\right)^{-\lambda_1}\sum_{\substack{\mathbf{n}_1\in\mathbb{Z}^d\setminus\mathbf{0}:\\ |L(\mathbf{n}_1)|=\lambda_1}}|a_{\mathbf{n}_1}|^2 \\
&\le (V-U)^2\left(1+\frac{1}{2d}\right)^d\cdot 2\sum_{\lambda_2=1}^{d}\sum_{\lambda_1=1}^{\lambda_2-1}\left(3de^{-2\pi^2U^2}\right)^{\lambda_2-\lambda_1}\sum_{\substack{\mathbf{n}_1\in\mathbb{Z}^d\setminus\mathbf{0}:\\ |L(\mathbf{n}_1)|=\lambda_1}}|a_{\mathbf{n}_1}|^2 \\
&\le 4(V-U)^2d\left(\sum_{j=1}^{\infty}\left(3de^{-2\pi^2U^2}\right)^j\right)\sum_{\mathbf{n}\in\mathbb{Z}^d\setminus\mathbf{0}}|a_{\mathbf{n}}|^2 \\
&= 4(V-U)^2d\frac{3de^{-2\pi^2U^2}}{1-3de^{-2\pi^2U^2}}\sum_{\mathbf{n}\in\mathbb{Z}^d\setminus\mathbf{0}}|a_{\mathbf{n}}|^2, \qquad (9.27)
\end{aligned}$$

where we used the substitution $j=\lambda_2-\lambda_1$.

Using (9.25) in (9.27), we have

$$\begin{aligned}
\text{Part One} &\le 12(V-U)^2\frac{d^2e^{-2\pi^2U^2}}{1-3de^{-2\pi^2U^2}}\sum_{\mathbf{n}\in\mathbb{Z}^d\setminus\mathbf{0}}|a_{\mathbf{n}}|^2 \\
&\le 12(V-U)^2\frac{3^{-4}U^{-2}}{1-3de^{-2\pi^2U^2}}\sum_{\mathbf{n}\in\mathbb{Z}^d\setminus\mathbf{0}}|a_{\mathbf{n}}|^2 \\
&\le \frac{1}{6}\sum_{\mathbf{n}\in\mathbb{Z}^d\setminus\mathbf{0}}|a_{\mathbf{n}}|^2, \qquad (9.28)
\end{aligned}$$

since $V-U\le U$.

Next we use (9.18) in (9.21):

$$\begin{aligned}
\text{Part Two} &\le \sum_{\lambda_1=1}^{d}\int_{t_1=U}^{V}\sum_{\substack{\mathbf{n}_1\in\mathbb{Z}^d\setminus\mathbf{0}:\\ |L(\mathbf{n}_1)|=\lambda_1}}|a_{\mathbf{n}_1}|^2 \\
&\quad\times\left(\sum_{L_{1,2}\subset L(\mathbf{n}_1)}\sum_{\lambda_2=\max\{|L_{1,2}|,1\}}^{\lambda_1-1}\cdot e^{-2\pi^2(\lambda_1-\lambda_{1,2})U^2}(V-U)\right.
\end{aligned}$$

$$\times \left(3de^{-2\pi^2U^2}\right)^{\lambda_2-\lambda_{1,2}} \left(1+3e^{-\pi^2U^2/2}\right)^{\lambda_{1,2}} \Bigg) \, dt_1$$

$$\leq (V-U)^2 \max_{\lambda_{1,2}\leq d} \left(1+3e^{-\pi^2U^2/2}\right)^{\lambda_{1,2}} \sum_{\lambda_1=1}^{d} \sum_{\substack{\mathbf{n}_1\in\mathbb{Z}^d\setminus\mathbf{0}: \\ |L(\mathbf{n}_1)|=\lambda_1}} |a_{\mathbf{n}_1}|^2$$

$$\cdot \sum_{\lambda_2=1}^{\lambda_1-1} \sum_{\lambda_{1,2}=0}^{\lambda_2} \binom{\lambda_1}{\lambda_{1,2}} \left(3de^{-2\pi^2U^2}\right)^{\lambda_2-\lambda_{1,2}} e^{-2\pi^2(\lambda_1-\lambda_{1,2})U^2}$$

$$\leq (V-U)^2 \left(1+\frac{1}{2d}\right)^d \sum_{\lambda_1=1}^{d} \sum_{\lambda_2=1}^{\lambda_1-1} \sum_{\lambda_{1,2}=0}^{\lambda_2}$$

$$\cdot \left(3de^{-2\pi^2U^2}\right)^{\lambda_2-\lambda_{1,2}} \left(d\cdot e^{-2\pi^2U^2}\right)^{\lambda_1-\lambda_{1,2}} \sum_{\mathbf{n}\in\mathbb{Z}^d\setminus\mathbf{0}} |a_{\mathbf{n}}|^2$$

$$\leq (V-U)^2\cdot 2\cdot \sum_{\lambda_1=1}^{d} \sum_{\lambda_2=1}^{\lambda_1-1} \sum_{\lambda_{1,2}=0}^{\lambda_2} \left(3de^{-2\pi^2U^2}\right)^{\lambda_1+\lambda_2-2\lambda_{1,2}} \sum_{\mathbf{n}\in\mathbb{Z}^d\setminus\mathbf{0}} |a_{\mathbf{n}}|^2, \tag{9.29}$$

where in the last steps we used (9.25) and the trivial upper bound

$$\binom{\lambda_1}{\lambda_{1,2}} = \binom{\lambda_1}{\lambda_1-\lambda_{1,2}} \leq \lambda_1^{\lambda_1-\lambda_{1,2}} \leq d^{\lambda_1-\lambda_{1,2}}. \tag{9.30}$$

Applying (9.25) in (9.29), we obtain

$$\text{Part Two} \leq 2\left(\sum_{\mathbf{n}\in\mathbb{Z}^d\setminus\mathbf{0}} |a_{\mathbf{n}}|^2\right)(V-U)^2\cdot$$

$$\cdot \sum_{\lambda_1=1}^{d} \sum_{\lambda_2=1}^{\lambda_1-1} \sum_{\lambda_{1,2}=0}^{\lambda_2} \left(3\cdot 3^{-4}d^{-1}U^{-2}\right)^{\lambda_1+\lambda_2-2\lambda_{1,2}}$$

$$\leq 2\left(\sum_{\mathbf{n}\in\mathbb{Z}^d\setminus\mathbf{0}} |a_{\mathbf{n}}|^2\right)(V-U)^2 \sum_{j=1}^{\infty} d(j+1)j\left(3^{-3}d^{-1}U^{-2}\right)^j$$

$$= 2\left(\sum_{\mathbf{n}\in\mathbb{Z}^d\setminus\mathbf{0}} |a_{\mathbf{n}}|^2\right)(V-U)^2\cdot d\cdot 2\frac{3^{-3}d^{-1}U^{-2}}{(1-3^{-3}d^{-1}U^{-2})^3}$$

$$= \frac{4}{27}\left(\sum_{\mathbf{n}\in\mathbb{Z}^d\setminus\mathbf{0}} |a_{\mathbf{n}}|^2\right)(V-U)^2U^{-2}\left(1-3^{-3}d^{-1}U^{-2}\right)^{-3}$$

$$\leq \frac{1}{4}\sum_{\mathbf{n}\in\mathbb{Z}^d\setminus\mathbf{0}} |a_{\mathbf{n}}|^2, \tag{9.31}$$

where we used the substitution $j = \lambda_1+\lambda_2-2\lambda_{1,2}$, the assumption $V-U \leq U$, and the simple fact

$$\sum_{j=1}^{\infty} j(j+1)x^j = \frac{2x}{(1-x)^3} \quad \text{for all } |x|<1.$$

Next we apply (9.18) in (9.22):

$$\text{Part Three} \leq \sum_{\lambda_1=1}^{d}\int_{t_1=U}^{V}\sum_{\substack{\mathbf{n}_1\in\mathbb{Z}^d\setminus\mathbf{0}:\\ |L(\mathbf{n}_1)|=\lambda_1}} |a_{\mathbf{n}_1}|^2$$

$$\times\left(\sum_{\substack{L_{1,2}\subset L(\mathbf{n}_1):\\ L_{1,2}\neq L(\mathbf{n}_1)}} \cdot e^{-2\pi^2(\lambda_1-\lambda_{1,2})U^2}(V-U)\right.$$

$$\left.\times\left(3de^{-2\pi^2U^2}\right)^{\lambda_1-\lambda_{1,2}}\left(1+3e^{-\pi^2U^2/2}\right)^{\lambda_{1,2}}\right)dt_1$$

$$\leq (V-U)^2\max_{\lambda_{1,2}\leq d}\left(1+3e^{-\pi^2U^2/2}\right)^{\lambda_{1,2}}\sum_{\lambda_1=1}^{d}\sum_{\substack{\mathbf{n}_1\in\mathbb{Z}^d\setminus\mathbf{0}:\\ |L(\mathbf{n}_1)|=\lambda_1}} |a_{\mathbf{n}_1}|^2$$

$$\cdot\sum_{\lambda_{1,2}=0}^{\lambda_1-1}\binom{\lambda_1}{\lambda_{1,2}}\left(3de^{-(2+2)\pi^2U^2}\right)^{\lambda_1-\lambda_{1,2}}. \tag{9.32}$$

Using (9.30) in (9.32), we have

$$\text{Part Three} \leq (V-U)^2\max_{\lambda_{1,2}\leq d}\left(1+3e^{-\pi^2U^2/2}\right)^{\lambda_{1,2}}\sum_{\lambda_1=1}^{d}\sum_{\substack{\mathbf{n}_1\in\mathbb{Z}^d\setminus\mathbf{0}:\\ |L(\mathbf{n}_1)|=\lambda_1}} |a_{\mathbf{n}_1}|^2$$

$$\cdot\sum_{\lambda_{1,2}=0}^{\lambda_1-1}\left(3d^2e^{-4\pi^2U^2}\right)^{\lambda_1-\lambda_{1,2}}. \tag{9.33}$$

Using (9.25) in (9.33), we obtain

$$\text{Part Three} \le (V-U)^2\left(1+3(dU)^{-1}\right)^d$$

$$\times \sum_{\lambda_1=1}^{d}\sum_{\lambda_{1,2}=0}^{\lambda_1-1}(3(3^4dU)^{-2})^{\lambda_1-\lambda_{1,2}} \cdot \sum_{\mathbf{n}\in\mathbb{Z}^d\setminus\mathbf{0}}|a_{\mathbf{n}}|^2$$

$$\le (V-U)^2\left(1+\frac{1}{2d}\right)^d \cdot d\left(\sum_{j=1}^{\infty}(3^7d^2U^2)^{-j}\right)\sum_{\mathbf{n}\in\mathbb{Z}^d\setminus\mathbf{0}}|a_{\mathbf{n}}|^2$$

$$\le (V-U)^2\cdot 2\cdot d\frac{(3^7d^2U^2)^{-1}}{1-(3^7d^2U^2)^{-1}}\sum_{\mathbf{n}\in\mathbb{Z}^d\setminus\mathbf{0}}|a_{\mathbf{n}}|^2$$

$$= (V-U)^2U^{-2}\frac{2(3^7d)^{-1}}{1-(3^7d^2U^2)^{-1}}\sum_{\mathbf{n}\in\mathbb{Z}^d\setminus\mathbf{0}}|a_{\mathbf{n}}|^2$$

$$\le \frac{1}{100d}\sum_{\mathbf{n}\in\mathbb{Z}^d\setminus\mathbf{0}}|a_{\mathbf{n}}|^2, \tag{9.34}$$

where we used the substitution $j=\lambda_1-\lambda_{1,2}$ and the hypothesis $V-U\le U$.

Finally we estimate (9.23). We have

$$\text{Part Four} = \text{Part Four A} + \text{Part Four B}, \tag{9.35}$$

where

$$\text{Part Four A} = \int_{t_1=U}^{V}\sum_{\mathbf{n}_1\in\mathbb{Z}^d\setminus\mathbf{0}}|a_{\mathbf{n}_1}|^2$$

$$\cdot\left(\int_{t_2=U}^{V}\sum_{\substack{\mathbf{n}_2\in\mathbb{Z}^d\setminus\mathbf{0}:\\ L(\mathbf{n}_2)=L(\mathbf{n}_1),\ h_0(t_2;\mathbf{n}_2)<|L(\mathbf{n}_1)|}} e^{-2\pi^2|t_1\mathbf{n}_1-t_2\mathbf{n}_2|^2}\,dt_2\right)dt_1, \tag{9.36}$$

and

$$\text{Part Four B} = \int_{t_1=U}^{V}\sum_{\mathbf{n}_1\in\mathbb{Z}^d\setminus\mathbf{0}}|a_{\mathbf{n}_1}|^2$$

$$\cdot\left(\int_{t_2=U}^{V}\sum_{\substack{\mathbf{n}_2\in\mathbb{Z}^d\setminus\mathbf{0}:\\ L(\mathbf{n}_2)=L(\mathbf{n}_1),\ h_0(t_2;\mathbf{n}_2)=|L(\mathbf{n}_1)|}} e^{-2\pi^2|t_1\mathbf{n}_1-t_2\mathbf{n}_2|^2}\,dt_2\right)dt_1. \tag{9.37}$$

We start with Part Four A, and use the notation $\lambda_1 = |L(\mathbf{n}_1)|$. In this special case the argument of (9.13)–(9.18) simplifies to the following upper bound (note that here $\lambda_1 = \lambda_2 = \lambda_{1,2}$)

$$\text{Part Four A} \leq \int_{t_1=U}^{V} \sum_{\mathbf{n}_1 \in \mathbb{Z}^d \setminus \mathbf{0}} |a_{\mathbf{n}_1}|^2 (V-U)$$

$$\cdot \left(\sum_{\substack{(h_0,h_1,\ldots,h_r) \text{ with } r\geq 0:\ h_0<\lambda_1 \\ h_0+h_1+\cdots+h_r=\lambda_1, h_j\geq 0, 0\leq j\leq r-1 \quad \text{and} \quad h_r\geq 1}} \binom{\lambda_1}{h_0} \right.$$

$$\cdot \binom{\lambda_1 - h_0}{h_1} 2^{h_1} e^{-2\pi^2 h_1 (U/2)^2} \binom{\lambda_1 - h_0 - h_1}{h_2} 2^{h_2} e^{-2\pi^2 h_2 (3U/2)^2}$$

$$\left. \cdot \binom{\lambda_1 - h_0 - h_1 - h_2}{h_3} 2^{h_3} e^{-2\pi^2 h_3 (5U/2)^2} \cdots \right) dt_1$$

$$= \int_{t_1=U}^{V} \sum_{\mathbf{n}_1 \in \mathbb{Z}^d \setminus \mathbf{0}} |a_{\mathbf{n}_1}|^2 (V-U)$$

$$\cdot \left(\left(1 + 2e^{-\pi^2 U^2/2} + 2e^{-\pi^2 3^2 U^2/2} + 2e^{-\pi^2 5^2 U^2/2} + \cdots \right)^{\lambda_1} - 1 \right) dt_1, \tag{9.38}$$

where the term -1 at the end is explained by the restriction $h_0 < \lambda_1$, i.e., the term -1 is due to the fact that in the application of the multinomial theorem we have an *almost* complete sum, where the single missing case is $h_0 = \lambda_1$.

By (9.38)

$$\begin{aligned} \text{Part Four A} &\leq \sum_{\mathbf{n}_1 \in \mathbb{Z}^d \setminus \mathbf{0}} |a_{\mathbf{n}_1}|^2 (V-U)^2 \left(\left(1 + 3e^{-\pi^2 U^2/2} \right)^d - 1 \right) \\ &\leq \sigma_0^2(f)(V-U)^2 \left(\exp\left(3de^{-\pi^2 U^2/2} \right) - 1 \right) \\ &\leq \sigma_0^2(f)(V-U)^2 6de^{-\pi^2 U^2/2}, \end{aligned} \tag{9.39}$$

since

$$de^{-\pi^2 U^2/2} \leq \frac{1}{3}. \tag{9.40}$$

Note that in (9.39) we used Parseval's formula and the simple inequalities $1+x \le e^x \le 1+2x$ for $0 \le x \le 1$; and of course (9.40) follows from (9.25).

Applying (9.25) in (9.39), we have

$$\text{Part Four A} \le \sigma_0^2(f)(V-U)^2 \frac{6}{3U} \le 2\sigma_0^2(f)(V-U), \tag{9.41}$$

since $V - U \le U$.

To estimate (9.37)

$$\text{Part Four B} = \int_{t_1=U}^{V} \sum_{\mathbf{n}_1 \in \mathbb{Z}^d \setminus \mathbf{0}} |a_{\mathbf{n}_1}|^2$$
$$\cdot \left(\int_{t_2=U}^{V} \sum_{\substack{\mathbf{n}_2 \in \mathbb{Z}^d \setminus \mathbf{0}: \\ L(\mathbf{n}_2)=L(\mathbf{n}_1),\ h_0(t_2;\mathbf{n}_2)=|L(\mathbf{n}_1)|}} e^{-2\pi^2|t_1\mathbf{n}_1 - t_2\mathbf{n}_2|^2}\, dt_2 \right) dt_1, \tag{9.42}$$

we make use of the fact that for every given triple $(t_1, \mathbf{n}_1, t_2)$ with $t_1 \in [U,V]$, $\mathbf{n}_1 \in \mathbb{Z}^d \setminus \mathbf{0}$ and $t_2 \in [U,V]$, there is *at most one* term in the endsum

$$\sum_{\substack{\mathbf{n}_2 \in \mathbb{Z}^d \setminus \mathbf{0}: \\ L(\mathbf{n}_2)=L(\mathbf{n}_1),\ h_0(t_2;\mathbf{n}_2)=|L(\mathbf{n}_1)|}} e^{-2\pi^2|t_1\mathbf{n}_1 - t_2\mathbf{n}_2|^2}$$

of (9.42). Indeed, it follows from the definition of $h_0(t_2;\mathbf{n}_2)$ that for every given triple $(t_1, \mathbf{n}_1, t_2)$ the inequality $|t_1 n_{1,j} - t_2 n_{2,j}| < U/2$ has at most one integer solution $n_{2,j}$, and because $h_0(t_2;\mathbf{n}_2) = |L(\mathbf{n}_1)|$ and $L(\mathbf{n}_2) = L(\mathbf{n}_1)$, there is at most one $\mathbf{n}_2 \in \mathbb{Z}^d \setminus \mathbf{0}$ satisfying the requirements. Let $\mathbf{n}_2^* \in \mathbb{Z}^d \setminus \mathbf{0}$, $\mathbf{n}_2^* = \mathbf{n}_2^*(t_1, \mathbf{n}_1, t_2)$ denote this single integer lattice point, *if it exists.* Thus we can rewrite (9.42) as follows:

$$\text{Part Four B} = \int_{t_1=U}^{V} \sum_{\mathbf{n}_1 \in \mathbb{Z}^d \setminus \mathbf{0}} |a_{\mathbf{n}_1}|^2$$
$$\times \left(\int_{t_2=U}^{V} \left\{ e^{-2\pi^2|t_1\mathbf{n}_1 - t_2\mathbf{n}_2^*|^2} \text{ or } 0 \right\} dt_2 \right) dt_1, \tag{9.43}$$

where $\mathbf{n}_2^* = \mathbf{n}_2^*(t_1, \mathbf{n}_1, t_2)$, if it exists.

Let $j_0 = j_0(\mathbf{n}_1)$ be an arbitrary but fixed element of the set $L(\mathbf{n}_1)$; then we have the following trivial upper bound for (9.43):

$$\text{Part Four B} \leq \int_{t_1=U}^{V} \sum_{\mathbf{n}_1 \in \mathbb{Z}^d \setminus \mathbf{0}} |a_{\mathbf{n}_1}|^2 \times \left(\int_{t_2=U}^{V} \left\{ e^{-2\pi^2 |t_1 n_{1,j_0} - t_2 n^*_{2,j_0}|^2} \text{ or } 0 \right\} dt_2 \right) dt_1. \tag{9.44}$$

Given $(t_1, \mathbf{n}_1)$, we consider the following decomposition of the interval $U \leq t_2 \leq V (\leq 2U)$:

$$I_\ell(t_1, \mathbf{n}_1) = \left\{ t_2 \in [U, V] : \ \ell - 1 \leq |t_1 n_{1,j_0} - t_2 n^*_{2,j_0}| < \ell \right\}, \quad \ell = 1, 2, 3, \ldots, \tag{9.45}$$

where $j_0 = j_0(\mathbf{n}_1)$ and $\mathbf{n}^*_2 = \mathbf{n}^*_2(t_1, \mathbf{n}_1, t_2)$.

Using decomposition (9.45) in (9.44), we obtain the upper bound

$$\text{Part Four B} \leq \int_{t_1=U}^{V} \sum_{\mathbf{n}_1 \in \mathbb{Z}^d \setminus \mathbf{0}} |a_{\mathbf{n}_1}|^2 \left(\sum_{\ell=1}^{\infty} \text{length}\,(I_\ell)\, e^{-2\pi^2 (\ell-1)^2} \right) dt_1, \tag{9.46}$$

where $I_\ell = I_\ell(t_1, \mathbf{n}_1)$. Applying Lemma 9.3 with $C = t_1 n_{1,j_0}$ and $C' = \ell$, we obtain the upper bound

$$\text{length}\,(I_\ell) < 6\ell,$$

and using it in (9.46), we have

$$\begin{aligned} \text{Part Four B} &\leq \int_{t_1=U}^{V} \sum_{\mathbf{n}_1 \in \mathbb{Z}^d \setminus \mathbf{0}} |a_{\mathbf{n}_1}|^2 \left(\sum_{\ell=1}^{\infty} 6\ell e^{-2\pi^2 (\ell-1)^2} \right) dt_1 \\ &\leq 7(V-U) \sum_{\mathbf{n}_1 \in \mathbb{Z}^d \setminus \mathbf{0}} |a_{\mathbf{n}_1}|^2. \end{aligned} \tag{9.47}$$

Using (9.28), (9.31), (9.34), (9.41), (9.47) and Parseval's formula in (9.19), we conclude

$$\begin{aligned} |\Delta^2_f(\text{Gauss}; U, V)| &\leq \left(\frac{1}{6} + \frac{1}{4} + \frac{1}{100d} + 2(V-U) + 7(V-U) \right) \sigma_0^2(f) \\ &\leq (9(V-U)+1)\sigma_0^2(f). \end{aligned}$$

This completes the proof of Theorem 9.1.

To complete the proof of Theorem 4.2, we have to eliminate the technical restriction $U < V \le 2U$ in Theorem 9.1. It goes by a routine application of the Cauchy–Schwarz inequality as follows.

Deduction of Theorem 4.2 from Theorem 9.1. Given $1 \le U < W$, let $k \ge 1$ denore the integer such that $2^{k-1}U < W \le 2^k U$. It is easy to find a sequence $1 \le W_0 = U < W_1 < \cdots < W_k = W$ such that $W_i < W_{i+1} \le 2W_i$ for all $0 \le i < k$. By using the Cauchy–Schwarz inequality and Theorem 9.1, we have

$$\begin{aligned}
&\Delta_f^2(\text{Gauss}; U, W) \\
&= (2\pi)^{-d/2} \int_{\mathbf{v}\in\mathbb{R}^d} \left(\text{TimeDiscrep}_f(\mathbf{v}; U, W)\right)^2 e^{-|\mathbf{v}|^2/2}\, d\mathbf{v} \\
&= (2\pi)^{-d/2} \int_{\mathbf{v}\in\mathbb{R}^d} \left(\sum_{j=0}^{k-1} \text{TimeDiscrep}_f(\mathbf{v}; W_j, W_{j+1})\right)^2 e^{-|\mathbf{v}|^2/2}\, d\mathbf{v} \\
&\le (2\pi)^{-d/2} \int_{\mathbf{v}\in\mathbb{R}^d} \left(\sum_{j=0}^{k-1} \left(\text{TimeDiscrep}_f(\mathbf{v}; W_j, W_{j+1})\right)^2\right) \\
&\quad \times \left(\sum_{j=0}^{k-1} 1\right) e^{-|\mathbf{v}|^2/2}\, d\mathbf{v} \\
&= \left(\sum_{j=0}^{k-1} (2\pi)^{-d/2} \int_{\mathbf{v}\in\mathbb{R}^d} \left(\text{TimeDiscrep}_f(\mathbf{v}; W_j, W_{j+1})\right)^2 e^{-|\mathbf{v}|^2/2}\, d\mathbf{v}\right) \\
&\quad \times \left(\sum_{j=0}^{k-1} 1\right) \\
&\le \sigma_0^2(f) \left(\sum_{j=0}^{k-1} (9(W_{j+1} - W_j) + 1)\right) \cdot \left(\sum_{j=0}^{k-1} 1\right) \\
&\le k\sigma_0^2(f)\,(10(W-U) + 10)\,.
\end{aligned}$$

Since $k = \lceil \log_2(W/U) \rceil$ (upper integral part), the proof of Theorem 4.2 is complete. □

Finally, we turn to Theorem 4.3.

Proof of Theorem 4.3. Clearly

$$\Delta^2_{f0}(\text{Gauss}; U, V) = \int_U^V \left(\int_U^V e^{-2\pi^2(t_1-t_2)^2}\, dt_2 \right) dt_1, \tag{9.48}$$

since the off-diagonal part is empty by definition. We give a lower bound to the inner integral in (9.48) (t_1 is fixed, and t_2 is the variable):

$$\begin{aligned}
\int_{t_2=U}^{V} e^{-2\pi^2(t_2-t_1)^2|}\, dt_2 &= \int_{U-t_1}^{V-t_1} e^{-2\pi^2 y^2}\, dy = \frac{1}{\sqrt{2}\pi} \int_{\sqrt{2}\pi(U-t_1)}^{\sqrt{2}\pi(V-t_1)} e^{-x^2}\, dx \\
&= \frac{1}{\sqrt{2}\pi} \int_{-\infty}^{\infty} e^{-x^2}\, dx - \frac{1}{\sqrt{2}\pi} \int_{-\infty}^{\sqrt{2}\pi(U-t_1)} e^{-x^2}\, dx \\
&\quad - \frac{1}{\sqrt{2}\pi} \int_{\sqrt{2}\pi(V-t_1)}^{\infty} e^{-x^2}\, dx \\
&\geq \frac{1}{\sqrt{2}\pi}\sqrt{\pi} - \left(\frac{1}{\sqrt{2}\pi} - (t_1 - U) \right)^+ \\
&\quad - \frac{1}{2\sqrt{2}\pi} e^{-2\pi^2(t_1-U)^2} - \left(\frac{1}{\sqrt{2}\pi} - (V - t_1) \right)^+ \\
&\quad - \frac{1}{2\sqrt{2}\pi} e^{-2\pi^2(V-t_1)^2},
\end{aligned} \tag{9.49}$$

where we used the simple facts

$$\int_L^{\infty} e^{-x^2}\, dx \leq \int_L^{\infty} x e^{-x^2}\, dx = \frac{1}{2} e^{-L^2} \quad \text{for} \quad L > 1$$

and

$$\int_L^1 e^{-x^2}\, dx \leq 1 - L \quad \text{for} \quad L \leq 1$$

(z^+ denotes the positive part of z, i.e., $z^+ = z$ if $z > 0$ and 0 otherwise).

Using (9.49) in (9.48), we have

$$\begin{aligned}
\Delta^2_{f0}(\text{Gauss}; U, V) &= \int_U^V \left(\int_U^V e^{-2\pi^2(t_1-t_2)^2}\, dt_2 \right) dt_1 \\
&\geq \int_{t_1=U}^{V} \left(\frac{1}{\sqrt{2\pi}} - \left(\frac{1}{\sqrt{2}\pi} - (t_1 - U) \right)^+ \right.
\end{aligned}$$

$$- \frac{1}{2\sqrt{2}\pi} e^{-2\pi^2(U-t_1)^2} - \left(\frac{1}{\sqrt{2}\pi} - (V-t_1)\right)^+$$

$$\left. - \frac{1}{2\sqrt{2}\pi} e^{-2\pi^2(V-t_1)^2}\right) dt_1$$

$$\geq \frac{V-U}{\sqrt{2}\pi} - 2\int_0^{\pi^{-1}2^{-1/2}} x\, dx - \frac{1}{2\sqrt{2}\pi}\int_{-\infty}^{\infty} e^{-2\pi^2 x^2}\, dx$$

$$= \frac{V-U}{\sqrt{2}\pi} - \frac{1}{2\pi^2} - \frac{1}{2\sqrt{2}\pi}\frac{1}{\sqrt{2\pi}} > \frac{V-U}{\sqrt{2}\pi} - \frac{1}{5}, \tag{9.50}$$

which completes the proof of Theorem 4.3. □

Chapter 2

General Models

10. General Model: Unique Ergodicity via Typical Rotations

In Chapter 1 we studied the simplest model of large systems: the torus-billiard system. Here we study a far reaching generalization of the torus-billiard model. Assume that every particle has its own *type* of trajectory described by a parametrized space curve in the h-space $\mathbb{R}^h$, $h \geq 2$. It means that the actual trajectory $\mathbf{x}_k(t)$ of the kth particle in the torus model is in fact a *translated* and *rotated* copy of $\mathbf{r}_k(t)$ *modulo one*, where $\mathbf{r}_k(0) = \mathbf{0}$, $1 \leq k \leq N$ (see (10.2) below). We assume that the parametrized space curves are continuous, but beyond that they can be (almost) arbitrary: straight lines, bending lines, twisted lines, any kind of general helix, spiral, wave, and so on (of course, in the torus model we take the space curves modulo one). The only condition that we require is that the parametrized space curves $\mathbf{r}_k(t) \subset \mathbb{R}^h$ have a tendency of "drifting away from the start".

A constant speed motion on a straight line is obviously drifting away from the start rather fast. What we require here is much less. A natural quantitative way to define "drifting away from the start" is the following. Let $I^h + \mathbf{w}$, $\mathbf{w} \in \mathbb{R}^h$ be an arbitrary but fixed axis-parallel h-dimensional cube of side length one (i.e., a shifted copy of the unit cube $I^h = [0,1)^h$), and consider the total time t in the interval $0 \leq t \leq T$ that the curve $\mathbf{r}_k(t)$ spends in $I^h + \mathbf{w}$:

$$\text{time}_T \left\{ \mathbf{r}_k(t) \in I^h + \mathbf{w} \right\} = \text{length}\{0 \leq t \leq T : \; \mathbf{r}_k(t) \in I^h + \mathbf{w}\}.$$

We say that the system is *drifting away* if there exist two constants $\alpha > 0$ and $0 \le \beta < 1$ such that

$$\sup_{\mathbf{w}\in\mathbb{R}^h} \text{time}_T \left\{\mathbf{r}_k(t) \in I^h + \mathbf{w}\right\} \le \alpha T^\beta \tag{10.1}$$

uniformly for all $1 \le k \le N$ and all $T \ge 1$.

Here we interrupt our mathematical discussion, and give some real life motivation for introducing arbitrary space curves $\mathbf{r}_k(t)$. In a constant magnetic field $\mathbf{B}$, there is a force on a moving charged particle, say a proton. The force is the cross product $\mathbf{F} = \mathbf{v} \times \mathbf{B}$, where $\mathbf{v}$ is the velocity of the proton. Since $\mathbf{F}$ is perpendicular to $\mathbf{v}$, the proton travels on a helix like $\mathbf{r}(t) = (\cos t, \sin t, t)$, instead of a straight line. Of course, an uncharged particle still travels on a straight line.

A second motivation is that particle–particle collision leads to random-walk-like chaotic zig-zag trajectories, and by allowing arbitrary space curves $\mathbf{r}_k(t)$, we attempt to imitate particle–particle collision. Of course a random-walk-like trajectory is completely different from a straight line; nevertheless, it still exhibits a tendency of "drifting away from the start" (see (10.1)), which is included in our general model.

A third real life motivation for arbitrary space curves is that it mimicks the basic mechanism of the Second Law of Thermodynamics: "energy dispersal" on a microscopic level. "Energy dispersal" is far more complicated than particle–particle collision for gas-in-a-box; it also applies for fluids and solids, it applies for all real life many-particle systems. We feel that our general torus model with arbitrary space curves is a reasonable mathematical model for "energy dispersal" on a microscopic level.

Let us return to mathematics: by using the Weyl's criterion, we can prove a result describing the typical long-term time evolution of a general torus model. It is an analog of the Kronecker–Weyl equidistribution theorem. We may call it a "Weak Uniformity" Theorem (by contrast, Theorems 4.1-2 are "strong uniformity" theorems).

We use $\mathbf{x}_k(t)$ to denote the trajectory of the kth particle in the torus $I^h = [0,1)^h$ $(1 \le k \le N)$. We allow arbitrary initial point configuration $\mathbf{x}_k(0) = \mathbf{y}_k \in I^h$, $1 \le k \le N$. Let $\mathbf{r}_k(t) \subset \mathbb{R}^h$, $1 \le k \le N$ be an arbitrary family of N continuous parametrized space curves in the h-dimensional space with $\mathbf{r}_k(0) = \mathbf{0}$, and satisfying requirement (10.1), i.e., "the system is drifting away" with parameters $\alpha > 0$ and $0 \le \beta < 1$.

We know that $\mathbf{r}_k(t)$ describes the *type* of the trajectory of the kth particle, and $\mathbf{x}_k(0) = \mathbf{y}_k \in I^h$ is the starting point, but we also need the "initial direction", which actually means a *rotation* of the h-space $\mathbb{R}^h$. Consider e.g., the case $h = 3$: a general space curve — a helix, say — is a proper 3-dimensional object: unlike a straight line, a helix spans the 3-space. So, in dimension $h \geq 3$, for a general h-dimensional space curve $\mathbf{r}_k(t) \subset \mathbb{R}^h$ we must choose a rotation $\vartheta_k \in \mathrm{SO}(h)$ ("special orthogonal group") of the h-space $(1 \leq k \leq N)$.

For comparison note that the 2-dimensional case is simpler: the rotation of the plane is the same as choosing a point on the unit circle $\mathbf{S}^1$, or equivalently, to choose an angle $0 \leq \theta < 2\pi$. Formally, SO(2) is the additive group of the reals modulo 2π, so SO(2) is commutative. On the other hand, SO(h) is noncommutative for every $h \geq 3$. (This fundamental difference between SO(2) and SO(h), $h \geq 3$ is the source of many technical problems; see later.)

The choice of the N rotations ϑ_k, $1 \leq k \leq N$ in the 3-space makes the initial condition complete: the *real* trajectory of the kth particle in the general torus model is

$$\mathbf{x}_k(t) = \mathbf{y}_k + \vartheta_k \mathbf{r}_k(t) \quad \text{modulo one,} \tag{10.2}$$

which describes the time evolution of the whole N-particle system for all time $0 \leq t < \infty$.

To talk about *typical* directions we need a measure. We assume that $\vartheta_1, \vartheta_2, \ldots, \vartheta_N$ are N independent random variables, where $\vartheta_1, \vartheta_2, \ldots, \vartheta_N$ are all uniformly distributed in the rotation group SO(h) (= special orthogonal group of all proper rotations of the h-space) with respect to the Haar measure. This means we use the corresponding product measure on the product space

$$\omega = (\vartheta_1, \vartheta_2, \ldots, \vartheta_N) \in \Omega = \mathrm{SO}(h) \times \cdots \times \mathrm{SO}(h) = (\mathrm{SO}(h))^N. \tag{10.3}$$

Let ProdMeas denote the normalized version of this product measure on the set Ω, and for notational simplicity write $d\omega = d\mathrm{ProdMeas}$. That is,

$$\int_\Omega 1\, d\omega = \mathrm{ProdMeas}(\Omega) = 1.$$

We refer to the huge product space Ω as the "initial direction space" (somewhat vaguely speaking).

The trajectory $\Gamma(t) = \vec{\Gamma}_{\mathcal{Y},\omega}(t)$ of the whole system is

$$\begin{aligned}\Gamma(t) &= \vec{\Gamma}_{\mathcal{Y},\omega}(t) = \vec{\Gamma}(\mathcal{Y};\vec{\mathcal{R}};\omega;t) = (\mathbf{x}_1, \mathbf{x}_2, \ldots, \mathbf{x}_N) \\ &= (\mathbf{y}_1 + \vartheta_1\mathbf{r}_1(t), \mathbf{y}_2 + \vartheta_2\mathbf{r}_2(t), \ldots, \mathbf{y}_N + \vartheta_N\mathbf{r}_N(t)) \quad \text{modulo one,}\end{aligned} \tag{10.4}$$

where t is the time,

$$\omega = (\vartheta_1, \vartheta_2, \ldots, \vartheta_N) \in \Omega = (\mathrm{SO}(h))^N,$$

$$\mathcal{Y} = \{\mathbf{y}_1, \mathbf{y}_2, \ldots, \mathbf{y}_N\} \in [0,1]^h$$

is the initial point configuration,

$$\vec{\mathcal{R}} = (\mathbf{r}_1(t), \mathbf{r}_2(t), \ldots, \mathbf{r}_N(t))$$

is a (vector) family of parametrized h-dimensional curves "drifting away" with parameters $\alpha > 0$ and $0 \le \beta < 1$ (see (10.1)). Since $\vartheta_k\mathbf{r}_k(t) \in \mathbb{R}^h$, $1 \le k \le N$, $\vec{\Gamma}_{\mathcal{Y},\omega}(t)$ in (10.4) is a curve in the hN-dimensional unit torus (= configuration space).

Theorem 10.1. (Weak Uniformity Theorem: "kind of" unique ergodicity via typical rotations) *Consider the h-dimensional general torus model $\vec{\Gamma}_{\mathcal{Y},\omega}(t) = \vec{\Gamma}(\mathcal{Y};\vec{\mathcal{R}};\omega;t)$ with $h \ge 2$, drifting away from the start; see definitions* (10.1)–(10.4). *For almost every vector of rotations*

$$\omega = (\vartheta_1, \vartheta_2, \ldots, \vartheta_N) \in \Omega = (\mathrm{SO}(h))^N,$$

$$\lim_{T\to\infty} \frac{1}{T}\text{length}\left\{0 \le t \le T : \ \vec{\Gamma}_{\mathcal{Y},\omega}(t) \in \mathcal{S}\right\} = \mathrm{vol}_d(\mathcal{S}) \tag{10.5}$$

for all Jordan measurable subsets $\mathcal{S}$ of the $d = hN$-dimensional unit cube $I^d = [0,1]^d$ (= configuration space), and in general, for every Riemann integrable function f defined on I^d,

$$\lim_{T\to\infty} \frac{1}{T}\int_0^T f(\vec{\Gamma}_{\mathcal{Y},\omega}(t))\,dt = \int_{I^d} f\,d\mathrm{vol}_d, \tag{10.6}$$

where length is the 1-dimensional Lebesgue measure, and vol_d *is the d-dimensional Lebesgue measure.*

Remarks. The Weak Uniformity Theorem is similar to the Kronecker–Weyl theorem. There is, however, a crucial difference: in the Kronecker–Weyl theorem we have an explicit characterization of the set of "good" directions (linear independence of the direction coordinates over the rationals); in the Weak Uniformity Theorem, on the other hand, we do not know any criterion to decide that a given explicit vector of rotations

$$\omega = (\vartheta_1, \vartheta_2, \ldots, \vartheta_N) \in \Omega = (\mathrm{SO}(h))^N \tag{10.7}$$

is "good" or "bad". This is certainly a disadvantage. It is compensated by the fact that the Weak Uniformity Theorem is far more general: it extends the limited class of constant speed motions on straight lines to (basically) arbitrary parametrized curves.

It is interesting to point out that, for a fixed

$$\vec{\mathcal{R}} = (\mathbf{r}_1(t), \mathbf{r}_2(t), \ldots, \mathbf{r}_N(t)) \tag{10.8}$$

and

$$\omega = (\vartheta_1, \vartheta_2, \ldots, \vartheta_N) \in (\mathrm{SO}(h))^N, \tag{10.9}$$

the *time flow* defined by (10.4) *does not necessarily* satisfy the composition property ("group action") $T_{t_1}T_{t_2} = T_{t_1+t_2}$, or equivalently,

$$T_{t_1}(T_{t_2}(\mathbf{x})) = T_{t_1+t_2}(\mathbf{x}).$$

The composition property ("group action") holds in the special case of constant speed motion of the particles on straight lines, but it fails for varying speed motions, especially for nonlinear curves. The violation of the group action means that in the general case we are beyond the framework of traditional dynamical systems (= describing the solution of autonomous systems of differential equations) and ergodic theory. Nevertheless, we keep the "ergodic viewpoint".

The "ergodic viewpoint" explains why we refer to Theorem 10.1 as a result exhibiting "kind of" *unique ergodicity*. We recall a key result in topological ergodic theory (= topological dynamical systems), often called Unique Ergodicity Theorem. For simplicity we discuss the discrete time evolution (the continuous case of a time-flow is a straightforward analog). Let X be a compact metric space, let $\mathcal{B}$ denote the σ-algebra of Borel subsets of X, let $T: X \to X$ be a continuous map of X to itself, and let μ be a measure such that $(X, \mathcal{B}, \mu)$ is a probability space, and T is μ-measure-preserving. We say that $x_0 \in X$ is a *generic* point for the quadruple $(X, \mathcal{B}, \mu, T)$ if

$$\lim_{n\to\infty} \frac{1}{n} \sum_{k=0}^{n-1} f(T^k x_0) = \int_X f\, d\mu \tag{10.10}$$

for every continuous function f defined on X. Birkhoff's ergodic theorem implies that, if T is ergodic then μ-almost-every point $x_0 \in X$ is generic. It raises the natural question: under what condition can we upgrade *μ-almost-every* to *every*? It turns out that this question has a surprisingly elegant answer.

Unique Ergodicity Theorem. *The following four statements are equivalent:*

(a) *every point $x \in X$ is generic for the quadruple $(X, \mathcal{B}, \mu, T)$;*
(b) *μ is the only probability measure on the Borel σ-algebra $\mathcal{B}$ such that $(X, \mathcal{B}, \mu)$ is a probability space and T is μ-measure-preserving;*
(c) *for every continuous function f defined on X we have the pointwise convergence to a constant*

$$\lim_{n\to\infty} \frac{1}{n} \sum_{k=0}^{n-1} f(T^k x) = c(f), \tag{10.11}$$

where the constant $c(f)$ depends only on f (i.e., it is independent of x);
(d) *for every continuous function f defined on X*

$$\frac{1}{n} \sum_{k=0}^{n-1} f(T^k x) \to \int_X f\, d\mu \tag{10.12}$$

uniformly in $x \in X$.

A pair (X, T) (= topological dynamical system) is called *uniquely ergodic* if (b) holds, i.e., if there is only one probability measure μ such that $(X, \mathcal{B}, \mu)$ is a probability space and T is μ-measure-preserving. The bad news is that it is not easy to decide whether or not a topological dynamical system (X, T) is uniquely ergodic.

Note that the proof of the Unique Ergodicity Theorem is substantially simpler than that of Birkhoff's ergodic theorem. Unfortunately both theorems are "soft": they do not say anything about the speed of convergence. This is why we do not use any of them in the rest of the book.

Notice that (10.6) is a "kind of" time-flow analog of part (c) in the Unique Ergodicity Theorem (where μ is of course the ordinary Lebesgue measure). Thus, Theorem 10.1 implies that, under the "drifting away" condition (10.1)–(10.4), the time-flow (10.4) on the torus is "kind of" *uniquely ergodic* for almost every vector of rotations

$$\omega = (\vartheta_1, \vartheta_2, \ldots, \vartheta_N) \in (\mathrm{SO}(h))^N.$$

We can summarize this book from the viewpoint of ergodic theory as follows. We prove and apply "short-time" ergodic type theorems for "kind of" uniquely ergodic time-flows in the configuration space. Such a time-flow describes the typical short-term time evolution of a large system of non-interacting particles, starting from off-equilibrium.

Proof of Theorem 10.1. Write $d = hN$. According to the curve Weyl's criterion (2.5), it suffices to verify (10.6) for the complex exponential functions $f(\mathbf{z}) = f_{\mathbf{n}}(\mathbf{z}) = e^{2\pi i \mathbf{n}\cdot\mathbf{z}}$ with $\mathbf{n} \in \mathbb{Z}^d \setminus \mathbf{0}$. We show that, given any $\varepsilon > 0$, there is a subset $\Omega(\varepsilon) \subset \Omega$ such that $\mathrm{ProdMeas}(\Omega(\varepsilon)) < \varepsilon$, and for every $\omega \in \Omega \setminus \Omega(\varepsilon)$

$$\limsup_{T\to\infty} \frac{1}{T} \left| \int_0^T f_{\mathbf{n}}(\vec{\Gamma}_{\mathcal{Y},\omega}(t))\, dt \right| < \varepsilon \tag{10.13}$$

holds for all

$$f_{\mathbf{n}}(\vec{\Gamma}_{\mathcal{Y}\omega}(t)) = e^{2\pi i \mathbf{n}\cdot\vec{\Gamma}_{\mathcal{Y},\omega}(t)}, \ \mathbf{n} \in \mathbb{Z}^d \setminus \mathbf{0}. \tag{10.14}$$

Choosing $\varepsilon \to 0$, (10.13)–(10.14) imply that (10.6) holds for all $f(\mathbf{z}) = f_{\mathbf{n}}(\mathbf{z})$, $\mathbf{n} \in \mathbb{Z}^d \backslash \mathbf{0}$ for almost every vector of rotations $\omega \in \Omega$ — this completes the proof of the theorem.

We start the proof of (10.13)–(10.14) by introducing a notation: for $T > 0$ write

$$D_{f_{\mathbf{n}}}(\mathcal{Y}; \vec{\mathcal{R}}; \omega; T) = \int_0^T f_{\mathbf{n}}(\vec{\Gamma}_{\mathcal{Y},\omega}(t))\, dt. \tag{10.15}$$

We study the big average of the square of the absolute value of $D_{f_{\mathbf{n}}}(\mathcal{Y}; \vec{\mathcal{R}}; \omega; T)$:

$$\begin{aligned}\Delta^2_{f_{\mathbf{n}}}(\mathcal{Y}; \vec{\mathcal{R}}; T) &= \int_{\omega\in\Omega} \left| D_{f_{\mathbf{n}}}(\mathcal{Y}; \vec{\mathcal{R}}; \omega; T)\right|^2 d\omega \\ &= \int_{\vartheta_1\in\mathrm{SO}(3)} \cdots \int_{\vartheta_N\in\mathrm{SO}(h)} \left| D_{f_{\mathbf{n}}}(\mathcal{Y}; \vec{\mathcal{R}}; \omega; T)\right|^2 d\vartheta_1 \ldots d\vartheta_N,\end{aligned} \tag{10.16}$$

where $d\vartheta$ denotes the normed Haar measure on the special orthogonal group $\mathrm{SO}(h)$, i.e., $\int_{\mathrm{SO}(h)} d\vartheta = 1$.

First we prove the theorem in the special case $h = 3$. It is based on the following lemma.

Lemma 10.1. *For every $T > 0$ and every complex exponential function $f_{\mathbf{n}}(\mathbf{z}) = e^{2\pi \mathrm{i}\mathbf{n}\cdot\mathbf{z}}$, where $\mathbf{n} = (n_1, \ldots, n_d) \in \mathbb{Z}^d \setminus \mathbf{0}$ is a nonzero d-dimensional lattice point and $d = 3N$, we have*

$$\Delta^2_{f_{\mathbf{n}}}(\mathcal{Y}; \vec{\mathcal{R}}; T) = \int_0^T \int_0^T \prod_{k=1}^{N} \frac{\sin\left(2\pi\left|\mathbf{r}_k(t_1) - \mathbf{r}_k(t_2)\right|\left|\mathbf{n}(k)\right|\right)}{2\pi\left|\mathbf{r}_k(t_1) - \mathbf{r}_k(t_2)\right|\left|\mathbf{n}(k)\right|}\, dt_1\, dt_2$$

with $\mathbf{n}(k) = (n_{3k-2}, n_{3k-1}, n_{3k})$.

Proof of Lemma 10.1. Using the trivial general fact

$$\int_a^b h(x)\, dx \int_c^d g(y)\, dy = \int_a^b \int_c^d h(x)g(y)\, dx\, dy$$

in (10.15), we have

$$\left|D_{f_{\mathbf{n}}}(\mathcal{Y}; \vec{\mathcal{R}}; \omega; T)\right|^2 = \int_0^T \int_0^T e^{2\pi \mathrm{i}(\vec{\Gamma}_{\mathcal{Y},\omega}(t_1)\cdot\mathbf{n} - \vec{\Gamma}_{\mathcal{Y},\omega}(t_2)\cdot\mathbf{n})}\, dt_1\, dt_2,$$

and so we obtain

$$\begin{aligned}
\Delta^2_{f_{\mathbf{n}}}(\mathcal{Y}; \vec{\mathcal{R}}; T) &= \int_{\vartheta_1\in\mathrm{SO}(3)} \cdots \int_{\vartheta_N\in\mathrm{SO}(3)} \\
&\quad \cdot \left(\int_0^T \int_0^T e^{2\pi \mathrm{i}(\vec{\Gamma}_{\mathcal{Y},\omega}(t_1) - \vec{\Gamma}_{\mathcal{Y},\omega}(t_2))\cdot\mathbf{n}}\, dt_1\, dt_2\right) d\vartheta_1 \ldots d\vartheta_N \\
&= \int_0^T \int_0^T \left(\int_{\vartheta_1\in\mathrm{SO}(3)} \cdots \right. \\
&\quad \left. \int_{\vartheta_1\in\mathrm{SO}(3)} e^{2\pi \mathrm{i}(\vec{\Gamma}_{\mathcal{Y},\omega}(t_1) - \vec{\Gamma}_{\mathcal{Y},\omega}(t_2))\cdot\mathbf{n}}\, d\vartheta_1 \ldots d\vartheta_N\right) dt_1\, dt_2.
\end{aligned} \tag{10.17}$$

Since

$$e^{2\pi \mathrm{i}(\vec{\Gamma}_{\mathcal{Y},\omega}(t_1) - \vec{\Gamma}_{\mathcal{Y},\omega}(t_2))\cdot\mathbf{n}} = \prod_{k=1}^{N} e^{2\pi \mathrm{i}(\vartheta_k\mathbf{r}_k(t_1) - \vartheta_k\mathbf{r}_k(t_2))\cdot\mathbf{n}(k)},$$

we have

$$\int_{\vartheta_1 \in \mathrm{SO}(3)} \cdots \int_{\vartheta_N \in \mathrm{SO}(3)} e^{2\pi \mathrm{i}(\vec{\Gamma}_{\mathcal{Y},\omega}(t_1) - \vec{\Gamma}_{\mathcal{Y},\omega}(t_2)) \cdot \mathbf{n}} \, d\vartheta_1 \ldots d\vartheta_N$$

$$= \prod_{k=1}^{N} \int_{\vartheta_k \in \mathrm{SO}(3)} e^{2\pi \mathrm{i}(\vartheta_k \mathbf{r}_k(t_1) - \vartheta_k \mathbf{r}_k(t_2)) \cdot \mathbf{n}(k)} \, d\vartheta_k$$

$$= \prod_{k=1}^{N} \int_{\vartheta_k \in \mathrm{SO}(3)} e^{2\pi \mathrm{i}(\mathbf{r}_k(t_1) - \mathbf{r}_k(t_2)) \cdot \vartheta_k^{-1} \mathbf{n}(k)} \, d\vartheta_k, \tag{10.18}$$

where $\vartheta_k^{-1} \in \mathrm{SO}(3)$ denotes the inverse of ϑ_k.

Inserted Remark. In view of (10.17)–(10.18) the initial point configuration $\mathcal{Y} = \{\mathbf{y}_1, \ldots, \mathbf{y}_N\}$ cancels out in the proof. This is perfectly natural, since the torus is translation invariant.

To evaluate the last integral in (10.18), we need the following lemma.

Lemma 10.2. *Let* $\mathbf{w}$ *be an arbitrary* 3*-dimensional vector. Then*

$$\frac{1}{4\pi} \int_{\mathbf{v} \in \mathbf{S}^2} e^{\mathrm{i}\mathbf{w} \cdot \mathbf{v}} \, d\mathrm{SA}(\mathbf{v}) = \frac{\sin(|\mathbf{w}|)}{|\mathbf{w}|}, \tag{10.19}$$

where $d\mathrm{SA}(\mathbf{v})$ *represents the surface area.*

Proof of Lemma 10.2. The special case $\mathbf{w} = \mathbf{0}$ is trivial: the left side is equal to 1, and the right side tends to 1 as $\mathbf{w}$ tends to $\mathbf{0}$; so we may assume that $\mathbf{w} \neq \mathbf{0}$. We need a well-known geometric fact about the surface area of some special spherical domains. Let $\mathbf{e} \in \mathbf{S}^2$ be an arbitrary but fixed unit vector, and let $0 < \delta < 1$ be an arbitrary real number. Consider the "spherical belt" (often called the "spherical zone"):

$$\mathbf{S}^2(\mathbf{e}; \delta) = \left\{ \mathbf{v} \in \mathbf{S}^2 : \ |\mathbf{e} \cdot \mathbf{v}| \leq \delta \right\}.$$

On one hand, it is trivial that the surface area of the "belt" $\mathbf{S}^2(\mathbf{e}; \delta)$ is independent of the choice of the unit vector $\mathbf{e}$; on the other hand, it is a very surprising nontrivial fact — discovered by Archimedes — that the dependence on δ is just plain linear

$$\mathrm{SurfaceArea}\left(\mathbf{S}^2(\mathbf{e}; \delta)\right) = \mathrm{SurfaceArea}\left(\mathbf{S}^2(\mathbf{e}; \delta)\right) = 4\pi\delta. \tag{10.20}$$

(10.20) implies that, for any nonzero 3-dimensional vector $\mathbf{w}$ and any real number ϱ with $0 < \varrho < |\mathbf{w}|$,

$$\text{SurfaceArea}\left(\{\mathbf{v} \in \mathbf{S}^2 : \ |\mathbf{w} \cdot \mathbf{v}| \le \varrho\}\right) = 4\pi \cdot \frac{\varrho}{|\mathbf{w}|}. \tag{10.21}$$

By using (10.21), we have

$$\frac{1}{4\pi}\int_{\mathbf{S}^2} e^{\mathrm{i}\mathbf{w}\cdot\mathbf{v}}\, d\mathrm{SA}(\mathbf{v}) = \frac{1}{2}\int_{-1}^{1} e^{\mathrm{i}|\mathbf{w}|u}\, du = \frac{\sin(|\mathbf{w}|)}{|\mathbf{w}|},$$

which proves (10.19). □

By (10.19),

$$\begin{aligned}\int_{\vartheta_k \in \mathrm{SO}(3)} & e^{2\pi \mathrm{i}(\mathbf{r}_k(t_1)-\mathbf{r}_k(t_2))\cdot\vartheta_k^{-1}\mathbf{n}(k)}\, d\vartheta_k \\ &= \frac{1}{4\pi}\int_{\mathbf{u}\in\mathbf{S}^2} e^{\mathrm{i}\mathbf{w}\cdot\mathbf{u}}\, d\mathrm{SA}(\mathbf{u}) = \frac{\sin w}{w}\end{aligned} \tag{10.22}$$

with

$$w = |\mathbf{w}| = 2\pi\, |\mathbf{r}_k(t_1) - \mathbf{r}_k(t_2)|\, |\mathbf{n}(k)|.$$

Lemma 10.1 follows from (10.17), (10.18) and (10.22). □

We are now ready to complete the proof of the theorem for $h = 3$. Let $\mathbf{n} = (n_1, \ldots, n_d) \in \mathbb{Z}^d \setminus \mathbf{0}$ be fixed. Then there is a j such that $\mathbf{n}(j) = (n_{3j-2}, n_{3j-1}, n_{3j}) \neq \mathbf{0}$, so $|\mathbf{n}(j)| \ge 1$. Thus by Lemma 10.1 we have the trivial upper bound

$$\begin{aligned}\Delta^2_{f_{\mathbf{n}}}(\mathcal{Y}; \vec{\mathcal{R}}; T) &\le \int_0^T \int_0^T \min\left\{\frac{1}{2\pi\,|\mathbf{r}_j(t_1) - \mathbf{r}_j(t_2)|}, 1\right\} dt_1\, dt_2 \\ &= \int_{t_1=0}^T \left(\int_{t_2=0}^T \min\left\{\frac{1}{2\pi\,|\mathbf{r}_j(t_1) - \mathbf{r}_j(t_2)|}, 1\right\} dt_2\right) dt_1.\end{aligned} \tag{10.23}$$

We want to estimate the inner integral in the last line of (10.23) (so t_1 is fixed, and t_2 is a variable). Let A_1 denote the aligned unit size cube $I^3 + \mathbf{w}$ centered at $\mathbf{r}_j(t_1)$, and consider the tiling of the 3-space $\mathbb{R}^3$ with aligned unit size cubes where A_1 is a member of the tiling. Let $A_2, A_3, \ldots, A_{27}$ be the first neighborhood of A_1 in this tiling, let $A_{28}, A_{29}, \ldots, A_{125}$ be the second neighborhood of A_1, and in general, let

$$A_{(2k-1)^3+1}, A_{(2k-1)^3+2}, \ldots, A_{(2k+1)^3}$$

be the kth neighborhood of A_1 in this tiling ($k \ge 1$).

If A_i is in the kth neighborhood of A_1 and $\mathbf{r} \in A_i$, then $|\mathbf{r}_j(t_1) - \mathbf{r}| \ge (2k-1)/2$, and by (10.1),

$$\text{length}\,\{0 \le t \le T : \; \mathbf{r}_j(t) \in A_i\} \le \alpha T^\beta.$$

Let K be the largest integer with $(2K+1)^3 \le T^{1-\beta}$. We have the following trivial upper bound for the inner integral in the last line of (10.23):

$$\begin{aligned}
&\int_{t_2=0}^{T} \min\left\{\frac{1}{2\pi\,|\mathbf{r}_j(t_1) - \mathbf{r}_j(t_2)|}, 1\right\} dt_2 \\
&\le \alpha T^\beta + \frac{(27-1)\alpha T^\beta}{2\pi \cdot \frac{1}{2}} + \frac{(125-27)\alpha T^\beta}{2\pi \cdot \frac{3}{2}} \\
&\quad + \cdots + \frac{((2k+1)^3 - (2k-1)^3)\alpha T^\beta}{2\pi \cdot \frac{2k-1}{2}} \\
&\quad + \cdots + \frac{((2K+1)^3 - (2K-1)^3)\alpha T^\beta}{2\pi \cdot \frac{2K-1}{2}} + \frac{T}{2\pi \cdot \frac{2K+1}{2}},
\end{aligned} \tag{10.24}$$

where k indicates the running index $1 \le k \le K$.

We rewrite (10.24) in the shorter form

$$\begin{aligned}
&\int_{t_2=0}^{T} \min\left\{\frac{1}{2\pi\,|\mathbf{r}_j(t_1) - \mathbf{r}_j(t_2)|}, 1\right\} dt_2 \\
&\le \alpha T^\beta \left(1 + \frac{1}{\pi} \sum_{1 \le k \le K} \frac{((2k+1)^3 - (2k-1)^3)}{2k-1}\right) + \frac{T}{\pi(2K+1)},
\end{aligned} \tag{10.25}$$

where K is the largest integer with $(2K+1)^3 \le T^{1-\beta}$.

Clearly

$$\begin{aligned}
&\sum_{1 \le k \le K} \frac{((2k+1)^3 - (2k-1)^3)}{2k-1} \\
&= 2 \sum_{1 \le k \le K} \frac{(2k+1)^2 + (2k+1)(2k-1) + (2k-1)^2}{2k-1}
\end{aligned}$$

$$\leq 6 \sum_{1\leq k\leq K} \frac{(2k+1)^2}{2k-1} \leq 6 \sum_{1\leq k\leq K} 3(2k-1)$$

$$= 18(K+1)^2 < 18T^{2(1-\beta)/3}. \tag{10.26}$$

Moreover,

$$\frac{T}{\pi(2K+1)} \leq \frac{T}{T^{(1-\beta)/3}} = T^{(2+\beta)/3}, \tag{10.27}$$

since $(2K+3)^3 > T^{1-\beta}$ and $\pi(2K+1) > 2K+3$.

Combining (10.23) and (10.25)–(10.27), we have

$$\Delta^2_{f_{\mathbf{n}}}(\mathcal{Y};\vec{\mathcal{R}};T) \leq \int_{t_1=0}^{T} \left(\int_{t_2=0}^{T} \min\left\{ \frac{1}{2\pi\, |\mathbf{r}_j(t_1) - \mathbf{r}_j(t_2)|}, 1 \right\} dt_2 \right) dt_1$$

$$\leq T\left(\alpha T^{\beta} \left(1 + \frac{1}{\pi} 18 T^{2(1-\beta)/3} \right) + T^{(2+\beta)/3} \right)$$

$$< (7\alpha+1)T^{(5+\beta)/3}. \tag{10.28}$$

Now it is just a routine calculation to finish the proof of the theorem, i.e., to prove (10.13)–(10.14) for $h=3$. Given $\varepsilon > 0$, define the sparse sequence

$$T_n = \left(1 + \frac{\varepsilon}{2}\right)^n, \quad n = 1, 2, 3, \ldots. \tag{10.29}$$

Let

$$\Omega_n(\text{bad}) = \left\{ \omega \in \Omega : \ |D_{f_{\mathbf{n}}}(\mathcal{Y};\vec{\mathcal{R}};\omega;T_n)| > n\sqrt{7\alpha+1}\,T_n^{(5+\beta)/6} \right\}. \tag{10.30}$$

Combining (10.15)–(10.16) and (10.28)–(10.30), we have

$$\text{ProdMeas}(\Omega_n(\text{bad})) < \frac{1}{n^2}. \tag{10.31}$$

Let

$$\Omega(\text{good}) = \Omega \setminus \bigcup_{n > C_0} \Omega_n(\text{bad}), \tag{10.32}$$

where the threshold $C_0 = C_0(\varepsilon, \alpha, \beta)$ will be specified later.

Now let

$$T_n = \left(1+\frac{\varepsilon}{2}\right)^n \le T < T_{n+1} = \left(1+\frac{\varepsilon}{2}\right)^{n+1}.$$

Then

$$\begin{aligned}|D_{f_{\mathbf{n}}}(\mathcal{Y};\vec{\mathcal{R}};\omega;T)| &= \left|\int_0^T e^{2\pi i\mathbf{n}\cdot\vec{\Gamma}_{\mathcal{Y},\omega}(t)}\,dt\right| \\ &\le |D_{f_{\mathbf{n}}}(\mathcal{Y};\vec{\mathcal{R}};\omega;T_n)| + (T-T_n) \\ &\le |D_{f_{\mathbf{n}}}(\mathcal{Y};\vec{\mathcal{R}};\omega;T_n)| + \frac{\varepsilon T}{2}. \qquad (10.33)\end{aligned}$$

If $\omega \in \Omega(\text{good})$ then by (10.30)

$$|D_{f_{\mathbf{n}}}(\mathcal{Y};\vec{\mathcal{R}};\omega;T_n)| \le n\sqrt{7\alpha+1}\,T_n^{(5+\beta)/6},$$

and using it in (10.33), we have (log stands for the natural logarithm)

$$\begin{aligned}|D_{f_{\mathbf{n}}}(\mathcal{Y};\vec{\mathcal{R}};\omega;T)| &\le |D_{f_{\mathbf{n}}}(\mathcal{Y};\vec{\mathcal{R}};\omega;T_n)| + \frac{\varepsilon T}{2} \le n\sqrt{7\alpha+1}\,T_n^{(5+\beta)/6} + \frac{\varepsilon T}{2} \\ &\le \frac{\log T}{\log(1+\varepsilon/2)}\sqrt{7\alpha+1}\,T^{(5+\beta)/6} + \frac{\varepsilon T}{2}, \qquad (10.34)\end{aligned}$$

where

$$n \le \frac{\log T}{\log(1+\varepsilon/2)} \quad \text{follows from } T_n = \left(1+\frac{\varepsilon}{2}\right)^n \le T.$$

Since $\beta < 1$, the inequality

$$\frac{\log T}{\log(1+\varepsilon/2)}\sqrt{7\alpha+1}\,T^{(5+\beta)/6} < \frac{\varepsilon T}{2} \qquad (10.35)$$

holds for all sufficiently large integers n, i.e., for all $n > C_0$, where $C_0 = C_0(\varepsilon,\alpha,\beta)$ is some appropriate absolute constant. We also assume that $C_0 > 1/\varepsilon$. Since

$$\sum_{n>1/\varepsilon} \frac{1}{n^2} < \varepsilon,$$

by (10.31)–(10.32) we have

$$\text{ProdMeas}(\Omega(\text{good})) > 1-\varepsilon. \qquad (10.36)$$

Finally, by (10.34) and (10.35), for all $\omega \in \Omega(\text{good})$,

$$\limsup_{T\to\infty} \frac{1}{T}|D_{f_{\mathbf{n}}}(\mathcal{Y};\vec{\mathcal{R}};\omega;T)| < \frac{\varepsilon}{2} + \frac{\varepsilon}{2} = \varepsilon. \tag{10.37}$$

Notice that (10.36)–(10.37) prove (10.13)–(10.14) for $h = 3$, completing the proof of the Weak Uniformity Theorem in the special case $h = 3$.

The proof of the cases $h = 2$ and $h \geq 4$ goes along the same lines. The only change is that we need the analog of Lemma 10.2 for arbitrary dimension.

Lemma 10.3. *Let $h \geq 2$ be an integer, and let* SA *denote the surface area in the h-space $\mathbb{R}^h$. For every h-dimensional vector $\mathbf{w}$, we obtain with $w = |\mathbf{w}|$,*

$$\frac{1}{\mathrm{SA}(\mathbf{S}^{h-1})}\int_{\mathbf{u}\in\mathbf{S}^{h-1}} e^{\mathrm{i}\mathbf{w}\cdot\mathbf{u}}\, d\mathrm{SA}(\mathbf{u}) = (h-2)!!\frac{J_{(h-2)/2}(w)}{w^{(h-2)/2}} \cdot \left\{1 \text{ or } \sqrt{\frac{\pi}{2}}\right\},$$

depending on whether $h \geq 2$ is even or odd, accordingly. Here $J_\nu(z)$ is the well-known Bessel function, and, as usual, $k!! = k(k-2)(k-4)\cdots$, where the last factor is 1 or 2 depending on the parity of k, and $0!! = 1$.

Remarks. Note that the Bessel functions arise as solutions of some important differential equations, but here we define the Bessel function (of the first kind) $J_\nu(z)$ (where ν is called the order) in an equivalent shortcut way via power series

$$J_\nu(z) = \sum_{m=0}^{\infty} \frac{(-1)^m}{m!\mathrm{Gamma}(m+\nu+1)}\left(\frac{z}{2}\right)^{2m+\nu},$$

where $\mathrm{Gamma}(z)$ is Euler's Gamma function.

If $\nu \geq 0$ is a half-integer, then $J_\nu(z)$ has a finite form, e.g.,

$$J_{\frac{1}{2}}(z) = \sqrt{\frac{2}{\pi z}}\sin z \quad \text{and} \quad J_{\frac{3}{2}}(z) = \sqrt{\frac{2}{\pi z}}\left(-\cos z + \frac{\sin z}{z}\right),$$

but for $\nu \geq 0$ integer the Bessel function $J_\nu(z)$ does not have a finite expression. What we can still use is an asymptotic expression — called Hankel's formula — that works for all $\nu \geq 0$; see later.

The unusual notation $\mathrm{Gamma}(s)$ for the Gamma function is explained by our preference to save letter $\Gamma = \Gamma(t)$ for the trajectory in the configuration space, describing the time evolution of a system.

Proof of Lemma 10.3. We need information about the surface area of a "spherical belt" in arbitrary dimension. We recall that $\mathbf{S}^m = \{\mathbf{u} \in \mathbb{R}^{m+1} : |\mathbf{u}| = 1\}$ denotes the unit sphere in the $(m+1)$-dimensional space. Let $\mathbf{v} \in \mathbf{S}^m$ be an arbitrary but fixed unit vector, and let $\beta < \gamma$ be real. We need to know the surface area of the "spherical belt"

$$\mathbf{S}^m(\mathbf{v}; \beta, \gamma) = \{\mathbf{u} \in \mathbf{S}^m : \beta \leq \mathbf{v} \cdot \mathbf{u} \leq \gamma\}. \tag{10.38}$$

The special case $m = 2$ is particularly simple: we have the elegant formula

$$\text{SurfaceArea}\left(\mathbf{S}^2(\mathbf{v}; \beta, \gamma)\right) = \frac{\gamma - \beta}{2} \cdot 4\pi \tag{10.39}$$

for $-1 \leq \beta < \gamma \leq 1$. Formula (10.39) was used in the proof of Lemma 10.2.

Unfortunately, the elegant linear formula (10.39) breaks down in dimensions $m \neq 2$. Nevertheless, we have a (less elegant) general integral formula, which suffices for our purposes. Since the surface area of $\mathbf{S}^m(\mathbf{v}; \beta, \gamma)$ is independent of the choice of the unit vector $\mathbf{v}$, in the following lemma we drop $\mathbf{v}$, and simply write $\mathbf{S}^m(\beta, \gamma)$.

Lemma 10.4. *If $m \geq 1$ then for any $-1 \leq \beta < \gamma \leq 1$, we have*

$$\frac{\text{SurfaceArea}\left(\mathbf{S}^m(\beta, \gamma)\right)}{\text{SurfaceArea}\left(\mathbf{S}^m\right)} = \frac{\int_\beta^\gamma (1-x^2)^{\frac{m}{2}-1}\, dx}{\int_{-1}^1 (1-x^2)^{\frac{m}{2}-1}\, dx}.$$

Unfortunately this technical result is not well known, so for the sake of completeness, we include a short proof at the end of the section.

Let $h \geq 2$ be an arbitrary dimension with $h \neq 3$, and let $\mathbf{w} \in \mathbb{R}^h$ be an arbitrary but fixed vector. By using Lemma 10.4 with $m = h - 1$, we are ready now to evaluate the critical integral (let $w = |\mathbf{w}|$)

$$\frac{1}{\text{SA}(\mathbf{S}^{h-1})} \int_{\mathbf{u} \in \mathbf{S}^{h-1}} e^{\mathrm{i}\mathbf{w}\cdot\mathbf{u}}\, d\text{SA}(\mathbf{u}) = \int_0^1 e^{iwx} \frac{(1-x^2)^{(h-3)/2}}{\int_0^1 (1-y^2)^{(h-3)/2}\, dy}\, dx. \tag{10.40}$$

The good news is that integral (10.40) can be expressed in terms of the classical Bessel function $J_\nu(z)$. Indeed, we make use of the Poisson representation formula (see e.g. formula (8) on p. 48 in Watson [Wat58])

$$J_\nu(z) = \frac{(z/2)^\nu}{\text{Gamma}(\nu + \frac{1}{2})\sqrt{\pi}} \int_{-1}^1 e^{\mathrm{i}zx}(1-x^2)^{\nu - \frac{1}{2}}\, dx, \tag{10.41}$$

which holds for all $\nu > -1/2$ and complex z.

Combining (10.4) and (10.5) with $\nu = (h-2)/2$ and $z = w = |\mathbf{w}|$, we obtain

$$\frac{1}{\mathrm{SA}(\mathbf{S}^{h-1})} \int_{\mathbf{u} \in \mathbf{S}^{h-1}} e^{\mathbf{i}\mathbf{w}\cdot\mathbf{u}}\, d\mathrm{SA}(\mathbf{u})$$

$$= J_{(h-2)/2}(w) \cdot \frac{\sqrt{\pi}\mathrm{Gamma}(\frac{h-1}{2})}{(w/2)^{(h-2)/2} \int_{-1}^{1} (1-y^2)^{(h-3)/2}\, dy}. \quad (10.42)$$

It remains to evaluate the factor

$$C(h) = \frac{\sqrt{\pi} 2^{(h-2)/2} \mathrm{Gamma}(\frac{h-1}{2})}{\int_{-1}^{1} (1-y^2)^{(h-3)/2}\, dy}.$$

The integral $\int_{-1}^{1}(1-y^2)^{(h-3)/2}\,dy$ in the denominator can be easily evaluated by a simple recurrence formula. Indeed, let

$$A(k) = \int_{-1}^{1} (1-y^2)^{(k-1)/2}\, dy, \quad k \geq 0 \text{ integer},$$

then $A(0) = \pi$, $A(1) = 2$, $A(2) = \pi/2$, and

$$A(k) = \frac{k-1}{k} A(k-2) \quad \text{for all } k \geq 2.$$

Furthermore, we have

$$\mathrm{Gamma}(s) = (s-1)\mathrm{Gamma}(s-1), \ \mathrm{Gamma}(1/2) = \sqrt{\pi}, \ \mathrm{Gamma}(1) = 1.$$

Combining these formulas, we obtain the recurrence relation $C(h) = (h-2)!!C(h-2)$. Since $C(2) = 1$ and $C(3) = \sqrt{\pi/2}$, for $h \geq 3$ we obtain

$$C(h) = (h-2)!! \cdot \left\{ 1 \text{ or } \sqrt{\frac{\pi}{2}} \right\} \quad (10.43)$$

depending on whether h is even or odd, accordingly. Using (10.43) in (10.42), Lemma 10.3 follows. □

Switching from $h = 3$ to an arbitrary dimension $h \geq 2, h \neq 3$ the crucial change is that the key function $\frac{\sin z}{z}$ in Lemma 10.1 is replaced by

$$(h-2)!! \frac{J_{(h-2)/2}(z)}{z^{(h-2)/2}} \cdot \left\{ 1 \text{ or } \sqrt{\frac{\pi}{2}} \right\}, \quad (10.44)$$

depending on whether $h \geq 2$ is even or odd, accordingly.

If $\nu \geq 0$ is a half-integer, then $J_\nu(z)$ has a finite form. For $\nu \geq 0$ integer the Bessel function $J_\nu(z)$ does not have a finite expression, but we can still use an asymptotic expression — called Hankel's formula — that works for all $\nu \geq 0$, covering both cases. For our purposes it suffices to apply the first order approximation in Hankel's formula

$$J_\nu(z) = \sqrt{\frac{2}{\pi z}} \cos\left(z - \frac{(2\nu+1)\pi}{4}\right) + O(z^{-3/2}) \quad \text{for } z \geq 1 \tag{10.45'}$$

and

$$J_\nu(z) = O(1) \quad \text{for } 0 \leq z < 1, \tag{10.45''}$$

where the implicit constants in the O-notations depend only on the value of $\nu \geq 0$.

By using Lemma 10.3, (10.45′) and (10.45″), we can complete the proof of the general case $h \geq 2, h \neq 3$ exactly the same way as we completed the case of $h = 3$ via Lemma 10.1.

As we promised, we conclude Sec. 10 by including a proof of Lemma 10.4.

Proof of Lemma 10.4. For notational convenience, we use $\mathbf{v} = \mathbf{e}_1 = (1, 0, \ldots, 0)$ in (10.38). The special case $m = 1$ gives the unit circle $\mathbf{S}^1$, and then we have the elementary result

$$\text{ArcLength}(\mathbf{S}^1(\beta, \gamma)) = 2(\arcsin(\gamma) - \arcsin(\beta)) = 2\int_\beta^\gamma \frac{dx}{\sqrt{1-x^2}}. \tag{10.46}$$

Since

$$\int_{-1}^{1} \frac{dx}{\sqrt{1-x^2}} = \pi,$$

(10.46) proves Lemma 10.4 in the special case $m = 1$.

It remains to prove the case $m \geq 3$. In higher dimension we use an analytic formula. If an m-dimensional surface ("manifold") is given in the graph $x_{m+1} = f(x_1, \ldots, x_m)$, then the ($m$-dimensional) surface area is expressed in terms of the partial derivatives f_{x_j} of f as a multiple integral:

$$m\text{-dimensional surface area} = \int \cdots \int \left(1 + f_{x_1}^2 + \cdots + f_{x_m}^2\right)^{1/2} dx_1\, dx_2 \ldots dx_m. \tag{10.47}$$

By symmetry we can assume that $0 \leq \beta < \gamma \leq 1$. For the upper hemisphere we use the particular function

$$f(x_1, \ldots, x_m) = \left(1 - x_1^2 - x_2^2 - \cdots - x_m^2\right)^{1/2},$$

which, via simple calculations, gives the following form of (10.47):

m-dimensional surface area of the spherical belt

$$= \int \cdots \int \left(1 - x_1^2 - x_2^2 - \cdots - x_m^2\right)^{-1/2} dx_1\, dx_2 \ldots dx_m. \tag{10.48}$$

More precisely, for $0 \leq \beta < \gamma \leq 1$ we have

$$\begin{aligned} \text{SA}\left(\mathbf{S}^m(\mathbf{e}_1; \beta, \gamma)\right) &= \int \cdots \int_{\substack{x_1^2+\cdots+x_m^2 \leq 1 \\ \beta \leq x_1 \leq \gamma}} \frac{dx_1 \ldots dx_m}{\left(1 - x_1^2 - x_2^2 - \cdots - x_m^2\right)^{1/2}} \\ &= \int_{x_1=\beta}^{\gamma} \left(\int \cdots \int_{x_2^2+\cdots+x_m^2 \leq 1-x_1^2} \frac{dx_2 \ldots dx_m}{\left((1 - x_1^2) - x_2^2 - \cdots - x_m^2\right)^{1/2}} \right) dx_1. \end{aligned} \tag{10.49}$$

The inner integral

$$\int \cdots \int_{x_2^2+\cdots+x_m^2 \leq 1-x_1^2} \frac{dx_2 \ldots dx_m}{\left((1 - x_1^2) - x_2^2 - \cdots - x_m^2\right)^{1/2}}$$

in (10.49) is defined on the $(m-1)$-dimensional ball (= solid sphere) of radius $R = \sqrt{1 - x_1^2}$. By decomposing this ball into concentric spheres of radius r, $0 < r < R$, we have

$$\begin{aligned} &\int \cdots \int_{x_2^2+\cdots+x_m^2 \leq R^2 = 1-x_1^2} \frac{dx_2 \ldots dx_m}{\left(R^2 - x_2^2 - \cdots - x_m^2\right)^{1/2}} \\ &= \int_{r=0}^{R=\sqrt{1-x_1^2}} \frac{\text{SA}(\mathbf{S}^{m-2}) \cdot r^{m-2}}{\sqrt{R^2 - r^2}}\, dr \\ &= \text{SA}(\mathbf{S}^{m-2}) \cdot R^{m-2} \int_{\theta=0}^{\pi/2} \sin^{m-2}(\theta)\, d\theta \\ &= (1 - x_1^2)^{(m-2)/2} \cdot \text{SA}(\mathbf{S}^{m-2}) \cdot \int_{\theta=0}^{\pi/2} \sin^{m-2}(\theta)\, d\theta, \end{aligned} \tag{10.50}$$

where SA means the surface area, and we applied the substitution $r = R \sin\theta$ with $R = \sqrt{1 - x_1^2}$.

Combining (10.49) and (10.50), for $0 \le \beta < \gamma \le 1$ we have

$$\mathrm{SA}\left(\mathbf{S}^m(\mathbf{e}_1;\beta,\gamma)\right) = \mathrm{SA}(\mathbf{S}^{m-2}) \cdot \int_{\theta=0}^{\pi/2} \sin^{m-2}(\theta)\, d\theta \cdot \int_{x_1=\beta}^{\gamma} (1-x_1^2)^{(m-2)/2}\, dx_1. \tag{10.51}$$

Using (10.51) in the special case $\beta = 0$ and $\gamma = 1$ ("hemisphere"):

$$\frac{1}{2}\mathrm{SA}(\mathbf{S}^m) = \mathrm{SA}\left(\mathbf{S}^m(\mathbf{e}_1;0,1)\right) = \mathrm{SA}(\mathbf{S}^{m-2}) \cdot \int_{\theta=0}^{\pi/2} \sin^{m-2}(\theta)\, d\theta \cdot \int_{x_1=0}^{1} (1-x_1^2)^{(m-2)/2}\, dx_1. \tag{10.52}$$

Taking the ratio of (10.51) and (10.52), the first factors cancel out, and only the integrals with respect to x_1 survive. This proves Lemma 10.4 for $0 \le \beta < \gamma \le 1$. The general case follows by symmetry.

Thus the proof of Theorem 10.1 is complete. □

11. Asymptotic Time-Lapse Randomness

Consider now the 3-dimensional *general* torus model introduced in (10.1)–(10.4), and let us go back to the proof of Theorem 3.2. Replacing the application of the Kronecker–Weyl theorem to the characteristic function of a set $S^*(\Delta) \subset [0,1]^{3N}$ with an application of Theorem 10.1 ("Weak Uniformity Theorem"), we obtain the following analog of Theorem 3.2.

Theorem 11.1. (another asymptotic binomial distribution) *Consider the 3-dimensional general N-particle torus model drifting away from the start; see* (10.1)–(10.4). *Then for almost every vector of rotations*

$$\omega = (\vartheta_1, \vartheta_2, \ldots, \vartheta_N) \in \Omega = (\mathrm{SO}(3))^N,$$

$$\lim_{T\to\infty} \frac{1}{T}\mathrm{length}\left\{t \in [0,T]:\ N_S(\mathcal{Y};\vec{\mathcal{R}};\omega;t) = n\right\} = \binom{N}{n} p^n q^{N-n},$$

where

$$N_S(t) = N_S(\mathcal{Y};\vec{\mathcal{R}};\omega;t) = \sum_{\substack{1\le k\le N:\\ \mathbf{x}_k(t)\in S}} 1$$

is the corresponding point-counting function, the perfect analog of (3.11).

□

Applying the classical DeMoivre–Laplace theorem with error term, Theorem 11.1 implies the following result (CLT stand for the central limit theorem).

Theorem 11.2. (asymptotic CLT) (a) *Under the conditions of Theorem* 11.1, *with* $N_S(t) = N_S(\mathcal{Y}; \vec{\mathcal{R}}; \omega; t)$,

$$\sup_{\infty < y < \infty} \left| \lim_{T \to \infty} \frac{1}{T} \text{length} \left\{ t \in [0, T] : \frac{N_S(t) - pN}{\sqrt{pqN}} \leq y \right\} - \frac{1}{\sqrt{2\pi}} \int_{-\infty}^{y} e^{-u^2/2} \, du \right| \leq \frac{1}{\sqrt{pqN}}, \tag{11.1}$$

where $p = \text{vol}(S)$ *and* $q = 1 - p$. □

Note that the error term $O(1/\sqrt{pqN})$ in (11.1) is the inevitable discrepancy between the (continuous) standard normal distribution and the (discrete) normalized binomial distribution with parameters N and $0 < p < 1$.

These are asymptotic results, describing the (very!) long-term time evolution of a large N-particle system, starting from an arbitrary initial point configuration (e.g., "Big Bang", where all particles start from the same point). Theorem 11.2 has the message of *asymptotic time-lapse randomness* as $t \to \infty$.

The same argument proves a far-reaching generalization of Theorems 11.1 and 11.2, extending the binomial distribution to the multinomial distribution.

Theorem 11.3. (multinomial version) *Let* $k \geq 2$ *be an integer, and let* $S_1, S_2, \ldots, S_k$ *be arbitrary pairwise disjoint Jordan measurable subsets of the unit cube* $[0,1]^3$. *Let* $0 < p_i = \text{vol}(S_i) < 1$, $1 \leq i \leq k$.

(a) *Under the condition of Theorem* 3.2, *with* $N_{S_i}(t) = N_{S_i}(\mathcal{Y}; \mathcal{V}; t)$,

$$\lim_{T \to \infty} \frac{1}{T} \text{length} \{ t \in [0, T] : \ N_{S_i}(t) = n_i \quad \textit{for all } 1 \leq i \leq k \}$$
$$= \frac{N!}{n_1! \cdots n_k! (N - n_1 - \cdots - n_k)!} \times p_1^{n_1} \cdots p_k^{n_k} (1 - p_1 - \cdots - p_k)^{N - n_1 - \cdots - n_k}. \tag{11.2}$$

(b) *Similarly, under the condition of Theorem* 11.1, *with* $N_{S_i}(t) = N_{S_i}(\mathcal{Y}; \vec{\mathcal{R}}; \omega; t)$, *we have the perfect analog of* (11.2). □

Indeed, to prove Theorem 11.3 we just apply either the Kronecker–Weyl theorem (via unfolding) or the Weak Uniformity Theorem in the configuration space.

The same argument gives the generalization of Theorem 11.1 where different particles may have different masses. Let μ_k, $1 \le k \le N$ denote the mass of the kth particle. For simplicity we assume that average mass is one, i.e., $\sum_{k=1}^{N} \mu_k = N$. Then the point-counting function $N_S(t)$ is replaced by the mass-counting function $N_S(\text{mass}; t)$:

$$N_S(\text{mass}; t) = \sum_{\substack{1 \le k \le N: \\ \mathbf{x}_k(t) \in S}} \mu_k. \tag{11.3}$$

It is natural to expect that, when the time evolution of the system reaches "equilibrium", the mass-counting function $N_S(\text{mass}; t)$ gets very close to pN, and the discrepancy exhibits "random fluctuation" with "standard deviation" $\sigma\sqrt{pqN}$ as t runs in a long time interval; here

$$\sigma^2 = \frac{1}{N}\sum_{k=1}^{N} \mu_k^2, \quad \text{and for simplicity } \frac{1}{N}\sum_{k=1}^{N} \mu_k = 1. \tag{11.4}$$

Again the Kronecker–Weyl theorem and the Weak Uniformity Theorem justify this intuition; see (11.7) below. The first step is the following perfect analog of Theorem 11.3 (the proof is the same).

Theorem 11.4. (mass-counting version) *Consider the general case of arbitrary masses; see* (11.3)– (11.4).

(a) *Under the conditions of Theorem* 3.2, *for every real number* $-\infty < y < \infty$,

$$\lim_{T\to\infty} \frac{1}{T}\text{length}\left\{t \in [0,T] : \frac{N_S(\text{mass}; t) - pN}{\sigma\sqrt{pqN}} \le y\right\}$$
$$= \sum_{\substack{I \subset \{1,2,\dots,N\}: \\ \sum_{i\in I} \mu_k \le pN + y\sigma\sqrt{pqN}}} p^{|I|} q^{N-|I|}, \tag{11.5}$$

where $p = \text{vol}(S)$, $q = 1 - p$, *and* $|I|$ *denotes the number of elements of a finite set* I.

(b) *Under the conditions of Theorem* 11.1, *we have the perfect analog of* (11.5). □

The right-hand side of (11.5) is the distribution of the sum $Y = \sum_{k=1}^{N} X_i$ of N independent random variables X_k, $1 \le k \le N$ such that $\Pr[X_k = \mu_k] = p$ and $\Pr[X_k = 0] = q$. Then the expectation of Y is pN, and the variance of Y is $\sigma^2 pqN$. The well-known Berry–Esseen inequality in probability theory (see e.g., in [F71]) says that

$$\sup_{-\infty<y<\infty} \left| \sum_{\substack{I\subset\{1,2,\ldots,N\}: \\ \sum_{i\in I}\mu_k \le pN + y\sigma\sqrt{pqN}}} p^{|I|}q^{N-|I|} - \frac{1}{\sqrt{2\pi}}\int_{-\infty}^{y} e^{-u^2/2}\,du \right|$$

$$\le \frac{\rho^3/\sigma^3}{\sqrt{pqN}}, \quad \text{where } \rho^3 = \frac{1}{N}\sum_{k=1}^{N}\mu_k^3. \tag{11.6}$$

Combining (11.5) and (11.6), we obtain

$$\lim_{T\to\infty}\frac{1}{T}\text{length}\left\{t\in[0,T]: \ \frac{N_S(\text{mass};t) - pN}{\sqrt{pqN}} \le y\right\}$$

$$= \frac{1}{\sqrt{2\pi}}\int_{-\infty}^{y} e^{-u^2/2}\,du + O\left(\frac{\rho^3/\sigma^3}{\sigma\sqrt{pqN}}\right). \tag{11.7}$$

Let us return to Theorem 11.2. If the volume $p = \text{vol}(S)$ of the test set S is very small — say, it is around $1/N$ — then of course (11.1) becomes useless (since the error term $O(1/\sqrt{Npq})$ is in the range of 1). However, Theorem 11.1 still works, and implies the following analog of Theorem 3.4.

Theorem 11.5. (asymptotic Poisson law) *Write $\lambda = pN = \text{vol}(S)N$. Under the conditions of Theorem* 11.1, *for every integer $\ell \ge 0$,*

$$\lim_{T\to\infty}\frac{1}{T}\text{length}\{t\in[0,T]: \ N_S(t) = \ell\} = \frac{\lambda^\ell e^{-\lambda}}{\ell!} + O(\lambda^2/N). \tag{11.8}$$

Note that (11.8) immediately follows from Theorem 11.1 combined with the simple approximation

$$\binom{N}{\ell}(\lambda/N)^\ell(1-\lambda/N)^{N-\ell} = \frac{\lambda^\ell e^{-\lambda}}{\ell!} + O(\lambda^2/N).$$

If $\lambda = pN = \text{vol}(S)N$ is in the "constant range" (i.e., the volume of the test set S is very small), then by (11.8), the point-counting function $N_S(t)$ has Poisson distribution with negligible error (since the error term $O(\lambda^2/N)$ is very small).

In the general case, where there are several sets $S_1, S_2, \ldots, S_k$, $k \geq 2$, the multinomial Theorem 11.3 gives the following result.

Theorem 11.6. (multinomial Poisson) *Let $k \geq 2$ be an integer, and let $S_1, S_2, \ldots, S_k$ be k small pairwise disjoint Jordan measurable subsets of the unit cube $[0,1]^3$ such that $0 < \lambda_i = p_i N = \mathrm{vol}(S_i)N$, $1 \leq i \leq k$ are all in the constant range.*

(a) *Under the conditions of Theorem* 3.2,

$$\begin{aligned}\lim_{T\to\infty} &\frac{1}{T}\mathrm{length}\left\{t \in [0,T]:\ N_{S_i}(t) = \ell_i \quad \text{for all } 1 \leq i \leq k\right\}\\ &= \frac{N!}{\ell_1!\cdots\ell_k!(N-\ell_1-\cdots-\ell_k)!}(\lambda_1/N)^{\ell_1}\cdots\\ &\quad \times(\lambda_k/N)^{\ell_l}(1-(\lambda_1+\cdots+\lambda_k)/N)^{N-\ell_1-\cdots-\ell_k}\\ &= \prod_{i=1}^{k}\frac{\lambda_i^{\ell}e^{-\lambda_i}}{\ell!}+O(1/N),\end{aligned} \tag{11.9}$$

where the implicit constant in the error term $O(1/N)$ depends only on the given constants $\lambda_1, \ldots, \lambda_k$.

(b) *Similarly, under the conditions of Theorem* 11.3, *we have the perfect analog of* (11.9). □

The product at the beginning of the last line in (11.9) tells us that the long-term time evolution of the system exhibits "asymptotic independence" of "very small" disjoint test sets S_i, $1 \leq i \leq k$ (of course we are talking about statistical independence).

Notice that "asymptotic independence" certainly fails for "large" subsets S_i. The simplest example is when we have two sets, and they are complements $S_2 = [0,1]^3 \setminus S_1$. Then we have a deterministic relation between the number of point billiards in S_1 and in S_2 (the sum is constant), which contradicts (statistical) independence.

What happens if the disjoint test sets S_i, $1 \leq i \leq k$ are still "small" but not "very small"?

Here is a positive result. Assume that the volumes $p_i = \mathrm{vol}(S_i)$, $1 \leq i \leq k$ are all in the range of $1/\sqrt{N}$ (i.e., each test set is expected to contain around $\sqrt{N}$ point billiards), then

$$\lim_{T\to\infty}\frac{1}{T}\text{length}\left\{t\in[0,T]:\ \frac{N_{S_i}(t)-p_iN}{\sqrt{p_i(1-p_i)N}}\le y_i\quad\text{for all } 1\le i\le k\right\}$$

$$=\prod_{i=1}^{k}\mathcal{N}(y_i)+O\left(N^{-1/4}\log N\right) \tag{11.10}$$

for all fixed constants $-\infty<y_1,y_2<\infty$, where log stands for the natural (base e) logarithm and $\mathcal{N}(y)$ denotes the normal distribution:

$$\mathcal{N}(y)=\frac{1}{\sqrt{2\pi}}\int_{-\infty}^{y}e^{-u^2/2}\,du.$$

Note that (11.10) follows easily from Theorem 11.3 via routine estimation. We outline this simple — but somewhat long — estimation in the special case of two disjoint midsize sets S_1 and S_2 (i.e., $k=2$) with $p_i=\text{vol}(S_i)\approx 1/\sqrt{N}$, $i=1,2$. In view of Theorem 11.3 we have to estimate the ratio

$$\frac{\frac{N!}{n_1!n_2!(N-n_1-n_2)!}p_1^{n_1}p_2^{n_2}(1-p_1-p_2)^{N-n_1-n_2}}{\binom{N}{n_1}p_1^{n_1}(1-p_1)^{N-n_1}\cdot\binom{N}{n_2}p_2^{n_2}(1-p_2)^{N-n_2}}, \tag{11.11}$$

where the product in the denominator of (11.11) represents independence. To prove (11.10) (for $k=2$), it suffices to show that the ratio (11.11) is

$$1+O\left(N^{-1/4}\log N\right)\ \text{for}$$

$$p_iN-\log N\sqrt{p_iN}\le n_i\le p_iN+\log N\sqrt{p_iN},\quad i=1,2. \tag{11.12}$$

Indeed, the binomial distribution is close to the normal distribution if Np_i is "large" (see the denominator of (11.11)), and $|n_i-p_iN|>\log N\sqrt{p_iN}$ represents a "large deviation" with probability around

$$e^{-(\log N)^2/2}=N^{-\log N/2},$$

which is *superpolynomially small* in terms of N, so it is totally negligible.

We have

$$\text{ratio (11.11)}=\frac{(N-n_1)!(N-n_2)!(1-p_1-p_2)^{N-n_1-n_2}}{(N-n_1-n_2)!N!(1-p_1)^{N-n_1}(1-p_2)^{N-n_2}}$$

$$=\frac{(1-p_1-p_2)^{N-n_1-n_2}}{(1-p_1)^{N-n_1}(1-p_2)^{N-n_2}}\prod_{\ell=1}^{n_1}\left(1-\frac{n_2}{N-n_1+\ell}\right)$$

$$= \frac{\left(\frac{1-p_1-p_2}{(1-p_1)(1-p_2)}\right)^{N-n_1-n_2}}{(1-p_1)^{n_2}(1-p_2)^{n_1}} \prod_{\ell=1}^{n_1}\left(1-\frac{n_2}{N-n_1+\ell}\right)$$

$$= \frac{\left(1-\frac{p_1p_2}{(1-p_1)(1-p_2)}\right)^{N-n_1-n_2}}{(1-p_1)^{n_2}(1-p_2)^{n_1}} \prod_{\ell=1}^{n_1}\left(1-\frac{n_2}{N-n_1+\ell}\right)$$

$$= \exp\left((N-n_1-n_2)\log\left(1-\frac{p_1p_2}{(1-p_1)(1-p_2)}\right)\right.$$

$$-n_1\log(1-p_2)-n_2\log(1-p_1)$$

$$\left.+\sum_{\ell=1}^{n_1}\log\left(1-\frac{n_2}{N-n_1+\ell}\right)\right),$$

where of course $\exp(x) = e^x$. Using the Taylor series

$$\log(1-x) = -(x+x^2/2+x^3/3+\cdots) = -x+O(x^2),$$

we have

$$\text{ratio (11.11)} = \exp\left(-(N-n_1-n_2)\left(\frac{p_1p_2}{(1-p_1)(1-p_2)}+O(p_1^2p_2^2)\right)\right.$$

$$+n_1p_2+O(n_1p_2^2)+n_2p_1+O(n_2p_1^2)$$

$$\left.-\sum_{\ell=1}^{n_1}\frac{n_2}{N-n_1+\ell}+O(n_1n_2^2N^{-2})\right). \tag{11.13}$$

Estimating the Riemann sum with the integral, we have

$$\sum_{\ell=1}^{n_1}\frac{n_2}{N-n_1+\ell} = n_2\int_0^{n_1}\frac{1}{N-x}\,dx+O(n_2/N)$$

$$= n_2\log\left(\frac{N}{N-n_1}\right)+O(n_2/N)$$

$$= -n_2\log\left(1-\frac{n_1}{N}\right)+O(n_2/N)$$

$$= \frac{n_1n_2}{N}+O(n_1^2n_2N^{-2})+O(n_2/N). \tag{11.14}$$

Using (11.14) in (11.13), we have

$$\text{ratio (11.19)} = \exp\Bigg(\left(\frac{Np_1p_2}{(1-p_1)(1-p_2)} - \frac{n_1n_2}{N}\right)$$
$$+ (n_1p_2 - Np_1p_2) + (n_2p_1 - Np_2p_1)$$
$$+ O(Np_1^2p_2^2) + O(n_1p_2^2) + O(n_2p_1^2) + O(n_1n_2^2N^{-2})$$
$$+ O(n_1^2n_2N^{-2}) + O(n_2/N)\Bigg). \tag{11.15}$$

By using the condition

$$n_i = p_iN \pm \log N\sqrt{p_iN},\ i = 1,2$$

(see (11.12)) in (11.15), we conclude

$$\text{ratio (11.11)} = \exp\left(O\left(\log N\sqrt{Np_1}p_2\right) + O\left(\log N\sqrt{Np_2}p_1\right)\right)$$
$$= 1 + O\left(N^{-1/4}\log N\right),$$

which is exactly what we wanted to prove (see (11.12)).

Consider now the 3-dimensional *general* torus model introduced in (10.1)–(10.4). Replacing the application of the Kronecker theorem in the proof of Theorem 3.1 with an application of Theorem 10.1 ("Weak Uniformity Theorem"), we obtain the following analog of Theorem 3.1.

Theorem 11.7. (another superdiscrepancy result) *Consider the 3-dimensional general N-particle torus model drifting away from the start; see* (10.1)–(10.4). *Then for almost every vector of rotations*

$$\omega = (\vartheta_1, \vartheta_2, \ldots, \vartheta_N) \in \Omega = (\mathrm{SO}(3))^N,$$

the time evolution of the system exhibits the $(B_1, \ldots, B_N)$-type superdiscrepancy described in Theorem 3.1. □

The next result shows that some application of the superdiscrepancy result Theorem 3.1 *does* require ridiculously large time, i.e., in some cases unrealistic time is inevitable. Assume that the $3N$ initial velocity coordinates of an N-particle point-billiard system are linearly independent over the rationals; let $B \subset [0,1]^3$ be an arbitrary box in the unit cube; and, finally, let $I \subset \{1,2,\ldots,N\}$ be an arbitrary subset. Here is a simple corollary of Theorem 3.1: starting from an arbitrary initial point configuration ($t = 0$), it will eventually happen at some time instant $t > 0$ that the ith particle is in B if and only if $i \in I$. That is, the index-set I is exactly the set of particles in B at time t.

For simplicity assume that the box $B \subset [0,1]^3$ has volume $1/2$. For a typical time instant t the number of particles in B is expected to be close to $N/2$. This leads us to the following natural question: How long does it take to have *all possible* $N/2$-element subsets to show up as the set of particles in B? We prove that, under very mild conditions, the required time is *exponentially large* in terms of N.

More precisely, we say that the time evolution of an N-particle billiard system is $(B; N/2)$-*complete* in a time interval $[0, T]$ (where $t = 0$ is the start), if for every $N/2$-element subset $I \subset \{1, 2, \ldots, N\}$ there exists a time instant $0 \le t(I) \le T$ such that the ith particle is in B at time $t(I)$ if and only if $i \in I$.

Let B_0 denote the cube of volume $1/2$ centered at $(1/2, 1/2, 1/2)$

$$B_0 = \left[\frac{1}{2} - 2^{-4/3}, \frac{1}{2} + 2^{-4/3}\right]. \tag{11.16}$$

Theorem 11.8. *Assume that at least* 99% *of the speeds of an* N*-particle point-billiard system* $\{\mathbf{x}_k(t) : 1 \le k \le N\}$ *are less than* 100, *and the time evolution is* $(B_0; N/2)$*-complete in a time interval* $[0, T]$, *where* B_0 *is the cube defined in* (11.16). *Then*

$$T \ge \frac{2^{99N/100}}{10^3 N^2 e^{N/9800}} = \exp\left(\left(\frac{99\log 2}{100} - \frac{1}{9800}\right) N\right) 10^{-3} N^{-2}.$$

The same holds for torus lines instead of billiard paths.

Remarks. The crucial fact here is that the lower bound is exponential in N. So, if N is in the range of the Avogadro number $N = 10^{24}$, or larger, then the lower bound gives a totally unrealistic, absurd time scale.

The choice of the constants "99%" and "100" are accidental. We could choose basically any other constants, and the proof still works providing an exponential lower bound.

On an intuitive level Theorem 11.8 is perfectly plausible, almost trivial. Indeed, since there are $\binom{N}{N/2} \approx 2^N N^{-1/2}$ $N/2$-element subsets of the set of N particles, and the particles have constant speed, the exponential (in terms of N) lower bound for the necessary time T is not surprising at all. The precise proof of Theorem 11.8 is also simple; see below.

It is a trivial by-product of Theorem 11.8 that *perfect time-lapse randomness*, or even *nearly perfect time-lapse randomness*, requires the same totally unrealistic, exponential time scale. Indeed, let $S \subset [0,1]^3$ be a Jordan measurable subset in the unit cube with volume 1/2. If the $3N$ initial velocity coordinates of the point-billiard system are linearly independent over the rationals, then for every $1 > \varepsilon > 0$ there is a finite threshold T_0 — depending on the initial velocity coordinates, on S, and on ε — such that

$$1-\varepsilon \le \frac{\text{length}\,\{t \in [0,T] :\ \mathbf{x}_k(t) \in S \iff k \in I_1\}}{\text{length}\,\{t \in [0,T] :\ \mathbf{x}_k(t) \in S \iff k \in I_2\}} \le \frac{1}{1-\varepsilon} \tag{11.17}$$

holds for any two different $N/2$-element subsets $I_i \subset \{1,2,\ldots,N\}$, $i=1,2$ and for every $T \ge T_0$. We say that the time evolution of a billiard system exhibits *nearly perfect time-lapse randomness* if (11.17) holds for some $1 > \varepsilon > 0$ and T_0. The message of Theorem 11.8 is that it is *impossible* to have nearly perfect time-lapse randomness on a realistic time scale. Indeed, for every $1 > \varepsilon > 0$, Theorem 11.8 applies and gives the lower bound

$$T_0 \ge \exp\left(\left(\frac{99\log 2}{100} - \frac{1}{9800}\right)N\right)10^{-3}N^{-2},$$

which is exponential in N, i.e., totally unrealistic.

This negative result raises the natural question: what kind of randomness can we prove on a realistic time scale? In the physical world most systems reach equilibrium very fast; if nearly perfect randomness is unrealistic, how can we still justify the physicist's intuition *equilibrium = randomness*? We will address this very interesting problem with rigorous theorems later in Chap. 3. For example, Theorem 19.1 can be interpreted as a "realistic time version" of the CLT-result Theorem 11.2. We also prove that large billiard systems demonstrate time-lapse randomness in the form of exhibiting a "multiple mixing" type property on a constant time scale (independent of N); see *multiple mixedupness* in Sec. 12.

Proof of Theorem 11.8. We show that, under the condition

$$T < \frac{2^{99N/100}}{10^3N^2}e^{-N/9800}$$

there exists a time interval $[a,b] \subset [0,T]$ with $b-a \le 10^{-3}$ and a particle — the ith particle, say — that has speed less than 100, and in the time interval

$a \le t \le b$ the ith particle makes the following in-out-in motion: at time $t = a$ it is in B_0, later at some time $t = c \in (a,b)$ it is outside of B_0, and finally at time $t = b$ it is again in B_0. A slow particle (= billiard) with speed less than 100 simply cannot do this in-out-in motion in such a short time, which is a contradiction. Indeed, B_0 is relatively far from the six faces ("walls") of the unit cube, the distance is $> 1/10$ — so, leaving B_0 at some instant t with $t_{j_1} \le t \le t(I_{j_2})$, and moving toward the "walls", a slow billiard does not have the time to bounce back from the walls to be in B_0 again (since $b - a \le 10^{-3}$).

First consider the special case where every particle has speed ≤ 100. Assume that the particles are labelled $1, 2, \ldots, N$, let I be an arbitrary $N/2$-element subset of $\{1, 2, \ldots, N\}$ and let $t(I) \in [0, T]$ be a time instant such that at time $t = t(I)$ the index set of particles in B_0 is exactly I. We write these times in increasing order

$$0 \le t_1 = t(I_1) < t_2 = t(I_2) < \cdots < t_m = t(I_m) \le T \quad \text{with} \quad m = \binom{N}{N/2}, \tag{11.18}$$

and accordingly we define the following 0,1 matrix $\mathcal{A}$ of m rows and N columns. The entry $a_{i,j} \in \mathcal{A}$ of the matrix is 1 if at time $t = t_i$ the jth particle is in B_0, and 0 otherwise. A *vertical* 1-*block* in $\mathcal{A}$ is a maximal line-segment of consequtive 1's in a column; formally, $a_{i,j} = 1$ for some interval $k \le i \le \ell$ such that either $a_{k-1,j} = 0, a_{\ell+1,} = 0$, or $k = 1$, or $\ell = m$ (j is arbitrary). It is easy to see that the number of *vertical* 1-*blocks* in $\mathcal{A}$ is $\ge m = \binom{N}{N/2}$. Indeed, for every row i there is an entry $a_{i,j} = 1$ such that either $a_{i-1,j} = 0$ or $i = 1$, so $a_{i,j} = 1$ is the top member of a *vertical* 1-*block* in $\mathcal{A}$. Since there are N columns, there is a column i_0 that contains at least $\binom{N}{N/2}/N$ *vertical* 1-*blocks*. If $T < 10^{-3}\binom{N}{N/2}/N$ then by the pigeonhole principle there exist $1 \le j_1 < j_2 \le m = \binom{N}{N/2}$ with $j_2 - j_1 \ge 2$ such that (see (11.18)) $t_{j_2} - t_{j_1} \le 10^{-3}$, $a_{i_0,j_1} = 1$ is the top member of a *vertical* 1-*block* in $\mathcal{A}$, and $a_{i_0,j_2} = 1$ is the top member of another *vertical* 1-*block* in $\mathcal{A}$. It follows that the i_0th particle makes the following in-out-in motion: at time $t = t_{j_1}$ it is in B_0, later at time $t = t_{j_2-1}$ it is outside of B_0, and finally at time $t = t_{j_2}$ it is again in B_0 — a contradiction. The contradiction proves the lower bound $T \ge 10^{-3}\binom{N}{N/2}/N$, which is exponential in N.

Next we explain how to modify the argument in the general case. By hypothesis, there are at least $N_1 \ge 99N/100$ "slow" particles with speed ≤ 100. Throwing out the "fast" particles, we can repeat the previous argument with $m = \binom{N_1}{N/2}$, and obtain the lower bound

$$T \geq \frac{\binom{N_1}{N/2}}{10^3 N_1}. \tag{11.19}$$

Clearly

$$\begin{aligned}\binom{N_1}{N/2} &\geq \binom{N}{N/2} \prod_{j=0}^{N-N_1+1} \frac{N-j-N/2}{N-j} \geq \binom{N}{N/2}\left(\frac{N_1-N/2}{N_1}\right)^{N-N_1}\\ &\geq \binom{N}{N/2}\left(\frac{1}{2}\left(\frac{98}{99}\right)\right)^{N-N_1} \geq \frac{2^N}{N}2^{-(N-N_1)}\left(1+\frac{1}{98}\right)^{-(N-N_1)}\\ &\geq \frac{2^{N_1}}{N}e^{-N/9800} \geq \frac{2^{99N/100}}{N}e^{-N/9800},\end{aligned} \tag{11.20}$$

where in the last steps we applied the elementary facts $\binom{n}{n/2} \geq 2^n/n$ and $1 + x \leq e^x$. Using (11.20) in (11.19), Theorem 11.8 follows. □

We conclude Sec. 11 with a generalization of Theorem 11.8. A key step in the proof of Theorem 11.8 was the impossibility of the "in-out-in motion" by a point billiard. This argument breaks down if the point billiards are replaced by tiny solid balls ("hardcore model"), since the balls can collide, and the trajectory of a single billiard ball is more like a chaotic zig-zag than a curve of wall-to-wall line segments. The average speed is one, so in unit time a typical point billiard moves from wall to wall on a line segment. A tiny billiard ball, on the other hand, has many particle–particle collisions in unit time. In gas models, the number of collisions in unit time is subtantially less than N (= the number of particles in the box). This fact motivates the following definition. Assume that every particle has a piecewise linear continuous trajectory in the unit cube $[0, 1]^3$ such that

(1) the particle moves with constant speed on a linear piece, where the speed depends on the piece, and

(2) the particle travels less than N^2 linear pieces in unit time.

We refer to a system with properties (1) and (2) as a *collision-like* system. When a particle moves from one linear piece to another linear piece, we call it a *break* (it imitates a particle–particle collision).

Of course, the choice of N^2 is accidental; we could also write N^{10}, say. Note that the upper bound N^2 is already "very generous".

Generalization of Theorem 11.8. *Consider a collision-like N-particle system. Assume that the time evolution of the system in a time interval $[0,T]$ is $(B_0; N/2)$-complete, where B_0 is the cube of volume $1/2$ in Theorem 11.8. Moreover, assume that for every time instant t in $0 \le t \le T$ at least 99% of the speeds are less than $100N^2$. Then*

$$T \geq \frac{2^{99N/100}}{10^3 N^7} e^{-N/9800} = \exp\left(\left(\frac{99\log 2}{100} - \frac{1}{9800}\right) N\right) 10^{-3} N^{-7}.$$

Remark. Again the choices of $100N^2$ and 99% are accidental; we could choose other constants and also a higher power of N. Again the lower bound is ridiculously large if N is in the range of the Avogadro number (or larger).

Proof. We repeat the proof of Theorem 11.8 with some minor modifications at the end. The first modification is that we change $b - a \le 10^{-3}$ to $b - a \le 10^{-3}N^{-2}$.

Since the system is *collision-like* satisfying properties (1) and (2) above, and the length of the time-interval $a \le t \le b$ is less than one, there are at most $H < N \cdot N^2 = N^3$ *breaking points*

$$a \le t(1) < t(2) < \cdots < t(H) \le b$$

such that at time $t = t(h)$ the trajectory of some particle breaks. This explains the first extra factor N^3 in the denominator of the lower bound. The second extra factor N^2 in the denominator comes from replacing $b-a \le 10^{-3}$ to $b - a \le 10^{-3}N^{-2}$.

This is why we have the extra factor $N^5 = N^3 \cdot N^2$, replacing N^2 in the denominator of the lower bound in Theorem 11.8 with $N^7 = N^5 \cdot N^2$.

Again we apply the pigeonhole principle, noting that in "break-free" time intervals we can save the basic idea: a slow particle with speed less than $100N^2$ simply cannot do the in-out-in motion in a short "break-free" time interval. □

12. Short-Term Time-Lapse Randomness: Multiple Mixedupness (I)

We discuss a "time-lapse randomness" type result motivated by (statistical) independence (= product rule); we call it *multiple ε-mixedupness*. For comparison, first we recall the concept of *ε-perfect time-lapse randomness* introduced in Sec. 11; see (11.17). The Kronecker–Weyl equidistribution theorem implies (via unfolding) the following result: If the initial velocity coordinates of an N-particle point-billiard system $\{\mathbf{x}_k(t) : 1 \le k \le N\}$ are linearly independent over the rationals, and $S \subset [0,1]^3$ is a Jordan measurable subset of the unit cube with volume $0 < p = \mathrm{vol}(S) < 1$, then for every $1 > \varepsilon > 0$ there is a finite threshold T_0 — depending on the initial velocity coordinates, on S, and on ε — such that

$$1 - \varepsilon \le \frac{\mathrm{length}\,\{t \in [0,T] : \mathbf{x}_k(t) \in S \iff k \in I\}}{p^{|I|}(1-p)^{N-|I|}} \le \frac{1}{1-\varepsilon} \tag{12.1}$$

holds for every subset $I \subset \{1, 2, \ldots, N\}$ and for every $T \ge T_0$. Equation (12.1) has the interpretation that the time evolution of a billiard system exhibits *ε-perfect time-lapse randomness* for all $T \ge T_0$.

Theorem 11.8 implies that it is impossible to have ε-perfect time-lapse randomness on a realistic time scale. Indeed, for every $1 > \varepsilon > 0$, Theorem 11.8 applies, and yields the lower bound

$$T_0 \ge \exp\left(\left(\frac{99\log 2}{100} - \frac{1}{9800}\right)N\right) 10^{-3} N^{-2},$$

which is exponential in N, i.e., totally unrealistic.

In sharp contrast with this "extremely long-term time-lapse randomness" result (exponential time in terms of N), in this section we prove a "short-term time-lapse randomness" result that goes as follows. Starting from an arbitrary initial configuration, the majority of the time evolutions of a point billiard system exhibits *multiple ε-mixedupness* in *constant time* independent of the number of particles N (it depends only on $\varepsilon > 0$). Constant time is certainly realistic.

The formal definition of *multiple ε-mixedupness* is the following. For simplicity we start with the special case where the test sets are axis-parallel boxes; we switch to the general case of arbitrary measurable test sets at the end of the section. Let $k \ge 1$ be an arbitrary integer, let $B_j \subset [0,1]^3$, $1 \le j \le k$ be an arbitrary sequence of k axis-parallel boxes in the unit cube such that $0 < \mathrm{vol}(B_j) = p_j < 1$, $1 \le j \le k$, and let $0 < t_1 < t_2 < \cdots < t_k$ be k times in increasing order. Given an $\varepsilon > 0$, we say that the time evolution of a point billiard system, starting from a given initial configuration $\mathcal{Y} = \{\mathbf{y}_1, \ldots, \mathbf{y}_N\}$ at $t = 0$ with some given initial velocities, exhibits *multiple ε-mixedupness* at $0 < t_1 < t_2 < \cdots < t_k$ with respect to the sequence $B_j \subset [0,1]^3$, $1 \le j \le k$ of the given test sets, if the number of particles that are simultaneously in B_j at $t = t_j$ for every $j = 1, 2, \ldots, k$ is between

$$\left(\prod_{j=1}^{k} p_j - \varepsilon\right) N \quad \text{and} \quad \left(\prod_{j=1}^{k} p_j + \varepsilon\right) N. \tag{12.2}$$

The concept of *multiple ε-mixedupness* is motivated by the physical act of stirring a liquid, and the product form of the density $\prod_{j=1}^{k} p_j$ represents "perfect stirring": perfect time-lapse independence via the product rule.

Notice that (12.1) demonstrates ε-approximation of the product rule (= time-lapse statistical independence) in a very strong sense. On the other hand, (12.2) demonstrates ε-approximation of the product rule (= time-lapse statistical independence) in a different *weaker* sense. This explains the dramatic drop from exponential time to constant time in the required time scale.

What we can actually prove is the *typicality* of multiple ε-mixedupness. Here is the precise statement of what we prove. Let ε be a positive real with $0 < \varepsilon < 1$ and $k \ge 1$ be an integer. Then there exist absolute constants $C_1(\varepsilon; k)$ and $C_2(\varepsilon; k)$, both depending only on ε and k, such that, given an arbitrary real $T \ge C_1(\varepsilon; k)$, given an arbitrary sequence $B_j \subset [0,1]^3$, $1 \le j \le k$ of k axis-parallel boxes satisfying

$$\prod_{j=1}^{k} p_j = \prod_{j=1}^{k} \mathrm{vol}(B_j) > \varepsilon > 0,$$

given an arbitrary initial configuration $\mathcal{Y} = \{\mathbf{y}_1, \ldots, \mathbf{y}_N\} \subset [0,1]^3$ with $N \ge C_2(\varepsilon; k)$, and given an arbitrary real $T_0 \ge 0$, for at least $1 - \varepsilon$ part of all choices of k times $t_1 < \cdots < t_k$ with

$$\begin{aligned} T_0 < t_1 < T_0 + T < t_2 < T_0 + 2T < t_3 < T_0 + 3T \\ < \cdots < T_0 + (k-1)T < t_k < T_0 + kT, \end{aligned} \tag{12.3}$$

the time evolution for at least $1 - \varepsilon$ part of the initial velocities of an N-particle billiard system — starting from $\mathcal{Y}$ at $t = 0$ — exhibits *multiple ε-mixedupness* at $t_1, t_2, \ldots, t_k$ (see (12.3)) with respect to the given sequence $B_j \subset [0,1]^3$, $1 \le j \le k$ of test sets.

Here *at least* $1-\varepsilon$ *part of all choices of* k *times* $t_1 < \cdots < t_k$ *with*

$$T_0 < t_1 < T_0 + T < t_2 < T_0 + 2T < t_3 < T_0 + 3T$$
$$< \cdots < T_0 + (k-1)T < t_k < T_0 + kT$$

precisely means that we choose a point in the k-dimensional cube $[0,T]^k$ with uniform distribution, and its jth coordinate gives $t_j - T_0 - (j-1)T$, $1 \le j \le k$. Of course, *at least* $1-\varepsilon$ *part* refers to the normalized k-dimensional volume (= Lebesgue measure), i.e., we divide by T^k.

We picked the name mixedupness because *multiple mixedupness* shows some formal similarity to the concept of *multiple mixing* measure-preserving transformation in ergodic theory. But they are not the same: *mixing* is a "soft" long-term property, which is perfectly meaningful even for one particle; *mixedupness*, on the other hand, is a "hard" quantitative property, which is meaningful for many-particle systems only, and it is used to describe the realistic short term time evolution of a (large) system. At the end of the section — after formulating Theorems 12.1 and 12.2 — we return to the question of comparing our mixedupness to the concept of mixing in ergodic theory. We show that the time-flow of our model is *not* mixing; nevertheless it exhibits multiple mixedupness in a constant time scale.

As usual, we apply unfolding (see Sec. 3) to convert the time evolution of a billiards-in-a-box system to that of a torus-billiard system. We use the general continuous Erdős–Turán–Koksma inequality (see (3.21)) in the special case where the continuous curves are torus lines. That is, we study the straight line $\mathbf{r}(t) = t\mathbf{v}$, where $\mathbf{v} = (v_1, \ldots, v_d) \in \mathbb{R}^d$ is the direction vector (and of course we take the straight line modulo one).

Erdős–Turán–Koksma Inequality for torus lines: Let $t\mathbf{v}$, $0 \le t \le T$ ($\mathbf{v} \in \mathbb{R}^d \setminus \mathbf{0}$ is fixed) be a straight line and $H \ge 1$ be an arbitrary integer; then with $f = \chi_B$, where B denotes axis-parallel boxes and, as usual, f is defined on the unit torus $I^d = [0,1)^d$ (i.e., f is periodic),

$$\max_{B = I_1 \times \cdots \times I_d \subset I^d} \left| \frac{1}{T} \int_0^T f(t\mathbf{v})\, dt - \int_{I^d} f\, d\mathrm{V} \right|$$
$$\le \left(\frac{3}{2}\right)^d \left(\frac{2}{H+1} + \sum_{\substack{\mathbf{h} \in \mathbb{Z}^d: \\ 1 \le \|\mathbf{h}\|_\infty \le H}} \frac{1}{\Lambda(\mathbf{h})} \left| \frac{1}{T} \int_0^T e^{2\pi i t \mathbf{h} \cdot \mathbf{v}}\, dt \right| \right), \tag{12.4}$$

where $\mathrm{i} = \sqrt{-1}$, $H \geq 1$ is an arbitrary integer,

$$\|\mathbf{h}\|_\infty = \max_{1 \leq j \leq d} |h_j| \tag{12.5}$$

is the maximum norm of the vector $\mathbf{h} = (h_1, \ldots, h_d) \in \mathbb{Z}^d$, and

$$\Lambda(\mathbf{h}) = \prod_{1 \leq j \leq d} (1 + |h_j|) \,. \tag{12.6}$$

In the special case of torus lines we can explicitly evaluate the key integral

$$\int_0^T e^{2\pi \mathrm{i} t \mathbf{h}\cdot\mathbf{v}}\, dt = \frac{e^{2\pi \mathrm{i} T \mathbf{h}\cdot\mathbf{v}} - 1}{2\pi \mathrm{i} \mathbf{h}\cdot\mathbf{v}},$$

which implies the upper bound

$$\left| \int_0^T e^{2\pi \mathrm{i} t \mathbf{h}\cdot\mathbf{v}}\, dt \right| \leq \min\left\{ \frac{1}{\pi |\mathbf{h}\cdot\mathbf{v}|}, T \right\}. \tag{12.7}$$

Using (12.7) in (12.4), we have with $f = \chi_B$,

$$\max_{B = I_1 \times \cdots \times I_d \subset I^d} \left| \frac{1}{T} \int_0^T \chi_B(t\mathbf{v})\, dt - \int_{I^d} \chi_B \, d\mathrm{V} \right|$$

$$\leq \left(\frac{3}{2}\right)^d \left(\frac{2}{H+1} + \sum_{1 \leq \|\mathbf{h}\|_\infty \leq H} \frac{1}{\Lambda(\mathbf{h})} \min\left\{ \frac{1}{T\pi |\mathbf{h}\cdot\mathbf{v}|}, 1 \right\} \right), \tag{12.8}$$

where of course

$$\frac{1}{T} \int_0^T \chi_B(t\mathbf{v})\, dt = \frac{1}{T} \text{length}\{t \in [0,T] : \ t\mathbf{v} \in B \text{ modulo one}\} \tag{12.9}$$

is the *relative time* the torus line spends in box B, and

$$\int_{I^d} \chi_B \, d\mathrm{V} = \mathrm{vol}_d(B) \tag{12.10}$$

is the (d-dimensional) volume of the box, representing the *expected relative time.*

We recall Archimedes's theorem (10.21): for any 3-dimensional unit vector $\mathbf{w}$ and any real number ϱ with $0 < \varrho < 1 = |\mathbf{w}|$,

$$\text{SurfaceArea}\left(\{\mathbf{v} \in \mathbf{S}^2 : \ |\mathbf{w}\cdot\mathbf{v}| \leq \varrho\}\right) = 4\pi \cdot \varrho. \tag{12.11}$$

It follows from (12.11) that, if $\mathbf{w} \in \mathbf{S}^2$ is fixed and $\mathbf{v}^* \in \mathbf{S}^2$ is a random variable uniformly distributed on the unit sphere, then the dot product $\mathbf{w} \cdot \mathbf{v}^*$ is uniformly distributed in the interval $[-1, 1]$. We refer to this remarkable geometric fact as the "uniformity property of the sphere".

Let $\mathbf{w} \in \mathbf{S}^2$ be a fixed unit vector, and let $\mathbf{v}_1^*, \mathbf{v}_2^*, \ldots, \mathbf{v}_N^* \in \mathbf{S}^2$ be N independent random variables uniformly distributed on the unit sphere. For a given integer ℓ, let $E_1(\mathbf{w}; \ell; 6 \log N)$ denote the event that the interval

$$\left[\frac{\ell - \frac{1}{2}}{N}, \frac{\ell + \frac{3}{2}}{N}\right] \tag{12.12}$$

contains at least $6 \log N$ dot products $\mathbf{w} \cdot \mathbf{v}_k^*$, $1 \le k \le N$; and for given integers ℓ and r with $2^r \ge N^{1/3}$, let $E_2(\mathbf{w}; \ell; r)$ denote the event that the interval

$$\left[\frac{\left(\ell - \frac{1}{4}\right) 2^r}{N}, \frac{\left(\ell + \frac{5}{4}\right) 2^r}{N}\right] \tag{12.13}$$

contains at least 2^r dot products $\mathbf{w} \cdot \mathbf{v}_k^*$, $1 \le k \le N$.

Combining independence with the "uniformity property of the sphere", we can estimate the probability of event $E_1(\mathbf{w}; \ell; 6 \log N)$ by using the binomial distribution as follows:

$$\begin{aligned}
\Pr[E_1(\mathbf{w}; \ell; 6 \log N)] &= \sum_{m \ge 6 \log N} \binom{N}{m} \left(\frac{1}{N}\right)^m \left(1 - \left(\frac{1}{N}\right)\right)^{N-m} \\
&\le \sum_{m \ge 6 \log N} \left(\frac{eN}{m}\right)^m \left(\frac{1}{N}\right)^m \le \sum_{m \ge 6 \log N} \left(\frac{e}{m}\right)^m \\
&\le \sum_{m \ge 6 \log N} \left(\frac{e}{6 \log N}\right)^m \le \frac{\left(\frac{e}{6 \log N}\right)^{6 \log N}}{1 - \frac{e}{\log N}},
\end{aligned} \tag{12.14}$$

where we applied the weak form of Stirling's formula $m! \ge (m/e)^m$.

Similarly, we can estimate the probability of event $E_2(\mathbf{w}; \ell; r)$ from above as follows:

$$\Pr[E_2(\mathbf{w}; \ell; r)] \le \sum_{m \ge 2^r} \binom{N}{m} \left(\frac{3 \cdot 2^{r-2}}{N}\right)^m \left(1 - \left(\frac{3 \cdot 2^{r-2}}{N}\right)\right)^{N-m}. \tag{12.15}$$

To estimate (12.15), we use the well-known large deviation inequality for the binomial distribution: if $X_1, \ldots, X_n$ are independent random variables with common distribution $\Pr[X_i = 1 \text{ or } 0] = p$ or $q = 1 - p$, then for every positive γ,

$$\Pr\left[\left|\sum_{i=1}^{n}(X_i - p)\right| \geq \gamma\sqrt{npq}\right]$$
$$= \sum_{\substack{0\leq k\leq n:\\ k\geq pn+\gamma\sqrt{pqn}}} \binom{n}{k} p^k q^{n-k} \leq \exp\left(-\frac{\gamma^2/2}{1+\frac{\gamma}{3\sqrt{npq}}}\right), \tag{12.16}$$

where $\sqrt{npq}$ is the standard deviation of the binomial distribution. We apply (12.16) with $n = N$, $p = 3 \cdot 2^{r-2}/N$ (so $p \leq 1/2$, implying $p \leq 1/2 \leq q = 1 - p$), and

$$\gamma = \frac{2^{r-2}}{\sqrt{Npq}}, \tag{12.17}$$

which implies (note that $2^r \geq N^{1/3}$)

$$\gamma > \frac{2^{r-2}}{\sqrt{Np}} = \frac{2^{(r-2)/2}}{\sqrt{3}} = \gamma_1 \geq \frac{N^{1/6}}{2\sqrt{3}} = \gamma_2 \tag{12.18}$$

and

$$\gamma \leq \frac{2^{r-2}}{\sqrt{Np/2}} = \frac{2^{(r-1)/2}}{\sqrt{3}} = \gamma_3. \tag{12.19}$$

Using (12.16)–(12.19), we obtain the following upper bound for the end sum of (12.15)

$$\sum_{m\geq 2^r} \binom{N}{m} \left(\frac{3\cdot 2^{r-2}}{N}\right)^m \left(1 - \left(\frac{3\cdot 2^{r-2}}{N}\right)\right)^{N-m}$$
$$\leq \exp\left(-\frac{\gamma_1^2/2}{1+\frac{\gamma_3}{3\sqrt{Np/2}}}\right)$$
$$= \exp\left(-\frac{\gamma_1^2/2}{1+\frac{2^{(r-1)/2}/\sqrt{3}}{3\sqrt{3}2^{(r-3)/2}}}\right) = \exp\left(-\frac{\gamma_1^2/2}{1+\frac{2}{9}}\right)$$
$$< \exp\left(-\frac{\gamma_2^2/2}{11/9}\right) < \exp\left(-\frac{N^{1/3}}{30}\right). \tag{12.20}$$

Assume that $\mathbf{w}$ runs over the special unit vectors of the form

$$\mathbf{w} = \frac{\mathbf{n}}{|\mathbf{n}|} = \frac{(n_1, n_2, n_3)}{\sqrt{n_1^2 + n_2^2 + n_3^2}} \quad \text{with } \mathbf{n} \in \mathbb{Z}^3 \setminus \mathbf{0},\ |n_i| \leq M, 1 \leq i \leq 3, \tag{12.21}$$

where $M = M(N)$ will be specified later. Consider the two union events

$$E_1(\text{bad}; 6 \log N; M) = \bigcup_{\mathbf{w}\mathbf{w} \text{ has the form } (12.21)} \bigcup_{-N \leq \ell < N} E_1(\mathbf{w}; \ell; 6 \log N) \tag{12.22}$$

and

$$E_2(\text{bad}; M) = \bigcup_{\mathbf{w} \text{ has the form } (12.21)} \bigcup_{-N^{2/3} \leq \ell < N^{2/3}} \bigcup_{\substack{r \in \mathbb{Z}: \\ N \geq 2^r \geq N^{1/3}}} E_2(\mathbf{w}; \ell; r). \tag{12.23}$$

By using (12.14) in (12.22), we have the upper bound

$$\Pr[E_1(\text{bad}; 6 \log N; M] \leq \sum_{\mathbf{w} \text{ has the form } (12.21)} \sum_{-N \leq \ell < N} \Pr[E_1(\mathbf{w}; \ell; 6 \log N)]$$

$$\leq (2M+1)^3\, 2N \frac{\left(\frac{e}{6 \log N}\right)^{6 \log N}}{1 - \frac{e}{6 \log N}}. \tag{12.24}$$

Similarly, by using (12.15) and (12.20) in (12.23), we have

$$\Pr[E_2(\text{bad}; M)] \leq \sum_{\mathbf{w} \text{ has the form } (12.21)} \sum_{-N^{2/3} \leq \ell < N^{2/3}}$$

$$\times \sum_{\substack{r \in \mathbb{Z}: \\ N \geq 2^r \geq N^{1/3}}} \Pr[E_2(\mathbf{w}; \ell; r)]$$

$$\leq (2M+1)^3 2N \exp\left(-\frac{N^{1/3}}{30}\right). \tag{12.25}$$

We say that a sequence $\mathbf{v}_1, \mathbf{v}_2, \ldots, \mathbf{v}_N \in \mathbf{S}^2$ of N unit vectors on the unit sphere satisfies the *Separation Condition* if for every nonzero lattice point $\mathbf{h} \in \mathbb{Z}^3 \setminus \mathbf{0}$ we have the following two properties: **(1)** for every integer $-N \leq \ell \leq N-1$ the interval

$$\left[\frac{\ell|\mathbf{h}|}{N}, \frac{(\ell+1)|\mathbf{h}|}{N}\right] \tag{12.26}$$

contains less than $6 \log N$ dot products $\mathbf{h} \cdot \mathbf{v}_k$, and **(2)** for every integer ℓ and integer r with $2^r \geq N^{1/3}$, the interval

$$\left[\frac{\ell 2^r |\mathbf{h}|}{N}, \frac{(\ell+1)2^r|\mathbf{h}|}{N}\right] \tag{12.27}$$

contains less than 2^r dot products $\mathbf{h} \cdot \mathbf{v}_k$, where $1 \leq k \leq N$.

Lemma 12.1. *Let $N \geq 10^{15}$; then the set of unit vectors $\mathbf{v}_1, \mathbf{v}_2, \ldots, \mathbf{v}_N \in \mathbf{S}^2$ that satisfy the Separation Condition (12.26)–(12.27) represents an overwhelming majority: it is at least $1 - N^{-10}$ part of the total.*

Proof. Let $\mathbf{w} = (w_1, w_2, w_3) \in \mathbf{S}^2$ be an arbitrary unit vector. By Dirichlet's simultaneous approximation theorem, there exist integers n_1, n_2, n_3 such that

$$\left| w_i - \frac{n_i}{N} \right| \leq \frac{1}{N^{4/3}}, \quad 1 \leq i \leq 3. \tag{12.28}$$

Let $\mathbf{v}_1, \mathbf{v}_2, \ldots, \mathbf{v}_N \in \mathbf{S}^2$ be an arbitrary sequence of N unit vectors. With $\mathbf{n} = (n_1, n_2, n_3)$ we have for every $1 \leq k \leq N$,

$$\left| \mathbf{w} \cdot \mathbf{v}_k - \frac{1}{N}\mathbf{n} \cdot \mathbf{v}_k \right| \leq \frac{3}{N^{4/3}}, \tag{12.29}$$

where we used (12.28).

Again by (12.28),

$$\frac{3}{N^{4/3}} \geq \left| |\mathbf{w}| - \frac{1}{N}|\mathbf{n}| \right| = \left| 1 - \frac{1}{N}|\mathbf{n}| \right|. \tag{12.30}$$

By (12.28) and (12.30),

$$\begin{aligned}
\left| \mathbf{w} \cdot \mathbf{v}_k - \frac{1}{|\mathbf{n}|}\mathbf{n} \cdot \mathbf{v}_k \right| &\leq \left| \mathbf{w} \cdot \mathbf{v}_k - \frac{1}{N}\mathbf{n} \cdot \mathbf{v}_k \right| + \left| \frac{1}{N}\mathbf{n} \cdot \mathbf{v}_k - \frac{1}{|\mathbf{n}|}\mathbf{n} \cdot \mathbf{v}_k \right| \\
&\leq 3N^{-4/3} + \left| \frac{1}{|\mathbf{n}|}|\mathbf{n}| - \frac{1}{N}|\mathbf{n}| \right| \\
&= 3N^{-4/3} + \left| 1 - \frac{1}{N}|\mathbf{n}| \right| \\
&\leq 3N^{-4/3} + 3N^{-4/3} = 6N^{-4/3}.
\end{aligned} \tag{12.31}$$

Notice that $\mathbf{n}/|\mathbf{n}|$ in (12.31) has the special form of (12.21) with $M = 2N$. Thus by (12.24)

$$\begin{aligned}\Pr[E_1 \text{bad}; 6\log N; 2N] &\le (4N+1)^3 2N \frac{\left(\frac{e}{6\log N}\right)^{6\log N}}{1-\frac{e}{6\log N}} \\ &\le N^{-11} \quad \text{for } N \ge 10^{15}. \end{aligned} \tag{12.32}$$

Moreover, by (12.25)

$$\begin{aligned}\Pr[E_2(\text{bad}; M)] &\le (4N+1)^3 2N \exp\left(-\frac{N^{1/3}}{30}\right) \\ &\le N^{-11} \quad \text{for } N \ge 10^{15}. \end{aligned} \tag{12.33}$$

Combining (12.32) and (12.33), we have

$$\Pr[E(\text{bad})] < N^{-10}, \tag{12.34}$$

where

$$E(\text{bad}) = E_1(\text{bad}; 6\log N; 2N) \cup E_2(\text{bad}; 2N).$$

Finally, by (12.31)

$$\left| \mathbf{w}\cdot\mathbf{v}_k - \frac{1}{|\mathbf{n}|}\mathbf{n}\cdot\mathbf{v}_k \right| \le 6N^{-4/3} \le \frac{1}{2N},$$

and so we have the implication

$$\mathbf{w}\cdot\mathbf{v}_k \in \left[\frac{\ell}{N}, \frac{\ell+1}{N}\right] \Longrightarrow \frac{1}{|\mathbf{n}|}\mathbf{n}\cdot\mathbf{v}_k \in \left[\frac{\ell-\frac{1}{2}}{N}, \frac{\ell+\frac{3}{2}}{N}\right]. \tag{12.35}$$

Combining (12.12)–(12.13) with (12.35), and also using (12.34), Lemma 12.1 follows. □

We now *fix* a sequence $\mathbf{v}_1, \mathbf{v}_2, \ldots, \mathbf{v}_N \in \mathbf{S}^2$ of N unit vectors on the unit sphere that satisfies the Separation Condition (12.26)–(12.27). Let $J \subseteq \{1, 2, 3, \ldots, N\}$ be a subset, and consider the torus-billiard system

$$\begin{aligned}\mathcal{Z}_J(t) &= \mathcal{Z}_J(\mathbf{v}_j, j \in J; t) \\ &= \{\mathbf{z}_j + t\mathbf{v}_j \text{ modulo one}: \ j \in J\}, \quad 0 \le t \le U, \end{aligned} \tag{12.36}$$

where of course modulo one means the fractional part applied coordinate-wise, $0 \le t \le U$ refers to the time evolution, and

$$\mathcal{Z}_J(0) = \{\mathbf{z}_j \text{ modulo one}: \ j \in J\} \tag{12.37}$$

is the initial configuration (i.e., a $|J|$-element point set in the unit torus $I^3 = [0,1)^3$). Let $B \subset [0,1)^3$ be an axis-parallel box; we study the square integral ("variance")

$$\begin{aligned}
&\int_0^U \left(|\mathcal{Z}_J(t) \cap B| - |J|\text{vol}(B)\right)^2 dt \\
&\quad = \int_0^U \left(|\mathcal{Z}_J(\mathbf{v}_j, j \in J; t) \cap B| - |J|\text{vol}(B)\right)^2 dt \\
&\quad = \int_0^U \left(\sum_{j\in J} \chi_B(j;t) - |J|\text{vol}(B)\right)^2 dt,
\end{aligned} \tag{12.38}$$

where

$$\chi_B(j;t) = \begin{cases} 1 & \text{if } \mathbf{z}_j + t\mathbf{v}_j \in B \quad \text{modulo one}, \\ 0 & \text{otherwise}. \end{cases} \tag{12.39}$$

We need the following key lemma.

Lemma 12.2. *Let $\mathbf{v}_1, \mathbf{v}_2, \ldots, \mathbf{v}_N \in \mathbf{S}^2$ be a sequence of N unit vectors on the unit sphere that satisfies the Separation Condition* (12.26)–(12.27). *Then for every subset $J \subseteq \{1,2,3,\ldots,N\}$, every initial configuration* (12.37), *every axis-parallel box $B \subset [0,1)^3$, and every real number $U \ge 10$,*

$$\begin{aligned}
&\int_0^U \left(|\mathcal{Z}_J(\mathbf{v}_j, j \in J; t) \cap B| - |J|\text{vol}(B)\right)^2 dt \\
&\quad \le 10^{14} N|J|(\log U)^2 + 10^{16} N|J| \log U + 54|J|^2 U N^{-1/3}.
\end{aligned}$$

The proof is basically a straightforward application of the Erdős–Turán–Koksma Inequality for torus lines (see (12.4)). Despite the simplicity of the basic idea of the proof, the technical details are rather long and complicated, so we decided to postpone it to Sec. 33.

Next we repeatedly apply Lemma 12.2 the same way as we usually apply Chebyshev's inequality in probability theory. We choose an arbitrary sequence $\mathbf{v}_1, \mathbf{v}_2, \ldots, \mathbf{v}_N \in \mathbf{S}^2$ of N unit vectors on the unit sphere that satisfies the Separation Condition (12.26)–(12.27); in applications of Lemma 12.2 we use the short notation (see (12.36))

$$\mathcal{Z}_J(t) = \mathcal{Z}_J(\mathbf{v}_j, j \in J; t).$$

We choose a sequence $B_1, \ldots, B_k$ of k axis-parallel boxes in the unit cube $[0,1)^3$; choose a positive ε such that

$$0 < \varepsilon < p_0 = \prod_{i=1}^{k} p_i \quad \text{with } p_i = \text{vol}(B_i), \tag{12.40}$$

choose an initial configuration

$$\mathcal{Y} = \{\mathbf{y}_1, \ldots, \mathbf{y}_N\} \subset [0,1]^3,$$

and choose a real number $T_0 > 0$. We apply Lemma 12.2 with $B = B_1$,

$$\begin{aligned} J = J_0 &= \{1, 2, 3, \ldots, N\}, \\ \mathcal{Z}_{J_0}(0) &= \{\mathbf{y}_j + T_0\mathbf{v}_j \text{ modulo one}: \ j \in J_0\} \end{aligned} \tag{12.41}$$

representing the initial configuration, and obtain the upper bound for every real $U_1 \geq 10$

$$\begin{aligned} &\int_0^{U_1} \left(|\mathcal{Z}_{J_0}(t) \cap B_1| - Np_1\right)^2 \, dt \\ &\quad \leq 10^{14} N^2 (\log U_1)^2 + 10^{16} N^2 \log U_1 + 54 U_1 N^{5/3}. \end{aligned} \tag{12.42}$$

For every $\eta_1 > 0$ let

$$S_1(U_1; \eta_1) = \left\{ t \in [0, U_1] : \ \left| \frac{1}{N} |\mathcal{Z}_{J_0}(\tau) \cap B_1| - p_1 \right| \geq \eta_1 \right\}. \tag{12.43}$$

Combining (12.42) and (12.43), we have

$$\text{length}\,(S_1(U_1; \eta_1)) \leq \frac{1}{\eta_1^2} \left(10^{14} (\log U_1)^2 + 10^{16} \log U_1 + 54 U_1 N^{-1/3} \right). \tag{12.44}$$

Next we choose an arbitrary $\tau_1 \in [0, U_1] \setminus S_1(U_1; \eta_1)$, and define

$$J_1(\tau_1) = \{ j \in J_0 : \ \mathbf{y}_j + (T_0 + \tau_1)\mathbf{v}_j \in B_1 \text{ modulo one} \}. \tag{12.45}$$

By definition

$$|J_1(\tau_1)| = |\mathcal{Z}_{J_0}(\tau_1) \cap B_1| \,. \tag{12.46}$$

By (12.43) and (12.46),

$$||J_1(\tau_1)| - p_1 N| \le \eta_1 N. \tag{12.47}$$

Next we apply Lemma 12.2 with the initial configuration

$$\mathcal{Z}_{J_1(\tau_1)}(0) = \{\mathbf{y}_j + (T_0 + \tau_1)\mathbf{v}_j \text{ modulo one}: \ j \in J_1(\tau_1)\}, \ B = B_2,$$

and obtain the upper bound for every real $U_2 \ge 10$

$$\begin{aligned} &\int_0^{U_2} \left(\left|\mathcal{Z}_{J_1(\tau_1)}(t) \cap B_2\right| - |J_1(\tau_1)|p_2\right)^2 dt \\ &\quad \le 10^{14} N|J_1(\tau_1)|(\log U_2)^2 + 10^{16} N|J_1(\tau_1)| \log U_2 + 54|J_1(\tau_1)|^2 U_2 N^{-1/3}. \end{aligned} \tag{12.48}$$

For every $\eta_2 > 0$ let

$$S_2(\tau_1; U_2; \eta_2) = \left\{ t \in [0, U_2] : \ \left| \frac{1}{|J_1(\tau_1)|} \left|\mathcal{Z}_{J_1(\tau_1)}(t) \cap B_2\right| - p_2 \right| \ge \eta_2 \right\}. \tag{12.49}$$

Combining (12.48) and (12.49), we have

$$\begin{aligned} &\text{length}\,(S_2(\tau_1; U_2; \eta_2)) \\ &\quad \le \frac{1}{\eta_2^2} \left(10^{14} \frac{N}{|J_1(\tau_1)|} (\log U_2)^2 + 10^{16} \frac{N}{|J_1(\tau_1)|} \log U_2 + 54 U_2 N^{-1/3} \right). \end{aligned} \tag{12.50}$$

Next we choose an arbitrary $\tau_2 \in [0, U_2] \setminus S_1(\tau_1 : U_2; \eta_2)$, and define

$$\begin{aligned} J_2(\tau_1; \tau_2) &= \{j \in J_1(\tau_1) : \ \mathbf{y}_j + (T_0 + \tau_1 + \tau_2)\mathbf{v}_j \in B_2 \text{ modulo one}\} \\ &= \left\{ j \in J_0 : \ \mathbf{y}_j + \left(T_0 + \sum_{i=1}^{h} \tau_i\right) \mathbf{v}_j \in B_h \text{ modulo one for } h = 1, 2 \right\}, \end{aligned} \tag{12.51}$$

where we used (12.45). By definition

$$|J_2(\tau_1;\tau_2)| = \left|\mathcal{Z}_{J_1(\tau_1)}(\tau_2) \cap B_2\right|. \tag{12.52}$$

By (12.49) and (12.52),

$$||J_2(\tau_1;\tau_2)| - p_2|J_1(\tau_1)|| \le \eta_2|J_1(\tau_1)|,$$

and combining it with (12.47), we have

$$\begin{aligned}||J_2(\tau_1;\tau_2)| - p_1p_2N| &\le \eta_2\,(p_1N + \eta_1 N) + p_2\eta_1 N\\ &\le (\eta_1 + \eta_2 + \eta_1\eta_2)\,N = ((1+\eta_1)\,(1+\eta_2) - 1)\,N.\end{aligned} \tag{12.53}$$

Next we apply Lemma 12.2 with the initial configuration

$$\begin{aligned}\mathcal{Z}_{J_2(\tau_1;\tau_2)}(0) &= \{\mathbf{y}_j + (T_0 + \tau_1 + \tau_2)\mathbf{v}_j \text{ modulo one} : j \in J_2(\tau_1;\tau_2)\},\\ B &= B_3,\end{aligned} \tag{12.54}$$

and obtain the upper bound for every real $U_3 \ge 10$

$$\begin{aligned}&\int_0^{U_3} \left(\left|\mathcal{Z}_{J_2(\tau_1;\tau_2)}(t) \cap B_3\right| - |J_2(\tau_1;\tau_2)|p_3\right)^2 dt\\ &\quad \le 10^{14}N|J_2(\tau_1;\tau_2)|(\log U_3)^2 + 10^{16}N|J_2(\tau_1;\tau_2)|\log U_3\\ &\qquad + 54|J_2(\tau_1;\tau_2)|^2U_3N^{-1/3}.\end{aligned} \tag{12.55}$$

For every $\eta_3 > 0$ let

$$\begin{aligned}S_3(\tau_1;\tau_2;U_3;\eta_3) = \Bigg\{t \in [0,U_3] :\\ \left|\frac{1}{|J_2(\tau_1;\tau_2)|}\left|\mathcal{Z}_{J_2(\tau_1;\tau_2)}(t) \cap B_3\right| - p_3\right| \ge \eta_3\Bigg\}.\end{aligned} \tag{12.56}$$

Combining (12.55) and (12.56), we have

$$\begin{aligned}\text{length}\,(S_3(\tau_1;\tau_2;U_3;\eta_3)) \le \frac{1}{\eta_3^2}\Bigg(10^{14}\frac{N}{|J_2(\tau_1;\tau_2)|}(\log U_3)^2\\ + 10^{16}\frac{N}{|J_2(\tau_1;\tau_2)|}\log U_3 + 54U_3N^{-1/3}\Bigg).\end{aligned} \tag{12.57}$$

Next we choose an arbitrary $\tau_3 \in [0, U_3] \setminus S_2(\tau_1; \tau_2; U_3; \eta_3)$, and define

$$\begin{aligned} &J_3(\tau_1; \tau_2; \tau_3) \\ &\quad = \{j \in J_2(\tau_1; \tau_2) : \ \mathbf{y}_j + (T_0 + \tau_1 + \tau_2 + \tau_3)\mathbf{v}_j \in B_3 \text{ modulo one}\} \\ &\quad = \left\{ j \in J_0 : \ \mathbf{y}_j + \left(T_0 + \sum_{i=1}^{h} \tau_i \right) \mathbf{v}_j \in B_h \text{ modulo one} \ \text{ for } h = 1, 2, 3 \right\}, \end{aligned} \tag{12.58}$$

where we used (12.51). By definition

$$|J_3(\tau_1; \tau_2; \tau_3)| = \left| \mathcal{Z}_{J_2(\tau_1;\tau_2)}(\tau_3) \cap B_3 \right|. \tag{12.59}$$

By (12.56) and (12.59),

$$||J_3(\tau_1; \tau_2; \tau_3)| - p_3 |J_2(\tau_1; \tau_2)|| \leq \eta_3 |J_2(\tau_1; \tau_2)|,$$

and combining it with (12.53), we have

$$\begin{aligned} &||J_3(\tau_1; \tau_2; \tau_3)| - p_1 p_2 p_3 N| \leq \eta_3 \left(p_1 p_2 N + \left((1 + \eta_1)(1 + \eta_2) - 1 \right) N \right) \\ &\quad + p_3 \left((1 + \eta_1)(1 + \eta_2) - 1 \right) N \leq \eta_3 N + \eta_3 \left((1 + \eta_1)(1 + \eta_2) - 1 \right) N \\ &\quad + \left((1 + \eta_1)(1 + \eta_2) - 1 \right) N = \left((1 + \eta_1)(1 + \eta_2)(1 + \eta_3) - 1 \right) N, \end{aligned} \tag{12.60}$$

which is the analog of (12.53).

We keep repeating this argument until we cover all k given test sets $B_1, \ldots, B_k$. At the end of the process we obtain a subset of $J_0 = \{1, 2, 3, \ldots, N\}$

$$\begin{aligned} J_k(\tau_1; \tau_2; \ldots; \tau_k) = \Bigg\{ j \in J_0 : \ \mathbf{y}_j + \left(T_0 + \sum_{i=1}^{h} \tau_i \right) \mathbf{v}_j \in B_h \text{ modulo one} \\ \text{for } h = 1, 2, \ldots, k \Bigg\} \end{aligned} \tag{12.61}$$

with

$$\left| |J_k(\tau_1; \tau_2; \ldots; \tau_k)| - N \prod_{i=1}^{k} p_i \right| \leq \left(\prod_{i=1}^{k} (1 + \eta_i) - 1 \right) N \tag{12.62}$$

(a generalization of (12.60)), where $p_i = \text{vol}(B_i)$ (see (12.40)),

$$\tau_i \in [0, U_i] \setminus S_i(\tau_1; \ldots; \tau_{i-1}; U_i; \eta_i) \quad \text{for } i \geq 2, \quad \text{and}$$
$$\tau_1 \in [0, U_1] \setminus S_1(U_1; \eta_1), \tag{12.63}$$

and (see (12.57) for $i = 3$)

$$\text{length}\,(S_i(\tau_1; \ldots; \tau_{i-1}; U_i; \eta_i)) \leq \frac{1}{\eta_i^2} \left(\frac{10^{14} N}{|J_{i-1}(\tau_1; \ldots; \tau_{i-1})|} (\log U_i)^2 + \frac{10^{16} N}{|J_{i-1}(\tau_1; \ldots; \tau_{i-1})|} \log U_i + 54 U_i N^{-1/3} \right) \tag{12.64}$$

for $i \geq 2$, and (see (12.44))

$$\text{length}\,(S_1(U_1; \eta_1)) \leq \frac{1}{\eta_1^2} \left(10^{14} (\log U_1)^2 + 10^{16} \log U_1 + 54 U_1 N^{-1/3} \right). \tag{12.65}$$

Since

$$\left| |J_h(\tau_1; \ldots; \tau_h)| - N \prod_{i=1}^{h} p_i \right| \leq \left(\prod_{i=1}^{h} (1 + \eta_i) - 1 \right) N$$

holds for every $1 \leq h \leq k$, by (12.64) we have for $i \geq 2$

$$\text{length}\,(S_i(\tau_1; \ldots; \tau_{i-1}; U_i; \eta_i)) \leq \frac{1}{\eta_i^2} \left(\frac{10^{14} (\log U_i)^2 + 10^{16} \log U_i}{\prod_{\ell=1}^{i-1} p_\ell - \left(\prod_{\ell=1}^{i-1} (1 + \eta_\ell) - 1 \right)} + \frac{54 U_i}{N^{1/3}} \right). \tag{12.66}$$

Next we specify the parameters U_i and η_i: let

$$U_i = kT \quad \text{and} \quad \eta_i = \frac{\varepsilon}{4k}, \quad 1 \leq i \leq k. \tag{12.67}$$

Since $0 < \varepsilon < 1$, we have

$$\prod_{i=1}^{k} (1 + \eta_i) - 1 = \left(1 + \frac{\varepsilon}{4k} \right)^k - 1 \leq e^{\varepsilon/4} - 1 < \varepsilon/2. \tag{12.68}$$

Using (12.68) in (12.62), we have

$$\left| |J_k(\tau_1;\tau_2;\ldots;\tau_k)| - N\prod_{i=1}^{k} p_i \right| \le \varepsilon N/2. \tag{12.69}$$

Next we use (12.67) and (12.68) in (12.66), and also use

$$0<\varepsilon<p_0=\prod_{i=1}^{k} p_i$$

(see (12.40)); thus we obtain

$$\begin{aligned}&\text{length}\,(S_i(\tau_1;\ldots;\tau_{i-1};U_i;\eta_i))\\ &\le \frac{16k^2}{\varepsilon^2}\left(\frac{10^{14}(\log(kT))^2+10^{16}\log(kT)}{\varepsilon-\frac{\varepsilon}{2}}+\frac{54kT}{N^{1/3}}\right)\\ &= \frac{32k^2}{\varepsilon^2}\left(\frac{10^{14}(\log(kT))^2+10^{16}\log(kT)}{\varepsilon}+\frac{27kT}{N^{1/3}}\right).\end{aligned} \tag{12.70}$$

Finally, simple calculation gives that the inequality

$$\frac{32k^2}{\varepsilon^2}\left(\frac{10^{14}(\log(kT))^2+10^{16}\log(kT)}{\varepsilon}+\frac{27kT}{N^{1/3}}\right)\le\frac{\varepsilon}{k}T \tag{12.71}$$

holds if

$$T\ge C_1(\varepsilon;k)=\frac{10^{30}k^3\,(\log(1+k))^3}{\varepsilon^4}, \tag{12.72}$$

and

$$N\ge C_2(\varepsilon;k)=\frac{10^{15}k^{12}}{\varepsilon^9}. \tag{12.73}$$

Thus by (12.70)–(12.73) we have

$$\text{length}\,(S_i(\tau_1;\ldots;\tau_{i-1};U_i;\eta_i))\le\frac{\varepsilon}{k}T \quad \text{for } 1\le i\le k,$$

which implies

$$\frac{1}{T}\sum_{i=1}^{k}\text{length}\,(S_i(\tau_1;\ldots;\tau_{i-1};U_i;\eta_i))\le\varepsilon. \tag{12.74}$$

Combining the Separation Condition (see (12.26)–(12.27)) with (12.61), (12.63), (12.69) and (12.74), we obtain the following theorem with $T_0 < t_1 = \tau_1 < T_0 + T$, $T_0 + T < t_2 = \tau_1 + \tau_2 < T_0 + 2T$,

$T_0 + 2T < t_3 = \tau_1 + \tau_2 + \tau_2 < T_0 + 3T$, and so on with

$$T_0 + (k-1)T < t_k = \sum_{i=1}^{k} \tau_i < t_0 + kT$$

at the end.

Theorem 12.1. *Let ε be a positive real with $0 < \varepsilon < 1$ and $k \geq 1$ be an integer. Given an arbitrary real*

$$T \geq C_1(\varepsilon; k) = \frac{10^{30}k^3 (\log(1+k))^3}{\varepsilon^4},$$

given an arbitrary sequence $B_j \subset [0,1]^3$, $1 \leq j \leq k$ of k axis-parallel boxes satisfying

$$p_0 = \prod_{j=1}^{k} \text{vol}(B_j) > \varepsilon > 0,$$

given an arbitrary initial configuration $\mathcal{Y} = \{\mathbf{y}_1, \ldots, \mathbf{y}_N\} \subset [0,1]^3$ with

$$N \geq C_2(\varepsilon; k) = \frac{10^{15}k^{12}}{\varepsilon^9},$$

and given an arbitrary real $T_0 \geq 0$, for at least $1 - \varepsilon$ part of all choices of k times $t_1 < \cdots < t_k$ with

$$T_0 < t_1 < T_0 + T < t_2 < T_0 + 2T < t_3 < T_0 + 3T < \cdots < T_0$$
$$+ (k-1)T < t_k < T_0 + kT,$$

the time evolution of an N-particle billiard system, starting from $\mathcal{Y}$ at $t = 0$, and with initial velocities $\mathbf{v}_1, \mathbf{v}_2, \ldots, \mathbf{v}_N \in \mathbf{S}^2$ satisfying the Separation Condition (12.26)–(12.27), *exhibits multiple ε-mixedupness at $t_1, t_2, \ldots, t_k$ with respect to the given sequence $B_j \subset [0,1]^3$, $1 \leq j \leq k$ of test sets. That is, the number of particles that are simultaneously in B_j at $t = t_j$ for every $j = 1, 2, \ldots, k$ is between*

$$(p_0 - \varepsilon) N \quad \textit{and} \quad (p_0 + \varepsilon) N,$$

ε-approximating the product rule of statistical independence. Moreover, "at least $1-\varepsilon$ part of all choices of $t_1 < \cdots < t_k$" precisely means that we choose a point in the k-dimensional cube $[0,T]^k$ with uniform distribution, the jth coordinate of this point gives $t_j - T_0 - (j-1)T$, $1 \leq j \leq k$, and we take the normalized k-dimensional volume (i.e., we divide by T^k). Finally, by Lemma 12.1, *the set of initial velocities $\mathbf{v}_1, \mathbf{v}_2, \ldots, \mathbf{v}_N \in \mathbf{S}^2$ that satisfy the Separation Condition* (12.26)–(12.27) *represents an overwhelming majority: it is at least $1 - N^{-10}$ part of the total.* □

Remarks. The main point here is that the thresholds $C_1(\varepsilon; k)$ and $C_2(\varepsilon; k)$ are constants in the sense that they are independent of the number N of the particles. It is very likely that the explicit formulas given in (12.72)–(12.73) can be substantially improved.

Theorem 12.1 is about N-particle billiard systems where the initial velocities

$$\mathbf{v}_1, \mathbf{v}_2, \ldots, \mathbf{v}_N \in \mathbf{S}^2$$

are unit vectors, i.e., the particles have the same constant speed (like photons). Note that we did not really use this fact in the proof. The proof of Theorem 12.1 actually works for any other isotropic initial velocity distribution (for example, the 3-dimensional normal, or Gaussian, distribution; physicists often call it the Maxwellian distribution).

Next we give a result that can be considered — in some sense — a far-reaching extension of Theorem 12.1. We are motivated by the following natural question: What happens if one replaces the "nice" test sets of k axis-parallel boxes $B_i \subset [0,1]^3$, $1 \le i \le k$ with "ugly" test sets $A_i \subset [0,1]^3$, $1 \le i \le k$, where A_i are *arbitrary* Lebesgue measurable sets? Can one still prove an analog of Theorem 12.1? In general — without any extra condition — the answer is no, but we can still prove multiple ε-mixedupness in the following modified form. ε-mixedupness holds with respect to the *majority* of the translated sequences $A_i + \mathbf{x}_i$, $1 \le i \le k$, $\mathbf{x}_i \in [0,1)^3$ (where of course the translation is taken modulo one, i.e., we are on the unit torus).

Theorem 12.2. *Let ε be a positive real with $0 < \varepsilon < 1$ and $k \ge 1$ be an integer. Given an arbitrary real*

$$T \ge C_3(\varepsilon; k) = \frac{10^4 k^5}{\varepsilon^5},$$

given an arbitrary sequence $A_j \subset [0,1]^3 = I^3$, $1 \le j \le k$ of k Lebesgue measurable sets satisfying

$$p_0 = \prod_{j=1}^{k} \mathrm{vol}(A_j) > \varepsilon > 0,$$

given an arbitrary initial configuration $\mathcal{Y} = \{\mathbf{y}_1, \ldots, \mathbf{y}_N\} \subset [0,1]^3$ with

$$N \ge C_4(\varepsilon; k) = \frac{10^6 k^6}{\varepsilon^3},$$

and given an arbitrary real $T_0 \ge 0$, for at least $1 - \varepsilon$ part of all choices of the k pairs $(t_j, \mathbf{w}_j)$, $1 \le j \le k$ consisting of translations $\mathbf{w}_j \in I^3$ and times

$t_1 < \cdots < t_k$ *with*

$$T_0 < t_1 < T_0 + T < t_2 < T_0 + 2T < t_3 < T_0 + 3T < \cdots$$
$$< T_0 + (k-1)T < t_k < T_0 + kT,$$

the time evolution of an N*-particle billiard system, starting from* $\mathcal{Y}$ *at* $t = 0$, *and with initial velocities* $\mathbf{v}_1, \mathbf{v}_2, \ldots, \mathbf{v}_N \in \mathbf{S}^2$ *satisfying the Separation Condition* (12.26)–(12.27), *exhibits multiple* ε*-mixedupness at* $t_1, t_2, \ldots, t_k$ *with respect to the translated sequence* $A_j + \mathbf{w}_j$, $1 \le j \le k$. *That is, the number of particles that are simultaneously in* $A_j + \mathbf{w}_j$ *at* $t = t_j$ *for every* $j = 1, 2, \ldots, k$ *is between*

$$(p_0 - \varepsilon)N \quad \text{and} \quad (p_0 + \varepsilon)N,$$

ε*-approximating the product rule of statistical independence. Moreover, "at least* $1 - \varepsilon$ *part of all choices of the* k *pairs* $(t_j, \mathbf{w}_j)$, $1 \le j \le k$*" precisely means that we choose a point in the* $4k$*-dimensional box* $[0, T]^k \times (I^3)^k$ *with uniform distribution, the* j*th coordinate of this point with* $1 \le j \le k$ *gives* $t_j - T_0 - (j-1)T$, *the remaining* $3k$ *coordinates give* $\mathbf{w}_1, \ldots, \mathbf{w}_k$ *in this order, and we take the normalized* $4k$*-dimensional volume (i.e., we divide by* T^k*). Finally, by Lemma* 12.1, *the set of initial velocities* $\mathbf{v}_1, \mathbf{v}_2, \ldots, \mathbf{v}_N \in \mathbf{S}^2$ *that satisfy the Separation Condition* (12.26)–(12.27) *represents an overwhelming majority: it is at least* $1 - N^{-10}$ *part of the total.*

Remarks. We postpone the proof of Theorem 12.2 to Sec. 34.

Similarly to Theorem 12.1, the main point here is that the thresholds $C_3(\varepsilon; k)$ and $C_4(\varepsilon; k)$ are constants in the sense that they are independent of the number N of the particles. It is very likely that the explicit formulas for $C_3(\varepsilon; k)$ and $C_4(\varepsilon; k)$ can be substantially improved.

Similarly to Theorem 12.1, Theorem 12.2 is about N-particle billiard systems where the initial velocities

$$\mathbf{v}_1, \mathbf{v}_2, \ldots, \mathbf{v}_N \in \mathbf{S}^2$$

are unit vectors, i.e., the particles have the same constant speed (like photons). Note that we did not really use this fact in the proof. The proof of Theorem 12.2 actually works for any other isotropic initial velocity distribution.

A major difference between Theorems 12.1 and 12.2 is the appearance of translations in the latter. We emphasize that without translation one cannot expect any positive result, not even for a single test set A_1, due to the overwhelming generality of the Lebesgue measure. Indeed, choosing an arbitrary but fixed initial configuration $\mathcal{Y} = \{\mathbf{y}_1, \ldots, \mathbf{y}_N\} \subset [0,1]^3$ and choosing an arbitrary but fixed set of initial velocities $\mathbf{v}_1, \mathbf{v}_2, \ldots, \mathbf{v}_N \in \mathbf{S}^2$ satisfying the Separation Condition (12.26)–(12.27), the time evolution of this N-particle billiard system is described by N infinite smooth curves in the unit cube I^3. The union of these N infinite smooth curves has volume (= 3-dimensional Lebesgue measure) zero, so choosing A_1 to be the complement of this zero set, we obtain a shocking "anti-uniformity" property of the time evolution of the system in the following sense. No particle will ever enter $A_1 \subset I^3$, despite the fact that its measure is one! This simple argument explains why we need some extra condition (e.g., to involve typical translation, what we did in Theorem 12.2).

Both Theorems 12.1 and 12.2 are about mixedupness (in constant time). What about *mixing*? What is the relation of the two concepts? We start with the definition; for notational simplicity we just consider the case of discrete time evolution (the continuous case is a straightforward analog). Given a probability space $(X, \mathcal{A}, \mu)$ and a μ-measure-preserving transformation $T : X \to X$, we call T *mixing* (or strongly mixing in some books) if for any two measurable sets $A, B \in \mathcal{A}$

$$\lim_{n \to \infty} \mu\left(T^{-n} A \cap B\right) = \mu(A)\mu(B).$$

In this section we were studying the translation-flow on the d-dimensional torus $I^d = [0,1)^d$. The discrete version of the translation-flow is $T_{\mathbf{a}} : \mathbf{x} \to \mathbf{x} + \mathbf{a}$ modulo one (of course we mean it coordinatewise); and so $T_{\mathbf{a}}^k : \mathbf{x} \to \mathbf{x} + k\mathbf{a}$ modulo one for every integer $k \in \mathbb{Z}$. We know that $T_{\mathbf{a}}$ is ergodic if and only if $1, \alpha_1, \ldots, \alpha_d$ are linearly independent over the rationals, where $\mathbf{a} = (\alpha_1, \ldots, \alpha_d)$. We show that $T_{\mathbf{a}}$ is *not* mixing for any of the vectors $\mathbf{a} = (\alpha_1, \ldots, \alpha_d)$. Indeed, consider two disjoint congruent balls in $I^d = [0,1)^d$; for example, let

$$B_1 = B(1/4; 1/5) = \left\{ \mathbf{x} \in I^d : \left(\sum_{i=1}^{d} \left(x_i - \frac{1}{4} \right)^2 \right)^{1/2} \le \frac{1}{5} \right\}$$

and

$$B_2 = B(3/4; 1/5) = \left\{ \mathbf{x} \in I^d : \left(\sum_{i=1}^{d} \left(x_i - \frac{3}{4} \right)^2 \right)^{1/2} \le \frac{1}{5} \right\}.$$

By Dirichlet's simultaneous approximation theorem for every $\varepsilon > 0$ there exists an infinite set S of positive integers n such that

$$\|n\alpha_i\| < \varepsilon \text{ for every } 1 \le i \le d \text{ and every } n \in S.$$

Note that the distance between B_1 and B_2 is $1/10$; so, if $\varepsilon > 0$ is sufficiently small, for $n \in S$ the balls $T^{-n}B_1$ and B_1 almost coincide, implying that $T^{-n}B_1$ and B_2 are disjoint, where $T = T_{\mathbf{a}}$. It follows that

$$\liminf_{n\to\infty} \mathrm{vol}_d\left(T^{-n}B_1 \cap B_2\right) = 0 \ne \mathrm{vol}_d(B_1)\mathrm{vol}_d(B_2),$$

which contradicts mixing.

In the traditional theory of dynamical systems the concepts of *mixing* and *hyperbolicity* play a key role. The message of Sec. 12 is that, despite the lack of mixing, the time evolution of our systems exhibit multiple mixedupness, and other forms of short-term advanced randomness, including snapshot randomness and time-lapse randomness (see the next chapters). We may say that, mixing — a "soft" asymptotic property — is not necessary in the study of realistic short-term time evolution of large systems toward "advanced randomness". Similarly, as our results demonstrate, we do not necessarily need *hyperbolicity* to demonstrate "advanced randomness" in the short-term time evolution of large systems.

13. Extensions of Theorem 4.2 beyond the Gaussian Case

To generalize Theorem 4.2 for arbitrary initial velocities beyond the Gaussian, we have to point out a simple necessary condition for Configuration Equilibrium. It comes from a "dimension argument", which provides a trivial obstacle that prevents us from proving Configuration Equilibrium for some speed distributions. We call this obstacle the *dimension deficit* for the family of corresponding curves in the configuration space. For example, consider the simplest case, where the particles move on straight lines (modulo one, i.e., the torus-billiard model), and the velocity distribution for each particle is isotropic with constant unit speed (like photons). In this case the family of all possible system-trajectories in the configuration space (starting from a fixed initial point configuration, e.g., from the origin of the configuration space) has dimension $2N + 1$. Indeed, the initial directions are determined by a vector

$$\omega \in \mathbf{S}^2 \times \cdots \times \mathbf{S}^2 = \left(\mathbf{S}^2\right)^N,$$

where $(\mathbf{S}^2)^N$ is a $2N$-dimensional manifold, and the time evolution $0 \le t < \infty$ gives the last extra dimension. On the other hand, the configuration space $[0,1)^{3N}$ is of course $3N$-dimensional, and the inequality $2N + 1 < 3N$ represents a clear-cut case of dimension deficit. The union of the corresponding family of curves (= all system-trajectories) in the configuration space (describing the time evolution of the system starting from a fixed initial point configuration) has dimension (at most) $2N + 1$, so the $3N$-dimensional Lebesgue measure of this union set is zero. By choosing the test set S to be the complement of this union set in $[0,1]^{3N}$, it is obvious why Configuration Equilibrium is impossible: this particular test set has $3N$-dimensional Lebesgue measure one, so the system-trajectory *should be* in S most of the time. But what we have here is the complete opposite: the system-trajectory will never ever enter this particular test set!

We conclude, therefore, that "photon like" systems with isotropic constant speed cannot reach Configuration Equilibrium — there is no analog of Theorem 4.2. In a typical "photon like" system with isotropic constant speed the initial velocity coordinates are linearly independent, so the Kronecker–Weyl theorem still applies. But it is impossible to upgrade this "weak uniformity result" to a "strong uniformity result" in the configuration space.

On the other hand, if every particle has an initial speed distribution with "sufficiently smooth" density function, then there is no dimension deficit, and we can indeed prove an analog of Theorem 4.2. In our first extension of Theorem 4.2 we consider a general torus model where the particles move on *torus lines with (possibly) varying speeds*, and every particle has its own (almost arbitrary) smooth initial speed distribution. (In Sec. 14 we will discuss a generalization of Theorem 4.2 for the case where the particles move on nonlinear trajectories — that case is more complicated.) The curve of the whole system in the configuration space is denoted by

$$\begin{aligned}
&\vec{\mathcal{Y}}(\mathcal{R};\rho_1,\mathbf{e}_1,\ldots,\rho_N,\mathbf{e}_N;t)\\
&\quad= \vec{\mathcal{Y}} + \vec{\Gamma}(\mathcal{R};\rho_1,\mathbf{e}_1,\ldots,\rho_N,\mathbf{e}_N;t) \text{ modulo one}\\
&\quad= (\mathbf{y}_1 + \rho_1 r_1(t)\mathbf{e}_1, \mathbf{y}_2 + \rho_2 r_2(t)\mathbf{e}_2, \ldots, \mathbf{y}_N + \rho_N r_N(t)\mathbf{e}_N) \text{ modulo one,}
\end{aligned} \tag{13.1}$$

where t is the time,

$$\vec{\mathcal{Y}} = (\mathbf{y}_1, \mathbf{y}_2, \ldots, \mathbf{y}_N) \tag{13.2}$$

is the vector (= point in the configuration space) formed from the set

$$\mathcal{Y} = \{\mathbf{y}_1, \mathbf{y}_2, \ldots, \mathbf{y}_N\}$$

of the initial point configuration at the start $t = 0$,

$$\mathcal{R} = (r_1(t), r_2(t), \ldots, r_N(t)) \tag{13.3}$$

represents a family of functions such that

$$\mathcal{R}^{(\text{inf})}(t) = \min_{1 \le k \le N} \inf_{x \ge t} r_k(x) \ge t^{\alpha} \text{ for all } t \ge \alpha_0 \tag{13.4}$$

and

$$\mathcal{R}^{(\text{sup})}(t) = \max_{1 \le k \le N} \sup_{x \le t} r_k(x) \le t^{\alpha_1} \text{ for all } t \ge \alpha_0, \tag{13.5}$$

where $\alpha_0 \ge 1$, $0 < \alpha \le \alpha_1$ are some absolute constants. (We also need a family of probability density functions

$$\mathcal{G} = (g_1, g_2, \ldots, g_N),$$

where g_k defines the distribution of ρ_k by the integral $\Pr[\rho_k \le u] = \int_0^u g_k(y)\, dy$; we elaborate on $\mathcal{G}$ later.)

Moreover, let $\mathcal{I}$ denote an arbitrary finite interval of positive length. Assume that

$$\max_{1 \le k \le N} \sup_{\mathcal{I} \subset \mathbb{R}} \frac{\text{length}\{0 \le t \le T : \ r_k(t) \in \mathcal{I}\}}{\text{length}(\mathcal{I})} \le T^{\beta} \tag{13.6}$$

uniformly for all $T \ge \beta_0$, where $\beta_0 \ge 1$ and $0 \le \beta < 1$ are absolute constants.

Note that it is easy to satisfy conditions (13.4)–(13.6). For example, consider the forward motions where the speeds $r_k'(t)$ of the particles satisfy the inequalities

$$\min\left\{\kappa_1 t^{-\zeta_1}, \kappa_1\right\} \le r_k'(t) \le \max\left\{\kappa_2 t^{\zeta_2}, \kappa_2\right\}, \ 1 \le k \le N, \ t > 0,$$

which hold for some absolute constants $0 \le \zeta_1 < 1$ and $\zeta_2 \ge 0$.

Of course, there are many other ways, mixing forward and backward motions of the particles, such that (13.4)–(13.6) are still satisfied.

We focus on the most natural case where the particle space (= "gas container") is 3-dimensional, i.e., $d = 3N$. Notice that the curve (13.1) of the whole system is in the d-dimensional unit torus $I^d = [0,1)^d$, which is the configuration space.

The trajectory of the kth particle is $\mathbf{y}_k + \rho_k r_k(t)\mathbf{e}_k \in \mathbb{R}^3$, $1 \le k \le N$ modulo one, representing a (possibly) varying speed motion on a torus line in the direction $\mathbf{e}_k$. For the *set* of N particles at time t we use the notation

$$\begin{aligned}&\mathcal{Y}(\mathcal{R};\rho_1,\mathbf{e}_1,\ldots,\rho_N,\mathbf{e}_N;t)\\&\quad=\{\mathbf{y}_1+\rho_1 r_1(t)\mathbf{e}_1,\ldots,\mathbf{y}_N+\rho_N r_N(t)\mathbf{e}_N\}\text{ modulo one.}\end{aligned}\tag{13.7}$$

Comparing (13.7) to (13.1), we see that the switch from the point set in the particle space (= "gas container") to the point in the configuration space is indicated by an extra arrow $\vec{\mathcal{Y}}$ on the top of $\mathcal{Y}$.

Let $d = 3N$, and let $f \in L_2\left(I^d\right)$, $I^d = [0,1)^d$ be a complex-valued Lebesgue square-integrable function in the d-dimensional unit torus (i.e., we extend f over the whole d-space $\mathbb{R}^d$ periodically). Write

$$\sigma_0^2(f) = \int_{I^d} |f|^2\, d\mathrm{V} - \left|\int_{I^d} f\, d\mathrm{V}\right|^2 = \int_{I^d}\left|f(\mathbf{y}) - \int_{I^d} f\, d\mathrm{V}\right|^2 d\mathbf{y}. \tag{13.8}$$

Since the torus is translation invariant, it suffices to study the time discrepancy of the $\vec{\mathcal{Y}}$-independent part $\vec{\Gamma}(t)$ in (13.1):

$$D_f(\mathcal{R};\rho_1,\mathbf{e}_1,\ldots,\rho_N,\mathbf{e}_N;T_1,T_2) = \int_{T_1}^{T_2} f(\vec{\Gamma}(t))\,dt - (T_2 - T_1)\int_{I^d} f\, d\mathrm{V}, \tag{13.9}$$

where $\vec{\Gamma}(t) = \vec{\Gamma}(\mathcal{R};\rho_1,\mathbf{e}_1,\ldots,\rho_N,\mathbf{e}_N;t)$ and $0 < T_1 < T_2$.

Let us discuss now the family

$$\mathcal{G} = (g_1, g_2, \ldots, g_N)$$

of probability density functions. The kth function g_k defines the initial speed distribution $\Pr[\rho_k \le u] = \int_0^u g_k(y)\,dy$ of the kth particle ($1 \le k \le N$). We assume that

$$\infty > g_k(u) \ge 0 \text{ for } 0 \le u < \infty,\ \int_0^\infty g_k(u)\,du = 1 \text{ and } \int_0^\infty u g_k(u)\,du < \infty, \tag{13.10}$$

where the last requirement in (13.10) means that the "expectation = mean-speed is finite". The corresponding initial velocity space is

$$\omega = (\rho_1, \mathbf{e}_1, \rho_2, \mathbf{e}_2, \ldots, \rho_N, \mathbf{e}_N) \in \Omega_{\mathcal{G}} = \left([0,\infty) \times \mathbf{S}^2\right)^N, \tag{13.11}$$

equipped with the product measure $\text{ProdMeas}_{\mathcal{G}}$, where the kth factor $[0,\infty) \times \mathbf{S}^2$ in (13.11) has the normed surface area for the unit sphere $\mathbf{S}^2 \subset \mathbb{R}^3$ and has the probability density function g_k for $[0,\infty)$.

By using (13.11), we can rewrite the time discrepancy (13.9) as follows:

$$\begin{aligned} D_f(\mathcal{R};\omega;T_1,T_2) &= D_f(\mathcal{R};\rho_1,\mathbf{e}_1,\ldots,\rho_N,\mathbf{e}_N;T_1,T_2) \\ &= \int_{T_1}^{T_2} f(\vec{\Gamma}_\omega(t))\,dt - (T_2 - T_1)\int_{I^d} f\,d\text{V}, \end{aligned} \tag{13.12}$$

where

$$\vec{\Gamma}_\omega(t) = \vec{\Gamma}(\mathcal{R};\omega;t) = \vec{\Gamma}(\mathcal{R};\rho_1,\mathbf{e}_1,\ldots,\rho_N,\mathbf{e}_N;t),$$
$$\omega = (\rho_1,\mathbf{e}_1,\ldots,\rho_N,\mathbf{e}_N) \in \Omega_{\mathcal{G}}$$

and $0 < T_1 < T_2$.

To extend Lemma 9.1 in the general case, we need to define the so-called 3-dimensional "isotropic Fourier transform" of $g = g_k$, $1 \le k \le N$:

$$\begin{aligned} \mathbf{F}_g(\mathbf{w}) &= \int_{u=0}^{\infty} \frac{1}{4\pi}\int_{\mathbf{v}\in\mathbf{S}^2} e^{\mathrm{i}u\mathbf{v}\cdot\mathbf{w}} g(u)\,du\,d\text{SA}(\mathbf{v}) \\ &= \int_{u=0}^{\infty} \frac{1}{4\pi}\int_{\mathbf{v}\in\mathbf{S}^2} \cos(u\mathbf{v}\cdot\mathbf{w}) g(u)\,du\,d\text{SA}(\mathbf{v}), \end{aligned} \tag{13.13}$$

where $\mathbf{w}$ is an arbitrary 3-dimensional vector, $d\text{SA}(\mathbf{v})$ represents the surface area. $\mathbf{F}_g(\mathbf{w})$ defines a real-valued function depending only on the length $w = |\mathbf{w}|$ of the input vector, and we have

$$\mathbf{F}_g(0) = \mathbf{F}_g(\mathbf{0}) = 1 \text{ and } |\mathbf{F}_g(w)| = |\mathbf{F}_g(\mathbf{w})| < 1 \text{ for } w = |\mathbf{w}| > 0.$$

By the Riemann–Lebesgue theorem $\mathbf{F}_g(w) \to 0$ as $w \to \infty$.

Write

$$\mathbf{F}_g^{(\text{sup})}(w) = \sup_{x \ge w} |\mathbf{F}_g(x)| \tag{13.14}$$

and

$$\mathbf{F}_{\mathcal{G}}^{(\text{sup})}(w) = \max_{1 \le k \le N} \mathbf{F}_{g_k}^{(\text{sup})}(w). \tag{13.14$'$}$$

So both $\mathbf{F}_{g}^{(\text{sup})}(w)$ and $\mathbf{F}_{\mathcal{G}}^{(\text{sup})}(w)$ are monotone decreasing positive functions tending to 0 as $w \to \infty$.

Technical Condition: polynomial decay. Assume that there exist real numbers $\gamma > 3$ and $\gamma_0 \ge 1$ such that

$$\mathbf{F}^{(\text{sup})}(w) = \mathbf{F}_{\mathcal{G}}^{(\text{sup})}(w) = \max_{1 \le k \le N} \sup_{x \ge w} |\mathbf{F}_{g_k}(x)| \le w^{-\gamma} \text{ for all} w \ge \gamma_0. \tag{13.15}$$

(We need the condition $\gamma > 3$ to avoid a "divergence problem" later in the proof of Theorem 13.1; see below.)

Condition (13.15) with $\gamma > 3$ is easily satisfied if $g = g_k$ is "sufficiently smooth". This is the message of the following lemma.

Lemma 13.1. *Let $g(u)$ be a probability density function in $[0, \infty)$ with $g(0) = 0$, and extend it over the whole real line as an even function: $g(-u) = g(u)$. Let $k \ge 2$, assume that $g(u)$, $-\infty < u < \infty$ is k-times continuously differentiable, also assume that for every integer $1 \le j \le k$, the jth derivative $g^{(j)}(u) \to 0$ as $u \to \infty$, and finally assume*

$$\int_{-\infty}^{\infty} |g^{(j)}(u)|\, du \le C_0 < \infty.$$

Then the 3-dimensional "isotropic Fourier transform"

$$\mathbf{F}_g(w) = \int_0^{\infty} \frac{1}{4\pi} \int_{\mathbf{v} \in \mathbf{S}^2} e^{iu\mathbf{v}\cdot\mathbf{w}} g(u)\, du\, d\text{SA}(\mathbf{v})$$

with $w = |\mathbf{w}|$ satisfies the upper bound

$$|\mathbf{F}_g(w)| \le \frac{C_0}{2(k-1)w^k}.$$

Note that $d\text{SA}(\mathbf{v})$ represents the surface area.

Proof of Lemma 13.1. Let $\widehat{g}(x)$ denote the (ordinary) Fourier transform of g. By hypothesis we can apply integration by parts k times in the definition

$$\widehat{g}(x) = \frac{1}{\sqrt{2\pi}} \int_{-\infty}^{\infty} e^{-ixu} g(u)\, du$$

of the Fourier transform, and thus we have

$$\widehat{g}(x) = \frac{1}{\sqrt{2\pi}} \int_{-\infty}^{\infty} e^{-\mathrm{i}xu} g(u)\, du = \frac{1}{\sqrt{2\pi}} \int_{-\infty}^{\infty} e^{-\mathrm{i}xu} \frac{(-1)^k g^{(k)}(u)}{(-\mathrm{i}x)^k}\, du$$
$$= \frac{1}{\sqrt{2\pi}} \frac{1}{(\mathrm{i}x)^k} \int_{-\infty}^{\infty} e^{-\mathrm{i}xu} g^{(k)}(u)\, du,$$

and so

$$|\widehat{g}(x)| \le \frac{1}{\sqrt{2\pi}|x|^k} \int_{-\infty}^{\infty} \left| g^{(k)}(u) \right| du \le \frac{C_0}{\sqrt{2\pi}|x|^k}. \tag{13.16}$$

We also need the following lemma.

Lemma 13.2. *Let $g(u)$ be a probability density function in $[0, \infty)$, let us extend g over the whole real line as an even function $g(-u) = g(u)$. The 3-dimensional "isotropic Fourier transform" $\mathbf{F}_g(\mathbf{w})$ (defined in (13.13)) can be expressed in terms of the Fourier transform $\widehat{g}(x)$ of g as follows:*

$$\mathbf{F}_g(w) = \mathbf{F}_g(\mathbf{w}) = \frac{\sqrt{\pi}}{2\sqrt{2}w} \int_{-w}^{w} \widehat{g}(x)\, dx = \sqrt{\frac{\pi}{2}} \frac{1}{w} \int_{0}^{w} \widehat{g}(x)\, dx.$$

Moreover, if $g(y)$ is continuous at $y = 0$ then

$$g(0) = \frac{1}{\sqrt{2\pi}} \int_{-\infty}^{\infty} \widehat{g}(x)\, dx,$$

and so we can rewrite $\mathbf{F}_g(w)$ as

$$\mathbf{F}_g(w) = \frac{\sqrt{\pi}}{2\sqrt{2}w} \left(g(0)\sqrt{2\pi} - \int_{|x| \ge w} \widehat{g}(x)\, dx \right).$$

We are now ready to finish the proof of Lemma 13.1. By Lemma 13.2,

$$\mathbf{F}_g(w) = -\frac{\sqrt{\pi}}{\sqrt{2}w} \int_{x \ge w} \widehat{g}(x)\, dx,$$

and by using (13.16), we conclude

$$|\mathbf{F}_g(w)| \le \frac{\sqrt{\pi}}{\sqrt{2}w} \int_{x \ge w} |\widehat{g}(x)|\, dx \le \frac{\sqrt{\pi}}{\sqrt{2}w} \int_{x \ge w} \frac{C_0}{\sqrt{2\pi} x^k}\, dx \le \frac{C_0}{2(k-1)w^k},$$

completing the proof of Lemma 13.1. □

It remains to prove Lemma 13.2. We need another lemma.

Lemma 13.3. *Let* $\mathbf{w}$ *be an arbitrary 3-dimensional vector. Then*

$$\frac{1}{4\pi}\int_{\mathbf{v}\in\mathbf{S}^2} e^{i\mathbf{w}\cdot\mathbf{v}}\, d\mathrm{SA}(\mathbf{v}) = \frac{\sin(|\mathbf{w}|)}{|\mathbf{w}|}.$$

Proof of Lemma 13.3. The special case $\mathbf{w} = \mathbf{0}$ is trivial: the left side is equal to 1, and the right side tends to 1 as $\mathbf{w}$ tends to $\mathbf{0}$; so we may assume that $\mathbf{w} \neq \mathbf{0}$. We need a well-known geometric fact about the surface area of some special spherical domains. Let $\mathbf{e} \in \mathbf{S}^2$ be an arbitrary but fixed unit vector, and let $0 < \delta < 1$ be an arbitrary real number. Consider the "spherical belt" (often called the "spherical zone"):

$$\mathbf{S}^2(\mathbf{e};\delta) = \left\{\mathbf{v}\in\mathbf{S}^2 :\ |\mathbf{e}\cdot\mathbf{v}| \le \delta\right\}.$$

On one hand, it is trivial that the surface area of the "belt" $\mathbf{S}^2(\mathbf{e};\delta)$ is independent of the choice of the unit vector $\mathbf{e}$; it is a very surprising fact, on the other hand, that the dependence on δ is just plain linear (discovered by Archimedes)

$$\mathrm{SurfaceArea}\left(\mathbf{S}^2(\mathbf{e};\delta)\right) = \mathrm{SurfArea}\left(\mathbf{S}^2(\mathbf{e};\delta)\right) = 4\pi\delta. \tag{13.17}$$

(13.17) implies that, for any nonzero 3-dimensional vector $\mathbf{w}$ and any real number ϱ with $0 < \varrho < |\mathbf{w}|$,

$$\mathrm{SurfArea}\left(\left\{\mathbf{v}\in\mathbf{S}^2 :\ |\mathbf{w}\cdot\mathbf{v}| \le \varrho\right\}\right) = 4\pi\cdot\frac{\varrho}{|\mathbf{w}|}. \tag{13.18}$$

By using (13.18), we have

$$\frac{1}{4\pi}\int_{\mathbf{S}^2} e^{i\mathbf{w}\cdot\mathbf{v}}\, d\mathrm{SA}(\mathbf{v}) = \frac{1}{2}\int_{-1}^{1} e^{i|\mathbf{w}|u}\, du = \frac{\sin(|\mathbf{w}|)}{|\mathbf{w}|},$$

proving Lemma 13.3. □

We are now ready to finish the proof of Lemma 13.2. By Lemma 13.3,

$$\frac{1}{4\pi}\int_{\mathbf{v}\in\mathbf{S}^2} e^{iu\mathbf{v}\cdot\mathbf{w}}\, d\mathrm{SA}(\mathbf{v}) = \frac{\sin(uw)}{uw} \quad \text{with } w = |\mathbf{w}|,$$

implying

$$\mathbf{F}_g(\mathbf{w}) = \int_0^\infty g(u)\frac{\sin(uw)}{uw}\, du. \tag{13.19}$$

Since $g(-x) = g(x)$, we can rewrite (13.19):

$$\mathbf{F}_g(\mathbf{w}) = \frac{1}{2}\int_{-\infty}^{\infty} g(u)\frac{\sin(uw)}{uw}\, du. \tag{13.20}$$

We recall some well-known facts from the theory of Fourier transform:

$$\widehat{f}(t) = \frac{1}{\sqrt{2\pi}} \int_{-\infty}^{\infty} e^{-\mathrm{i}xt} f(x)\, dx \tag{13.21}$$

denotes the Fourier transform of f, and we have the similar inversion formula

$$f(x) = \frac{1}{\sqrt{2\pi}} \int_{-\infty}^{\infty} e^{\mathrm{i}xt} \widehat{f}(t)\, dt. \tag{13.22}$$

It follows from (13.21)–(13.22) that the Fourier inverse of $f(x)$ is $\widehat{f}(-t)$.

We also need the fact

$$\int_{-\infty}^{\infty} f_1(x)\overline{f_2(x)}\, dx = \int_{-\infty}^{\infty} \widehat{f_1}(t)\overline{\widehat{f_2}(t)}\, dt, \tag{13.23}$$

which is a corollary of the Parseval–Plancherel formula. Note that (13.21)–(13.23) hold under mild conditions; for example, they hold if the functions are in the space $L_2(\mathbb{R})$.

If $\chi_w(x)$ denotes the 0, 1 valued characteristic function of the interval $[-w, w]$, then (13.21) gives via simple calculations

$$\widehat{\chi_w}(u) = \sqrt{\frac{2}{\pi}} \frac{\sin(uw)}{u}. \tag{13.24}$$

Since $\chi_w(x)$ is an even function, the Fourier transform of

$$f(u) = \sqrt{\frac{2}{\pi}} \frac{\sin(uw)}{u}$$

is $\chi_w(x)$. Combining this fact with (13.23), we have

$$\int_{-\infty}^{\infty} g(u)\frac{\sin(uw)}{uw}\, du = \int_{-\infty}^{\infty} \widehat{g}(x)\sqrt{\frac{\pi}{2}}\frac{\chi_w(x)}{w}\, dx = \frac{\sqrt{\pi}}{\sqrt{2}w} \int_{-w}^{w} \widehat{g}(x)\, dx. \tag{13.25}$$

Note that g is not necessarily in the $L_2(\mathbb{R})$-space; nevertheless, (13.23) applies in (13.25), since $\int_{-w}^{w} \widehat{g}(x)\, dx$ is a "safe" finite integral.

By (13.20) and (13.25),

$$\mathbf{F}_g(\mathbf{w}) = \frac{\sqrt{\pi}}{2\sqrt{2}w} \int_{-w}^{w} \widehat{g}(x)\, dx.$$

Thus the proof of Lemma 13.2 is complete. □

Let us return to (13.12); we take the square of the absolute value of the time discrepancy

$$D_f(\mathcal{R};\omega;T_1,T_2) = D_f(\mathcal{R};\rho_1,\mathbf{e}_1,\ldots,\rho_N,\mathbf{e}_N;T_1,T_2),$$

and study the following average ($d = 3N$)

$$\begin{aligned}\Delta_f^2(\mathcal{G};\mathcal{R};T_1,T_2) &= \int_{\omega\in\Omega_{\mathcal{G}}} |D_f(\mathcal{R};\omega;T_1,T_2)|^2 \, d\mathrm{ProdMeas}_{\mathcal{G}}(\omega)\\ &= \int_{\rho_1=0}^{\infty}\int_{\mathbf{e}_1\in\mathbf{S}^2}\cdots\int_{\rho_N=0}^{\infty}\int_{\mathbf{e}_N\in\mathbf{S}^2}\\ &\quad\cdot |D_f(\mathcal{R};\rho_1,\mathbf{e}_1,\ldots,\rho_N,\mathbf{e}_N;T_1,T_2)|^2\\ &\quad\cdot g_1(\rho_1)\,d\rho_1\,d\mathrm{SA}^*(\mathbf{e}_1)\ldots g_N(\rho_N)\,d\rho_N\,d\mathrm{SA}^*(\mathbf{e}_N),\end{aligned} \tag{13.26}$$

where $d\mathrm{SA}^*(\mathbf{e})$ means the normed surface area on the unit sphere $\mathbf{S}^2$, i.e., $\mathrm{SA}^*\left(\mathbf{S}^2\right) = 1$.

Here we make a brief detour to explain that $\Delta_f^2(\mathrm{Gauss};T_1,T_2)$ in (6.29) represents the special case of (13.26) where every particle moves on some torus line with some constant speed (both depending on the particle), and the particles have the same speed distribution with density $g(u) = \sqrt{\frac{2}{\pi}}u^2e^{-u^2/2}$, $0 \le u < \infty$ (which is the density of the speed distribution of the 3-dimensional Gaussian velocity distribution). In this special case the trajectory of the kth particle is $\mathbf{y}_k + \rho_k t\mathbf{e}_k \in \mathbb{R}^3$, $1 \le k \le N$ modulo one, where $\Pr[\rho_k \le u] = \sqrt{\frac{2}{\pi}}\int_0^u z^2e^{-z^2/2}\,dz$; and the *set* of the N particles at time t is (see (13.7))

$$\begin{aligned}\mathcal{Y}(\mathrm{Gauss};\omega;t) &= \mathcal{Y}(\mathrm{Gauss};\rho_1,\mathbf{e}_1,\ldots,\rho_N,\mathbf{e}_N;t)\\ &= \{\mathbf{y}_1+\rho_1 t\mathbf{e}_1,\ldots,\mathbf{y}_N+\rho_N t\mathbf{e}_N\} \text{ modulo one.}\end{aligned} \tag{13.27}$$

The fact that (6.29) is a special case of (13.26) can formally be expressed as follows:

$$\Delta_f^2(\mathcal{G};\mathcal{R};T_1,T_2) = \Delta_f^2(\mathrm{Gauss};T_1,T_2) \text{ with the choices of}$$

$$r_1(t) = \cdots = r_N(t) = t \quad\text{and}\quad g_1(u) = \cdots = g_N(u) = \sqrt{\tfrac{2}{\pi}}u^2e^{-u^2/2} \tag{13.28}$$

in the definitions of $\mathcal{R}$ and $\mathcal{G}$, respectively.

In this special case the curve $\vec{\mathcal{Y}}(\omega;t)$, $t \geq 0$ of the whole system in the configuration space is a *straight line* in the d-space (with $d = 3N$) modulo one:

$$\begin{aligned}\vec{\mathcal{Y}}(\omega;t) &= \vec{\mathcal{Y}}(\text{Gauss};\omega;t) = \vec{\mathcal{Y}} + \vec{\Gamma}_\omega(t) \text{ modulo one}\\ &= \vec{\mathcal{Y}} + \vec{\Gamma}(\text{Gauss};\omega;t) \text{ modulo one}\\ &= \vec{\mathcal{Y}} + \vec{\Gamma}(\text{Gauss};\rho_1,\mathbf{e}_1,\ldots,\rho_N,\mathbf{e}_N;t) \text{ modulo one}\\ &= (\mathbf{y}_1 + \rho_1 t\mathbf{e}_1, \mathbf{y}_2 + \rho_2 t\mathbf{e}_2, \ldots, \mathbf{y}_N + \rho_N t\mathbf{e}_N) \text{ modulo one}\\ &= \vec{\mathcal{Y}} + t\mathbf{v}(\omega) \text{ modulo one,}\end{aligned} \tag{13.29}$$

where (see (13.11))

$$\omega = (\rho_1, \mathbf{e}_1, \rho_2, \mathbf{e}_2, \ldots, \rho_N, \mathbf{e}_N) \in \Omega_{\text{Gauss}} = ([0,\infty) \times \mathbf{S}^2)^N, \tag{13.30}$$

$$\mathbf{v}(\omega) = (\rho_1\mathbf{e}_1, \rho_2\mathbf{e}_2, \ldots, \rho_N\mathbf{e}_N), \tag{13.31}$$

and the product space Ω_{Gauss} is equipped with the product measure $\text{ProdMeas}_{\text{Gauss}}$; here the half-line $[0,\infty)$ has the probability density function $g(u) = \sqrt{\frac{2}{\pi}}u^2e^{-u^2/2}$ (see (13.28)). We refer to this special case as the *simplest Gaussian torus-billiard model* (in the 3-space).

We leave now the Gaussian case, and formulate the first extension of Theorem 4.2 in the case of the "polynomial decay".

Theorem 13.1. *Assume* $\mathcal{R}$ *satisfies* (13.4)–(13.6), $\mathcal{G}$ *satisfies the "polynomial decay" condition* (13.15) *for the* 3*-dimensional "isotropic Fourier transforms"* $\mathbf{F}_{g_k}$, *and*

$$W > U \geq \max\left\{\alpha_0, \beta_0, \left(\frac{\gamma_0}{2\pi}\right)^{1/\alpha}\right\}, \quad \gamma > 3 \quad \textit{and} \quad \alpha\gamma > 1.$$

Then

$$\Delta_f^2(\mathcal{G};\mathcal{R};U,W) \leq \lceil \log_2(W/U)\rceil \sigma_0^2(f)(W-U)$$
$$\cdot\left(25N^2\pi^{-\gamma}U^{1-\alpha\gamma}\frac{2^{5-\gamma}}{1-2^{3-\gamma}}\right.$$

$$+ 5N^3\pi^{-2\gamma}U^{1-2\alpha\gamma}\left(\frac{2^{5-\gamma}}{1-2^{3-\gamma}}\right)^2$$

$$+ 2W^{\beta}\left(1 + (\alpha_1 - \alpha)\log W + \alpha_1 \log 2 + \log 3\right)$$

$$\times \left(1 + \gamma_0 + (2\pi)^{-\gamma}\frac{\gamma_0^{1-\gamma}}{\gamma - 1}\right)\Bigg). \tag{13.32}$$

The upper bound (13.32) is quite complicated; nevertheless, it is very simple to apply Theorem 13.1. In the applications N is very large (say, N is around the Avogadro number), so we have to choose U in such a way that $N^2U^{1-\alpha\gamma} = O(1)$.

We will use Theorem 13.1 to extend the application of Theorem 4.2 about advanced randomness and its long-term stability for large classes of smooth initial velocity distributions far beyond the Gaussian case.

The proof of Theorem 13.1 goes along the same lines as that of Theorem 4.2, but the details are more complicated; see Secs. 27–29.

The Class of Exponential Decay. The technical condition (13.15) represents a polynomial decay of tending to zero (for the tail of the "isotropic Fourier transform").

A simple sufficient condition — formulated in Lemma 13.1 — is that the density function of the speed distribution is "relatively smooth".

On the other hand, the Gaussian distribution represents a superexponential decay tending to zero (for the tail of the "isotropic Fourier transform"); see Lemma 9.1. We show that there is a whole class of probability distributions such that the corresponding "isotropic Fourier transform" tends to zero (at least) exponentially fast. For this "exponential decay class" we can prove a stronger version of Theorem 13.1.

We need the following "Paley–Wiener type" result in complex variables.

Lemma 13.4. *Let $f(z)$ be a function analytic in the horizontal strip on the complex plane where the imaginary part of z has absolute value $\leq a$, and let*

$$\sup_{|y|\leq a}\int_{-\infty}^{\infty}|f(x+\mathrm{i}y)|^2\,dx < \infty.$$

Then the tail of the (ordinary) Fourier transform has the decay

$$\left|\widehat{f}(z)\right| \leq C_b e^{-b|z|} \text{ for every } z \in \mathbb{C} \quad \text{and} \quad 0 < b < a, \tag{13.33}$$

where $C_b = c(f; b)$ is a positive constant independent of $z \in \mathbb{C}$.

For a proof, we refer the reader to Reed and Simon, Methods of Modern Mathematical Physics, Vol. 2, Theorem IX.14.

For illustration, consider the function $f(z) = e^{-z^4}$, which is analytic on the whole complex plane. (We picked this function, because, unlike the case of e^{-z^2}, for $f(z) = e^{-z^4}$ we do not know the explicit form of the (ordinary) Fourier transform.) With $z = x + iy$ the function

$$f(z) = e^{-z^4} \text{ has absolute value } |f(z)| = e^{-x^4+6x^2y^2-y^4},$$

and so

$$\sup_{|y|\leq a} \int_{-\infty}^{\infty} |f(x+iy)|^2\, dx = \sup_{|y|\leq a} e^{-2a^4} \int_{-\infty}^{\infty} e^{-2x^4+12a^2x^2}\, dx = C(a) < \infty$$

for every $a > 0$. Since (13.33) holds for *arbitrarily large* a, it follows that $|\widehat{f}(z)|$ tends to zero exponentially fast as $|z| \to \infty$ on the complex plane. (In fact, we may call it superexponentially fast since a is arbitrarily large.)

The ordinary Fourier transform is just an intermediate step; what we are really interested in is the 3-dimensional "isotropic Fourier transform" (see Lemma 13.2)

$$\mathbf{F}_g(w) = \frac{\sqrt{\pi}}{2\sqrt{2}w}\left(g(0)\sqrt{2\pi} - \int_{|x|\geq w} \widehat{g}(x)\, dx\right) \tag{13.34}$$

of the probability density function $g(u)$ defined in $[0, \infty)$ (noting that on the right-hand side of (13.34) g is extended over the whole real line as an even function $g(-u) = g(u)$).

If $g(y)$ is continuous at $y = 0$ and $g(0) = 0$, then (13.34) simplifies to

$$\mathbf{F}_g(w) = -\frac{\sqrt{\pi}}{\sqrt{2}w} \int_{x=w}^{\infty} \widehat{g}(x)\, dx.$$

Suppose that g also satisfies the condition of Lemma 13.4 with some $a > 0$; then by Lemma 13.4 for every $0 < b < a$,

$$|\mathbf{F}_g(w)| \le \frac{\sqrt{\pi}}{\sqrt{2w}} \int_{x=w}^{\infty} |\widehat{g}(x)|\, dx$$

$$\le \frac{\sqrt{\pi}}{\sqrt{2w}} \int_{x=w}^{\infty} c(g;b)e^{-bx}\, dx = \frac{\sqrt{\pi}}{\sqrt{2b}w} c(g;b)e^{-bw},$$

which implies

$$|\mathbf{F}_g(w)| \le \min\left\{ \frac{\sqrt{\pi}c(g;b)}{\sqrt{2b}w} e^{-bw}\, dx, 1 \right\}. \tag{13.35}$$

Consider now the following class of even functions $g(y)$ with $g(0) = 0$. The class consists of

$$y^2e^{-y^2},\ y^4e^{-y^2},\ y^6e^{-y^2},\ldots, y^2e^{-y^4},\ y^4e^{-y^4},\ y^6e^{-y^4},\ldots,$$
$$y^2e^{-y^6},\ y^4e^{-y^6},\ y^6e^{-y^6},\ldots,$$

all finite products formed from these functions, and finally,

all finite sums formed from these products taken with positive

coefficients such that we obtain a probability density function on $[0,\infty)$. (13.36)

Notice that (13.36) is an "exponential decay class". Indeed, the application of Lemma 13.4 for $f(z) = e^{-z^4}$ above can be repeated for every member $g(y)$ of the class (13.36). Thus we obtain that the ordinary Fourier transform $|\widehat{g}(z)|$ tends to zero exponentially fast as $|z| \to \infty$ on the complex plane. Combining this fact with (13.35), we conclude that for every member $g(y)$ of the class (13.36), the "isotropic Fourier transform"

$$\mathbf{F}_g(w) = -\frac{\sqrt{\pi}}{2\sqrt{2}w} \int_{|x|\ge w} \widehat{g}(x)\, dx$$

also tends to zero exponentially fast as $w \to \infty$.

Theorem 13.2. *Assume $\mathcal{R}$ satisfies (13.4)–(13.6), $\mathcal{G}$ satisfies the "exponential decay" condition*

$$\mathbf{F}^{(\text{sup})}(w) = \mathbf{F}_{\mathcal{G}}^{(\text{sup})}(w) \le e^{-\gamma w} \text{ for all } w \ge \gamma_0$$

for the 3-dimensional "isotropic Fourier transforms" $\mathbf{F}_{g_k}$, *and*

$$W > U \geq \max\left\{\alpha_0, \beta_0, \left(\frac{\gamma_0}{2\pi}\right)^{1/\alpha}, (2\pi\gamma\alpha)^{-1/\alpha}\right\}.$$

Then

$$\begin{aligned}\Delta_f^2(\mathcal{G}; \mathcal{R}; U, W) \leq & \lceil \log_2(W/U) \rceil \sigma_0^2(f)(W-U) \\ & \cdot \left(25N^2 \frac{Ue^{-2\pi\gamma U^\alpha}}{1-e^{-2\pi\gamma U^\alpha}} + 5N^3 \frac{Ue^{-4\pi\gamma U^\alpha}}{(1-e^{-2\pi\gamma U^\alpha})^2}\right. \\ & + 2W^\beta \left(1+\gamma_0+\frac{e^{-2\pi\gamma\gamma_0}}{2\pi\gamma}\right) \\ & \left.\times (1+(\alpha_1-\alpha)\log W + \alpha_1 \log 2 + \log 3)\right).\end{aligned}$$

Next we switch from the case of "exponential decay" to the case of generalized Gaussian case, i.e., we assume that $\mathcal{G}$ exhibits "superexponential decay". The advantage of the next result over Theorem 4.2 is that here the particles may move with varying speeds on their own torus lines.

Theorem 13.3. *Assume* $\mathcal{R}$ *satisfies* (13.4)–(13.6), $\mathcal{G}$ *satisfies the "super-exponential decay" condition*

$$\mathbf{F}^{(\mathrm{sup})}(w) = \mathbf{F}_{\mathcal{G}}^{(\mathrm{sup})}(w) \leq e^{-\gamma w^2} \text{ for all } w \geq \gamma_0$$

for the 3-dimensional "isotropic Fourier transforms" $\mathbf{F}_{g_k}$, *and*

$$W > U \geq \max\left\{\alpha_0, \beta_0, \left(\frac{\gamma_0}{2\pi}\right)^{1/\alpha}, (8\pi^2\gamma\alpha)^{-1/(2\alpha)}\right\}.$$

Then

$$\begin{aligned}\Delta_f^2(\mathcal{G}; \mathcal{R}; U, W) \leq & \lceil \log_2(W/U) \rceil \cdot \sigma_0^2(f)(W-U) \\ & \cdot \left(25N^2 \frac{Ue^{-4\pi^2\gamma U^{2\alpha}}}{1-e^{-4\pi^2\gamma U^{2\alpha}}} + 5N^3 \frac{Ue^{-8\pi^2\gamma U^{2\alpha}}}{(1-e^{-4\pi^2\gamma U^{2\alpha}})^2}\right. \\ & + 2W^\beta(1+\gamma_0+e^{-4\pi^2\gamma\gamma_0^2})(1+(\alpha_1-\alpha)\log W \\ & \left. + \alpha_1 \log 2 + \log 3)\right).\end{aligned}$$

Next we switch from the "superexponential decay" to "vanishing", which implies *instant conf-space equilibrium*; for the details, see Sec. 25.

The Class of Vanishing Property. The probability density function $g(y)$ (defined on $0 \le y \le \infty$) satisfies the "vanishing property" if there exists a constant $C > 0$ such that the "isotropic Fourier transform" $\mathbf{F}_g(w)$ is zero for $w > C$. In view of Lemma 13.2, it suffices to guarantee that $g(y)$ is even, $g(y)$ is continuous at $y = 0$ with $g(0) = 0$, and $\widehat{g}(x) = 0$ for $x > C$.

We explain now how to construct such a probability density functions $g(y)$; see (13.49). The first step is to introduce the so-called "roof function"

$$f_2(y) = \begin{cases} \dfrac{1}{2} - \dfrac{|y|}{4}, & \text{if } |y| \le 2; \\ 0, & \text{if } |y| > 2. \end{cases} \tag{13.37}$$

The roof function $f_2(y)$ is familiar from probability theory: it is the density function of the convolution (denoted as usual by $*$) of the uniform distribution in $[-1, 1]$ with itself; formally,

$$f_2(y) = f_1 * f_1(y) = \int_{-\infty}^{\infty} f_1(y - z) f_1(z)\, dz,$$

where

$$f_1(y) = \begin{cases} \dfrac{1}{2}, & \text{if } |y| \le 1; \\ 0, & \text{if } |y| > 1, \end{cases} \tag{13.38}$$

is the probability density function of the uniform distribution in $[-1, 1]$.

It is a standard exercise in probability theory to compute the higher convolution powers of the uniform distribution in $[-1, 1]$. For example, a routine calculation gives

$$f_3(y) = f_1 * f_1 * f_1(y) = \begin{cases} (3 - y^2)/8, & \text{if } |y| \le 1; \\ (3 - |y|)^2/16, & \text{if } 1 \le |y| \le 3; \\ 0, & \text{if } |y| > 3, \end{cases} \tag{13.39}$$

and

$$f_4(y) = f_1 * f_1 * f_1 * f_1(y)$$
$$= \begin{cases} (3|y|^3 - 12y^2 + 32)/96, & \text{if } |y| \le 2; \\ (-|y|^3 + 12y^2 - 48|y| + 64)/96, & \text{if } 2 \le |y| \le 4; \\ 0, & \text{if } |y| > 4. \end{cases} \tag{13.40}$$

In general, let $\kappa \ge 2$ be an arbitrary integer, and consider the κth convolution power of f_1:

$$f_\kappa(y) = f_1 * f_1 * \cdots * f_1(y) \quad (\kappa \text{ times convolution}). \tag{13.41}$$

It is not very hard to prove by induction on κ the following generalization of (13.37) and (13.39)–(13.40) (see e.g. Rényi's book [Re62]):

$$f_\kappa(y) = \frac{1}{2^\kappa(\kappa-1)!} \sum_{j=0}^{\lfloor(\kappa+|y|)/2\rfloor} (-1)^j \binom{\kappa}{j} (\kappa + |y| - 2j)^{\kappa-1} \quad \text{if } |y| \le \kappa, \tag{13.42}$$

and 0 if $|y| > \kappa$. Note that $f_\kappa(y)$ is $(\kappa-2)$-times differentiable, and consists of a few generalized parabola arcs of degree $\kappa - 2$ (due to the jumps of the lower integral part function $\lfloor(\kappa+y)/2\rfloor$ as y runs in $0 \le y \le \kappa$) that smoothly fit together at the endpoints if $\kappa \ge 3$.

We need the Fourier transform of f_1:

$$\widehat{f_1}(u) = \frac{1}{\sqrt{2\pi}} \int_{-1}^{1} \frac{1}{2} e^{-iyu}\, dy = \frac{1}{\sqrt{2\pi}} \frac{\sin u}{u}. \tag{13.43}$$

We use the well-known fact that the Fourier transform of a convolution is the product of the Fourier transforms; formally,

$$\widehat{f * h} = \widehat{f}\, \widehat{h}. \tag{13.44}$$

By (13.43) and (13.44) we have

$$\widehat{f_\kappa}(u) = \left(\frac{1}{\sqrt{2\pi}} \frac{\sin u}{u} \right)^\kappa. \tag{13.45}$$

Write

$$g_\kappa(u) = (1 - \cos u)\left(\frac{1}{\sqrt{2\pi}}\frac{\sin u}{u}\right)^\kappa$$
$$= \left(1 - \frac{1}{2}e^{iu} - \frac{1}{2}e^{-iu}\right)\left(\frac{1}{\sqrt{2\pi}}\frac{\sin u}{u}\right)^\kappa. \tag{13.46}$$

Then by (13.45) and (13.46),

$$\widehat{g_\kappa}(y) = f_\kappa(y) - \frac{1}{2}f_\kappa(y+1) - \frac{1}{2}f_\kappa(y-1) \tag{13.47}$$

and so $\widehat{g_\kappa}(y) = 0$ for all $|y| \geq \kappa + 1$.

If $\kappa \geq 2$ is an even integer, then $g_\kappa(u) \geq 0$, so $c_\kappa g_\kappa(u)$ defines a probability density function in $0 \leq u < \infty$ with the constant factor c_κ satisfying

$$\frac{1}{c_\kappa} = \int_0^\infty g_\kappa(u)\,du = \frac{1}{2}\int_{-\infty}^\infty g_\kappa(u)\,du = \sqrt{\pi/2}\cdot\frac{1}{\sqrt{2\pi}}\int_{-\infty}^\infty g_\kappa(u)\,du$$
$$= \sqrt{\pi/2}\cdot\widehat{g_\kappa}(0) = \sqrt{\pi/2}\left(f_\kappa(0) - \frac{1}{2}f_\kappa(1) - \frac{1}{2}f_\kappa(-1)\right). \tag{13.48}$$

Let $\kappa = 4$; then by (13.40) we have

$$\frac{1}{c_4} = \sqrt{\pi/2}\left(f_4(0) - \frac{1}{2}f_4(1) - \frac{1}{2}f_4(-1)\right) = \frac{32-23}{96}\sqrt{\pi/2} = \frac{3}{32}\sqrt{\pi/2}.$$

Combining these facts we obtain that

$$g(u) = c_4 g_4(u) = \frac{32}{3}\sqrt{\frac{2}{\pi}}g_4(u)$$
$$= \frac{32}{3}\sqrt{\frac{2}{\pi}}(1-\cos u)\left(\frac{1}{\sqrt{2\pi}}\frac{\sin u}{u}\right)^4 = \frac{8\sqrt{2}}{3\pi^2\sqrt{\pi}}(1-\cos u)\left(\frac{\sin u}{u}\right)^4 \tag{13.49}$$

defines a probability density function in $0 \leq u < \infty$ such that $\widehat{g}(y) = 0$ for all $|y| \geq 5$, $g(y)$ is continuous at $y = 0$ with $g(0) = 0$, and $g(y)$ is even.

It easily follows from (13.49) that

$$\int_0^\infty ug(u)\,du < \infty \quad \text{and} \quad \int_0^\infty u^2 g(u)\,du < \infty, \tag{13.50}$$

that is, the first two moments of the corresponding random variable are both finite. (The reason why we did not choose $\kappa = 2$ is that the corresponding expected value $\int_0^\infty uc_2 g_2(u)\,du$ is infinite!)

By choosing $\kappa = 4$ in (13.49), the first two moments of the corresponding random variable are finite (see (13.50)), but the third moment is infinite. By choosing a larger even number κ (i.e., $g(u) = c_\kappa g_\kappa(u)$) we can guarantee that many more higher moments are finite. In fact, the first $\kappa - 2$ moments of the corresponding random variable are finite.

Theorem 13.4. *Assume $\mathcal{R}$ satisfies* (13.4)–(13.6), *$\mathcal{G}$ satisfies the "vanishing property"*

$$\max_{1\le k\le N} |\mathbf{F}_{g_k}(x)| = 0 \text{ for all } x \ge \gamma_0$$

for the 3*-dimensional "isotropic Fourier transforms" $\mathbf{F}_{g_k}$, and*

$$W > U \ge \max\left\{\alpha_0, \beta_0, \left(\frac{\gamma_0}{2\pi}\right)^{1/\alpha}\right\}.$$

Then

$$\Delta_f^2(\mathcal{G};\mathcal{R};U,W) \le 2\lceil\log_2(W/U)\rceil \cdot \sigma_0^2(f)(W-U)W^\beta$$
$$\cdot (1+\gamma_0)(1+(\alpha_1-\alpha)\log W + \alpha_1\log 2 + \log 3).$$

For the proofs of Theorems 13.1–13.4; see Secs. 27–29.

14. Extensions of Theorem 4.2 to Nonlinear Curves on the Plane

In Theorems 4.2 and 13.1–13.4 all particles stay on torus lines, and the only difference is whether the particles move with constant speeds or with varying speeds. This raises the natural question: does there exist a generalization where the orbits of the particles are arbitrary non-linear space curves (like in Theorem 10.1)?

The bad news is that we do not know how to extend Theorem 4.2 to the case of arbitrary space curves in the 3-space (or in higher dimensions), i.e., we do not know when the particle space (= "gas container") has dimension ≥ 3. The good news is that in the 2-dimensional case we can prove some positive results. If the particle space is the 2-dimensional unit torus

$[0,1)^2$, then the curve of the whole system in the configuration space has the form

$$\begin{aligned}&\vec{\mathcal{Y}}(\vec{\mathcal{R}};\rho_1,\vartheta_1,\ldots,\rho_N,\vartheta_N)\\&=\vec{\mathcal{Y}}+\vec{\Gamma}(\vec{\mathcal{R}};\rho_1,\vartheta_1,\ldots,\rho_N,\vartheta_N)\quad\text{modulo one}\\&=(\mathbf{y}_1+\rho_1\vartheta_1\mathbf{r}_1(t),\mathbf{y}_2+\rho_2\vartheta_2\mathbf{r}_2(t),\ldots,\mathbf{y}_N+\rho_N\vartheta_N\mathbf{r}_N(t)\quad\text{modulo one,}\end{aligned}\tag{14.1}$$

where t is the time,

$$\vec{\mathcal{Y}}=(\mathbf{y}_1,\mathbf{y}_2,\ldots,\mathbf{y}_N)\tag{14.2}$$

is the vector (= point in the configuration space) formed from the set

$$\mathcal{Y}=\{\mathbf{y}_1,\mathbf{y}_2,\ldots,\mathbf{y}_N\}$$

of the initial point configuration at the start $t=0$, $\vartheta_k\in\mathrm{SO}(2)$ are 2-dimensional rotations (i.e., we could also write $\vartheta_k\in[0,2\pi)$), and finally

$$\vec{\mathcal{R}}=(\mathbf{r}_1(t),\mathbf{r}_2(t),\ldots,\mathbf{r}_N(t))\tag{14.3}$$

represents a family of parametrized curves on the plane. Notice that the one-dimensional $r_k(t)$ in (13.3) is replaced by the 2-dimensional $\mathbf{r}_k(t)$ in (14.3); we indicate this dimension increase by putting an arrow on the top of $\mathcal{R}$.

(We also need to specify a family of probability density functions $\mathcal{G}=(g_1,g_2,\ldots,g_N)$, where g_k is the distribution of the "magnification factor" ρ_k by the integral $\Pr[\rho_k\le u]=\int_0^u g_k(y)\,dy$; see (14.7) later.)

Here we briefly interrupt the discussion to point out a basic geometric difference between magnifying/shrinking a straight line and a general non-linear curve from one of its own points. If we magnify/shrink a straight line from a point contained by the line, then the image of the line is the *same* line (as a set of points), but the image of a non-linear curve is a *different* (but similar) curve (as a set of points).

We assume that

$$\mathcal{R}^{(\mathrm{inf})}(t)=\min_{1\le k\le N}\inf_{x\ge t}|\mathbf{r}_k(x)|\ge t^\alpha\text{ for all }t\ge\alpha_0,\tag{14.4}$$

$$\sup_{x\le t}\max_{1\le k\le N}|\mathbf{r}'_k(x)|\le t^\kappa\text{ for all }t\ge\alpha_0\quad\text{and}$$

$$\max_{1\le k\le N}|\mathbf{r}'_k(x)|\le\kappa_0\text{ for }0<t<\alpha_0,\tag{14.5}$$

where $1 + \kappa > \alpha > 0$, $\alpha_0 \geq 1$, $\kappa \geq 0$, and $\mathbf{r}'_k(x)$ is the derivative of the parametrized curve $\mathbf{r}_k(x)$.

Moreover, let $C(\mathbf{c}; \varrho)$ denote an arbitrary circle centered at $\mathbf{c} \in \mathbb{R}^2$ with radius ϱ, and assume that

$$\sup_{\mathbf{c}\in\mathbb{R}^2, \varrho>0} \max_{1\leq k\leq N} \frac{\text{length}\{0 \leq t \leq T: \ \mathbf{r}_k(t) \in C(\mathbf{c}; \varrho)\}}{\varrho} \leq T^\beta \tag{14.6}$$

uniformly for all $T \geq \beta_0$ (note that $1 + \kappa > \alpha > 0$, $\alpha_0 \geq 1$, $\kappa \geq 0$, κ_0, $\beta_0 \geq 1$ and $0 \leq \beta < 1$ are absolute constants).

Since (14.1) is a curve in the $2N$-dimensional unit torus (= configuration space), here we work with $d = 2N$ instead of the usual $d = 3N$.

We need to specify a family of probability density functions $\mathcal{G} = (g_1, g_2, \ldots, g_N)$, where g_k is the distribution of the "magnification factor" ρ_k, i.e., $\Pr[\rho_k \leq u] = \int_0^u g_k(y)\, dy$. For simplicity consider first the simplest 2-dimensional Gaussian distribution, i.e.,

$$g_1(u) = \cdots = g_N(u) = ue^{-u^2/2}, \ 0 \leq u < \infty. \tag{14.7}$$

We indicate condition (14.7) by replacing $\mathcal{G}$ with "Gauss".

To formulate a curve analog of Lemma 9.2, we choose a Lebesgue square-integrable test function $f \in L_2(I^d)$ in the d-dimensional unit torus $I^d = [0, 1)^d$ (i.e., we extend f over the whole d-space $\mathbb{R}^d$ periodically), and consider the Fourier expansion of f:

$$f(\mathbf{u}) = \sum_{\mathbf{n}\in\mathbb{Z}^d} a_{\mathbf{n}} e^{2\pi i \mathbf{n}\cdot\mathbf{u}}, \tag{14.8}$$

where

$$a_{\mathbf{n}} = \int_{I^d} f(\mathbf{y}) e^{-2\pi i \mathbf{n}\cdot\mathbf{y}}\, d\mathbf{y},$$

and of course $\mathbf{v} \cdot \mathbf{w}$ denotes the dot product.

If f is real-valued, the Fourier coefficients corresponding to the pair $\mathbf{n}, -\mathbf{n}$ of indices are complex conjugates of each other.

Write

$$\sigma_0^2(f) = \int_{I^d} |f|^2\, d\mathrm{V} - \left| \int_{I^d} f\, d\mathrm{V} \right|^2 = \int_{I^d} \left| f(\mathbf{y}) - \int_{I^d} f\, d\mathrm{V} \right|^2 d\mathbf{y}.$$

Clearly

$$a_{\mathbf{0}} = \int_{I^d} f \, d\mathrm{V}, \quad \text{and so} \quad \sum_{\mathbf{n}\in\mathbb{Z}^d\setminus\mathbf{0}} |a_{\mathbf{n}}|^2 = \sigma_0^2(f),$$

where we used Parseval's formula.

Consider the integral

$$D_f(\vec{\mathcal{R}};\rho_1,\vartheta_1,\ldots,\rho_N,\vartheta_N;T_1,T_2)$$
$$= \int_{T_1}^{T_2} f(\vec{\Gamma}(t))\,dt - (T_2-T_1)\int_{I^d} f\,d\mathrm{V} = \int_{T_1}^{T_2} \sum_{\mathbf{n}\in\mathbb{Z}^d\setminus\mathbf{0}} a_{\mathbf{n}} e^{2\pi\mathrm{i}\mathbf{n}\cdot\vec{\Gamma}(t)}\,dt. \tag{14.9}$$

We study the (2-dimensional) Gaussian average of the square of the absolute value of

$$D_f(\vec{\mathcal{R}};\rho_1,\vartheta_1,\ldots,\rho_N,\vartheta_N;T_1,T_2),$$

which goes as follows:

$$\Delta_f^2(\text{Gauss};\vec{\mathcal{R}};T_1,T_2) = \int_{\rho_1=0}^{\infty}\int_{\vartheta_1\in\mathrm{SO}(2)}\cdots\int_{\rho_N=0}^{\infty}\int_{\vartheta_N\in\mathrm{SO}(2)}$$
$$\cdot\,|D_f(\vec{\mathcal{R}};\rho_1,\vartheta_1,\ldots,\rho_N,\vartheta_N;T_1,T_2)|^2$$
$$\times\,\rho_1 e^{-\rho_1^2/2}\,d\rho_1\,d\vartheta_1\ldots\rho_N e^{-\rho_N^2/2}\,d\rho_N\,d\vartheta_N, \tag{14.10}$$

where $d\vartheta$ refers to the normed measure on $\mathrm{SO}(2)=[0,2\pi)$, i.e., length$/(2\pi)$.

Using the trivial fact

$$\int_a^b h(x)\,dx\int_c^d g(y)\,dy = \int_a^b\int_c^d h(x)g(y)\,dx\,dy$$

in (14.9), we have

$$|D_f(\vec{\mathcal{R}};\rho_1,\vartheta_1,\ldots,\rho_N,\vartheta_N;T_1,T_2)|^2 = \int_{T_1}^{T_2}\int_{T_1}^{T_2}\sum_{\mathbf{n}_1\in\mathbb{Z}^d\setminus\mathbf{0}}\sum_{\mathbf{n}_2\in\mathbb{Z}^d\setminus\mathbf{0}} a_{\mathbf{n}_1}$$
$$\cdot\,\overline{a_{\mathbf{n}_2}}e^{2\pi\mathrm{i}(\vec{\Gamma}(t_1)\cdot\mathbf{n}_1-\vec{\Gamma}(t_2)\cdot\mathbf{n}_2)}\,dt_1\,dt_2, \tag{14.11}$$

where, as usual, $\overline{z}=x-\mathrm{i}y$ denotes the complex conjugate of a complex number $z=x+\mathrm{i}y$.

Applying (14.11) in (14.10), we obtain

$$\Delta_f^2(\text{Gauss}; \vec{\mathcal{R}}; T_1, T_2) = \int_{\rho_1=0}^{\infty} \int_{\vartheta_1 \in \mathrm{SO}(2)} \cdots \int_{\rho_N=0}^{\infty} \int_{\vartheta_N \in \mathrm{SO}(2)}$$

$$\cdot \left(\int_{T_1}^{T_2} \int_{T_1}^{T_2} \sum_{\mathbf{n}_1 \in \mathbb{Z}^d \setminus \mathbf{0}} \sum_{\mathbf{n}_2 \in \mathbb{Z}^d \setminus \mathbf{0}} \right.$$

$$\left. \times\, a_{\mathbf{n}_1} \overline{a_{\mathbf{n}_2}} e^{2\pi \mathrm{i}(\vec{\Gamma}(t_1) \cdot \mathbf{n}_1 - \vec{\Gamma}(t_2) \cdot \mathbf{n}_2)} \, dt_1 \, dt_2 \right)$$

$$\cdot\, \rho_1 e^{-\rho_1^2/2} \, d\rho_1 \, d\vartheta_1 \ldots \rho_N e^{-\rho_N^2/2} \, d\rho_N \, d\vartheta_N$$

$$= \sum_{\mathbf{n}_1 \in \mathbb{Z}^d \setminus \mathbf{0}} \sum_{\mathbf{n}_2 \in \mathbb{Z}^d \setminus \mathbf{0}} a_{\mathbf{n}_1} \overline{a_{\mathbf{n}_2}} \int_{T_1}^{T_2} \int_{T_1}^{T_2}$$

$$\times \left(\int_{\rho_1=0}^{\infty} \int_{\vartheta_1 \in \mathrm{SO}(2)} \cdots \int_{\rho_N=0}^{\infty} \int_{\vartheta_1 \in \mathrm{SO}(2)} \right.$$

$$\cdot\, e^{2\pi \mathrm{i}(\vec{\Gamma}(t_1) \cdot \mathbf{n}_1 - \vec{\Gamma}(t_2) \cdot \mathbf{n}_2)} \rho_1 e^{-\rho_1^2/2} \, d\rho_1 \, d\vartheta_1 \ldots$$

$$\left. \rho_N e^{-\rho_N^2/2} \, d\rho_N \, d\vartheta_N \right) dt_1 \, dt_2. \tag{14.12}$$

For $\mathbf{n} = (n_1, n_2, \ldots, n_d)$, $d = 2N$, let $\mathbf{n}(k) = (n_{2k-1}, n_{2k})$ denote the kth block of consecutive pairs of the coordinates ($1 \le k \le N = d/2$). Since

$$e^{2\pi \mathrm{i}(\vec{\Gamma}(t_1) \cdot \mathbf{n}_1 - \vec{\Gamma}(t_2) \cdot \mathbf{n}_2)} = \prod_{k=1}^{N} e^{2\pi \mathrm{i}(\rho_k \vartheta_k \mathbf{r}_k(t_1) \cdot \mathbf{n}_1(k) - \rho_k \vartheta_k \mathbf{r}_k(t_2) \cdot \mathbf{n}_2(k))},$$

we have

$$\int_{\rho_1=0}^{\infty} \int_{\vartheta_1 \in \mathrm{SO}(2)} \cdots \int_{\rho_N=0}^{\infty} \int_{\vartheta_N \in \mathrm{SO}(2)}$$

$$\cdot\, e^{2\pi \mathrm{i}(\vec{\Gamma}(t_1) \cdot \mathbf{n}_1 - \vec{\Gamma}(t_2) \cdot \mathbf{n}_2)} \rho_1 e^{-\rho_1^2/2} \, d\rho_1 \, d\vartheta_1 \ldots \rho_N e^{-\rho_N^2/2} \, d\rho_N \, d\vartheta_N$$

$$= \prod_{k=1}^{N} \int_{\rho_k=0}^{\infty} \int_{\vartheta_k \in \mathrm{SO}(2)} e^{2\pi \mathrm{i}(\rho_k \vartheta_k \mathbf{r}_k(t_1) \cdot \mathbf{n}_1(k) - \rho_k \vartheta_k \mathbf{r}_k(t_2) \cdot \mathbf{n}_2(k))}$$

$$\times\, \rho_k e^{-\rho_k^2/2} \, d\rho_k \, d\vartheta_k. \tag{14.13}$$

Given t_1 and t_2, write

$$\mathbf{e}_1 = \mathbf{e}_1(k; t_1) = \frac{\mathbf{r}_k(t_1)}{|\mathbf{r}_k(t_1)|} \quad \text{and} \quad \mathbf{e}_2 = \mathbf{e}_2(k; t_2) = \frac{\mathbf{r}_k(t_2)}{|\mathbf{r}_k(t_2)|},$$

where $\mathbf{e}_1, \mathbf{e}_2$ are 2-dimensional unit vectors. Thus we have $\mathbf{r}_k(t_1) = r_k(t_1)\mathbf{e}_1$ and $\mathbf{r}_k(t_2) = r_k(t_2)\mathbf{e}_2$, where $r_k(t_1) = |\mathbf{r}_k(t_1)|$ and $r_k(t_2) = |\mathbf{r}_k(t_2)|$. Let $\theta_{k;t_1\to t_2} \in \mathrm{SO}(2)$ denote the rotation of the plane that maps $\mathbf{e}_1$ to $\mathbf{e}_2$, i.e., $\mathbf{e}_2 = \theta_{k;t_1\to t_2}\mathbf{e}_1$.

Since the dot product is rotation invariant, by using $\mathbf{e}_2 = \theta_{k;t_1\to t_2}\mathbf{e}_1$, we have

$$\begin{aligned}
&\rho_k\vartheta_k\mathbf{r}_k(t_1)\cdot\mathbf{n}_1(k) - \rho_k\vartheta_k\mathbf{r}_k(t_2)\cdot\mathbf{n}_2(k)\\
&\quad= \rho_k\vartheta_k r_k(t_1)\mathbf{e}_1\cdot\mathbf{n}_1(k) - \rho_k\vartheta_k r_k(t_2)\mathbf{e}_2\cdot\mathbf{n}_2(k)\\
&\quad= \rho_k\vartheta_k r_k(t_1)\mathbf{e}_1\cdot\mathbf{n}_1(k) - \rho_k r_k(t_2)\vartheta_k\theta_{k;t_1\to t_2}\mathbf{e}_1\cdot\mathbf{n}_2(k)\\
&\quad= \rho_k\vartheta_k r_k(t_1)\mathbf{e}_1\cdot\mathbf{n}_1(k) - \rho_k r_k(t_2)\theta_{k;t_2\to t_1}\vartheta_k\theta_{k;t_1\to t_2}\mathbf{e}_1\cdot\theta_k[t_2\to t_1]\mathbf{n}_2(k)\\
&\quad= \rho_k\vartheta_k r_k(t_1)\mathbf{e}_1\cdot\mathbf{n}_1(k) - \rho_k r_k(t_2)\vartheta_k\mathbf{e}_1\cdot\theta_{k;t_2\to t_1}\mathbf{n}_2(k)\\
&\quad= \rho_k\vartheta_k\mathbf{e}_1\cdot(r_k(t_1)\mathbf{n}_1(k) - r_k(t_2)\theta_{k;t_2\to t_1}\mathbf{n}_2(k)),
\end{aligned} \tag{14.14}$$

where $\theta_{k;t_2\to t_1} \in \mathrm{SO}(2)$ denotes the inverse of $\theta_{k;t_1\to t_2} \in \mathrm{SO}(2)$. We emphasize the key step in (14.14):

$$\theta_{k;t_2\to t_1}\vartheta_k\theta_{k;t_1\to t_2} = \vartheta_k,$$

which follows from the fact that SO(2) is commutative. This step breaks down in higher dimension, since $\mathrm{SO}(h)$ is *not* commutative for $h \geq 3$.

Using (14.14) in (14.13), we have

$$\begin{aligned}
&\int_{\rho_1=0}^{\infty}\int_{\vartheta_1\in\mathrm{SO}(2)}\cdots\int_{\rho_N=0}^{\infty}\int_{\vartheta_N\in\mathrm{SO}(2)}\\
&\qquad\cdot e^{2\pi\mathrm{i}(\vec{\Gamma}(t_1)\cdot\mathbf{n}_1-\vec{\Gamma}(t_2)\cdot\mathbf{n}_2)}\rho_1 e^{-\rho_1^2/2}\,d\rho_1\,d\vartheta_1\ldots\rho_N e^{-\rho_N^2/2}\,d\rho_N\,d\vartheta_N\\
&\quad=\prod_{k=1}^{N}\int_{\rho_k=0}^{\infty}\int_{\vartheta_k\in\mathrm{SO}(2)} e^{2\pi\mathrm{i}\rho_k\vartheta_k\mathbf{e}\cdot(r_k(t_1)\mathbf{n}_1(k)-r_k(t_2)\theta_{k;t_2\to t_1}\mathbf{n}_2(k))}\\
&\qquad\times\rho_k e^{-\rho_k^2/2}\,d\rho_k\,d\vartheta_k\\
&\quad=\prod_{k=1}^{N} e^{-2\pi^2|r_k(t_1)\mathbf{n}_1(k)-r_k(t_2)\theta_{k;t_2\to t_1}\mathbf{n}_2(k)|^2},
\end{aligned} \tag{14.15}$$

where $\mathbf{e}$ is an arbitrary but fixed 2-dimensional unit vector, and in the last step we used the 2-dimensional special case of Lemma 9.1.

Combining (14.12)–(14.15), we obtain the following lemma.

Lemma 14.1. *For 2-dimensional curves*

$$\vec{\mathcal{R}} = (\mathbf{r}_1(t), \mathbf{r}_2(t), \ldots, \mathbf{r}_N(t))$$

we have

$$\begin{aligned}\Delta_f^2(\text{Gauss}; \vec{\mathcal{R}}; T_1, T_2) = &\sum_{\mathbf{n}_1 \in \mathbb{Z}^d \setminus \mathbf{0}} \sum_{\mathbf{n}_2 \in \mathbb{Z}^d \setminus \mathbf{0}} a_{\mathbf{n}_1} \overline{a_{\mathbf{n}_2}} \cdot \int_{T_1}^{T_2} \int_{T_1}^{T_2} \\ &\times \exp\left(- \sum_{k=1}^{N} 2\pi^2 |r_k(t_1)\mathbf{n}_1(k) \right. \\ &\left. - r_k(t_2)\theta_{k;t_2 \to t_1} \mathbf{n}_2(k)|^2 \right) dt_1\, dt_2. \qquad \square\end{aligned}$$

Lemma 14.1 is a curve analog of Lemma 9.2 in the 2-dimensional case. We do not know how to generalize Lemma 14.1 in higher dimensions.

By using Lemma 14.1 we can prove the following 2-dimensional *curve* version of Theorem 4.2.

Theorem 14.1. *Let $f \in L_2(I^d)$ be a test function, where $I^d = [0,1)^d$ with $d = 2N$. Under the condition of (14.4)–(14.6) for 2-dimensional curves*

$$\vec{\mathcal{R}} = (\mathbf{r}_1(t), \mathbf{r}_2(t), \ldots, \mathbf{r}_N(t)),$$

and also assuming (14.7) and $\exp(\pi^2 U^{2\alpha}/2) \geq 10UN$, for $W > U \geq \max\{\alpha_0, \beta_0, 1\}$ we have

$$\begin{aligned}\Delta_f^2(\text{Gauss}; \vec{\mathcal{R}}; U, W) \leq k \cdot \sigma_0^2(f) \Bigg(&2k + (W-U) \\ &\times \Bigg(2 + \max_{x \geq U} x \exp(-x^{2\alpha}\pi^2 2^{-8}) \\ &+ 9 \left(3 + 2\log_2 \left(1 + \frac{(2W)^{1+\kappa} + \alpha_0\kappa_0}{W^\alpha} \right)\right) \\ &\times (2W)^{1+\kappa+\beta-\alpha} \Bigg)\Bigg)\end{aligned}$$

with $k = \lceil \log_2(W/U) \rceil$ (upper integral part).

For the proof of Theorem 14.1; see Secs. 30–31.

Next we switch from the 2-dimensional Gaussian distribution (14.7) to the general 2-dimensional case of probability density functions $\mathcal{G} = (g_1, g_2, \ldots, g_N)$; here g_k defines the distribution of the "magnification factor" ρ_k by the integral $\Pr[\rho_k \le u] = \int_0^u g_k(y)\, dy$.

We need the 2-dimensional "isotropic Fourier transform" of $g = g_k$, $1 \le k \le N$; it is the perfect analog of (13.13):

$$\mathbf{F}_g(\mathbf{w}) = \int_{u=0}^{\infty} \frac{1}{2\pi} \int_{\mathbf{v}\in\mathbf{S}^1} e^{iu\mathbf{v}\cdot\mathbf{w}} g(u)\, du\, d\mathbf{v}, \tag{14.16}$$

where $\mathbf{w}$ is an arbitrary 2-dimensional vector, $\mathbf{S}^1$ is the unit circle of circumference 2π, and $\int_{\mathbf{v}\in\mathbf{S}^1} \ldots d\mathbf{v}$ stands for the arclength. $\mathbf{F}_g(\mathbf{w})$ defines a real-valued function depending only on the length $w = |\mathbf{w}|$ of the input vector, and we have

$$\mathbf{F}_g(0) = \mathbf{F}_g(\mathbf{0}) = 1 \quad \text{and} \quad |\mathbf{F}_g(w)| = |\mathbf{F}_g(\mathbf{w})| < 1 \quad \text{for } w = |\mathbf{w}| > 0.$$

By the Riemann–Lebesgue theorem $\mathbf{F}_g(w) \to 0$ as $w \to \infty$.

The 2-dimensional case of Lemma 9.1 states that the 2-dimensional "isotropic Fourier transform" of $g(u) = ue^{-u^2/2}$ is $\mathbf{F}_g(w) = e^{-w^2/2}$. This explicit formula was the key step in the proof of (14.15), and Theorem 14.1 is based on (14.15) via Lemma 14.1. Besides $g(u) = ue^{-u^2/2}$ there are many more "Gaussian type" probability density functions in $0 \le u < \infty$ for which the 2-dimensional "isotropic Fourier transform" has a similar explicit form of superexponential decay. For illustration, consider the probability density function $g_3(u) = \frac{1}{2}u^3 e^{-u^2/2}$, $0 \le u < \infty$. To determine the 2-dimensional "isotropic Fourier transform" of g_3, we basically repeat the proof of Lemma 9.1. We have

$$\begin{aligned}
\mathbf{F}_{g_3}(w) &= \int_{\rho=0}^{\infty} \frac{1}{2\pi} \int_{\mathbf{e}\in\mathbf{S}^1} e^{i\mathbf{w}\cdot\rho\mathbf{e}} \frac{1}{2}\rho^3 e^{-\rho^2/2}\, d\rho\, d\mathbf{e} \\
&= \frac{1}{4\pi} \int_{\mathbf{v}\in\mathbb{R}^2} e^{i\mathbf{w}\cdot\mathbf{v}} e^{-|\mathbf{v}|^2/2} |\mathbf{v}|^2\, d\mathbf{v} \\
&= \frac{1}{2}\left(\frac{1}{\sqrt{2\pi}} \int_{-\infty}^{\infty} e^{iw_1 v_1} e^{-v_1^2/2} v_1^2\, dv_1 \right)
\end{aligned}$$

$$\times\left(\frac{1}{\sqrt{2\pi}}\int_{-\infty}^{\infty} e^{\mathrm{i}w_2v_2}e^{-v_2^2/2}\,dv_2\right)$$
$$+\frac{1}{2}\left(\frac{1}{\sqrt{2\pi}}\int_{-\infty}^{\infty} e^{\mathrm{i}w_1v_1}e^{-v_1^2/2}\,dv_1\right)$$
$$\times\left(\frac{1}{\sqrt{2\pi}}\int_{-\infty}^{\infty} e^{\mathrm{i}w_2v_2}e^{-v_2^2/2}v_2^2\,dv_2\right), \qquad (14.17)$$

where the vector $\rho\mathbf{e} = \mathbf{v} = (v_1, v_2)$ has 2-dimensional standard normal (= Gaussian) distribution, and we used the well-known fact that the coordinates v_1, v_2 of $\mathbf{v}$ are independent random variables having standard normal distribution each. Note that

$$\frac{1}{\sqrt{2\pi}}\int_{-\infty}^{\infty} e^{\mathrm{i}wv}e^{-v^2/2}\,dv = \text{Fourier transform of } e^{-v^2/2} \text{ at } w = e^{-w^2/2}. \qquad (14.18)$$

Moreover,

$$\frac{1}{\sqrt{2\pi}}\int_{-\infty}^{\infty} v^2e^{\mathrm{i}wv}e^{-v^2/2}\,dv = -\text{second derivative of the Fourier transform of } e^{-v^2/2} \text{ at } w = e^{-w^2/2}(1-w^2). \qquad (14.19)$$

Combining (14.17)–(14.19), we have

$$\begin{aligned}\mathbf{F}_{g_3}(w) &= \int_{\rho=0}^{\infty}\frac{1}{2\pi}\int_{\mathbf{e}\in\mathbf{S}^1} e^{\mathrm{i}\mathbf{w}\cdot\rho\mathbf{e}}\frac{1}{2}\rho^3e^{-\rho^2/2}\,d\rho\,d\mathbf{e}\\ &= \frac{1}{4\pi}\int_{\mathbf{v}\in\mathbb{R}^2} e^{\mathrm{i}\mathbf{w}\cdot\mathbf{v}}e^{-|\mathbf{v}|^2/2}|\mathbf{v}|^2\,d\mathbf{v}\\ &= \frac{1}{2}e^{-w_1^2/2}(1-w_1^2)e^{-w_2^2/2} + \frac{1}{2}e^{-w_1^2/2}e^{-w_2^2/2}(1-w_2^2)\\ &= e^{-(w_1^2+w_2^2)/2}\left(1-\frac{w_1^2+w_2^2}{2}\right) = e^{-|\mathbf{w}|^2/2}\left(1-\frac{|\mathbf{w}|^2}{2}\right). \qquad (14.20)\end{aligned}$$

Similarly, we can replace the function $\frac{1}{2}u^3e^{-u^2/2}$ with any probability density function of the form

$$g(u) = (c_1u + c_2u^3 + c_3u^5 + \cdots + c_ku^{2k+1})e^{-u^2/2}. \qquad (14.21)$$

Of course we need $g(u) \geq 0$, which can be guaranteed e.g. if all coefficients c_i are nonnegative. The cases in (14.21) are basically the same in the sense that they all have an explicit 2-dimensional "isotropic Fourier transform" (where the factor $(1 - |\mathbf{w}|^2/2)$ in (14.20) is replaced by another polynomial of $|\mathbf{w}|^2$), exhibiting superexponentially fast asymptotic decay.

Write

$$\mathbf{F}_g^{(\text{sup})}(w) = \sup_{x \geq w} |\mathbf{F}_g(x)|. \tag{14.22}$$

So $\mathbf{F}_g^{(\text{sup})}(w)$ is a monotone decreasing positive function tending to 0 as $w \to \infty$.

Next we switch from the superexponentially fast decay of $\mathbf{F}_g^{(\text{sup})}(w)$, $w \to \infty$ in the class (14.21) to the class of much slower polynomial decay.

Condition of polynomial decay. Assume that there exist real numbers $\gamma > 2$ and $\gamma_0 \geq 1$ such that

$$\begin{aligned} \mathbf{F}^{(\text{sup})}(w) = \mathbf{F}_{\mathcal{G}}^{(\text{sup})}(w) &= \max_{1 \leq k \leq N} \mathbf{F}_{g_k}^{(\text{sup})}(w) \\ &= \max_{1 \leq k \leq N} \sup_{x \geq w} |\mathbf{F}_{g_k}(x)| \leq w^{-\gamma} \quad \text{for all} \quad w \geq \gamma_0. \end{aligned} \tag{14.23}$$

(We need the condition $\gamma > 2$ to avoid a "divergence problem" later in the proof of Theorem 14.2; see below.)

We now face the following problem: under what condition about the probability density function g can we guarantee that its 2-dimensional "isotropic Fourier transform" $\mathbf{F}_g$ satisfies the polynomial decay condition (14.23)? The following lemma provides a sufficient condition. It is some sort of analog of Lemma 13.1.

Lemma 14.2. *Let $g(u)$ be a probability density function in $[0, \infty)$ such that $g^{(m)}(0) = 0$ for some integer $m \geq 2$ (where $g^{(m)}$ is the mth derivative of g). We extend g over the whole real line as an even function: $g(-u) = g(u)$. Let $\ell \geq 2$, assume that $g(u)$, $-\infty < u < \infty$ is ℓ-times continuously differentiable, also assume that for every integer $1 \leq j \leq \ell$, the j-th derivative $g^{(j)}(u) \to 0$ as $u \to \infty$, and finally assume*

$$\int_{-\infty}^{\infty} |g^{(j)}(u)|\, du \leq C_0 < \infty.$$

Then the 2-dimensional "isotropic Fourier transform" of g has the upper bound

$$|\mathbf{F}_g(w)| \le C_0 w^{-\gamma} \text{ for all } w \ge \gamma_0$$

with some explicit constant γ_0, where

$$\gamma = \gamma(m;\ell) = \begin{cases} \ell(m+2)/(m+2+\ell), & \text{if } m+2 < \ell; \\ \ell/2, & \text{otherwise.} \end{cases}$$

Remarks. The condition $\gamma > 2$ in (14.23) is easily satisfied; for example, let $m \ge 2$ and $\ell \ge 5$.

Notice that

$$\gamma = \gamma(m;\ell) \ge \min\{(m+2)/2, \ell/2\},$$

so, if both m and ℓ are large, then $\gamma = \gamma(m;\ell)$ is also large.

Proof of Lemma 14.2. The unit circle is formally defined as the set $\mathbf{S}^1 = \{\mathbf{u} \in \mathbb{R}^2 : |\mathbf{u}| = 1\}$. Let $\mathbf{v} \in \mathbf{S}^1$ be an arbitrary but fixed 2-dimensional unit vector, let $-1 \le \beta < \gamma \le 1$ be reals, and consider the subset

$$\mathbf{S}^1(\mathbf{v};\beta,\gamma) = \{\mathbf{u} \in \mathbf{S}^1 : \beta \le \mathbf{v}\cdot\mathbf{u} \le \gamma\}$$

of the unit circle. The arclength of $\mathbf{S}^1(\mathbf{v};\beta,\gamma)$ is independent of the choice of the unit vector $\mathbf{v}$, so, we can simply write ArcLength($\mathbf{S}^1(\beta,\gamma)$). We have the elementary result

$$\frac{\text{ArcLength}(\mathbf{S}^1(\beta,\gamma))}{\text{ArcLength}(\mathbf{S}^1)} = \frac{1}{\pi}\int_\beta^\gamma \frac{dx}{\sqrt{1-x^2}}. \tag{14.24}$$

The proof of (14.24) is based on the facts that

$$\text{ArcLength}(\mathbf{S}^1(\beta,\gamma)) = 2(\arcsin(\gamma) - \arcsin(\beta)) = 2\int_\beta^\gamma \frac{dx}{\sqrt{1-x^2}}$$

and

$$\int_{-1}^1 \frac{dx}{\sqrt{1-x^2}} = \pi.$$

Using (14.24) in (14.16), we have

$$\begin{aligned}\mathbf{F}_g(w) &= \int_0^\infty \frac{1}{2\pi}\int_{\mathbf{v}\in\mathbf{S}^1} e^{\mathrm{i}u\mathbf{v}\cdot\mathbf{w}} g(u)\,du\,d\mathbf{v}\\ &= \frac{1}{\pi}\int_{u=0}^\infty \left(\int_{x=-1}^1 \frac{e^{\mathrm{i}uwx}}{\sqrt{1-x^2}}\,dx\right) g(u)\,du\\ &= \frac{1}{\pi}\int_{u=0}^\infty \left(\int_{x=-1}^1 \frac{\cos(uwx)}{\sqrt{1-x^2}}\,dx\right) g(u)\,du\\ &= \frac{2}{\pi}\int_{x=0}^1 \left(\int_{u=0}^\infty \cos(uwx)g(u)\,du\right)\frac{dx}{\sqrt{1-x^2}}. \qquad (14.25)\end{aligned}$$

Similarly to Sec. 13 (see Lemma 13.1), we extend the probability density function $g(u)$, $0 \le u < \infty$ over the whole real line as an even function: $g(-u) = g(u)$. Using the extended even function $g(u)$, $-\infty < u < \infty$ in (14.25), we have

$$\begin{aligned}\mathbf{F}_g(w) &= \frac{2}{\pi}\int_{x=0}^1 \left(\int_{u=0}^\infty \cos(uwx)g(u)\,du\right)\frac{dx}{\sqrt{1-x^2}}\\ &= \frac{1}{\pi}\int_{x=0}^1 \left(\int_{u=-\infty}^\infty \cos(uwx)g(u)\,du\right)\frac{dx}{\sqrt{1-x^2}}\\ &= \frac{1}{\pi}\int_{x=0}^1 \left(\int_{u=-\infty}^\infty e^{\mathrm{i}uwx}g(u)\,du\right)\frac{dx}{\sqrt{1-x^2}}\\ &= \frac{\sqrt{2}}{\sqrt{\pi}}\int_{x=0}^1 \left(\frac{1}{\sqrt{2\pi}}\int_{u=-\infty}^\infty e^{\mathrm{i}uwx}g(u)\,du\right)\frac{dx}{\sqrt{1-x^2}}\\ &= \sqrt{\frac{2}{\pi}}\int_{x=0}^1 \frac{\widehat{g}(-wx)}{\sqrt{1-x^2}}\,dx = \sqrt{\frac{2}{\pi}}\int_{x=0}^1 \frac{\widehat{g}(wx)}{\sqrt{1-x^2}}\,dx,\end{aligned} \qquad (14.26)$$

since the Fourier transform $\widehat{g}$ of g is also even.

For every fixed $0 < c < 1$, the binomial series

$$\frac{1}{\sqrt{1-x^2}} = 1 + \sum_{k=1}^\infty \binom{-1/2}{k}(-1)^k x^{2k} \qquad (14.27)$$

is absolutely convergent in the interval $0 \le x \le c$. Note that

$$\left|\binom{-1/2}{k}\right| = \left|\frac{(-1/2)(-3/2)(-5/2)\cdots(-k+\frac{1}{2})}{1\cdot 2\cdot 3\cdots k}\right| \le 1. \tag{14.28}$$

By using (14.27) and (14.28), we have

$$\begin{aligned}\left|\int_{x=0}^{1}\frac{\widehat{g}(wx)}{\sqrt{1-x^2}}\,dx\right| &\le \left|\int_{x=0}^{c}\frac{\widehat{g}(wx)}{\sqrt{1-x^2}}\,dx\right| + \left|\int_{x=c}^{1}\frac{\widehat{g}(wx)}{\sqrt{1-x^2}}\,dx\right| \\ &\le \sum_{k=0}^{\infty}\left|\int_{x=0}^{c}\widehat{g}(wx)x^{2k}\,dx\right| \\ &\quad + \max_{cw\le y\le w}|\widehat{g}(y)|\int_{x=0}^{1}\frac{1}{\sqrt{1-x^2}}\,dx \\ &= \frac{1}{2}\sum_{k=0}^{\infty}\left|\int_{x=-c}^{c}\widehat{g}(wx)x^{2k}\,dx\right| + \frac{\pi}{2}\max_{cw\le y\le w}|\widehat{g}(y)|,\end{aligned} \tag{14.29}$$

since $\widehat{g}$ is an even function.

Applying the substitution $y = xw$, we have

$$\begin{aligned}\int_{x=-c}^{c}\widehat{g}(wx)x^{2k}\,dx &= w^{-2k-1}\int_{-cw}^{cw}\widehat{g}(y)y^{2k}\,dy \\ &= w^{-2k-1}\left(\int_{-\infty}^{\infty}\widehat{g}(y)y^{2k}\,dy - \int_{|y|\ge cw}\widehat{g}(y)y^{2k}\,dy\right),\end{aligned} \tag{14.30}$$

assuming the infinite integrals are convergent. We recall the well-known general fact from Fourier analysis

$$\frac{1}{\sqrt{2\pi}}\int_{-\infty}^{\infty}\widehat{g}(y)y^{2k}\,dy = (-1)^k g^{(2k)}(0), \tag{14.31}$$

where $g^{(j)}(u)$ denotes the jth derivative of $g(u)$. (Notice that (14.19) is a special case of (14.31) with $k = 1$.) Using (14.31) in (14.30), we obtain the upper bound

$$\left|\int_{x=-c}^{c}\widehat{g}(wx)x^{2k}\,dx\right| \le w^{-2k-1}\left(\sqrt{2\pi}\left|g^{(2k)}(0)\right| + 2\int_{cw}^{\infty}|\widehat{g}(y)|\,y^{2k}\,dy\right). \tag{14.32}$$

Moreover, we have the trivial upper bound

$$\left|\int_{x=-c}^{c} \widehat{g}(wx)x^{2k}\,dx\right| \le \int_{x=-c}^{c} x^{2k}\,dx = \frac{2}{2k+1}c^{2k+1}. \tag{14.33}$$

We recall that the probability density function $g(u)$, $u \in [0,\infty)$ is extended over the whole real line as an even function: $g(-u) = g(u)$. Let $\ell \ge 2$, assume that $g(u)$, $-\infty < u < \infty$ is ℓ-times continuously differentiable, also assume that for every integer $1 \le j \le \ell$, the jth derivative $g^{(j)}(u) \to 0$ as $u \to \infty$, and finally assume

$$\int_{-\infty}^{\infty} |g^{(j)}(u)|\,du \le C_0 < \infty.$$

Then by (13.16)

$$|\widehat{g}(x)| \le \frac{C_0}{\sqrt{2\pi}|x|^{\ell}}. \tag{14.34}$$

If $\ell \ge 2k+2$, then by (14.34)

$$\int_{cw}^{\infty} |\widehat{g}(y)|\, y^{2k}\,dy \le \frac{C_0}{\sqrt{2\pi}} \int_{cw}^{\infty} y^{2k}\,dy = \frac{C_0}{\sqrt{2\pi}} \cdot \frac{(cw)^{2k+1-l}}{\ell-2k-1}. \tag{14.35}$$

Assume that $g^{(m)}(0) = 0$, that is, g has the form

$$g(u) = u^{m+1}g_1(u) \quad \text{with} \quad g_1(0) \ne 0 \text{ for some integer } m \ge 2. \tag{14.36}$$

Using (14.35) and (14.36) in (14.32), for $0 \le k \le \min\{m/2, (\ell-2)/2\}$ we have

$$\begin{aligned}\left|\int_{x=-c}^{c} \widehat{g}(wx)x^{2k}\,dx\right| &\le 2w^{-2k-1}\int_{cw}^{\infty} |\widehat{g}(y)|y^{2k}\,dy \\ &\le \frac{2C_0}{\sqrt{2\pi}} \cdot \frac{c^{2k+1-l}}{(\ell-2k-1)w^{\ell}}.\end{aligned} \tag{14.37}$$

Combining (14.26), (14.29) and (14.37), we obtain

$$\begin{aligned}|\mathbf{F}_g(w)| &\le \frac{1}{\sqrt{2\pi}} \sum_{k=0}^{\infty} \left|\int_{x=-c}^{c} \widehat{g}(wx)x^{2k}\,dx\right| + \sqrt{\frac{\pi}{2}} \max_{cw\le y\le w} |\widehat{g}(y)| \\ &= \frac{1}{\sqrt{2\pi}} \sum_{0\le k\le \min\{m/2,(\ell-2)/2\}} \left|\int_{x=-c}^{c} \widehat{g}(wx)x^{2k}\,dx\right| + \frac{1}{\sqrt{2\pi}}\end{aligned}$$

$$
\begin{aligned}
&\times \sum_{k>\min\{m/2,(\ell-2)/2\}} \left| \int_{x=-c}^{c} \widehat{g}(wx)x^{2k}\, dx \right| + \sqrt{\frac{\pi}{2}} \max_{cw\le y\le w} |\widehat{g}(y)| \\
&\le \frac{C_0}{\pi} \sum_{0\le k\le \min\{m/2,(\ell-2)/2\}} \frac{c^{2k+1-l}}{(\ell-2k-1)w^{\ell}} \\
&\quad + \frac{1}{\sqrt{2\pi}} \sum_{k>\min\{m/2,(\ell-2)/2\}} \frac{2}{2k+1} c^{2k+1} + \sqrt{\frac{\pi}{2}} \cdot \frac{C_0}{\sqrt{2\pi}(cw)^{\ell}},
\end{aligned} \tag{14.38}
$$

where in the last line we used (14.33) and (14.34).

Let $n_0 = \min\{m/2, (\ell-2)/2\}$; then we can rewrite (14.38) as follows:

$$
\begin{aligned}
|\mathbf{F}_g(w)| &\le \frac{C_0}{\pi(cw)^{\ell}} \sum_{0\le k\le n_0} \frac{c^{2k+1}}{\ell-2k-1} + \frac{1}{\sqrt{2\pi}} \sum_{k>n_0} \frac{2}{2k+1} c^{2k+1} + \frac{C_0}{2(cw)^{\ell}} \\
&\le \frac{C_0}{\pi(cw)^{\ell}} \cdot \frac{c}{1-c^2} + \frac{1}{\sqrt{2\pi}(1+n_0)} c^{2n_0+2} + \frac{C_0}{2(cw)^{\ell}}.
\end{aligned} \tag{14.39}
$$

We choose the value of $c \in (0,1)$ in such a way that

$$
c^{2n_0+2} = \frac{1}{(cw)^{\ell}} \iff c = w^{-\ell/(2n_0+2+\ell)}. \tag{14.40}
$$

Using (14.40) in (14.39), we have

$$
\begin{aligned}
|\mathbf{F}_g(w)| &\le \frac{C_0}{\pi(cw)^{\ell}} \cdot \frac{c}{1-c^2} + \frac{1}{\sqrt{2\pi}(1+n_0)} w^{-2\ell(n_0+1)/(2n_0+2+\ell)} + \frac{C_0}{2(cw)^{\ell}} \\
&\le C_0 w^{-\ell(2n_0+2)/(2n_0+2+\ell)}
\end{aligned} \tag{14.41}
$$

if w is larger than some explicit constant γ_0. By definition $2n_0 + 2 = \min\{m+2, \ell\}$, so we can rewrite (14.41) in the form

$$
|\mathbf{F}_g(w)| \le C_0 w^{-\gamma} \quad \text{for all } w \ge \gamma_0, \tag{14.42}
$$

where

$$\gamma = \frac{\ell \min\{m+2,\ell\}}{\min\{m+2,\ell\}+\ell}$$
$$= \begin{cases} \ell(m+2)/(m+2+\ell), & \text{if } m+2<\ell; \\ \ell/2, & \text{otherwise.} \end{cases} \tag{14.43}$$

This completes the proof of Lemma 14.2. □

We conclude Sec. 14 with formulating an analog of Theorem 14.1, where the Gaussian superexponential decay is replaced by the polynomial decay (14.23).

Theorem 14.2. *Let $f \in L_2(I^d)$ be a test function, where $I^d = [0,1)^d$ with $d = 2N$. Under the condition of* (14.4)–(14.6) *for 2-dimensional curves*

$$\vec{\mathcal{R}} = (\mathbf{r}_1(t), \mathbf{r}_2(t), \ldots, \mathbf{r}_N(t)),$$

and also assuming the polynomial decay condition (14.23), *for*

$$W > U \geq \max\left\{\alpha_0, \beta_0, \left(\frac{\gamma_0}{2\pi}\right)^{1/\alpha}\right\}, \ \gamma > 2 \quad \text{and} \quad \alpha\gamma > 1$$

we have

$$\begin{aligned}\Delta_f^2(\mathcal{G};\vec{\mathcal{R}};U,W) &\leq \lceil \log_2(W/U)\rceil \cdot \sigma_0^2(f)(W-U) \\ &\quad\times \left(25N^2\pi^{-\gamma}U^{1-\alpha\gamma}\frac{2^{4-\gamma}}{1-2^{2-\gamma}}\right. \\ &\quad + 5N^3\pi^{-2\gamma}U^{1-2\alpha\gamma}\left(\frac{2^{4-\gamma}}{1-2^{2-\gamma}}\right)^2 \\ &\quad + 8\left(1+\gamma_0+(2\pi)^{-\gamma}\frac{\gamma_0^{1-\gamma}}{\gamma-1}\right) \\ &\quad \left.\times\left(3+2\log_2\left(1+\frac{(2W)^{1+\kappa}+\alpha_0\kappa_0}{W^\alpha}\right)\right)(2W)^{1+\kappa+\beta-\alpha}\right).\end{aligned}$$

For the proof of Theorem 14.2; see Sec. 32.

In Chap. 3 we show more applications of our "sort-time ergodic theorems". To make it as simple and as transparent as possible, we just apply the simplest such result, Theorem 4.2, and explain the general case later in Sec. 24.

Chapter 3

More Applications of Theorem 4.2

15. Snapshot Randomness (II): Central Limit Theorem

In Sec. 8 we discussed the Poisson partition test. Here we continue Sec. 8 by replacing the Poisson limit law with the central limit theorem.

CLT partition test. Bob makes a partition $[0,1]^3 = H_1 \cup H_2 \cup \ldots \cup H_n$ of the unit cube with $n = \sqrt{N}$ parts (for simplicity assume that N is a complete square), where each H_i is measurable and $\mathrm{vol}(H_i) = 1/n$. (We choose $n = \sqrt{N}$ for simplicity; any other choice $n = \lfloor N^c \rfloor$ with $0 < c < 1$ would be fine.) Let $h_i = h_i(\mathcal{Z})$ denote the number of points of a given N-set

$$\mathcal{Z} = \{\mathbf{z}_1, \ldots, \mathbf{z}_N\}$$

(representing Bob's N-set) that are in H_i, and consider the n real numbers

$$\frac{h_i(\mathcal{Z}) - n}{\sqrt{n}}, \quad 1 \le i \le n = \sqrt{N}. \tag{15.1}$$

We compare the empirical distribution function of the set (15.1) to the standard normal distribution:

$$\mathrm{CltError}(\mathcal{Z}) = \sup_{-\infty<\lambda<\infty} \left| \frac{1}{n} \left| \left\{ 1 \le i \le n : \frac{h_i(\mathcal{Z}) - n}{\sqrt{n}} \le \lambda \right\} \right| - \frac{1}{\sqrt{2\pi}} \int_{-\infty}^{\lambda} e^{-u^2/2}\, du \right|. \tag{15.2}$$

Since $n^{-1/2}$ is the "random error", if the error in (15.2) satisfies the upper bound

$$\mathrm{CltError}(\mathcal{Z}) < n^{-\frac{1}{2}+\epsilon} = N^{-\frac{1}{4}+\frac{\epsilon}{2}}, \tag{15.3}$$

then we say that the N-set $\mathcal{Z}$ is ϵ-close to exhibiting *maximum snapshot randomness* with respect to the CLT-partition test $\mathcal{H} = \{H_1, \ldots, H_n\}$.

Note that Alice's honest random N-set certainly satisfies (15.3) with probability extremely close to one. Since the two sisters (Alice and Wicky) do not know Bob's partition $\mathcal{H}$, and the error term (15.3) is nearly optimal, if Bob's N-set $\mathcal{Z}$ happens to satisfy (15.3), then Bob has every reason to believe that his N-set is "truly random". For example, if $N = 10^{100}$ and $\varepsilon = 4 \cdot 10^{-2}$ (say), then

$$N^{-\frac{1}{4}+\frac{\varepsilon}{2}} = 10^{-23}$$

is a very small error. Such a very small error indicates "complete CLT snapshot randomness".

For the sake of brief notation, write

$$\mathcal{N}(\lambda) = \frac{1}{\sqrt{2\pi}} \int_{-\infty}^{\lambda} e^{-u^2/2}\, du \tag{15.4}$$

for the standard normal distribution.

For the usual application of Theorem 4.2, we need information about the volume of the subset

$$\begin{aligned} S(\lambda; \mathcal{H}; \beta) &= S(\text{Error}_\lambda \geq \beta n^{-1/2}) \\ &= \left\{ \vec{\mathcal{Z}} \in I^d : \left| \frac{1}{n} \left| \left\{ 1 \leq i \leq n : \frac{h_i(\mathcal{Z}) - n}{\sqrt{n}} \leq \lambda \right\} \right| - p_1(\lambda) \right| \geq \beta n^{-1/2} \right\} \end{aligned} \tag{15.5}$$

in the configuration space $I^d = [0,1)^d$, $d = 3N$, where $\beta > 0$ is arbitrary. Our goal is to prove the upper bound (5.15); see below.

We need to estimate the d-dimensional volume

$$\text{vol}_d(S(\lambda; \mathcal{H}; \beta)) = \text{vol}_d(S(\text{Error}_\lambda \geq \beta n^{-1/2})). \tag{15.6}$$

Like in Sec. 8, the d-dimensional volume in (15.6) is exactly the analog probability about the distribution of Alice's honestly constructed random N-element point set in Bob's $\sqrt{N}$-partition of the unit cube. And again (15.6) can be expressed in terms of a large deviation type probability for dependent random variables (see (15.20) below).

We reduce (15.6) to a problem of the N-balls-to-N-bins setup introduced in Sec. 8, and a key step in the argument is again an application of the simple Lemma 8.1. In order to do that, we divide every set H_i of volume $1/n$ into

n measurable parts $H_{i,j}$, $1 \leq j \leq n$ such that $\mathrm{vol}(H_{i,j}) = n^{-2} = 1/N$. We rename the sets: let $B_{(i-1)n+j} = H_{i,j}$, and let $b_l = b_l(\mathcal{Z})$ denote the number of points of the given N-set $\mathcal{Z}$ that are contained in B_l. So

$$h_i(\mathcal{Z}) = \sum_{j=1}^{n} b_{(i-1)n+j}(\mathcal{Z}). \tag{15.7}$$

Since there are N sets $B_{(i-1)n+j}(1 \leq i, j \leq n = \sqrt{N})$ of the same volume $1/N$, we can use what we know about the N-balls-to-N-bins problem. We place N balls independently and uniformly into N bins. Again let Y_l denote the number of balls in the lth bin B_l. (This represents the number of elements of the N-set $\mathcal{Z}$ in $H_{i,j}$ where $l = (i-1)n + j$.) Again we make use of the auxiliary random variables W_l, $1 \leq l \leq N$ that are *independent*. Let W_l, $1 \leq l \leq N$ be independent random variables, all having Poisson distribution with parameter one, i.e., $\Pr[W_l = k] = (k!e)^{-1}$, $k = 0, 1, 2, 3, \ldots$.

For later applications we need information about the first three moments of the Poisson distribution. We have

$$\mathbf{E}W_l = \sum_{k=0}^{\infty} k \cdot (k!e)^{-1} = \sum_{k=1}^{\infty} ((k-1)!e)^{-1} = e/e = 1, \tag{15.8}$$

$$\mathbf{E}W_l^2 = \sum_{k=0}^{\infty} k^2 \cdot (k!e)^{-1} = \sum_{k=0}^{\infty} (k + k(k-1)) \cdot (k!e)^{-1}$$

$$= \sum_{k=1}^{\infty} ((k-1)!e)^{-1} + \sum_{k=2}^{\infty} ((k-2)!e)^{-1} = 1 + 1 = 2, \tag{15.9}$$

and

$$\mathbf{E}|W_l - 1|^3 \leq 1 + \mathbf{E}W_l^3 = 1 + \sum_{k=0}^{\infty} k^3 \cdot (k!e)^{-1}$$

$$= 1 + \sum_{k=0}^{\infty} (k + 3k(k-1) + k(k-1)(k-2)) \cdot (k!e)^{-1}$$

$$= 1 + (1 + 3 + 1) = 6. \tag{15.10}$$

Combining (15.8) and (15.9), we obtain the variance

$$\text{variance} = \mathbf{E}W_l^2 - (\mathbf{E}W_l)^2 = 2 - 1 = 1. \tag{15.11}$$

Let $\mathbf{Y}(i) = Y_{(i-1)n+1} + \cdots + Y_{in}$ denote the number of balls in the union of the n bins B_l, $(i-1)n + 1 \leq l \leq in$. For every $1 \leq i \leq n$, write

$\mathbf{W}(i) = W_{(i-1)n+1} + \cdots + W_{in}$; it is a sum of n independent, identically distributed random variables with Poisson distribution of parameter one. For every $-\infty < \lambda < \infty$ write

$$p(\lambda) = \Pr[\mathbf{W}(i) \leq n + \lambda\sqrt{n}]. \tag{15.12}$$

(Note that $p(\lambda)$ is independent of the value of i.) We need a quantitative form of the central limit theorem with an explicit error term. We formulate a slightly more general result where the random components are not necessarily identically distributed (the proof is almost the same; see e.g. in Feller's well-known book [F71]).

Central Limit Theorem: Berry–Esseen version. *Let $Z_1, \ldots, Z_n$ be independent random variables with zero expectation $\mathbf{E}Z_i = 0$, and finite first three moments $\mathbf{E}Z_i^2 < \infty$, $\mathbf{E}|Z_i|^3 < \infty$, $1 \leq i \leq n$, and write*

$$V = \text{variance} = \sum_{i=1}^{n} \mathbf{E}Z_i^2.$$

Then for every real λ

$$\left| \Pr\left[\frac{Z_1 + \cdots + Z_n}{\sqrt{V}} \leq \lambda \right] - \mathcal{N}(\lambda) \right| \leq \frac{1}{\sqrt{V}} \max_{1 \leq i \leq n} \frac{\mathbf{E}|Z_i|^3}{\mathbf{E}Z_i^2}, \tag{15.13}$$

where $\mathcal{N}(\lambda)$ is the standard normal distribution, see (15.4).

By (15.8)–(15.13),

$$\sup_{-\infty<\lambda<\infty} |\mathcal{N}(\lambda) - p(\lambda)| \leq \frac{6}{\sqrt{n}}. \tag{15.14}$$

By repeating the proof of Lemma 8.2, we show that for $\beta > 6$

$$\text{vol}_d\left(S(\text{Error}_\lambda \geq \beta n^{-1/2}) \right) \leq 6\sqrt{N} \exp\left(-\frac{2(\beta-6)^2}{1+\frac{2\beta}{\sqrt{N}}} \right). \tag{15.15}$$

To prove (15.15), first we introduce some notation: for every $1 \leq i \leq n$, let $\mathbf{Y}^*(i;\lambda) = 1$ or 0 depending on whether

$$\mathbf{Y}(i) = Y_{(i-1)n+1} + \cdots + Y_{in} \leq n + \lambda\sqrt{n} \text{ or not,}$$

and similarly, for every $1 \leq i \leq n$, let $\mathbf{W}^*(i;\lambda) = 1$ or 0 depending on whether

$$\mathbf{W}(i) = W_{(i-1)n+1} + \cdots + W_{in} \leq n + \lambda\sqrt{n} \text{ or not.}$$

Let

$$\mathbf{Y}^*(\lambda) = \sum_{i=1}^{n} \mathbf{Y}^*(i;\lambda) \quad \text{and} \quad \mathbf{W}^*(\lambda) = \sum_{i=1}^{n} \mathbf{W}^*(i;\lambda),$$

where $\mathbf{W}^*(\lambda)$ is a sum of n $0,1$ valued independent Bernoulli variables with $p = p(\lambda)$.

By using (15.14) and Lemma 8.1, for every λ and $\gamma > 0$,

$$\Pr\left[\left|\frac{1}{n}\mathbf{Y}^*(\lambda) - \mathcal{N}(\lambda)\right| \geq \gamma n^{-1/2} + 6n^{-1/2}\right]$$
$$\leq \Pr\left[\left|\frac{1}{n}\mathbf{Y}^*(\lambda) - p(\lambda)\right| \geq \gamma n^{-1/2}\right]$$
$$\leq \frac{N!}{(N/e)^N} \Pr\left[\left|\frac{1}{n}\mathbf{W}^*(\lambda) - p(\lambda)\right| \geq \gamma n^{-1/2}\right]. \tag{15.16}$$

Since $\mathbf{W}^*(\lambda) = \sum_{i=1}^n \mathbf{W}^*(i;\lambda)$ is a sum of n $0,1$ valued independent Bernoulli variables with $p = p(\lambda)$, by Bernstein's inequality (12.9)

$$\Pr[|\mathbf{W}^*(\lambda) - np(\lambda)| \geq \gamma\sqrt{n}] \leq 2\exp\left(-\frac{\gamma^2(2p(1-p))^{-1}}{1 + \frac{\gamma}{3p(1-p)\sqrt{N}}}\right)$$
$$\leq 2\exp\left(-\frac{2\gamma^2}{1 + \frac{2\gamma}{\sqrt{N}}}\right), \tag{15.17}$$

where we used the trivial fact $p(1-p) \leq 1/4$.

Moreover, by (8.8)

$$\frac{N!}{(N/e)^N} < e^{1/(12N)}\sqrt{2\pi N} < 3\sqrt{N}. \tag{15.18}$$

Combining (15.16)–(15.18),

$$\Pr\left[\left|\frac{1}{n}\mathbf{Y}^*(\lambda) - \mathcal{N}(\lambda)\right| \geq (6+\gamma)n^{-1/2}\right]$$
$$\leq \frac{N!}{(N/e)^N} \Pr\left[\left|\frac{1}{n}\mathbf{W}^*(\lambda) - p(\lambda)\right| \geq \gamma n^{-1/2}\right] \leq 6\sqrt{N}\exp\left(-\frac{2\gamma^2}{1 + \frac{2\gamma}{\sqrt{N}}}\right). \tag{15.19}$$

We recall (15.5)–(15.6)

$$\mathrm{vol}_d\left(S(\lambda;\mathcal{H};\beta)\right) = \mathrm{vol}_d\left(S(\mathrm{Error}_\lambda \geq \beta n^{-1/2})\right)$$
$$= \mathrm{vol}_d\left(\left\{\vec{\mathcal{Z}} \in I^d : \left|\frac{1}{n}\left|\{1 \leq i \leq n : h_i(\mathcal{Z}) \leq n + \lambda\sqrt{n}\}\right|\right.\right.\right.$$
$$\left.\left.\left. -\mathcal{N}(\lambda)\right| \geq \beta n^{-1/2}\right\}\right)$$

$$= \Pr\left[\left|\frac{1}{n}\left|\{1 \le i \le n\colon Y_{(i-1)n+1} + \cdots + Y_{in} \le n + \lambda\sqrt{n}\}\right| \right.\right.$$
$$\left.\left. -\mathcal{N}(\lambda)\right| \ge \beta n^{-1/2}\right]$$
$$= \Pr\left[\left|\frac{1}{n}\mathbf{Y}^*(\lambda) - \mathcal{N}(\lambda)\right| \ge \beta n^{-1/2}\right]. \tag{15.20}$$

Combining (15.19) and (15.20), (15.15) follows.

Applying Theorem 4.2 as a short-time ergodic theorem in the configuration space (in fact, it works as a "large deviation theorem"). Like in Sec. 8, let $\mathcal{Y}_{\omega;t}$, $\omega \in \Omega_{\text{Gauss}}$ denote the 3-dimensional Gaussian torus-billiard model, where the initial configuration is $\mathcal{Y} \subset I^3 = [0,1)^3$. Let $\mathcal{H} = \{H_1, \ldots, H_n\}$ be an arbitrary but fixed CLT-partition ($n = \sqrt{N}$). Assume that N is large; is it true that the overwhelming majority of the time evolutions of this system reaches "complete CLT snapshot randomness" superexponentially fast, and stays in this state for an incredibly long time (with the possible exception of a totally negligible set of t's)? By using Theorem 4.2 we give a positive answer. Again we basically repeat the argument of Sec. 5, or what is the same, the argument in Sec. 8. The only (trivial) novelty is that the new parameter λ runs through a finite set of values (to be specified below).

The family of time evolutions $\mathcal{Y}_{\omega;t}$, $\omega \in \Omega_{\text{Gauss}}$ of this Gaussian torus-billiard model is represented by the family of torus lines $\vec{\mathcal{Y}}_{\omega;t}$ in the configuration space $I^d = [0,1)^d$ with $d = 3N$, all starting from the same point $\vec{\mathcal{Y}} \in I^d$.

Since the torus I^d is translation invariant, we can apply Theorem 4.2 with $f = \chi_S$ where S is the translated copy

$$S = S(\text{Error}_\lambda \ge \beta N^{-1/2}) - \vec{\mathcal{Y}} \tag{15.21}$$

of $S(\lambda) = S(\text{Error}_\lambda \ge \beta N^{-1/2})$ in the torus I^d. Thus, we obtain the following analog of (8.17): if $U \ge 1$ and $e^{\pi^2 U^2/2} > 3Ud$ then for every integer $j \ge 1$,

$$\int_{\omega\in\Omega_{\text{Gauss}}} \left(\text{length}\left\{U \le t \le 2^j U : \vec{\mathcal{Y}}_{\omega;t} \in S(\lambda)\right\}\right.$$
$$\left. -\text{vol}_d(S(\lambda))(2^j - 1)U\right)^2 d\text{ProdMeas}_{\text{Gauss}}(\omega)$$
$$\le \text{vol}_d(S(\lambda))10j((2^j - 1)U + 1) \text{ where } S(\lambda) = S(\lambda; \mathcal{H}; \beta). \tag{15.22}$$

To illustrate the power of (15.15) and (15.22), we choose the numerical values $\beta = 36$, $U = 8$, $j = 340$ and $N = 10^{100}$; it follows that $d = 3 \cdot 10^{100}$. Then by (15.15) and (15.22), for every λ,

$$\int_{\omega \in \Omega_{\text{Gauss}}} \Big(\text{length}\left\{8 \le t \le 2^{343} :\ \vec{\mathcal{Y}}_{\omega;t} \in S(\lambda)\right\}$$

$$-\text{vol}_d(S(\lambda))8(2^{340}-1)\Big)^2\, d\text{ProdMeas}_{\text{Gauss}}(\omega)$$

$$\le 6 \cdot 10^{50} \exp\left(-\frac{2 \cdot 30^2}{1+\frac{2\cdot 36}{\sqrt{N}}}\right) \cdot 8 \cdot 340 \cdot 2^{340} \cdot 20 \le 10^{-661}. \tag{15.23}$$

Let $\Omega_{\text{Gauss}}(\lambda; \text{bad})$ be the set of those $\omega \in \Omega_{\text{Gauss}}$ for which

$$\text{length}\left\{8 \le t \le 2^{343} :\ \vec{\mathcal{Y}}_{\omega;t} \in S(\lambda)\right\} \ge 10^{-220}. \tag{15.24}$$

Repeating the proof in (12.18)–(12.19), we obtain

$$\text{ProdMeas}_{\text{Gauss}}(\Omega_{\text{Gauss}}(\lambda; \text{bad})) \le 10^{-220}. \tag{15.25}$$

So far λ was an unspecified parameter: an arbitrary but fixed real number. In the application λ runs through the members of the following discrete sequence. For every integer $1 \le l \le \log n$ and every integer m with $0 \le m \le 20\sqrt{n}/l^2$ let

$$\lambda_{l,m} = 2(l-1) + \frac{ml^2}{10\sqrt{n}} \quad \text{and} \quad \lambda_{-l,-m} = -\lambda_{l,m}. \tag{15.26}$$

There are (at most)

$$2\sum_{1 \le l \le \log n} (1 + 20\sqrt{n}l^{-2}) \le 2\log n + 40\sqrt{n}\sum_{l=1}^{\infty} l^{-2} \le 80\sqrt{n} \tag{15.27}$$

numbers $\lambda_{l,m}$ and $\lambda_{-l,-m}$ defined in (15.26). Arrange the numbers $\lambda_{l,m}$ and $\lambda_{-l,-m}$ defined in (15.26) in increasing order, including $\pm\infty$:

$$\lambda_0 = -\infty < \lambda_1 < \lambda_2 < \ldots < \lambda_M < \lambda_{M+1} = \infty, \tag{15.28}$$

where $M \le 80\sqrt{n}$ (see (15.27)). It follows from (15.26) and (15.4) that

$$\mathcal{N}(\lambda_{j+1}) - \mathcal{N}(\lambda_j) < n^{-1/2} \text{ for every } 0 \le j < M. \tag{15.29}$$

Now we return to (15.24) and (15.25). Let

$$\Omega_{\text{Gauss}}(\text{bad}) = \bigcup_{\substack{1 \le j \le M: \\ \text{defined in (15.28)}}} \Omega_{\text{Gauss}}(\lambda_j; \text{bad})$$

and

$$\Omega_{\text{Gauss}}(\text{good}) = \Omega_{\text{Gauss}} \setminus \Omega_{\text{Gauss}}(\text{bad}). \tag{15.30}$$

It follows that for every $\omega \in \Omega_{\text{Gauss}}(\text{good})$

$$\text{length}\left\{8 \le t \le 2^{343} : \vec{\mathcal{Y}}_{\omega;t} \in S(\lambda_j)\right\} \le 10^{-220} \quad \text{for every } 1 \le j \le M. \tag{15.31}$$

By (15.31) we have that for every $\omega \in \Omega_{\text{Gauss}}(\text{good})$

$$\text{length}\left\{8 \le t \le 2^{343} : \vec{\mathcal{Y}}_{\omega;t} \in S(\lambda_j) \text{ holds for some } 1 \le j \le M\right\}$$
$$\le 80\sqrt{n}10^{-220} \le 10^{-193}, \tag{15.32}$$

since $n = \sqrt{N} = 10^{50}$.

Moreover, by (15.25) and (15.27),

$$\text{ProdMeas}_{\text{Gauss}}(\Omega_{\text{Gauss}}(\text{good})) \ge 1 - 80\sqrt{n}10^{-220} > 1 - 10^{-193}. \tag{15.33}$$

We recall (15.5) (and the choice $\beta = 36$)

$$S(\lambda) = S(\lambda; \mathcal{H}; \beta) S(\text{Error}_\lambda \ge \beta n^{-1/2})$$
$$= \left\{\vec{\mathcal{Z}} \in I^d : \left|\frac{1}{n}\left|\left\{1 \le i \le n : \frac{h_i(\mathcal{Z}) - n}{\sqrt{n}} \le \lambda\right\}\right| - \mathcal{N}(\lambda)\right| \ge 36n^{-1/2}\right\}, \tag{15.34}$$

where $h_i(\mathcal{Z})$ denotes the number of points of the given N-set

$$\mathcal{Z} = \{\mathbf{z}_1, \ldots, \mathbf{z}_N\}$$

(representing Bob's N-set) that are in H_i (where $H_i \in \mathcal{H}$ is a member of the given CLT-partition $[0,1]^3 = H_1 \cup H_2 \cup \ldots \cup H_n$).

By using (15.34), we can rewrite (15.32) as follows: for every $\omega \in \Omega_{\text{Gauss}}(\text{good})$

$$\text{length}\left\{8 \le t \le 2^{343} : \left|\frac{1}{n}\left|\left\{1 \le i \le n : \frac{h_i(\mathcal{Y}_{\omega;t}) - n}{\sqrt{n}} \le \lambda_j\right\}\right| - \mathcal{N}(\lambda_j)\right|\right.$$
$$\left. \ge 36n^{-1/2} \text{ holds for some } 1 \le j \le M\right\} \le 10^{-193}. \tag{15.35}$$

Let $\lambda_j < \lambda < \lambda_{j+1}$; then clearly

$$\frac{1}{n}\left|\left\{1 \le i \le n : \frac{h_i(\mathcal{Y}_{\omega;t}) - n}{\sqrt{n}} \le \lambda\right\}\right| - \mathcal{N}(\lambda)$$
$$\le \frac{1}{n}\left|\left\{1 \le i \le n : \frac{h_i(\mathcal{Y}_{\omega;t}) - n}{\sqrt{n}} \le \lambda_{j+1}\right\}\right| - \mathcal{N}(\lambda_j)$$

$$\leq \frac{1}{n}\left|\left\{1 \leq i \leq n : \frac{h_i(\mathcal{Y}_{\omega;t}) - n}{\sqrt{n}} \leq \lambda_{j+1}\right\}\right| - \mathcal{N}(\lambda_{j+1})$$
$$+(\mathcal{N}(\lambda_{j+1}) - \mathcal{N}(\lambda_j))$$
$$\leq \left(\frac{1}{n}\left|\left\{1 \leq i \leq n : \frac{h_i(\mathcal{Y}_{\omega;t}) - n}{\sqrt{n}} \leq \lambda_{j+1}\right\}\right| - \mathcal{N}(\lambda_{j+1})\right) + n^{-1/2}, \tag{15.36}$$

where in the last step we used (15.29). Similarly,

$$\frac{1}{n}\left|\left\{1 \leq i \leq n : \frac{h_i(\mathcal{Y}_{\omega;t}) - n}{\sqrt{n}} \leq \lambda\right\}\right| - \mathcal{N}(\lambda)$$
$$\geq \left(\frac{1}{n}\left|\left\{1 \leq i \leq n : \frac{h_i(\mathcal{Y}_{\omega;t}) - n}{\sqrt{n}} \leq \lambda_j\right\}\right| - \mathcal{N}(\lambda_j)\right) - n^{-1/2}. \tag{15.37}$$

Using (15.36)–(15.37) in (15.35), we obtain that for every $\omega \in \Omega_{\text{Gauss}}(\text{good})$

$$\text{length}\left\{8 \leq t \leq 2^{343} : \sup_{-\infty<\lambda<\infty}\left|\frac{1}{n}\left|\left\{1 \leq i \leq n : \frac{h_i(\mathcal{Y}_{\omega;t}) - n}{\sqrt{n}} \leq \lambda\right\}\right|\right.\right.$$
$$\left.\left. -\mathcal{N}(\lambda)\right| \geq (36+1)n^{-1/2}\right\} \leq 10^{-193}. \tag{15.38}$$

Again we study the classical Bernoulli gas model where the gas molecules are represented by $N = 10^{100}$ point billiards. Unfolding reduces the billiards-in-a-box model to the torus-billiard model. The threshold $U = 8$ represents — roughly speaking — the necessary number of "jumps" per particle in the Gaussian torus-billiard model (which is the half of the number of bounces in the analog billiards-in-a-box model) to reach "complete CLT snapshot randomness". As usual, assume that the gas molecules have average speed 10^3 meter per second. For this system it takes only a few milliseconds to reach "complete CLT snapshot randomness". Now (15.33) and (15.38) have the following interpretation. Choosing an arbitrary (measurable) CLT partition $\mathcal{H} = \{H_1, H_2, \ldots, H_n\}$ of the unit cube $[0,1]^3 = H_1 \cup H_2 \cup \ldots \cup H_n$ such that $\text{vol}(H_i) = 1/n$, $1 \leq i \leq n = \sqrt{N} = 10^{50}$ and an arbitrary $N = 10^{100}$-element initial point configuration $\mathcal{Y} \subset [0,1)^3$, for the totally overwhelming majority of the initial velocities (Gaussian distribution), the normalized distribution of the particles in the partition $\mathcal{H}$ remains very close to the standard normal distribution for an extremely long time, with the possible exception of a totally negligible set of times t.

Indeed, for every CLT partition $\mathcal{H} = \{H_1, H_2, \ldots, H_n\}$ and every $N = 10^{100}$-element initial point configuration $\mathcal{Y}$, there exists a subset $\Omega_{\text{Gauss}}(\text{good})$ with

$$\text{ProdMeas}_{\text{Gauss}}(\Omega_{\text{Gauss}}(\text{good})) \geq 1 - 10^{-193}$$

(see (15.33)), representing a totally overwhelming majority, such that for every $\omega \in \Omega_{\text{Gauss}}(\text{good})$,

$$\text{CltError} = \sup_{-\infty<\lambda<\infty} \left| \frac{1}{n} \left| \left\{ 1 \leq i \leq n : \frac{h_i(\mathcal{Y}_{\omega;t}) - n}{\sqrt{n}} \leq \lambda \right\} \right| - \mathcal{N}(\lambda) \right|$$
$$\leq 37n^{-1/2} = 3.7 \cdot 10^{-24} \tag{15.39}$$

holds for every $8 \leq t \leq 2^{343}$ with the possible exception of a set of times t of total length $<10^{-193}$, see (15.38). The latter actually represents less than 10^{-196} seconds, which is a ridiculously short time.

Note that $8 \leq t \leq 2^{343}$ represents a time interval of length about 10^{97} seconds, which is a ridiculously long time (the estimated age of the universe, starting from Big Bang, is less than 10^{20} seconds).

Summarizing, the typical time evolution of this system reaches "complete CLT snapshot randomness" (see (15.39); noting that $\mathcal{N}(\lambda)$ denotes the left tail probability of the standard normal distribution, see (15.4)) in a few milliseconds (even starting from Big Bang!), and then it remains in the state of "complete CLT snapshot randomness" for an incredibly long time (with the possible exception of a totally negligible set of t's).

The approximation of the famous bell-shaped curve — standard normal distribution — with error term $3.7 \cdot 10^{-24}$ in (15.39) represents a striking, and nearly optimal, precision (here $10^{-25} = n^{-1/2} = N^{-1/4}$ is the inevitable "random error"). This is why we may say that the system is *completely random*.

Another illustration. We conclude this section with another choice of the key parameters; this example will be used later in Sec. 17. In the crucial inequalities (15.15) and (15.22) we choose $\beta = 6 + \log N$, $U = \sqrt{\log N}$, $j = 10 \log_2 N$ with $N \geq 10^{20}$ (of course $d = 3N$). Then by (15.15) and (15.22), for every λ,

$$\int_{\omega\in\Omega_{\text{Gauss}}} \Big(\text{length} \left\{ U \leq t \leq 2^j U : \vec{\mathcal{Y}}_{\omega;t} \in S(\lambda) \right\}$$
$$- \text{vol}_d(S(\lambda))(2^j - 1)U \Big)^2 \, d\text{ProdMeas}_{\text{Gauss}}(\omega)$$

$$\leq \text{vol}_d(S(\lambda))10j\left((2^j-1)U+1\right)$$
$$\leq 6\sqrt{N}\exp\left(-\frac{2(\log N)^2}{1+\frac{2(6+\log N)}{\sqrt{N}}}\right)\cdot\sqrt{\log N}$$
$$\cdot 10\log_2 N\cdot 2^{10\log_2 N}\cdot 20\leq N^{-73}, \tag{15.40}$$

where

$$\text{vol}_d(S(\lambda))=\text{vol}_d(S(\lambda;\mathcal{H};\beta))=\text{vol}_d\left(S(\text{Error}_\lambda\geq\beta n^{-1/2})\right)$$
$$\leq 6\sqrt{N}\exp\left(-\frac{2(\log N)^2}{1+\frac{2(6+\log N)}{\sqrt{N}}}\right)<N^{-90}, \tag{15.41}$$

noting that $n=\sqrt{N}$.

Let $\Omega_{\text{Gauss}}(\lambda;\text{bad})$ be the set of those $\omega\in\Omega_{\text{Gauss}}$ for which

$$\text{length}\left\{U\leq t\leq U2^j:\ \vec{\mathcal{Y}}_{\omega;t}\in S(\lambda)\right\}\geq N^{-24}. \tag{15.42}$$

Repeating the proof in (12.17)–(12.18), we obtain

$$\text{ProdMeas}_{\text{Gauss}}(\Omega_{\text{Gauss}}(\lambda;\text{bad}))\leq N^{-24}. \tag{15.43}$$

Again λ runs through the members of the sequence defined in (15.26). Repeating the argument between (15.27) and (15.37), we obtain the following analogue of (15.33) and (15.38): for every $\omega\in\Omega_{\text{Gauss}}(\text{good})$

$$\text{length}\left\{U\leq t\leq U2^j:\ \sup_{-\infty<\lambda<\infty}\left|\frac{1}{n}\right.\right.$$
$$\left.\left\{1\leq i\leq n:\frac{h_i(\mathcal{Y}_{\omega;t})-n}{\sqrt{n}}\leq\lambda\right\}\right|-\mathcal{N}(\lambda)\Bigg|$$
$$\left.\geq(7+\log N)n^{-1/2}\right\}\leq 80\sqrt{n}N^{-24}<N^{-23}, \tag{15.44}$$

where

$$\text{ProdMeas}_{\text{Gauss}}(\Omega_{\text{Gauss}}(\text{good}))\geq 1-80\sqrt{n}N^{-24}>1-N^{-23}. \tag{15.45}$$

Again we study the classical Bernoulli gas model where the gas molecules are represented by $N\geq 10^{20}$ point billiards, and we employ *unfolding* to reduce the billiards-in-a-box model to the torus-billiard model. The threshold $U=\sqrt{\log N}$ represents — roughly speaking — the necessary number of "jumps" per particle in the Gaussian torus-billiard model (which is the half of the number of bounces in the analog billiards-in-a-box model) to reach "complete CLT snapshot randomness". As usual, assume that the gas molecules have average speed 10^3 meter per second. In view of the threshold $U = \sqrt{\log N}$, this Gaussian system reaches "complete CLT snapshot randomness" superexponentially fast. Now (15.44) and (15.45) have the following interpretation. For every CLT partition $\mathcal{H} = \{H_1, H_2, \ldots, H_n\}$ and every N-element initial point configuration $\mathcal{Y}$, there exists a subset $\Omega_{\text{Gauss}}(\text{good})$ with

$$\text{ProdMeas}_{\text{Gauss}}(\Omega_{\text{Gauss}}(\text{good})) \geq 1 - N^{-23}$$

(see (15.45)), representing an overwhelming majority, such that for every $\omega \in \Omega_{\text{Gauss}}(\text{good})$,

$$\begin{aligned}\text{CltError} &= \sup_{-\infty<\lambda<\infty} \left| \frac{1}{n} \left| \left\{ 1 \leq i \leq n : \frac{h_i(\mathcal{Y}_{\omega;t}) - n}{\sqrt{n}} \leq \lambda \right\} \right| - \mathcal{N}(\lambda) \right| \\ &\leq (7 + \log N) n^{-1/2} = \frac{7 + \log N}{N^{1/4}} \qquad (15.46)\end{aligned}$$

holds for every t in the time interval $\sqrt{\log N} \leq t \leq \sqrt{\log N} N^{10}$ with the possible exception of a set of times t of total length $< N^{-23}$, see (15.44). The latter represents less than $10^{-3} N^{-23}$ seconds, which is an extremely short time. Notice that $\sqrt{\log N} \leq t \leq \sqrt{\log N} N^{10}$ is an extremely long time interval.

The approximation of the standard normal distribution with error term

$$\frac{7 + \log N}{N^{1/4}}$$

in (15.46) represents a nearly optimal precision. Indeed, $n^{-1/2} = N^{-1/4}$ is the inevitable "random error".

16. Snapshot Randomness (III) Case of Closed Orbits

Red-Blue-CLT test. In Sec. 8 and 15 we studied two aspects — Poisson and CLT — of nearly complete snapshot randomness of Gaussian (at least 2-dimensional) box/torus billiard systems. What about the snapshot randomness of the very different Gaussian closed orbit systems introduced in Secs. 6 and 7? Here we show how the CLT partition test can be adapted for these closed orbit systems. Since this case is substantially harder, for the sake of simplicity, we just consider the spherical GreatCircleDiameter model, for which the underlying measure is the (normalized) surface area.

First we point out a trivial difference between the (at least two dimensional) box/torus (billiard) models and the spherical models: for the spherical models CLT time-lapse randomness *fails*, unless there is some extra condition. Indeed, assume that all N particles start from the North Pole $(0,0,1)$ of the unit sphere $\mathbf{S}^2$ ("spherical Big Bang"), and the CLT partition $\mathcal{H} = \{H_1, \ldots, H_n\}$, $n = \sqrt{N}$ is defined in terms of the great circles passing through the two poles $(0,0,\pm 1)$ as follows: assume that for every i the boundary of H_i consists of two great circles passing through $(0,0,\pm 1)$. Then, for every choice of the initial velocities, the number of particles in H_i is a time-independent constant, which is the complete opposite of the fluctuation described by the central limit theorem.

To avoid this kind of trivial nuisance, we choose test sets that cannot contain a whole great circle. For example, if the test set is in the (open) upper hemisphere

$$\mathbf{S}^2(+) = \{(x,y,z) \in \mathbf{S}^2 : \; z > 0\}, \tag{16.1}$$

then of course the test set cannot contain more than half of a great circle. Here we focus on CLT snapshot randomness instead of time-lapse randomness; nevertheless, roughly speaking, "we stay in the upper hemisphere" in the sense that we distinguish between the two hemispheres in such a way that a particle is considered positive or negative depending on whether it is in the upper or lower hemisphere. What we study here is the distribution of the $\pm$-*discrepancy*, or equivalently, the red-blue–*discrepancy*. This way the surface area cancels out — to be explained below.

The definition of the Gaussian $\pm$*GreatCircleDiameter model* is almost the same as that of the ordinary GreatCircleDiameter model. Again let

$$\mathcal{Y} = \{\mathbf{y}_1, \ldots, \mathbf{y}_N\} \subset \mathbf{S}^2$$

be an arbitrary N-element point set on the unit sphere — it represents the initial configuration. As usual, $\mathbf{y}_k \in \mathcal{Y}$ is the starting point of the kth particle. Again for every point $\mathbf{y}_k$, $1 \le k \le N$ we choose a great circle in $\mathbf{S}^2$ passing through $\mathbf{y}_k$ by choosing an angle α_k with uniform distribution in the interval $[0, 2\pi)$. The kth particle moves on this great circle in such a way that its projection on the $(\mathbf{y}_k, -\mathbf{y}_k)$-diameter has constant speed u_k (positive or negative; $-\mathbf{y}$ is the antipodal point of $\mathbf{y}$). We refer to this constant speed as the projection-speed. Again we choose the projection-speed

u_k of the kth particle by the one-dimensional normal distribution, that is, the density function of the distribution of the projection-speed $u = u_k$ is $(2\pi)^{-1/2}e^{-u^2/2}$. The pair (α_k, u_k) of the angle and the projection-speed determine the initial velocity of the kth particle. Finally, we choose the initial velocites in such a way that α_k, u_k, $1 \le k \le N$ are $2N$ *independent* random variables. This means that the ±GreatCircleDiameter system (±GCD system, in short) of the first kind has the following initial velocity space and product measure:

$$\begin{aligned} \omega &= (\alpha_1, \ldots, \alpha_N, u_1, \ldots, u_N) \in \Omega_{\text{Gauss}}(\pm GCD) \\ &= [0, 2\pi)^N \times \mathbb{R}^N, \quad \text{and} \\ \text{ProdMeas}_{\text{Gauss},\pm GCD} &= (2\pi)^{-N}\text{vol}_N \times \text{Gauss}_N. \end{aligned}$$

The first novelty is that a particle is considered positive or negative depending on whether it is in the upper or lower hemisphere. This leads to the "red-blue Alice–Wicky–Bob game", where red represents 1 and blue represents −1. In the "red-blue Alice–Wicky–Bob game" both sisters have a 2-colored (red and blue) N-element point set in the upper hemisphere $\mathbf{S}^2(+)$. Again Alice's 2-colored N-set is "truly random" with Pr[red] = Pr[blue] = 1/2, and again Wicky claims that her 2-colored N-set is also "truly random", but, because she is a liar, nobody knows what Wicky actually did. Again we open the door between the two sisters: Alice and Wicky put their 2-colored N-sets into a black box, and we give the black box to Bob, who pulls out one of the two 2-colored N-sets (despite being superman, he does not know which one is due to Alice and which one is due to Wicky). Bob has to decide whether the 2-colored N-set he picked is "truly random" or not. He can perform one test (say).

The Red-Blue-CLT test in this spherical system means that Bob makes a partition $\mathbf{S}^2(+) = H_1 \cup H_2 \cup \ldots \cup H_n$ of the upper hemisphere with $n = \sqrt{N}$ parts, where each H_i is measurable and they have the same surface area $\text{SA}^*(H_i) = 1/2n$ (SA^* is the normalized surface area, and the sisters have zero information about the partition).

Consider a 2-colored N-set in the upper hemisphere: it is represented by a pair $(\mathcal{Z}; \kappa)$, where

$$\mathcal{Z} = \{\mathbf{z}_1, \ldots, \mathbf{z}_N\} \subset \mathbf{S}^2(+) \tag{16.2}$$

is the point set, and

$$\kappa(\mathbf{z}_k) \in \{1, -1\}, \quad 1 \le k \le N \tag{16.3}$$

is the 2-coloring. The pair $(\mathcal{Z};\kappa)$ represents the 2-colored N-set picked by Bob. Let $\Delta_i = \Delta_i(\mathcal{H};\mathcal{Z};\kappa)$ denote the *color discrepancy* (or combinatorial discrepancy) in H_i:

$$\Delta_i(\mathcal{H};\mathcal{Z};\kappa) = \text{red minus blue} = \sum_{\substack{k:\ 1\le k\le N\\ \mathbf{z}_k\in H_i}} \kappa(\mathbf{z}_k). \tag{16.4}$$

Consider the n real numbers

$$\frac{\Delta_i(\mathcal{H};\mathcal{Z};\kappa)}{\sqrt{n}}, \quad 1\le i\le n=\sqrt{N}. \tag{16.5}$$

We compare the empirical distribution function of the set (16.5) to the standard normal distribution:

$$\text{CltColorError}(\mathcal{Z};\kappa) = \sup_{-\infty<\lambda<\infty}\left|\frac{1}{n}\left|\left\{1\le i\le n:\ \frac{\Delta_i(\mathcal{H};\mathcal{Z};\kappa)}{\sqrt{n}}\le\lambda\right\}\right| - \mathcal{N}(\lambda)\right|, \tag{16.6}$$

where, similarly to Sec. 15,

$$\mathcal{N}(\lambda) = \frac{1}{\sqrt{2\pi}}\int_{-\infty}^{\lambda} e^{-u^2/2}\,du \tag{16.7}$$

denotes the left tail probability of the standard normal distribution.

The "random error" is $n^{-1/2}$. Unlike in Sec. 15, here we cannot prove the sharp upper bound for the error. What we can prove is about the square root of the conjectured optimal result:

$$\text{CltColorError}(\mathcal{Z};\kappa) < n^{-\frac{1}{4}+\varepsilon} = N^{-\frac{1}{8}+\frac{\varepsilon}{2}}. \tag{16.8}$$

Nevertheless, if (16.8) holds, then we can still safely say that the 2-colored N-set $\mathcal{Z}$ exhibits *advanced snapshot randomness* with respect to the Red-Blue-CLT test $\mathcal{H} = \{H_1,\ldots,H_n\}$ in the upper hemisphere $\mathbf{S}^2(+)$. It would be very interesting to prove the optimal result by doubling the exponent in (16.8).

Let $\mathcal{Y}_{\omega;t}$ denote the Gaussian $\pm$GCD model with initial configuration

$$\mathcal{Y} = \{\mathbf{y}_1,\ldots,\mathbf{y}_N\}\subset\mathbf{S}^2$$

and

$$\omega = (\omega^*,\mathbf{u})\in\Omega_{\text{Gauss}}(\pm GCD), \quad \text{where} \quad \omega^* = (\alpha_1,\ldots,\alpha_N)\in[0,2\pi)^N$$

is the vector of the angles of the orbits of the particles, and finally,

$$\mathbf{u} = (u_1,\ldots,u_N)\in\mathbb{R}^N,$$

where u_k denotes the projection-speed of the kth particle. If a particle is in the lower hemisphere, then its antipodal point is in the upper hemisphere; we call the antipodal a blue point, and it has weight -1. If a particle is in the upper hemisphere, then of course it is a red point and it has weight 1. This way in every time instant t the $\pm$GCD model of the first kind $\mathcal{Y}_{\omega;t}$ generates a 2-colored N-set in the upper hemisphere. We refer to this 2-coloring as the *North–South 2-coloring* $\kappa_0 = \pm 1$: the point is red, meaning $\kappa_0 = 1$, if the point is in the upper hemishere (i.e., it is closer to the North Pole), and the point is blue, meaning $\kappa_0 = -1$, if the point is in the lower hemishere (i.e., it is closer to the South Pole). Note that the Equator is a zero measure set, so its contribution is irrelevant.

Assume that N is large; is it true that the overwhelming majority of the time evolutions of the system with the *North-South 2-coloring* $\kappa_0 = \pm 1$ reaches advanced Red-Blue-CLT snapshot randomness superexponentially fast, and then it stays in that state for an extremely long time (with the possible exception of a totally negligible set of t's)? By using Theorem 4.2 and a critical volume estimation (see Lemma 16.1 below), we give a positive answer to this question.

Like in Sec. 6, for every $\omega^* \in [0, 2\pi)^N$, let $GC_k = GC_k(\omega^*)$ $(1 \le k \le N)$ denote the common orbit of the kth particles: it is a great circle containing the antipodal points $\mathbf{y}_k \in \mathcal{Y}$ and $-\mathbf{y}_k$ with angle α_k. Since the projection of the motion on the $(\mathbf{y}_k, -\mathbf{y}_k)$-diameter has constant speed, we replace the motion of the kth particle on the great circle $GC_k = GC_k(\omega)$ with the constant speed motion in the interval $[0, 4)$ interpreted as a 1-dimensional torus ("projection"). Again we divide by 4 in the sense that we replace the constant speed motion in the 1-dimensional torus $[0, 4)$ with the constant speed motion in the 1-dimensional unit torus $[0, 1)$. This means that, for every *fixed* $\omega^* \in [0, 2\pi)^N$, the family of time evolutions $\mathcal{Y}_{\omega;t}$, $\omega = (\omega^*, \mathbf{u}) \in \Omega_{\text{Gauss}}(\pm GCD)$ of the $\pm$GCD model of the first kind with initial configuration $\mathcal{Y} \subset \mathbf{S}^2$ is represented by the family of torus lines in the configuration space $I^N = [0, 1)^N$, all starting from the origin $\mathbf{0} \in I^N$.

Let $-\mathcal{H} = \{-H_1, \ldots, -H_n\}$ be the antipodal extension of the partition $\mathcal{H}$ of the (open) upper hemisphere $\mathbf{S}^2(+)$ to the (open) lower hemisphere $\mathbf{S}^2(-)$. Formally,

$$\mathbf{w} \in -H_i \iff -\mathbf{w} \in H_i. \tag{16.9}$$

The great circle $GC_k = GC_k(\omega^*)$ (i.e., the orbit of the kth particle) intersects the ith member $H_i \subset \mathbf{S}^2(+)$ (and similarly $-H_i \subset \mathbf{S}^2(-)$) of the given partition $\mathcal{H}$ (and similarly $-\mathcal{H}$) in a 1-dimensional *measurable* set for almost every $\omega^* \in [0, 2\pi)^N$ — this is Fubini's theorem. Consider the projection of the intersection $GC_k(\omega^*) \cap H_i$ (and similarly $GC_k(\omega^*) \cap (-H_i)$) on the double copy of the $(\mathbf{y}_k, -\mathbf{y}_k)$-diameter, where the latter is represented as the interval $[0, 4)$. Let $H_{k,i}^{(+)} = H_{k,i}^{(+)}(\omega^*) \subset [0, 4)$ (and similarly $H_{k,i}^{(-)} = H_{k,i}^{(-)}(\omega^*) \subset [0, 4)$) denote this projection, and again we apply the usual division by 4:

$$\frac{1}{4} H_{k,i}^{(\pm)}(\omega^*) \subset [0,1) \quad \text{where} \quad \frac{1}{4} H = \left\{ \frac{1}{4} x : \ x \in H \right\}. \tag{16.10}$$

Write $h_{k,i}(\omega^*) = \frac{1}{4}\text{length}(H_{k,i}^{(\pm)}(\omega^*))$ (1-dimensional Lebesgue measure). The mean value of $h_{k,i}(\omega^*)$, $\omega^* \in [0, 2\pi)^N$ is clearly equal to the normalized surface area of H_i; formally,

$$(2\pi)^{-N} \int_{[0,2\pi)^N} h_{k,i}(\omega^*)\, d\omega^* = \text{SA}^*(H_i) = \frac{1}{4\pi}\text{SurfaceArea}(H_i) = \frac{1}{2n} \tag{16.11}$$

($d\omega^*$ represents the N-dimensional Lebesgue measure).

Write

$$\text{BadSet}_i = \left\{ \omega^* \in [0, 2\pi)^N : \ \left| \sum_{k=1}^{N} h_{k,i}(\omega^*) - \text{SA}^*(H_i) N \right| > 40 N^{1/4} \right\}. \tag{16.12}$$

We apply Bernstein's large deviation inequality (11.8) for the sequence Z_k, where $0 \le Z_k \le 1$, $\mathbf{E}Z_k = \text{SA}(H_i) = n^{-1}/2 = N^{-1/2}/2$, $1 \le k \le N$; thus (24.34) gives the upper bound

$$(2\pi)^{-N} \text{vol}_N\,(\text{BadSet}_i) \le 2 \exp\left(- \frac{1600}{1 + 80 N^{-1/4}/3} \right). \tag{16.13}$$

Estimating a critical volume in the configuration space. Let

$$\omega^* \notin \text{BadSet} = \bigcup_{i=1}^{n} \text{BadSet}_i \tag{16.14}$$

be arbitrary but *fixed*. To apply Theorem 4.2, we need information about the volume of the subset

$$S(\text{ColorError} \geq \beta n^{-1/4}; \omega^*) = \Bigg\{ \vec{z} \in I^N : \sup_{-\infty<\lambda<\infty} \Bigg| \frac{1}{n} \Bigg| \Bigg\{ 1 \leq i \leq n :$$

$$\Bigg| \Bigg\{ 1 \leq k \leq N : z_k \in \frac{1}{4} H_{k,i}^{(+)}(\omega^*) \Bigg\} \Bigg| - \Bigg| \Bigg\{ 1 \leq k \leq N : z_k \in \frac{1}{4} H_{k,i}^{(-)}(\omega^*) \Bigg\} \Bigg| \Big/ \sqrt{n}$$

$$\leq \lambda \Bigg\} \Bigg| - \mathcal{N}(\lambda) \Bigg| \geq \beta n^{-1/4} \Bigg\} \tag{16.15}$$

in the configuration space $I^N = [0,1)^N$, where $\beta > 0$ is arbitrary. Our goal is to prove the following lemma (unfortunately its proof is surprisingly long).

Lemma 16.1. *Under the condition* (16.14), *for every* $\beta > 20\sqrt{6}$ *and* $N = n^2 \geq 10^{10}$,

$$\text{vol}_N(S(\text{ColorError} \geq \beta n^{-1/4}; \omega^*)) \leq 4n^3 e^{-((\beta^2/12)-200)^2/2}.$$

Proof of Lemma 16.1. Like in Sec. 8 (or Sec. 15), the N-dimensional volume

$$\text{vol}_N(S(\text{ColorError} \geq \beta n^{-1/4}; \omega^*))$$

can be expressed in terms of a large deviation type probability for dependent random variables (see (16.17) below). This time we make use of a 2-colored N-balls-to-$2\sqrt{N}$-bins occupation number problem. We place N balls independently into $n = \sqrt{N}$ red bins and $n = \sqrt{N}$ blue bins; another novelty is that the placement is not necessarily uniform.

More precisely, we study the following particular occupancy problem. We place the kth ball into the ith red bin (ith blue bin) with probability

$$p_{k,i} = h_{k,i}(\omega^*) = \frac{1}{4}\text{length}(H_{k,i}^{(\pm)}(\omega^*)). \tag{16.16}$$

Let $Z_k = i$ if the kth ball is in the ith red bin, and let $Z_k = -i$ if the kth ball is in the ith blue bin; we assume that $Z_1, \ldots, Z_N$ are N independent random variables. Let $Y_i^{(+)}$ denote the number of balls in the ith red bin, and let $Y_i^{(-)}$ denote the number of balls in the ith blue bin. Thus we have (see (16.15))

$$\text{vol}_N\Big(S(\text{ColorError} \geq \beta n^{-1/4}; \omega^*)\Big)$$

$$= \Pr\Bigg[\sup_{-\infty<\lambda<\infty} \Bigg| \frac{1}{n} \Bigg| \Bigg\{ 1 \leq i \leq n : \frac{Y_i^{(+)} - Y_i^{(-)}}{\sqrt{n}} \leq \lambda \Bigg\} \Bigg| - \mathcal{N}(\lambda) \Bigg| \geq \beta n^{-1/4} \Bigg]. \tag{16.17}$$

The estimation of (16.17) is a more difficult occupancy problem. Similarly to the proof of Lemma 8.3 at the end of Sec. 8, we make use of martingales, and apply the Azuma–Hoeffding inequality.

Let time t refer to the point at which the first t balls have been thrown. Let $\mathcal{F}_t$ be the σ-algebra generated by the random choice of bins for the first t balls, i.e., the σ-algebra generated by the independent random variables $Z_1, \ldots, Z_t$. For every real γ let

$$X_\gamma = \sum_{\substack{1 \le i \le n: \\ Y_i^{(+)} - Y_i^{(-)} \le \gamma\sqrt{n}}} \left(\gamma\sqrt{n} - Y_i^{(+)} + Y_i^{(-)}\right)$$

$$= \sum_{i=1}^{n} \left(\gamma\sqrt{n} - Y_i^{(+)} + Y_i^{(-)}\right)^+, \tag{16.18}$$

where $(w)^+$ denotes the positive part of a real w, i.e., $(w)^+ = w$ if $w > 0$ and 0 otherwise. We refer to X_γ as the γ-sum.

Next let $U_t = \mathbf{E}(X_\gamma | \mathcal{F}_t)$. The random variables $U_0, U_1, \ldots, U_N$ form a martingale with

$$U_0 = \mathbf{E}(X_\gamma) \quad \text{and} \quad U_N = X_\gamma.$$

Notice that moving the kth ball from one bin to another can change the γ-sum by at most one. This implies that $|U_{i+1} - U_i| \le 1$ for $0 \le i < N$. Applying the Azuma–Hoeffding inequality with $M = 1$, for every $\alpha > 0$

$$\Pr\left[|X_\gamma - \mathbf{E}X_\gamma| \ge \alpha\sqrt{N}\right] \le 2e^{-\alpha^2/2}. \tag{16.19}$$

Motivated by (16.17), we define the empirical distribution function

$$\Delta(u) = \frac{1}{n}\left|\left\{1 \le i \le n : \frac{Y_i^{(+)} - Y_i^{(-)}}{\sqrt{n}} \le u\right\}\right|. \tag{16.20}$$

By (16.18) and (16.20),

$$X_\gamma = n^{3/2}\int_{-\infty}^{\gamma} (\gamma - u)\, d\Delta(u) = n^{3/2}\int_{-\infty}^{\gamma} \Delta(u)\, du, \tag{16.21}$$

where in the last step we used integration by parts.

Next we focus on $\mathbf{E}X_\gamma$.

Lemma 16.2. *Under the condition of* (16.14), *for* $N = n^2 \ge 10^{10}$,

$$\sup_{-\infty<\gamma<\infty} \left|\mathbf{E}X_\gamma - n^{3/2}\int_{-\infty}^{\gamma} \mathcal{N}(u)\, du\right| \le 200n.$$

Proof of Lemma 16.2. For every $1 \le i \le n$ let

$$G_i(u) = \Pr\left[\frac{Y_i^{(+)} - Y_i^{(-)}}{\sqrt{n}} \le u\right]. \tag{16.22}$$

By (16.18), (16.22), and using the linearity of the expectation,

$$\mathbf{E}X_\gamma = \sum_{i=1}^{n} \sqrt{n} \int_{-\infty}^{\gamma} (\gamma - u) dG_i(u) = \sum_{i=1}^{n} \sqrt{n} \int_{-\infty}^{\gamma} G_i(u)\, du, \tag{16.23}$$

where in the last step we used integration by parts.

We study (16.23). For $1 \le i \le n$, let $Z_{k,i} = \pm 1$ if $Z_k = \pm i$ and 0 otherwise; then

$$Y_i^{(+)} - Y_i^{(-)} = \sum_{k=1}^{N} Z_{k,i} \tag{16.24}$$

is a sum of N independent $0, \pm 1$ valued random variables with expectation

$$\mathbf{E}\left(Y_i^{(+)} - Y_i^{(-)}\right) = \sum_{k=1}^{N} p_{k,i} - \sum_{k=1}^{N} p_{k,i} = 0.$$

Applying the Berry–Esseen version of the Central Limit Theorem (15.13) for the sum (16.24), we have

$$\begin{aligned} &\sup_{-\infty<\tau<\infty} \left| G_i(\tau\sqrt{V}/\sqrt{n}) - \mathcal{N}(\tau) \right| \\ &= \sup_{-\infty<\tau<\infty} \left| \Pr\left[Y_i^{(+)} - Y_i^{(-)} \le \tau\sqrt{V}\right] - \mathcal{N}(\tau) \right| \\ &\le \frac{1}{\sqrt{V}} \max_{1\le k\le N} \frac{\mathbf{E}|Z_{k,i}|^3}{\mathbf{E}Z_{k,i}^2} = \frac{1}{\sqrt{V}}, \end{aligned} \tag{16.25}$$

where

$$V = \text{variance} = \sum_{k=1}^{N} \mathbf{E}Z_{k,i}^2 = 2\sum_{k=1}^{N} p_{k,i}. \tag{16.26}$$

We recall (16.14)

$$\omega^* \notin \text{BadSet} = \bigcup_{i=1}^{n} \text{BadSet}_i. \tag{16.27}$$

Thus by (16.12) and (16.16),

$$\left| \sum_{k=1}^{N} p_{k,i} - \frac{1}{2} N^{1/2} \right| \le 40 N^{1/4}. \tag{16.28}$$

By (16.26) and (16.28),

$$N^{1/2} - 80N^{1/4} \leq V = 2\sum_{k=1}^{N} p_{k,i} \leq N^{1/2} + 80N^{1/4}. \tag{16.29}$$

By (16.29),

$$N^{1/4} - 50 < \sqrt{V} < N^{1/4} + 40, \tag{16.30}$$

assuming $N \geq 10^{10}$.

For every real number u, let $\tau_1 = \tau_1(u)$ and $\tau_2 = \tau_2(u)$ be defined as follows:

$$\tau_1(u)(N^{1/4} + 40) = uN^{1/4} \quad \text{and} \quad \tau_2(u)(N^{1/4} - 50) = uN^{1/4}. \tag{16.31}$$

Combining (16.22), (16.25), (16.30) and (16.31), for $u < 0$ we have

$$\begin{aligned} G_i(u) &= \Pr\left[Y_i^{(+)} - Y_i^{(-)} \leq uN^{1/4}\right] \\ &= \Pr\left[Y_i^{(+)} - Y_i^{(-)} \leq \tau_2(u)(N^{1/4} - 50)\right] \\ &\geq \Pr\left[Y_i^{(+)} - Y_i^{(-)} \leq \tau_2(u)\sqrt{V}\right] \geq \mathcal{N}(\tau_2(u)) - \frac{1}{\sqrt{V}} \\ &\geq \mathcal{N}(\tau_2(u)) - \frac{1}{N^{1/4} - 50}, \end{aligned} \tag{16.32}$$

and similarly, for $u < 0$,

$$\begin{aligned} G_i(u) &= \Pr\left[Y_i^{(+)} - Y_i^{(-)} \leq uN^{1/4}\right] \\ &= \Pr\left[Y_i^{(+)} - Y_i^{(-)} \leq \tau_1(u)(N^{1/4} + 40)\right] \\ &\leq \Pr\left[Y_i^{(+)} - Y_i^{(-)} \leq \tau_1(u)\sqrt{V}\right] \leq \mathcal{N}(\tau_1(u)) + \frac{1}{\sqrt{V}} \\ &\leq \mathcal{N}(\tau_1(u)) + \frac{1}{N^{1/4} - 50}. \end{aligned} \tag{16.33}$$

By (16.31), for $u < 0$,

$$\tau_1(u) = u\frac{N^{1/4}}{N^{1/4} + 40} = u - u\frac{40}{\sqrt{n} + 40},$$

which implies that,

$$\begin{aligned} |\mathcal{N}(\tau_1(u)) - \mathcal{N}(u)| &= \left|\mathcal{N}\left(u - u\frac{40}{\sqrt{n}+40}\right) - \mathcal{N}(u)\right| \\ &= \frac{1}{\sqrt{2\pi}} \int_u^{u-u\frac{40}{\sqrt{n}+40}} e^{-x^2/2}\, dx \\ &\le \frac{40|u|}{\sqrt{2\pi}(\sqrt{n}+40)} \exp\left(-\frac{u^2 n}{2(\sqrt{n}+40)^2}\right) \le \frac{20}{\sqrt{n}}, \end{aligned} \tag{16.34}$$

assuming $N \ge 10^{10}$. Of course (16.34) remains true for $u \ge 0$ as well.

Similarly,

$$|\mathcal{N}(\tau_2(u)) - \mathcal{N}(u)| \le \frac{20}{\sqrt{n}}. \tag{16.35}$$

Combining (16.32)–(16.35), we have for every real u,

$$\mathcal{N}(u) - \frac{20}{\sqrt{n}} - \frac{1}{\sqrt{n}-50} \le G_i(u) \le \mathcal{N}(u) + \frac{20}{\sqrt{n}} + \frac{1}{\sqrt{n}-50},$$

which implies

$$\sup_{-\infty<u<\infty} |G_i(u) - \mathcal{N}(u)| \le \frac{22}{\sqrt{n}}, \tag{16.36}$$

assuming $N \ge 10^{10}$.

Moreover, by (16.24), (16.31), (16.33) and Bernstein's inequality, for $u < 0$ and $N \ge 10^{10}$,

$$\begin{aligned} G_i(u) &= \Pr\left[Y_i^{(+)} - Y_i^{(-)} \le uN^{1/4}\right] \\ &= \Pr\left[Y_i^{(+)} - Y_i^{(-)} \le \tau_1(u)(N^{1/4}+40)\right] \\ &\le \Pr\left[Y_i^{(+)} - Y_i^{(-)} \le \tau_1(u)\sqrt{V}\right] \le 2\exp\left(-\frac{\tau_1^2(u)V/2}{V + (|\tau_1(u)|\sqrt{V}/3)}\right) \\ &= 2\exp\left(-\frac{\tau_1^2(u)/2}{1 + (|\tau_1(u)|V^{-1/2}/3)}\right) \le 2\exp\left(-\frac{u^2/3}{1 + (|u|/\sqrt{n})}\right). \end{aligned} \tag{16.37}$$

Similarly, for $u \ge 0$,

$$1 - G_i(u) \le 2\exp\left(-\frac{u^2/3}{1 + (u/\sqrt{n})}\right). \tag{16.38}$$

Furthermore, for $u < 0$ we clearly have

$$1 - \mathcal{N}(-u) = \mathcal{N}(u) \le 2e^{-u^2/2}. \tag{16.39}$$

Combining (16.36)–(16.39), for every $\gamma < 0$ we have

$$\left| \int_{-\infty}^{\gamma} (G_i(u) - \mathcal{N}(u))\, du \right| \le \int_{-\infty}^{\gamma} |G_i(u) - \mathcal{N}(u)|\, du \le \frac{100}{\sqrt{n}}, \tag{16.40}$$

and similarly, for every $\gamma \ge 0$,

$$\left| \int_{0}^{\gamma} (G_i(u) - \mathcal{N}(u))\, du \right| = \left| \int_{0}^{\gamma} ((1 - G_i(u)) - (1 - \mathcal{N}(u)))\, du \right| \le \frac{100}{\sqrt{n}}. \tag{16.41}$$

By (16.40) and (16.41),

$$\sup_{-\infty<\gamma<\infty} \left| \int_{-\infty}^{\gamma} (G_i(u) - \mathcal{N}(u))\, du \right| \le \frac{200}{\sqrt{n}}. \tag{16.42}$$

Using (16.42) in (16.23), we obtain

$$\sup_{-\infty<\gamma<\infty} \left| \mathbf{E}X_\gamma - n^{3/2} \int_{-\infty}^{\gamma} \mathcal{N}(u)\, du \right| \le n^{3/2} \frac{200}{\sqrt{n}} = 200n,$$

which completes the proof of Lemma 16.2. □

Let $\alpha > 200$, then by (16.19), (16.21) and Lemma 16.2,

$$\begin{aligned} &\Pr\left[\left| \int_{-\infty}^{\gamma} \Delta(u)\, du - \int_{-\infty}^{\gamma} \mathcal{N}(u)\, du \right| \ge \alpha n^{-1/2} \right] \\ &= \Pr\left[\left| n^{-3/2} X_\gamma - \int_{-\infty}^{\gamma} \mathcal{N}(u)\, du \right| \ge \alpha n^{-1/2} \right] \\ &\le \Pr\left[n^{-3/2} \left| X_\gamma - \mathbf{E}X_\gamma \right| \ge (\alpha - 200) n^{-1/2} \right] \le 2e^{-(\alpha-200)^2/2}. \end{aligned} \tag{16.43}$$

Let $\gamma_j = j/n$, where the numerator j runs over the integers in the interval $-nN = -n^3 < j < nN = n^3$. Since the common denominator n is "large", the sequence γ_j is "very dense" in the relevant interval $-N < \gamma_j < N$.

Consider the following events: let

$$A: \quad \sup_{-\infty<\lambda<\infty} |\Delta(\lambda) - \mathcal{N}(\lambda)| \ge \sqrt{12\alpha}\, n^{-1/4}, \tag{16.44}$$

and for every integer j in $-n^3 < j < n^3$, let

$$B_j : \left| \int_{-\infty}^{\gamma_j} (\Delta(u) - \mathcal{N}(u))\, du \right| \geq \alpha n^{-1/2}, \tag{16.45}$$

where $\gamma_j = j/n$. We show that

$$A \subset \bigcup_{-n^3<j<n^3} B_j. \tag{16.46}$$

Indeed, let

$$|\Delta(\lambda_0) - \mathcal{N}(\lambda_0)| = \sup_{-\infty<\lambda<\infty} |\Delta(\lambda) - \mathcal{N}(\lambda)| \geq \sqrt{12\alpha} n^{-1/4}.$$

Assume first that $\Delta(\lambda_0) > \mathcal{N}(\lambda_0)$. Let $\varepsilon = \sqrt{12\alpha} n^{-1/4}$, then

$$\Delta(\lambda_0) - \mathcal{N}(u) \geq \frac{1}{2}\varepsilon \quad \text{for all } \lambda_0 \leq u \leq \lambda_0 + \frac{1}{2}\varepsilon. \tag{16.47}$$

In the interval $\lambda_0 \leq u \leq \lambda_1 = \lambda_0 + \frac{1}{2}\varepsilon$ we can find $\gamma' = \gamma_{j_1} = j_1/n$ and $\gamma'' = \gamma_{j_2} = j_2/n$ such that $\gamma'' - \gamma' \geq \frac{1}{3}\varepsilon$, so by (16.47),

$$\int_{\gamma'}^{\gamma''} (\Delta(u) - \mathcal{N}(u))\, du \geq (\gamma'' - \gamma')\frac{1}{2}\varepsilon \geq \frac{1}{3}\varepsilon \cdot \frac{1}{2}\varepsilon = \frac{1}{6}\varepsilon^2. \tag{16.48}$$

By (16.48),

$$\max\left\{ \left| \int_{-\infty}^{\gamma'} (\Delta(u) - \mathcal{N}(u))\, du \right|, \left| \int_{-\infty}^{\gamma''} (\Delta(u) - \mathcal{N}(u))\, du \right| \right\} \geq \frac{1}{12}\varepsilon^2 = \alpha n^{-1/2},$$

which proves (16.46) for $\Delta(\lambda_0) > \mathcal{N}(\lambda_0)$. If $\Delta(\lambda_0) < \mathcal{N}(\lambda_0)$, then we repeat the argument with the interval $\lambda_0 - \frac{1}{2}\varepsilon \leq u \leq \lambda_0$.

Combining (16.43)–(16.46), we have for every $\alpha > 200$,

$$\begin{aligned} &\Pr\left[\sup_{-\infty<\lambda<\infty} |\Delta(\lambda) - \mathcal{N}(\lambda)| \geq \sqrt{12\alpha} n^{-1/4} \right] \\ &\leq \sum_{-n^3<j<n^3} \Pr\left[\left| \int_{-\infty}^{\gamma_j} (\Delta(u) - \mathcal{N}(u))\, du \right| \geq \alpha n^{-1/2} \right] \\ &\leq 2n^3 \cdot 2e^{-(\alpha-200)^2/2}. \end{aligned} \tag{16.49}$$

By (16.20) and (16.49), for $\beta > 20\sqrt{6}$

$$\Pr\left[\sup_{-\infty<\lambda<\infty}\left|\frac{1}{n}\left|\left\{1 \le i \le n: \frac{Y_i^{(+)} - Y_i^{(-)}}{\sqrt{n}} \le \lambda\right\}\right| - \mathcal{N}(\lambda)\right| \ge \beta n^{-1/4}\right]$$

$$\le 4n^3 e^{-((\beta^2/12)-200)^2/2}. \tag{16.50}$$

Combining (16.17) and (16.50), Lemma 16.1 follows. □

Applying Theorem 4.2 as a short-time ergodic theorem in the configuration space. We study the usual question: assume that N is large; is it true that the overwhelming majority of the time evolutions of the system with the *North-South* 2*-coloring* reaches advanced Red-Blue-CLT snapshot randomness superexponentially fast, and then it stays in that state for an extremely long time (with the possible exception of a totally negligible set of t's)? Applying Theorem 4.2 and Lemma 16.1, we give a positive answer.

As usual, we take advantage of the fact that, for every *fixed* $\omega^* \in [0, 2\pi)^N$, the family of time evolutions $\mathcal{Y}_{\omega;t}$, $\omega = (\omega^*, \mathbf{u}) \in \Omega_{\text{Gauss}}(\pm GCD)$ of the Gaussian ±GCD model with initial configuration $\mathcal{Y} \subset \mathbf{S}^2$ is represented by the family of torus lines in the configuration space $I^N = [0, 1)^N$, all starting from the origin $\mathbf{0} \in I^N$.

Let

$$\omega^* \notin \text{BadSet} = \bigcup_{i=1}^{n} \text{BadSet}_i \tag{16.51}$$

be arbitrary but fixed. We apply Theorem 4.2 with $f = \chi_S$ where S is

$$S = S(\beta) = S(\text{ColorError} \ge \beta n^{-1/4}; \omega^*). \tag{16.52}$$

Thus we obtain the following analog of (8.17) (or (15.22)): if $U \ge 1$ and $e^{\pi^2 U^2/2} > 3UN$ then for every integer $j \ge 1$,

$$(2\pi)^{-N/2} \int_{\mathbf{u}\in\mathbb{R}^N} \Big(\text{length}\left\{U \le t \le 2^j U : \vec{\mathcal{Y}}_{\omega;t} \in S(\beta)\right\}$$

$$- \text{vol}_N(S(\beta))(2^j - 1)U\Big)^2$$

$$\cdot e^{-|\mathbf{u}|^2/2}\, d\mathbf{u} \le \text{vol}_N(S(\beta)) 10j((2^j - 1)U + 1), \tag{16.53}$$

where

$$\omega = (\omega^*, \mathbf{u}) \in \Omega_{\text{Gauss}}(\pm GCD),$$

and by Lemma 16.1,

$$\mathrm{vol}_N(S(\beta)) \le 4n^3 e^{-((\beta^2/12)-200)^2/2} \tag{16.54}$$

holds for $\beta > 20\sqrt{6}$ and $N \ge 10^{10}$.

To illustrate the power of (16.53)–(16.54), let $U = 6$, $j = 340$, $N = 10^{200}$ and $\beta = 60$. Then by (16.53)–(16.54), for every $\omega^* \notin$ BadSet,

$$(2\pi)^{-N/2} \int_{\mathbf{u}\in\mathbb{R}^N} \Big(\text{length}\left\{6 \le t \le 2^{342} :\ \vec{\mathcal{Y}}_{\omega;t} \in S(\beta)\right\}$$
$$- \mathrm{vol}_N(S(\beta))4(2^{340}-1)\Big)^2$$
$$\cdot e^{-|\mathbf{u}|^2/2}\, d\mathbf{u} \le 4\cdot 10^{300} e^{-100^2/2} \cdot 4 \cdot 340 \cdot 2^{340} \cdot 20 \le 10^{-1201}. \tag{16.55}$$

Let $\mathbb{R}^N(\text{bad};\omega^*)$ be the set of those $\mathbf{u} \in \mathbb{R}^N$ for which

$$\text{length}\left\{6 \le t \le 2^{342} :\ \vec{\mathcal{Y}}_{\omega;t} \in S(\beta)\right\} \ge 10^{-400} \tag{16.56}$$

holds for $\omega = (\omega^*, \mathbf{u})$. Repeating the proof of (11.17)–(11.18), we obtain

$$\mathrm{Gauss}_N\left(\mathbb{R}^N(\text{bad};\omega^*)\right) \le 10^{-400}. \tag{16.57}$$

Again the threshold $U = 6$ represents — roughly speaking — the necessary number of "rounds" per particle in the Gaussian ±GCD model to reach advanced Red-Blue-CLT snapshot randomness (for the typical time evolution). As usual assume that the average projection-speed is 10^3 meter per second. For this system it takes only a few milliseconds to reach advanced snapshot randomness. Now (16.13) and (16.56)–(16.57) have the following interpretation. Choosing an arbitrary (measurable) CLT partition $\mathcal{H} = \{H_1, H_2, \ldots, H_n\}$ of the upper hemisphere $\mathbf{S}^2(+) = H_1 \cup H_2 \cup \ldots \cup H_n$ such that $\mathrm{SA}(H_i) = 1/2n$, $1 \le i \le n = \sqrt{N} = 10^{100}$ and an arbitrary $N = 10^{200}$-element initial point configuration $\mathcal{Y} \subset \mathbf{S}^2$, for the totally overwhelming majority of the initial velocities (Gaussian distribution), the normalized distribution of the *signed* particle counting number (1 if the particle is in the upper hemisphere and -1 if the antipodal is in the upper hemisphere) in the partition $\mathcal{H}$ remains very close to the standard normal distribution for an extremely long time, with the possible exception of a totally negligible set of times t.

Indeed, for every CLT partition $\mathcal{H} = \{H_1, H_2, \ldots, H_n\}$ and every $N = 10^{200}$-element initial point configuration $\mathcal{Y}$, there exists a subset

$$\Omega_{\text{Gauss}}(\pm\text{GCD};\text{good}) = \{\omega = (\omega^*, \mathbf{u}) \in \Omega_{\text{Gauss}}(\pm\text{GCD}) :$$
$$\omega^* \notin \text{BadSet} \quad \text{and} \quad \omega^* \notin \mathbb{R}^N(\text{bad};\omega^*)\}$$

with (see (16.13) and (16.57))

$$\begin{aligned}
&\text{ProdMeas}_{\text{Gauss},\pm\text{GCD}}(\Omega_{\text{Gauss}}(\text{good}))\\
&\quad \geq 1 - n\cdot 2\exp\left(-\frac{1600}{1+80N^{-1/4}/3}\right) - 10^{-400}\\
&\quad = 1 - 10^{100}\cdot 2\exp\left(-\frac{1600}{1+80N^{-1/4}/3}\right) - 10^{-400} > 1 - 2\cdot 10^{-400},
\end{aligned}$$

representing a totally overwhelming majority, such that for every $\omega \in \Omega_1(\pm\text{GCD};\text{good})$,

$$\text{CLTColorError} \leq \beta n^{-1/4} = 6\cdot 10^{-24} \tag{16.58}$$

holds for every $6 \leq t \leq 2^{342}$ with the possible exception of a set of times t of total length $< 10^{-400}$, see (16.56). The latter actually represents less than 10^{-403} seconds, which is a ridiculously short time.

Note that $6 \leq t \leq 2^{342}$ represents a time interval of length about 10^{97} seconds, which is a ridiculously long time. (The estimated age of the universe, starting from Big Bang, is less than 10^{20} seconds.)

Summarizing, the typical time evolution of this Gaussian system reaches "advanced Red-Blue-CLT snapshot randomness" (see (16.58)) in a few milliseconds (even starting from Big Bang!), and then it remains in the state of "advanced randomness" for an incredibly long time (with the possible exception of a totally negligible set of t's).

The approximation of the standard normal distribution — see CLT in (16.58) and (16.6)—with error $\leq 6\cdot 10^{-24}$ represents a striking precision, justifying the term "advanced randomness".

17. Time-Lapse Randomness vs. Snapshot Randomness (I): A Fundamental Difference

We already studied snapshot randomness on a realistic time scale. The common message of Secs. 8 and 15 can be summarized in a nutshell as follows. The overwhelming majority of the time evolutions of a Gaussian system reaches "complete snapshot randomness" *superexponentially* fast, and then it remains in the state of "complete randomness" for an incredibly long time (with the possible exception of a totally negligible set of times t). We talk about "complete randomness" in the sense that the error term is (nearly) optimal.

For illustration, we recall the example at the end of Sec. 15; see "Another illustration". We emphasize that in CLT snapshot randomness we must work with a partition that consists of a large number of parts. For a single test set, or test function, CLT snapshot randomness does not make any sense. On the other hand, as we demonstrate later, CLT time-lapse randomness is perfectly meaningful for a single test set, or test function.

As usual, we consider the classical Bernoulli gas model where the gas molecules are represented by $N \geq 10^{20}$ point billiards. By using *unfolding* we can reduce the billiards-in-a-box model to the simplest torus-billiard model $\mathcal{Y}_{\omega;t}$; and again we assume Gaussian initial velocity distribution. The threshold $U = \sqrt{\log N}$ represents the necessary number of "jumps" per particle in the Gaussian torus model (which is half of the number of bounces in the analog billiard model) to reach "complete CLT snapshot randomness" — it is superexponentially fast. As usual, we assume that the gas molecules have average speed 10^3 meter per second. Equations (15.44) and (15.45) have the following interpretation. For every CLT partition $\mathcal{H} = \{H_1, H_2, \ldots, H_n\}$ with $n = \sqrt{N}$ and every N-element initial point configuration $\mathcal{Y}$, there exists a subset $\Omega_{\text{Gauss}}(\text{good})$ with

$$\text{ProdMeas}_{\text{Gauss}}(\Omega_{\text{Gauss}}(\text{good})) \geq 1 - N^{-23} \tag{17.1}$$

(see (15.45)), representing an overwhelming majority, such that for every $\omega \in \Omega_{\text{Gauss}}(\text{good})$,

$$\begin{aligned} \text{CltError} = & \sup_{-\infty<\lambda<\infty} \left| \frac{1}{n} \left| \left\{ 1 \leq i \leq n : \frac{h_i(\mathcal{Y}_{\omega;t}) - n}{\sqrt{n}} \leq \lambda \right\} \right| - \mathcal{N}(\lambda) \right| \\ & \leq (7 + \log N) n^{-1/2} = \frac{7 + \log N}{N^{1/4}} \end{aligned} \tag{17.2}$$

holds for every t in the extremely long time interval $\sqrt{\log N} \leq t \leq \sqrt{\log N} N^{10}$ with the possible exception of a set of times t of total length $<N^{-23}$, see (15.44). The latter represents less than $10^{-3} N^{-23}$ seconds, which is an extremely short time.

Since $n^{-1/2} = N^{-1/4}$ is the inevitable "random error", the approximation of the standard normal distribution with error term

$$\frac{7 + \log N}{N^{1/4}}$$

in (17.2) is nearly optimal.

Similarly to snapshot randomness, our basic intuition for time-lapse randomness goes as follows. We say that the time evolution of the system exhibits "complete CLT time-lapse randomness" if the bell-shaped curve (standard normal distribution) shows up with striking precision. We want optimal, or nearly optimal, error term $N^{\varepsilon-1/2}$ with some small $\varepsilon > 0$.

This leads to the following natural question. What happens if we switch from snapshot randomness to time-lapse randomness? Is it true that the overwhelming majority of the time evolutions of a Gaussian system reaches "complete CLT time-lapse randomness" *superexponentially* fast?

We show that the answer to the question is *no*: we do not have an analogue of (17.1)–(17.2) for time-lapse randomness. The overwhelmimg majority — like in (17.1) — of the time evolutions of a system of the first kind reaches "complete CLT time-lapse randomness" (like in (17.2)) *very slowly.* This fact represents a fundamental difference between snapshot randomness and time-lapse randomness.

To prove such a "slow approach" result, we need a simple but important inequality; see (17.19) below.

For later application, we switch from the simplest torus-billiard model to the general torus model. That is, similarly to Sec. 10, we assume that every particle has its own *type* of trajectory described by a parametrized space curve: the trajectory of the kth particle is $\mathbf{r}_k(t)$ modulo one, where $\mathbf{r}_k(0) = \mathbf{0}$, $1 \leq k \leq N$.

We of course assume that each parametrized space curve is continuous and has arclength (rectifiable), but beyond that they are almost arbitrary (straight line, bending line, twisted line, any kind of general helix, spiral, etc.). We require, however, that the parametrized space curves $\mathbf{r}_k(t)$ have a tendency of "drifting away from the start $\mathbf{0}$". A parametrized curve modulo one describes the motion of a particle in the torus, and $|\mathbf{r}_k(t)|$ denotes the Euclidean distance in the whole space (not in the torus!) from the start $\mathbf{r}_k(0) = \mathbf{0}$ at time t. We emphasize that here we do *not* take $\mathbf{r}_k(t)$ modulo one yet; for example, on a straight line $t\mathbf{e}$, where $\mathbf{e}$ is a unit vector, the distance of $t\mathbf{e}$ from the start $\mathbf{0}$ is t.

In the rest, if we do not specify it, it should always be clear from the context of the sentence that the *distance* between two points $\mathbf{x}, \mathbf{y} \in \mathbb{R}^d$ is the usual R^d-distance (= Euclidean distance) $(\sum_{j=1}^{d}(x_j - y_j)^2)^{1/2}$ or the "unit torus distance" $(\sum_{j=1}^{d}\|x_j - y_j\|^2)^{1/2}$ (where $\|z\|$ is the distance of z from the nearest integer).

We use $\mathbf{x}_k(t)$ to denote the trajectory of the kth particle on the torus $I^3 = [0,1)^3$ ($1 \le k \le N$). For simplicity assume that every particle starts from the origin ("Big Bang"), i.e., $\mathbf{x}_k(0) = \mathbf{0} \in I^3$, $1 \le k \le N$. Let $\mathbf{r}_k(t)$, $1 \le k \le N$ be an arbitrary family of N continuous parametrized space curves with $\mathbf{r}_k(0) = \mathbf{0}$, and write

$$\text{drifting distance at time } t = \min_{1\le k\le N} |\mathbf{r}_k(t)|. \tag{17.3}$$

We know that $\mathbf{r}_k(t)$ describes the *type* of the trajectory of the kth particle (of course for the torus we take $\mathbf{r}_k(t)$ modulo one), and $\mathbf{x}_k(0) = \mathbf{0}$ is the starting point, but we also need the initial velocity. We assume Gaussian initial velocity distribution: we choose a scale ρ_k in $0 < \rho_k < \infty$ with distribution

$$\Pr[\rho_k \le u] = \sqrt{\frac{2}{\pi}} \int_0^u y^2 e^{-y^2/2}\, dy \tag{17.4}$$

for all positive $u > 0$ (the speed distribution of the 3-dimensional normal distribution; note that the factor y^2 comes from the surface area $4\pi y^2$ of a sphere of radius $y > 0$). We also choose a "direction", which here actually means a *rotation* of the 3-space. The reason is that a general space curve (say a helix) is a proper 3-dimensional object (unlike a straight line), so we must choose a rotation of the 3-space $\vartheta_k \in \mathrm{SO}(3)$.

(For comparison, note that the 2-dimensional case is simpler: the rotation of the plane simply means to choose a point on the unit circle $\mathbf{S}^1$, or equivalently, to choose an angle $0 \le \theta < 2\pi$.)

The choice of the scale ρ_k and the rotation ϑ_k in the 3-space make the initial condition complete ($1 \le k \le N$). So, starting from Big Bang, the *real* trajectory of the kth particle is

$$\mathbf{x}_k(t) = \vartheta_k(\rho_k \mathbf{r}_k(t)) \quad \text{modulo one}, \quad 1 \le k \le N, \tag{17.5}$$

which describes the time evolution of the whole N-point system.

To define *typical*, we need a measure. As usual, we assume that $\rho_1, \vartheta_1, \rho_2, \vartheta_2, \ldots, \rho_N, \vartheta_N$ are $2N$ independent random variables, where $\rho_1, \rho_2, \ldots, \rho_N$ have the same Gaussian distribution described by (17.4), and

$\vartheta_1, \vartheta_2, \ldots, \vartheta_N$ are all uniformly distributed in the rotation group SO(3) (= special orthogonal group of all 3×3 real matrices) with respect to the Haar measure. This means we use the corresponding product measure, denoted by $\text{ProdMeas}_{\text{Gauss}}$, on the product space

$$\omega = (\rho_1, \vartheta_1, \rho_2, \vartheta_2, \ldots, \rho_N, \vartheta_N) \in \Omega_{\text{Gauss}} = ([0, \infty) \times \text{SO}(3))^N. \quad (17.6)$$

We have $\int_{\Omega_{\text{Gauss}}} 1 \, d\omega = 1$, where for notational simplicity $d\omega = d\text{ProdMeas}_{\text{GAUSS}}(\omega)$.

We refer to this as the general 3-dimensional Gaussian torus model (starting from Big Bang). We refer to the huge product space Ω_{Gauss} as the "initial velocity space".

Note that the Haar measure on SO(3) is equivalent to a simple product measure: the product of the surface area on the unit sphere $\mathbf{S}^2$ and the arclength on the unit circle $\mathbf{S}^1$. This follows from the well-known fact that every nontrivial rotation in the 3-space can be specified by an axis of rotation together with an angle of rotation about this axis (more precisely, one also needs to specify orientations), which leads to a measure-theoretic isomorphism between SO(3) and $\mathbf{S}^2 \times \mathbf{S}^1$.

Given a family of parametrized curves

$$\mathcal{R} = (\mathbf{r}_1, \mathbf{r}_2, \ldots, \mathbf{r}_N),$$

and given an initial condition

$$\omega = (\rho_1, \vartheta_1, \rho_2, \vartheta_2, \ldots, \rho_N, \vartheta_N) \in \Omega_{\text{GAUSS}} = ([0, \infty) \times \text{SO}(3))^N,$$

write

$$\vec{\Gamma}(\mathcal{R}; \omega; t) = (\vartheta_1(\rho_1 \mathbf{r}_1(t)), \ldots, \vartheta_N(\rho_N \mathbf{r}_N(t))) \text{ modulo one} \quad (17.7)$$

for the curve of the (Gaussian general torus) system in the configuration space; (17.7) represents the time evolution of the system.

We use the test function

$$g_0(z_1, z_2, z_3) = \cos(2\pi z_1) = \frac{1}{2} e^{2\pi i \mathbf{n}_0 \cdot \mathbf{z}} + \frac{1}{2} e^{-2\pi i \mathbf{n}_0 \cdot \mathbf{z}}, \quad (17.8)$$

where $\mathbf{n}_0 = (1, 0, 0)$ and $\mathbf{z} = (z_1, z_2, z_3)$. Write

$$g_0 \left[\vec{\Gamma}(\mathcal{R}; \omega; t) \right] = \frac{1}{2} \sum_{k=1}^{N} e^{2\pi i \mathbf{n}_0 \cdot \vartheta_k(\rho_k \mathbf{r}_k(t))} + \frac{1}{2} \sum_{k=1}^{N} e^{-2\pi i \mathbf{n}_0 \cdot \vartheta_k(\rho_k \mathbf{r}_k(t))} \quad (17.9)$$

and

$$
\begin{aligned}
T(\mathcal{R};\omega;g_0;U,W) &= \int_U^W g_0\left[\vec{\Gamma}(\mathcal{R};\omega;t)\right]\,dt \\
&= \int_U^W \left(\frac{1}{2}\sum_{k=1}^N e^{2\pi i \mathbf{n}_0\cdot\vartheta_k(\rho_k\mathbf{r}_k(t))}\right. \\
&\quad \left. +\frac{1}{2}\sum_{k=1}^N e^{-2\pi i \mathbf{n}_0\cdot\vartheta_k(\rho_k\mathbf{r}_k(t))}\right)\,dt, \qquad (17.10)
\end{aligned}
$$

where in (17.10) we used (17.9).

By (17.10) we have

$$
\begin{aligned}
T^2(\mathcal{R};\omega;g_0;U,W) &= \left(\int_U^W g_0[\vec{\Gamma}(\mathcal{R};\omega;t_1)]dt_1\right)\left(\int_U^W g_0[\vec{\Gamma}(\mathcal{R};\omega;t_2)]dt_2\right) \\
&= \int_{t_1=U}^W\int_{t_2=U}^W \left(\frac{1}{2}\sum_{k=1}^N e^{2\pi i \mathbf{n}_0\cdot\vartheta_k(\rho_k\mathbf{r}_k(t_1))} + \frac{1}{2}\sum_{k=1}^N e^{-2\pi i \mathbf{n}_0\cdot\vartheta_k(\rho_k\mathbf{r}_k(t_1))}\right) \\
&\quad \cdot\left(\frac{1}{2}\sum_{l=1}^N e^{2\pi i \mathbf{n}_0\cdot\vartheta_l(\rho_l\mathbf{r}_l(t_2))} + \frac{1}{2}\sum_{l=1}^N e^{-2\pi i \mathbf{n}_0\cdot\vartheta_l(\rho_l\mathbf{r}_l(t_2))}\right) dt_1 dt_2 \\
&= \int_{t_1=U}^W\int_{t_2=U}^W \left(\frac{1}{4}\sum_{1\le k,l\le N} e^{2\pi i \mathbf{n}_0\,\cdot\,(\vartheta_k(\rho_k\mathbf{r}_k(t_1))-\vartheta_l(\rho_l\mathbf{r}_l(t_2)))}\right. \\
&\quad + \frac{1}{4}\sum_{1\le k,l\le N} e^{-2\pi i \mathbf{n}_0\cdot(\vartheta_k(\rho_k\mathbf{r}_k(t_1))-\vartheta_l(\rho_l\mathbf{r}_l(t_2)))} \\
&\quad + \frac{1}{4}\sum_{1\le k,l\le N} e^{2\pi i \mathbf{n}_0\cdot(\vartheta_k(\rho_k\mathbf{r}_k(t_1))+\vartheta_l(\rho_l\mathbf{r}_l(t_2)))} \\
&\quad \left. + \frac{1}{4}\sum_{1\le k,l\le N} e^{-2\pi i \mathbf{n}_0\cdot(\vartheta_k(\rho_k\mathbf{r}_k(t_1))+\vartheta_l(\rho_l\mathbf{r}_l(t_2)))}\right) dt_1\, dt_2. \qquad (17.11)
\end{aligned}
$$

Next we integrate (17.11) over the big "initial velocity space" $\Omega_{\text{GAUSS}} = ([0,\infty) \times \text{SO}(3))^N$ (see (17.6)); we have

$$\int_{\Omega_{\text{GAUSS}}} T^2(\mathcal{R};\omega;g_0;U,W)\,d\omega$$
$$= \sqrt{\frac{2}{\pi}}\int_{\rho_1=0}^{\infty}\int_{\vartheta_1\in\text{SO}(3)}\cdots\sqrt{\frac{2}{\pi}}\int_{\rho_N=0}^{\infty}\int_{\vartheta_N\in\text{SO}(3)} T^2(\mathcal{R};\omega;g_0;U,W)\rho_1^2$$
$$\cdot e^{-\rho_1^2/2}\,d\rho_1\,d\vartheta_1\cdots\rho_N^2 e^{-\rho_N^2/2}\,d\rho_N\,d\vartheta_N$$
$$= \sum_{1\le k,l\le N:k\ne l}\int_{t_1=U}^{W}\int_{t_2=U}^{W}\exp(-2(\pi|\mathbf{r}_k(t_1)|)^2)\exp(-2(\pi|\mathbf{r}_l(t_2)|)^2)dt_1\,dt_2$$
$$+\frac{1}{2}\sum_{k=1}^{N}\int_{t_1=U}^{W}\int_{t_2=U}^{W}\exp(-2(\pi|\mathbf{r}_k(t_1)+\mathbf{r}_k(t_2)|)^2)dt_1\,dt_2$$
$$+\frac{1}{2}\sum_{k=1}^{N}\int_{t_1=U}^{W}\int_{t_2=U}^{W}\exp(-2(\pi|\mathbf{r}_k(t_1)-\mathbf{r}_k(t_2)|)^2)dt_1\,dt_2, \tag{17.12}$$

where we used the fact that for every 3-dimensional vector $\mathbf{w}$ and unit vector $\mathbf{e}\in\mathbf{S}^2$,

$$\sqrt{\frac{2}{\pi}}\int_0^{\infty}\int_{\vartheta\in\text{SO}(3)} e^{-2\pi i\vartheta(\rho\mathbf{e})\cdot\mathbf{w}}\rho^2 e^{-\rho^2/2}\,d\rho\,d\vartheta$$
$$= \sqrt{\frac{2}{\pi}}\int_0^{\infty}\frac{1}{4\pi}\int_{\mathbf{u}\in\mathbf{S}^2} e^{-2\pi i\rho\mathbf{u}\cdot\mathbf{w}}\rho^2 e^{-\rho^2/2}\,d\rho\,d\text{SA}(\mathbf{u}) = e^{-2\pi^2|\mathbf{w}|^2}, \tag{17.13}$$

and the last step in (17.13) follows from Lemma 9.1 with $d=3$.

The crucial fact in (17.12) is that the terms on the right-hand side are all positive. Thus we obtain the trivial lower bound (we just keep the last line in (17.12))

$$\int_{\Omega_1} T^2(\mathcal{R};\omega;g_0;U,W)\,d\omega$$
$$\ge \frac{1}{2}\sum_{k=1}^{N}\int_{t_1=U}^{W}\int_{t_2=U}^{W}\exp(-2(\pi|\mathbf{r}_k(t_1)-\mathbf{r}_k(t_2)|)^2)dt_1\,dt_2. \tag{17.14}$$

Consider now the simplest torus model $\mathbf{r}_k(t) = t\mathbf{n_0}$ with $\mathbf{n}_0 = (1,0,0)$, $1\le k\le N$. Then (17.7) simplifies to the torus line

$$\vec{\Gamma}(\omega;t) = t(\vartheta_1(\rho_1\mathbf{n}_0),\ldots,\vartheta_N(\rho_N\mathbf{n}_0)) \text{ modulo one}, \tag{17.15}$$

where

$$\omega = (\rho_1, \vartheta_1, \rho_2, \vartheta_2, \ldots, \rho_N, \vartheta_N) \in \Omega_{\text{Gauss}} = ([0,\infty) \times \mathbf{S}^2)^N. \quad (17.16)$$

In this special case (17.14) gives (for notational simplicity $d\omega = d\text{ProdMeas}_{\text{Gauss}}(\omega)$)

$$\int_{\Omega_{\text{Gauss}}} T^2(\omega; g_0; U, W)\, d\omega \geq \frac{N}{2} \int_{t_1=U}^{W} \int_{t_2=U}^{W} \exp(-2(\pi|t_1 - t_2|)^2) dt_1\, dt_2. \quad (17.17)$$

By (14.21) (which is basically the proof of Theorem 4.3)

$$\int_U^W \left(\int_U^W e^{-2\pi^2 (t_1 - t_2)^2}\, dt_2 \right) dt_1 \geq \frac{W-U}{\sqrt{2\pi}} - \frac{1}{5}. \quad (17.18)$$

Combining (17.17) and (17.18),

$$\int_{\Omega_{\text{Gauss}}} T^2(\omega; g_0; U, W)\, d\omega \geq \frac{N}{2} \left(\frac{W-U}{\sqrt{2\pi}} - \frac{1}{5} \right). \quad (17.19)$$

By using (17.19), we can derive a lower bound as follows — see (17.23) below.

Theorem 17.1. *Consider the simplest N-particle ($N \geq 10^{20}$) Gaussian 3-dimensional torus model in $I^3 = [0,1)^3$ with all particles starting from the origin ("Big Bang"); see (17.15)–(17.16). We use the "nice" test function g_0 defined in (17.8). Assume that there is a time interval $U \leq t \leq W$ with $U \geq \sqrt{\log N}$ and $W - U \geq 1$ (where $t = 0$ represents the Big Bang in the origin) such that*

$$\text{for at least } 1 - N^{-2} \text{ part of the time evolutions } \omega \in \Omega_{\text{Gauss}} \quad (17.20)$$

(representing overwhelming majority in the polynomial sense of (17.1)) we have

$$\sup_{-\infty<\lambda<\infty} \left| \frac{1}{W-U} \text{length} \left\{ t \in [U,W] : \frac{g_0[\vec{\Gamma}(\omega;t)]}{\sigma_0(g_0)\sqrt{N}} \leq \lambda \right\} - \mathcal{N}(\lambda) \right| \leq N^{\varepsilon - 1/2} \quad (17.21)$$

for some constant $1/4 > \varepsilon > 0$, where

$$\sigma_0(g_0) = \left(\int_0^1 \cos^2(2\pi x)\, dx \right)^{1/2} = 1/\sqrt{2}. \quad (17.22)$$

Then

$$W - U \geq \frac{N^{1-2\varepsilon}}{100(\log N)^2}. \tag{17.23}$$

Remark. This shows that we do not have an analogue of (17.1)–(17.2) for time-lapse randomness. Indeed, for the overwhelmimg majority (see (17.20)) of the time evolutions of a system of the first kind it takes nearly linear time (see (17.23), assuming ε is small) to reach nearly perfect CLT time-lapse randomness (see (17.21)). Nearly linear time $N^{1-o(1)}$ is strikingly different from the much smaller threshold $\sqrt{\log N}$ in snapshot randomness.

Proof of Theorem 17.1. By using the first line in (17.10) in the special case of the simplest 3-dimensional torus model, we have

$$\begin{aligned}\frac{T(\omega; g_0; U, W)}{(W-U)\sqrt{N/2}} &= \frac{1}{W-U}\int_U^W g_0[\vec{\Gamma}(\omega;t)]\,dt \\ &= \int_{\lambda=-\infty}^{\infty} \lambda d\frac{1}{W-U}\text{length}\left\{t \in [U,W] : \frac{g_0[\vec{\Gamma}(\omega;t)]}{\sqrt{N/2}} \leq \lambda\right\} \\ &= \int_{\lambda=-\infty}^{\infty} \lambda d\left(\frac{1}{W-U}\text{length}\left\{t \in [U,W] : \right.\right. \\ &\qquad \left.\left.\frac{g_0[\vec{\Gamma}(\omega;t)]}{\sqrt{N/2}} \leq \lambda\right\} - \mathcal{N}(\lambda)\right), \end{aligned} \tag{17.24}$$

where the last two integrals with

$$\int_{\lambda=-\infty}^{\infty} \lambda\, d\ldots$$

are Lebesgue–Stieltjes integrals, and in the last step we used the well-known fact that the expectation of the standard normal distribution is zero, i.e.,

$$\int_{\lambda=-\infty}^{\infty} \lambda\, d\mathcal{N}(\lambda) = 0.$$

Applying integration by parts in (17.24), we have

$$\begin{aligned}&\frac{T(\omega; g_0; U, W)}{(W-U)\sqrt{N/2}} \\ &= \int_{\lambda=-\infty}^{\infty}\left(\frac{1}{W-U}\text{length}\left\{t \in [U,W] : \frac{g_0[\vec{\Gamma}(\omega;t)]}{\sqrt{N/2}} \leq \lambda\right\} - \mathcal{N}(\lambda)\right) d\lambda.\end{aligned} \tag{17.25}$$

By (17.25),

$$\frac{T(\omega;g_0;U,W)}{(W-U)\sqrt{N/2}} = \int_{\lambda=-2\log N}^{2\log N}\left(\frac{1}{W-U}\text{length}\left\{t\in[U,W]:\right.\right.$$

$$\left.\left.\frac{g_0\left[\vec{\Gamma}(\omega;t)\right]}{\sqrt{N/2}}\le\lambda\right\}-\mathcal{N}(\lambda)\right)\,d\lambda$$

$$+\int_{\lambda=-\infty}^{-2\log N}\left(\frac{1}{W-U}\text{length}\left\{t\in[U,W]:\right.\right.$$

$$\left.\left.\frac{g_0\left[\vec{\Gamma}(\omega;t)\right]}{\sqrt{N/2}}\le\lambda\right\}-\mathcal{N}(\lambda)\right)\,d\lambda$$

$$+\int_{\lambda=2\log N}^{\infty}\left((1-\mathcal{N}(\lambda))-\frac{1}{W-U}\text{length}\left\{t\in[U,W]:\right.\right.$$

$$\left.\left.\frac{g_0\left[\vec{\Gamma}(\omega;t)\right]}{\sqrt{N/2}}>\lambda\right\}\right)\,d\lambda. \tag{17.26}$$

To estimate the last two lines in (17.26), we basically repeat the argument of Sec. 5: it is the usual application of Theorem 4.2. We carry out the details in the next section; see (18.13)–(18.14).

18. Time-Lapse Randomness vs. Snapshot Randomness (II): A Fundamental Difference

We make use of the fact that the family of time evolutions (see (17.15); let $\mathbf{n}_0=(1,0,0)$)

$$\vec{\Gamma}(\omega;t) = t(\vartheta_1(\rho_1\mathbf{n}_0),\ldots,\vartheta_N(\rho_N\mathbf{n}_0)) \text{ modulo one}$$
$$= t\mathbf{v}(\omega) \text{ modulo one}, \tag{18.1}$$

$\omega\in\Omega_{\text{Gauss}}$ of the simplest 3-dimensional Gaussian torus model is represented by a family of torus lines in the configuration space I^d starting from the origin $\mathbf{0}\in I^d$ (since the system started from Big Bang $\mathbf{0}\in I^3$).

For an arbitrary $\gamma > 0$ write

$$S(g_0;\gamma) = \left\{\vec{\mathcal{Z}} \in I^d : \left|\sum_{k=1}^{N} g_0(\mathbf{z}_k)\right| \geq \gamma\sqrt{N/2}\right\}$$
$$= \left\{\vec{\mathcal{Z}} \in I^d : \left|\sum_{k=1}^{N} \cos(z_{3k-2})\right| \geq \gamma\sqrt{N/2}\right\}, \qquad (18.2)$$

where

$$\vec{\mathcal{Z}} = (z_1, \ldots, z_{3N})$$

and $\mathbf{z}_k = (z_{3k-2}, z_{3k-1}, z_{3k})$, $1 \leq k \leq N$.

By the general form of Bernstein's large deviation inequality (see Sec. 5) with $n = N$ and $\tau = \gamma\sqrt{N/2}$,

$$\mathrm{vol}_d(S(g_0;\gamma)) \leq 2\exp\left(-\frac{\tau^2/2}{(\sum_{i=1}^{n} \mathbf{E}X_i^2) + (\tau M/3)}\right)$$
$$= 2\exp\left(-\frac{\gamma^2 N/4}{(N/2) + (\gamma\sqrt{N/2}/3)}\right)$$
$$= 2\exp\left(-\frac{\gamma^2/2}{1 + \gamma/(3\sqrt{N/2})}\right). \qquad (18.3)$$

We apply Theorem 4.2 with $f = \chi_S$ where $S = S(g_0;\gamma)$. Thus we obtain

$$(2\pi)^{-d/2}\int_{\mathbf{v}\in\mathbb{R}^d} (\mathrm{TimeDiscrep}_S(\mathbf{v}; U, W))^2 e^{-|\mathbf{v}|^2/2}\, d\mathbf{v}$$
$$= (2\pi)^{-d/2}\int_{\mathbf{v}\in\mathbb{R}^d} \left(\int_U^W \chi_S(t\mathbf{v})\, dt - \mathrm{vol}_d(S)(W-U)\right)^2 e^{-|\mathbf{v}|^2/2}\, d\mathbf{v}$$
$$\leq 10k \cdot \mathrm{vol}_d(S)(1 - \mathrm{vol}_d(S))(W - U + 1)$$
$$\leq 2\exp\left(-\frac{\gamma^2/2}{1 + \frac{\gamma}{3\sqrt{N/2}}}\right) 10k(W - U + 1), \qquad (18.4)$$

where the integer $k \geq 1$ comes from $2^{k-1}U < W \leq 2^k U$, and in the last step we used (18.3).

Note that

$$(2\pi)^{-d/2}\int_{\mathbf{v}\in\mathbb{R}^d}\left(\int_U^W \chi_S(t\mathbf{v})\,dt - \mathrm{vol}_d(S)(W-U)\right)^2 e^{-|\mathbf{v}|^2/2}\,d\mathbf{v}$$

$$= \int_{\omega\in\Omega_{\mathrm{Gauss}}}\left(\mathrm{length}\left\{U\le t\le W : \left|\frac{g_0\left[\vec{\Gamma}(\omega;t)\right]}{\sqrt{N/2}}\right| \ge \gamma\right\}\right.$$

$$\left. - \mathrm{vol}_d(S(g_0;\gamma))(W-U)\right)^2 d\omega. \tag{18.5}$$

We recall that $U \ge \sqrt{\log N}$ and $N \ge 10^{20}$, and we choose

$$\gamma = \gamma_j = 2^j \log N,\ j \ge 1. \tag{18.6}$$

Then

$$\mathrm{vol}_d(S(g_0; 2^j\log N)) \le 2\exp\left(-\frac{2^{2j-1}\log^2 N}{1+2^j\log N/(3\sqrt{N/2})}\right) \le N^{-90j}, \tag{18.7}$$

and so by (18.4)–(18.7),

$$\int_{\omega\in\Omega_{\mathrm{Gauss}}}\left(\mathrm{length}\left\{U\le t\le W : \left|\frac{g_0\left[\vec{\Gamma}(\omega;t)\right]}{\sqrt{N/2}}\right| \ge 2^j\log N\right\}\right.$$

$$\left. - \mathrm{vol}_d(S(g_0;2^j\log N))(W-U)\right)^2 d\omega$$

$$\le 2\exp\left(-\frac{2^{2j-1}\log^2 N}{1+\frac{2^j\log N}{3\sqrt{N/2}}}\right)10k(W-U+1)$$

$$\le 40\lceil\log_2(W/U)\rceil(W-U)N^{-90j} \le (W-U)^2N^{-90j+1}. \tag{18.8}$$

For every integer $j \ge 1$, let $\Omega(\mathrm{bad};j)$ be the set of those $\omega\in\Omega_{\mathrm{Gauss}}$ for which

$$\mathrm{length}\left\{U\le t\le W : \left|\frac{g_0\left[\vec{\Gamma}(\omega;t)\right]}{\sqrt{N/2}}\right| \ge 2^j\log N\right\} \ge (W-U)N^{-30j}. \tag{18.9}$$

We claim that (18.8) implies

$$\mathrm{ProdMeas}_{\mathrm{Gauss}}(\Omega(\mathrm{bad};j)) \le N^{-25j}. \tag{18.10}$$

Indeed, otherwise

$$\int_{\omega\in\Omega_{\text{Gauss}}}\left(\text{length}\left\{U\le t\le W:\ \left|\frac{g_0\left[\vec{\Gamma}(\omega;t)\right]}{\sqrt{N/2}}\right|\ge 2^j\log N\right\}\right.$$

$$\left.-\ \text{vol}_d(S(g_0;2^j\log N))(W-U)\right)^2 d\omega$$

$$\ge\int_{\omega\in\Omega_1(\text{bad};j)}\left(\text{length}\left\{U\le t\le W:\ \left|\frac{N(\omega;g_0;t)}{\sqrt{N/2}}\right|\ge 2^j\log N\right\}\right.$$

$$\left.-\ \text{vol}_d(S(g_0;2^j\log N))(W-U)\right)^2 d\omega$$

$$\ge N^{-25j}\left((W-U)N^{-30j}-N^{-90j}(W-U)\right)^2$$

$$\ge N^{-25j}\cdot\frac{1}{2}(W-U)^2N^{-60j}=\frac{1}{2}(W-U)^2N^{-85j},$$

which contradicts (18.8). In the last step we used $\text{vol}_d(S(g_0;2^j\log N))\le N^{-90j}$ (see (18.7)).

Let

$$\Omega(\text{bad})=\bigcup_{j=1}^{\infty}\Omega(\text{bad};j). \tag{18.11}$$

Then by (18.10),

$$\text{ProdMeas}_{\text{Gauss}}(\Omega(\text{bad}))\le\sum_{j=1}^{\infty}N^{-25j}=\frac{N^{-25}}{1-N^{-25}}. \tag{18.12}$$

Let

$$\omega\notin\Omega(\text{bad}), \tag{18.13}$$

then by (18.9)

$$\text{length}\left\{U\le t\le W:\ \left|\frac{g_0\left[\vec{\Gamma}(\omega;t)\right]}{\sqrt{N/2}}\right|\ge 2^j\log N\right\}\le(W-U)N^{-30j}$$

$$\text{for all}\quad j\ge 1. \tag{18.14}$$

Now we are ready to estimate the last two lines in (17.26). Under the condition of (18.13), by (18.14) we have

$$\int_{\lambda=-\infty}^{-2\log N}\left(\frac{1}{W-U}\text{length}\left\{t\in[U,W]:\frac{g_0\left[\vec{\Gamma}(\omega;t)\right]}{\sqrt{N/2}}\leq\lambda\right\}+\mathcal{N}(\lambda)\right)d\lambda$$

$$+\int_{\lambda=2\log N}^{\infty}\left((1-\mathcal{N}(\lambda))+\frac{1}{W-U}\text{length}\right.$$

$$\left.\times\left\{t\in[U,W]:\frac{g_0\left[\vec{\Gamma}(\omega;t)\right]}{\sqrt{N/2}}>\lambda\right\}\right)d\lambda$$

$$\leq 2\int_{\lambda=-\infty}^{-2\log N}\mathcal{N}(\lambda)\,d\lambda+\sum_{j=1}^{\infty}2^{j+1}\log N\frac{1}{W-U}(W-U)N^{-30j}$$

$$\leq 2\int_{-\infty}^{-2\log N}e^{-\lambda^2/2}\,d\lambda+2\log N\frac{2N^{-30}}{1-2N^{-30}}$$

$$\leq 2e^{-2\log^2 N}+2\log N\frac{2N^{-30}}{1-2N^{-30}}\leq N^{-29}. \tag{18.15}$$

Moreover, by (17.20) and (17.21),

$$\int_{\lambda=-2\log N}^{2\log N}\left|\frac{1}{W-U}\text{length}\left\{t\in[U,W]:\frac{g_0\left[\vec{\Gamma}(\omega;t)\right]}{\sqrt{N/2}}\leq\lambda\right\}-\mathcal{N}(\lambda)\right|d\lambda$$

$$\leq\int_{\lambda=-2\log N}^{2\log N}N^{\varepsilon-1/2}\,d\lambda=\frac{4\log N}{N^{1/2-\varepsilon}}, \tag{18.16}$$

except for a set $\omega\in\Omega(\text{defective})$ with

$$\text{ProdMeas}_{\text{Gauss}}(\Omega(\text{defective}))\leq N^{-3}. \tag{18.17}$$

Combining (17.26) and (18.15)–(18.17), we obtain

$$\left|\frac{T(\omega;g_0;U,W)}{(W-U)\sqrt{N/2}}\right|\leq\frac{4\log N}{N^{1/2-\varepsilon}}+N^{-29}\leq\frac{5\log N}{N^{1/2-\varepsilon}} \tag{18.18}$$

for all

$$\omega\notin\Omega(\text{faulty})=\Omega(\text{bad})\cup\Omega(\text{defective}) \tag{18.19}$$

with (see also (18.12))

$$\text{ProdMeas}_{\text{Gauss}}(\Omega(\text{faulty})) \le \frac{N^{-25}}{1-N^{-25}} + N^{-3} < 2N^{-3}. \tag{18.20}$$

Write

$$\Omega(\text{good}) = \Omega_{\text{Gauss}} \setminus \Omega(\text{faulty}). \tag{18.21}$$

Combining (18.18)–(18.21), we have

$$\begin{aligned} \int_{\Omega_{\text{Gauss}}} T^2(\omega; g_0; U, W)\, d\omega &= \int_{\Omega(\text{good})} T^2(\omega; g_0; U, W)\, d\omega \\ &\quad + \int_{\Omega(\text{faulty})} T^2(\omega; g_0; U, W)\, d\omega \\ &\le \left(\frac{5\log N}{N^{1/2-\varepsilon}}(W-U)\sqrt{N/2}\right)^2 \\ &\quad + 2N^{-2}(W-U)^2 N^2 \\ &= 25(W-U)^2(\log N)^2 N^{2\varepsilon}/2 + 2(W-U)^2, \end{aligned} \tag{18.22}$$

where we used the trivial upper bound $|T(\omega; g_0; U, W)| \le (W-U)N$ (see (17.10)).

On the other hand, by (17.19),

$$\int_{\Omega_1} T^2(\omega; g_0; U, W)\, d\omega \ge \frac{N}{2}\left(\frac{W-U}{\sqrt{2\pi}} - \frac{1}{5}\right). \tag{18.23}$$

Comparing (18.22) and (18.23), we have

$$\begin{aligned} \frac{N}{2}\left(\frac{W-U}{\sqrt{2\pi}} - \frac{1}{5}\right) &\le 25(W-U)^2(\log N)^2 N^{2\varepsilon}/2 + 2(W-U)^2 \\ &\le 13(W-U)^2(\log N)^2 N^{2\varepsilon}, \end{aligned}$$

which implies (17.23)

$$W - U \ge \frac{N^{1-2\varepsilon}}{100(\log N)^2} \quad \text{for all} \quad 1/4 > \varepsilon > 0.$$

This completes the proof of Theorem 17.1. □

Probably the reader is wondering: Why does it take so long to reach nearly optimal CLT time-lapse randomness? It seems like a "plausible explanation" that perhaps the N torus lines (representing the trajectories of the individual particles in the simplest torus model) are "too simple", and if the particles move on more complicated random walk like zig-zag trajectories then perhaps it takes much shorter time to reach nearly optimal CLT time-lapse randomness. Well, it turns out that this "plausible explanation" is completely wrong.

In Theorem 17.1 we studied the simplest 3-dimensional torus model $\mathbf{r}_k(t) = (t, 0, 0)$, $1 \le k \le N$. Consider now the most general case where $\mathbf{r}_k(t)$, $1 \le k \le N$ are arbitrary with only one natural restriction

$$|\mathbf{r}_k(t_1) - \mathbf{r}_k(t_2)| \le |t_1 - t_2|. \tag{18.24}$$

(18.24) simply means that the *speed is bounded* (it certainly holds for particle–particle collision of hard-core particles). Under the condition (18.24), Eq. (17.14) implies the *same* lower bound as in (17.17)

$$\begin{aligned}\int_{\Omega_{\text{Gauss}}} T^2(\mathcal{R}; \omega; g_0; U, V)\, d\omega &\ge \frac{1}{2}\sum_{k=1}^{N}\int_{t_1=U}^{V}\int_{t_2=U}^{V} \exp(-2(\pi|\mathbf{r}_k(t_1)\\ &\qquad - \mathbf{r}_k(t_2)|)^2)\, dt_1\, dt_2\\ &\ge \frac{N}{2}\int_{t_1=U}^{V}\int_{t_2=U}^{V} \exp(-2(\pi|t_1 - t_2|)^2)\, dt_1\, dt_2.\end{aligned} \tag{18.25}$$

Since (18.25) is the same as (17.17), we can basically repeat the whole argument above in the general case with some necessary changes (we omit the details). Thus we obtain the same nearly linear lower bound for the time interval as in (17.23). In other words, complicated zig-zag trajectories do not speed up the approach to nearly optimal CLT time-lapse randomness.

We have to admit that we do not know any simple intuitive explanation why the approach to nearly optimal CLT time-lapse randomness is so time-consuming.

19. CLT Time-Lapse Randomness: Upper Bound

We recall that in computational complexity theory, an algorithm with (worst-case) polynomial running time (in terms of the input data) is considered *tractable*, and exponential running time is considered *intractable*.

We use the same distinction here: polynomial time (in terms of N) is still considered *realistic*, and exponential time — what we have e.g. for nearly perfect time-lapse randomness (see Theorem 11.8 and after in the Remarks) — is considered *unrealistic.*

We complement the lower bound (17.23) in Theorem 17.1 with a positive result showing that polynomial time — i.e., realistic time — is sufficient to exhibit optimal CLT time-lapse randomness for the overwhelming majority of the time evolutions. The proof is another application of Theorem 4.2; again we use it as a *short-time ergodic theorem* in the configuration space.

For simplicity, we formulate this upper bound theorem for test sets $B \subset [0,1)^3$ in the particle space instead of the most general case of test functions.

Theorem 19.1. *Let $\mathcal{Y}_{\omega;t}$, $\omega \in \Omega_{\text{Gauss}}$ be the simplest 3-dimensional Gaussian torus model with N-element initial point configuration $\mathcal{Y} \subset [0,1)^3$. Let $N \geq 10^{20}$, and let $B \subset [0,1)^3$ be an arbitrary but fixed measurable test where $p = \text{vol}(B)$ denotes the 3-dimensional Lebesgue measure.*

Then for every $\eta > 0$ there exists a subset $\Omega(\text{bad};\eta) \subset \Omega_{\text{Gauss}}$ such that

$$\text{ProdMeas}_{\text{Gauss}}(\Omega(\text{bad};\eta)) \leq 200\eta,$$

and for every $\omega \notin \Omega(\text{bad};\eta)$, and every real number $T \geq 2\sqrt{\log N} + 2$ (we also use the notation $q = 1 - p$),

$$\sup_{-\infty<\beta<\infty} \left| \frac{1}{T}\text{length}\left\{0 \leq t \leq T : \frac{|\mathcal{Y}_{\omega;t} \cap B| - pN}{\sqrt{pqN}} \leq \beta\right\} - \mathcal{N}(\beta)\right|$$
$$\leq \frac{1}{\sqrt{pqN}} + \frac{2}{N} + \frac{\sqrt{2}(2+\log_2 T)^{7/2}(2\log_2 N)^2}{\sqrt{\eta}\sqrt{T}} + \frac{2\sqrt{\log N}+3}{T}, \tag{19.1}$$

where of course $\mathcal{N}(y)$ denotes the normal distribution:

$$\mathcal{N}(y) = \frac{1}{\sqrt{2\pi}} \int_{-\infty}^{y} e^{-u^2/2}\, du.$$

Remarks. For example, let $\eta = N^{-4}$ and $T \geq N^7$. Then the upper bound

$$\text{ProdMeas}_{\text{Gauss}}(\Omega(\text{bad};\eta)) \leq 200\eta = 200N^{-4}$$

implies that

$$\omega \in \Omega(\text{good};\eta) = \Omega_{\text{Gauss}} \setminus \Omega(\text{bad};\eta)$$

represents an overwhelming majority of the initial velocities, and for this overwhelming majority of $\omega \in \Omega(\text{good};\eta)$, (19.1) gives the upper bound

$$\sup_{-\infty<\beta<\infty}\left|\frac{1}{T}\text{length}\left\{0\le t\le T:\ \frac{|\mathcal{Y}_{\omega;t}\cap B|-pN}{\sqrt{pqN}}\le\beta\right\}-\mathcal{N}(\beta)\right|$$

$$\le\frac{1}{\sqrt{pqN}}+\frac{2}{N}+\frac{\sqrt{2}(2+\log_2 T)^{7/2}(2\log_2 N)^2}{\sqrt{\eta}\sqrt{T}}+\frac{2\sqrt{\log N}+2}{T}$$

$$\le\frac{1}{\sqrt{pqN}}+\frac{3}{N},$$

where the last line is the optimal CLT error apart from an absolute constant factor (the error is the discrepancy between the discrete binomial distribution and the continuous standard normal distribution). The lower bound $T\ge N^7$ is polynomial in N, so it is still considered a realistic time scale.

On the other hand, if T is in the much smaller polylogarithmic range of $(\log N)^{4++o(1)}$, then the right-hand side of (19.1) (= error term) is already $o(1)$ (for an appropriate $\eta=o(1)$). That is, polylogarithmic threshold suffices to prove a non-trivial CLT type result.

Returning to polynomial time: as we said above, it is still interpreted as a realistic time scale; nevertheless, the approach to time-lapse randomness is undoubtedly very slow; see (17.23). In sharp contrast, the approach to snapshot randomness is extremely fast, reflected by the threshold $\sqrt{\log N}$, which holds even for starting from Big Bang.

We may call Theorem 19.1 a quantitative/realistic version of Theorem 3.3. Similarly, by a straightforward modification of the proof of Theorem 19.1, one can easily prove a "hard" analogue of the "weak" Poisson Law, Multinomial CLT, and Multinomial Poisson Law in Secs. 3 and 11. We could also have a CLT-partition like in Sec. 15, and a Poisson-partition like in Sec. 8. The basic idea is always the same: replace the application of the Kronecker–Weyl theorem in the configuration space with the use of Theorem 4.2.

Also, we can easily extend Theorem 19.1 (i.e., CLT time-lapse randomness in realistic time) for the closed orbit models introduced in Secs. 6 and 17.

Proof of Theorem 19.1. As usual, given an N-element point set $\mathcal{Y}=\{\mathbf{y}_1,\ldots,\mathbf{y}_N\}\subset I^3$, we use the vector notation

$$\vec{\mathcal{Y}}=(\mathbf{y}_1,\ldots,\mathbf{y}_N)\in I^d$$

for the corresponding point in the configuration space $I^d = [0,1)^{3N}$. The family of time evolutions $\mathcal{Y}_{\omega;t}$, $\omega \in \Omega_{\text{Gauss}}$ of the simplest 3-dimensional torus model of the first kind is represented by the family of torus lines

$$\vec{\mathcal{Y}}_{\omega;t} = t\mathbf{v}(\omega) + \vec{\mathcal{Y}} \text{ modulo one}$$

in the configuration space I^d, all starting from the same point $\vec{\mathcal{Y}} \in I^d$. For $-\infty \le \gamma_1 < \gamma_2 \le \infty$, write with $q = 1-p$,

$$S(B;\gamma_1,\gamma_2) = \left\{ \vec{\mathcal{Z}} \in I^d :\ \gamma_1 \le \frac{|\mathcal{Z} \cap B| - pN}{\sqrt{pqN}} \le \gamma_2 \right\}, \tag{19.2}$$

where

$$\vec{\mathcal{Z}} = (z_1,\ldots,z_{3N}) \quad \text{and} \quad \mathcal{Z} = \{\mathbf{z}_1,\ldots,\mathbf{z}_N\} \tag{19.3}$$

with $\mathbf{z}_k = (z_{3k-2}, z_{3k-1}, z_{3k})$, $1 \le k \le N$.

We have

$$\text{vol}_d(S(B;\gamma_1,\gamma_2)) = \sum_{0\le k\le N:\ \gamma_1 \le \frac{k-pN}{\sqrt{pqN}} \le \gamma_2} \binom{N}{k} p^k q^{N-k}, \tag{19.4}$$

where $p = \text{vol}(B)$ and $q = 1-p$.

By using the Berry–Esseen form of the Central Limit Theorem in the special case of the binomial distribution with parameter $p = \text{vol}(B)$, we have

$$\begin{aligned} |\text{vol}_d(S(B;-\infty,\gamma)) - \mathcal{N}(\gamma)| &= \left| \sum_{0\le k\le N:\ \frac{k-pN}{\sqrt{pqN}} \le \gamma} \binom{N}{k} p^k q^{N-k} - \mathcal{N}(\gamma) \right| \\ &\le \frac{1}{\sqrt{pqN}} \frac{q^3p + p^3q}{pq} \le \frac{1}{\sqrt{pqN}}. \end{aligned} \tag{19.5}$$

Since the torus I^d is translation invariant, we can apply Theorem 4.2 with $f = \chi_S$ where S is the translated copy

$$S = S(\gamma_1,\gamma_2) = S(B;\gamma_1,\gamma_2) - \vec{\mathcal{Y}}$$

of $S(B;\gamma_1,\gamma_2)$ in the torus I^d. Thus we obtain: if

$$U \ge 1,\ W - U \ge 1 \quad \text{and} \quad e^{\pi^2 U^2/2} > 3dU, \tag{19.6}$$

then

$$(2\pi)^{-d/2}\int_{\mathbf{v}\in\mathbb{R}^d}(\text{TimeDiscrep}_{S(\gamma_1,\gamma_2)}(\mathbf{v};U,W))^2 e^{-|\mathbf{v}|^2/2}\,d\mathbf{v}$$

$$= (2\pi)^{-d/2}\int_{\mathbf{v}\in\mathbb{R}^d}\left(\int_U^W \chi_{S(\gamma_1,\gamma_2)}(t\mathbf{v})\,dt - \text{vol}_d(S(\gamma_1,\gamma_2))(W-U)\right)^2$$

$$\times e^{-|\mathbf{v}|^2/2}\,d\mathbf{v} \le (1+\log_2(W/U))\cdot \text{vol}_d(S(\gamma_1,\gamma_2))10(W-U+1). \tag{19.7}$$

Note that

$$(2\pi)^{-d/2}\int_{\mathbf{v}\in\mathbb{R}^d}\left(\int_U^W \chi_{S(\gamma_1,\gamma_2)}(t\mathbf{v})\,dt - \text{vol}_d(S(\gamma_1,\gamma_2))(W-U)\right)^2 e^{-|\mathbf{v}|^2/2}d\mathbf{v}$$

$$= \int_{\omega\in\Omega_{\text{Gauss}}}\left(\text{length}\left\{U\le t\le W:\ \gamma_1\le \frac{|\mathcal{Y}_{\omega;t}\cap B|-pN}{\sqrt{pqN}}\le\gamma_2\right\}\right.$$

$$\left.- \text{vol}_d(S(\gamma_1,\gamma_2))(W-U)\right)^2\,d\omega, \tag{19.8}$$

where, for notational convenience, $d\omega = d\text{ProdMeas}_{\text{Gauss}}(\omega)$.

Let

$$N\ge 10^{20}\quad\text{and}\quad U\ge N_0=\sqrt{\log N}. \tag{19.9}$$

We apply a standard binary decomposition trick. We call an interval $[m, m+2^r)$ *special* if $m, r\ge 0$ are nonnegative integers and m is divisible by 2^r.

Let $T\ge 2\sqrt{\log N}+2 = 2N_0+2$ be an arbitrary integer, and consider its binary representation

$$T = 2^{r_1}+2^{r_2}+2^{r_3}+\cdots,\quad\text{where}$$

$$r_1>r_2>r_3>\cdots\ge 0\quad\text{and}\quad 2^{r_1}-N_0\ge 1. \tag{19.10}$$

Then the interval $[N_0, T)$ can be decomposed into at most $1+\log_2 T$ intervals as follows:

$$[N_0,T) = [N_0,2^{r_1})\cup[2^{r_1},2^{r_1}+2^{r_2})\cup[2^{r_1}+2^{r_2},2^{r_1}+2^{r_2}+2^{r_3})\cup\cdots. \tag{19.11}$$

Note that the intervals on the right-hand side of (19.11) are all special except the first one.

Similarly, let $0 \le \lambda \le 1+\log N$ be a binary rational number of the form

$$\lambda = 2^{l_1} + 2^{l_2} + 2^{l_3} + \cdots \quad \text{where } l_1 > l_2 > l_3 > \cdots \ge -2 - \log_2 N. \tag{19.12}$$

Then

$$\begin{aligned} S(B; -\infty, \lambda) &= S(B; -\infty, 0) \cup S(B; 0, 2^{l_1}) \\ &\cup S(B; 2^{l_1}, 2^{l_1} + 2^{l_2}) \cup S(B; 2^{l_1} + 2^{l_2}, 2^{l_1} + 2^{l_2} + 2^{l_3}) \cup \cdots . \end{aligned} \tag{19.13}$$

Let $(R, r; l)$ be a triple of integers such that

$$0 \le r \le R, \quad \log_2 N_0 \le R \quad \text{and} \quad -2 - \log_2 N \le l \le l_0 = \log_2(1+\log N). \tag{19.14}$$

For $\eta > 0$, let $\Omega(R, r; l; \text{bad}; \eta)$ be the set of those $\omega \in \Omega_{\text{Gauss}}$ for which

$$\begin{aligned} &\sum_{\substack{(U,W):\ \text{either } U=N_0<W=2^r\le 2^R \text{ or} \\ N_0<U,W\le 2^R, W-U=2^r, 2^{-r}U \text{ is an integer}}} \ \sum_{\substack{(\gamma_1,\gamma_2):\ \text{either } \gamma_1=-\infty,\gamma_2=0 \text{ or } 0\le\gamma_1, \\ \gamma_2-\gamma_1=2^l, \gamma_2\le 1+\log N, 2^{-l}\gamma_1 \text{ is an integer}}} \\ &\left(\text{length}\left\{U \le t \le W :\ \gamma_1 \le \frac{|\mathcal{Y}_{\omega;t} \cap B| - pN}{\sqrt{pqN}} \le \gamma_2\right\}\right. \\ &\left. - \text{vol}_d(S(\gamma_1,\gamma_2))(W-U)\right)^2 > (1+R)^3(1+r)^2(1+l_0-l)^2 \cdot 2^R \cdot \frac{1}{\eta}, \end{aligned} \tag{19.15}$$

where $l_0 = \log_2(1+\log N)$.

By using (19.7) and (19.8) (noting that in view of (19.9)–(19.10) condition (19.6) is satisfied), we obtain that for every triple $(R, r; l)$ of integers satisfying (19.14),

$$\begin{aligned} &\sum_{\substack{(U,W):\ \text{either } U=N_0<W=2^r\le 2^R \text{ or} \\ N_0<U,W\le 2^R, W-U=2^r, 2^{-r}U \text{ is an integer}}} \ \sum_{\substack{(\gamma_1,\gamma_2):\ \text{either } \gamma_1=-\infty,\gamma_2=0 \text{ or } 0\le\gamma_1, \\ \gamma_2-\gamma_1=2^l, \gamma_2\le 1+\log N, 2^{-l}\gamma_1 \text{ is an integer}}} \\ &\int_{\omega\in\Omega_{\text{Gauss}}} \left(\text{length}\left\{U \le t \le W :\ \gamma_1 \le \frac{|\mathcal{Y}_{\omega;t} \cap B| - pN}{\sqrt{pqN}} \le \gamma_2\right\}\right. \\ &\left. - \text{vol}_d(S(\gamma_1,\gamma_2))(W-U)\right)^2 d\omega \end{aligned}$$

$$\leq \sum_{\substack{(U,W):\text{ either } U=N_0<W=2^r\leq 2^R \text{ or } N_0<U,\\ W\leq 2^R, W-U=2^r, 2^{-r}U \text{ is an integer}}} \sum_{\substack{(\gamma_1,\gamma_2):\text{ either } \gamma_1=-\infty,\gamma_2=0 \text{ or } 0\leq\gamma_1,\\ \gamma_2-\gamma_1=2^l,\gamma_2\leq 1+\log N, 2^{-l}\gamma_1 \text{ is an integer}}}$$

$$(1+\log_2(W/U))\cdot \text{vol}_d(S(\gamma_1,\gamma_2))10(W-U+1)$$

$$\leq 20(1+R)\left(\sum_{\substack{(U,W):\text{ either } U=N_0<W=2^r\leq 2^R \text{ or}\\ N_0<U,W\leq 2^R,W-U=2^r,2^{-r}U \text{ is an integer}}} (W-U)\right)$$

$$\cdot \sum_{\substack{(\gamma_1,\gamma_2):\text{ either } \gamma_1=-\infty,\gamma_2=0 \text{ or } 0\leq\gamma_1,\\ \gamma_2-\gamma_1=2^l,\gamma_2\leq 1+\log N,2^{-l}\gamma_1 \text{ is an integer}}} \text{vol}_d(S(\gamma_1,\gamma_2)) \leq 20(1+R)2^R. \tag{19.16}$$

Combining (19.15) and (19.16), we obtain the upper bound

$$\begin{aligned}\text{ProdMeas}_{\text{Gauss}}(\Omega(R,r;l;\text{bad};\eta)) &\leq \frac{20\eta(1+R)}{(1+R)^3(1+r)^2(1+l_0-l)^2}\\ &= \frac{20\eta}{(1+R)^2(1+r)^2(1+l_0-l)^2}.\end{aligned} \tag{19.17}$$

By (19.17), with $N_0=\sqrt{\log N}$ and $l_0=\log_2(1+\log N)$,

$$\sum_{\substack{(R,r;l):\text{ triple of integers}\\ \text{satisfying (19.14)}}} \text{ProdMeas}_{\text{Gauss}}(\Omega(R,r;l;\text{bad};\eta))$$

$$\leq \sum_{\substack{(R,r;l):\text{ triple of integers}\\ \text{satisfying (19.14)}}} \frac{20\eta}{(1+R)^2(1+r)^2(1+l_0-l)^2}$$

$$\leq 20\eta\sum_{R=0}^{\infty}\sum_{0\leq r\leq R}\sum_{l:\ l\leq l_0}\frac{1}{(1+R)^2(1+r)^2(1+l_0-l)^2}$$

$$\leq 20\eta\left(\sum_{R=0}^{\infty}\frac{1}{(1+R)^2}\right)\left(\sum_{r=0}^{\infty}\frac{1}{(1+r)^2}\right)\left(\sum_{l:\ l\leq l_0}\frac{1}{(1+l_0-l)^2}\right)$$

$$\leq 20\eta\left(\sum_{n=1}^{\infty}\frac{1}{n^2}\right)^3 = 20\eta(\pi^2/6)^3 < 100\eta. \tag{19.18}$$

Let

$$\Omega(+;\text{bad};\eta) = \bigcup_{\substack{(R,r;l):\text{ triple of integers}\\ \text{satisfying (19.14)}}} \Omega(R,r;l;\text{bad};\eta). \tag{19.19}$$

Then by (19.18),

$$\text{ProdMeas}_{\text{Gauss}}(\Omega(+;\text{bad};\eta)) \leq 100\eta. \tag{19.20}$$

If $\omega \notin \Omega(+;\text{bad};\eta)$, then by (19.15),

$$\sum_{\substack{(U,W):\text{ either } U=N_0<W=2^r\leq 2^R \text{ or}\\ N_0<U,W\leq 2^R, W-U=2^r, 2^{-r}U \text{ is an integer}}} \sum_{\substack{(\gamma_1,\gamma_2):\text{ either } \gamma_1=-\infty,\gamma_2=0 \text{ or } 0\leq\gamma_1,\\ \gamma_2-\gamma_1=2^l,\gamma_2\leq 1+\log N, 2^{-l}\gamma_1 \text{ is an integer}}} \left(\text{length}\left\{U\leq t\leq W:\ \gamma_1\leq \frac{|\mathcal{Y}_{\omega;t}\cap B|-pN}{\sqrt{pqN}}\leq\gamma_2\right\}\right.$$
$$\left. - \text{vol}_d(S(\gamma_1,\gamma_2))(W-U)\right)^2 \leq (1+R)^3(1+r)^2(1+l_0-l)^2\cdot 2^R\cdot\frac{1}{\eta} \tag{19.21}$$

holds for every triple of integers $(R,r;l)$ satisfying (19.14).

We recall (19.10)–(19.13): Let $T\geq 2\sqrt{\log N}+2 = 2N_0+2$ be an arbitrary integer with binary representation

$$T = 2^{r_1}+2^{r_2}+2^{r_3}+\cdots \quad \text{where } r_1>r_2>r_3>\cdots\geq 0,$$

and let $0\leq\lambda\leq 1+\log N$ be a binary rational number of the form

$$\lambda = 2^{l_1}+2^{l_2}+2^{l_3}+\cdots \quad \text{where} \quad l_1>l_2>l_3>\cdots\geq -2-\log_2 N.$$

Then the interval $[N_0,T)$ has the decomposition

$$[N_0,T) = [N_0,2^{r_1})\cup[2^{r_1},2^{r_1}+2^{r_2})\cup[2^{r_1}+2^{r_2},2^{r_1}+2^{r_2}+2^{r_3})\cup\cdots,$$

and similarly,

$$S(-\infty,\lambda) = S(-\infty,0)\cup S(0,2^{l_1})$$
$$\cup S(2^{l_1},2^{l_1}+2^{l_2})\cup S(2^{l_1}+2^{l_2},2^{l_1}+2^{l_2}+2^{l_3})\cup\cdots,$$

where $S(\gamma_1,\gamma_2) = S(B;\gamma_1,\gamma_2)-\vec{\mathcal{Y}}$.

If $\omega\notin\Omega(+;\text{bad};\eta)$, then by using first the triangle inequality, using second the Cauchy–Schwarz inequality, and finally using (19.21), we obtain for every integer $T \geq 2N_0 + 2 = 2\sqrt{\log N} + 2$ and every binary rational λ of the form (19.12) in the interval $0 \leq \lambda \leq 1 + \log N$,

$$\Bigg| \text{length}\left\{ N_0 \leq t \leq T : \frac{|\mathcal{Y}_{\omega;t} \cap B| - pN}{\sqrt{pqN}} \leq \lambda \right\} - \text{vol}_d(S(-\infty, \lambda))(T - N_0) \Bigg|$$

$$\leq \sum_{\substack{(U,W):\text{ it shows up on the} \\ \text{right side of (19.11)}}} \sum_{\substack{(\gamma_1,\gamma_2):\text{ it shows up on the} \\ \text{right side of(19.13)}}} \Bigg| \text{length}\left\{ U \leq t \leq W : \gamma_1 \leq \frac{|\mathcal{Y}_{\omega;t} \cap B| - pN}{\sqrt{pqN}} \leq \gamma_2 \right\} - \text{vol}_d(S(\gamma_1, \gamma_2))(W - U) \Bigg|$$

$$\leq \left(\sum_{\substack{(U,W):\text{ it shows up on the} \\ \text{right side of (19.11)}}} \sum_{\substack{(\gamma_1,\gamma_2):\text{ it shows up} \\ \text{on the right side of (19.13)}}} \left(\text{length}\left\{ U \leq t \leq W : \gamma_1 \leq \frac{|\mathcal{Y}_{\omega;t} \cap B| - pN}{\sqrt{pqN}} \leq \gamma_2 \right\} - \text{vol}_d(S(\gamma_1, \gamma_2))(W - U) \right)^2 \right)^{1/2}$$

$$\cdot \left(\sum_{\substack{(U,W):\text{ it shows up on the} \\ \text{right side of (19.11)}}} \sum_{\substack{(\gamma_1,\gamma_2):\text{ it shows up on the} \\ \text{right side of (19.13)}}} 1 \right)^{1/2}$$

$$\leq \left(\sum_{0 \leq r \leq R = \lceil \log_2 T \rceil} \sum_{l_0 \geq l \geq -2 - \log_2 N} \sum_{\substack{(U,W):\ \{\text{either}\} U = N_0 < W = 2^r \leq 2^R \text{ or} \\ N_0 < U, W \leq 2^R, W - U = 2^r, 2^{-r}U \text{ is an integer}}} \sum_{\substack{(\gamma_1,\gamma_2):\text{ either } \gamma_1 = -\infty, \gamma_2 = 0 \text{ or} \\ 0 \leq \gamma_1, \gamma_2 - \gamma_1 = 2^l, \gamma_2 \leq 1 + \log N, 2^{-l}\gamma_1 \text{ is an integer}}} \left(\text{length}\left\{ U \leq t \leq W : \gamma_1 \leq \frac{|\mathcal{Y}_{\omega;t} \cap B| - pN}{\sqrt{pqN}} \leq \gamma_2 \right\} - \text{vol}_d(S(\gamma_1, \gamma_2))(W - U) \right)^2 \right)^{1/2}$$

$$\cdot\left(\sum_{\substack{(U,W):\text{ it shows up on the}\\ \text{right side of (19.11)}}}\ \sum_{\substack{(\gamma_1,\gamma_2):\text{ it shows up on the}\\ \text{right side of (19.13)}}} 1\right)^{1/2}$$

$$\leq\left(\sum_{0\leq r\leq R=\lceil\log_2 T\rceil}\ \sum_{l_0\geq l\geq -2-\log_2 N}\frac{R^3(1+r)^2(1+l_0-l)^2 2^R}{\eta}\right)^{1/2}$$

$$\cdot\left(\sum_{\substack{(U,W):\text{ it shows up on the}\\ \text{right side of (19.11)}}}\ \sum_{\substack{(\gamma_1,\gamma_2):\text{ it shows up on the}\\ \text{right side of (19.13)}}} 1\right)^{1/2}$$

$$\leq\left(\sum_{0\leq r\leq R=\lceil\log_2 T\rceil}\ \sum_{l_0\geq l\geq -2-\log_2 N}\frac{\begin{array}{c}(2+\log_2 T)^3(2+\log_2 T)^2\\ \times(2\log_2 N)^2 2^R\end{array}}{\eta}\right)^{1/2}$$

$$\cdot\left((1+\log_2 T)2\log_2 N\right)^{1/2}\leq\frac{(2+\log_2 T)^{7/2}(2\log_2 N)^2\sqrt{2T}}{\sqrt{\eta}}. \tag{19.22}$$

Using (19.3) in (19.22), we obtain that for $\omega\notin\Omega_1(+;\text{bad};\eta)$,

$$\left|\text{length}\left\{N_0\leq t\leq T:\ \frac{|\mathcal{Y}_{\omega;t}\cap B|-pN}{\sqrt{pqN}}\leq\lambda\right\}-\mathcal{N}(\lambda)(T-N_0)\right|$$
$$\leq\frac{(2+\log_2 T)^{7/2}(2\log_2 N)^2\sqrt{2T}}{\sqrt{\eta}}+\frac{T-N_0}{\sqrt{pqN}}. \tag{19.23}$$

For every real number $0\leq\beta\leq\log N$ there exist two binary rationals $0\leq\lambda_1\leq\beta\leq\lambda_2\leq 1+\log N$ of the form (19.12) such that

$$0\leq\lambda_2-\lambda_1\leq 1/N. \tag{19.24}$$

We have

$$\left|\text{length}\left\{N_0\leq t\leq T:\ \frac{|\mathcal{Y}_{\omega;t}\cap B|-pN}{\sqrt{pqN}}\leq\beta\right\}-\mathcal{N}(\beta)(T-N_0)\right|$$
$$\leq\max\left\{\left|\text{length}\left\{N_0\leq t\leq T:\ \frac{|\mathcal{Y}_{\omega;t}\cap B|-pN}{\sqrt{pqN}}\leq\lambda_1\right\}\right.\right.$$
$$\left.-\mathcal{N}(\lambda_2)(T-N_0)\right|,$$

$$\left|\text{length}\left\{N_0 \le t \le T : \frac{|\mathcal{Y}_{\omega;t} \cap B| - pN}{\sqrt{pqN}} \le \lambda_2\right\} - \mathcal{N}(\lambda_1)(T - N_0)\right|\Bigg\}$$

$$\le \max_{i=1,2}\left\{\left|\text{length}\left\{N_0 \le t \le T : \frac{|\mathcal{Y}_{\omega;t} \cap B| - pN}{\sqrt{pqN}} \le \lambda_i\right\}\right.\right.$$

$$\left.\left. - \mathcal{N}(\lambda_i)(T - N_0)\right|\right\} + \mathcal{N}(\lambda_1, \lambda_2)(T - N_0), \tag{19.25}$$

where

$$\mathcal{N}(\lambda_1, \lambda_2) = \frac{1}{\sqrt{2\pi}} \int_{\lambda_1}^{\lambda_2} e^{-u^2/2}\, du \le \frac{\lambda_2 - \lambda_1}{\sqrt{2\pi}} < \lambda_2 - \lambda_1 \le \frac{1}{N}, \tag{19.26}$$

and in the last step of (19.26) we used (19.24). Combining (19.23) and (19.25)–(19.26), we have for $\omega \notin \Omega(+; \text{bad}; \eta)$,

$$\sup_{0 \le \beta \le \log N}\left|\text{length}\left\{N_0 \le t \le T : \frac{|\mathcal{Y}_{\omega;t} \cap B| - pN}{\sqrt{pqN}} \le \beta\right\}\right.$$

$$\left. - \mathcal{N}(\beta)(T - N_0)\right|$$

$$\le \frac{(2 + \log_2 T)^{7/2}(2\log_2 N)^2\sqrt{2T}}{\sqrt{\eta}} + \frac{T - N_0}{\sqrt{pqN}}$$

$$+ \frac{T - N_0}{N} \quad \text{for all } T \ge 2N_0 + 2, \tag{19.27}$$

where $N_0 = \sqrt{\log N}$.

Moreover, we have

$$\sup_{\beta > \log N}\left|\text{length}\left\{N_0 \le t \le T : \frac{|\mathcal{Y}_{\omega;t} \cap B| - pN}{\sqrt{pqN}} \le \beta\right\} - \mathcal{N}(\beta)(T - N_0)\right|$$

$$= \sup_{\beta > \log N}\left|\text{length}\left\{N_0 \le t \le T : \frac{|\mathcal{Y}_{\omega;t} \cap B| - pN}{\sqrt{pqN}} > \beta\right\}\right.$$

$$\left. - (1 - \mathcal{N}(\beta))(T - N_0)\right|$$

$$\le \sup_{\beta > \log N}\left(\text{length}\left\{N_0 \le t \le T : \frac{|\mathcal{Y}_{\omega;t} \cap B| - pN}{\sqrt{pqN}} > \beta\right\}\right.$$

$$\left. + (1 - \mathcal{N}(\beta))(T - N_0)\right)$$

$$\leq \text{length}\left\{N_0 \leq t \leq T : \frac{|\mathcal{Y}_{\omega;t} \cap B| - pN}{\sqrt{pqN}} > \log N\right\}$$
$$+(1 - \mathcal{N}(\log N))(T - N_0)$$
$$\leq \Bigg|\text{length}\left\{N_0 \leq t \leq T : \frac{|\mathcal{Y}_{\omega;t} \cap B| - pN}{\sqrt{pqN}} > \log N\right\}$$
$$-(1 - \mathcal{N}(\log N))(T - N_0)\Bigg| + 2(1 - \mathcal{N}(\log N))(T - N_0)$$
$$= \Bigg|\text{length}\left\{N_0 \leq t \leq T : \frac{|\mathcal{Y}_{\omega;t} \cap B| - pN}{\sqrt{pqN}} \leq \log N\right\}$$
$$-\mathcal{N}(\log N)(T - N_0)\Bigg| + 2(1 - \mathcal{N}(\log N))(T - N_0). \tag{19.28}$$

Using (19.27) with $\beta = \log N$ in (19.28), we have for $\omega \notin \Omega(+; \text{bad}; \eta)$,

$$\sup_{\beta > \log N} \Bigg|\text{length}\left\{N_0 \leq t \leq T : \frac{|\mathcal{Y}_{\omega;t} \cap B| - pN}{\sqrt{pqN}} \leq \beta\right\} - \mathcal{N}(\beta)(T - N_0)\Bigg|$$
$$\leq \Bigg|\text{length}\left\{N_0 \leq t \leq T : \frac{|\mathcal{Y}_{\omega;t} \cap B| - pN}{\sqrt{pqN}} \leq \log N\right\}$$
$$-\mathcal{N}(\log N)(T - N_0)\Bigg| + 2(1 - \mathcal{N}(\log N))(T - N_0)$$
$$\leq \frac{(2 + \log_2 T)^{7/2}(2\log_2 N)^2\sqrt{2T}}{\sqrt{\eta}} + \frac{T - N_0}{\sqrt{pqN}} + \frac{T - N_0}{N}$$
$$+2e^{-\log^2 N/2}(T - N_0) \quad \text{for all} \quad T \geq 2N_0 + 2. \tag{19.29}$$

By (19.9),

$$e^{-\log^2 N/2} = N^{-\log N/2} < N^{-2}. \tag{19.30}$$

Combining (19.27) and (19.29)–(19.30), for $\omega \notin \Omega(+; \text{bad}; \eta)$,

$$\sup_{\beta \geq 0} \Bigg|\text{length}\left\{N_0 \leq t \leq T : \frac{|\mathcal{Y}_{\omega;t} \cap B| - pN}{\sqrt{pqN}} \leq \beta\right\} - \mathcal{N}(\beta)(T - N_0)\Bigg|$$
$$\leq \frac{(2 + \log_2 T)^{7/2}(2\log_2 N)^2\sqrt{2T}}{\sqrt{\eta}} + \frac{T - N_0}{\sqrt{pqN}} + \frac{2(T - N_0)}{N}, \tag{19.31}$$

where by (19.20),

$$\text{ProdMeas}_{\text{Gauss}}(\Omega(+;\text{bad};\eta)) \le 100\eta. \tag{19.32}$$

(19.31)–(19.32) settle the case of positive β's. Repeating the same argument, we can settle the missing case of negative β's. That is, there exists a subset

$$\Omega(-;\text{bad};\eta) \subset \Omega_{\text{Gauss}}$$

such that

$$\text{ProdMeas}_{\text{Gauss}}\Big(\Omega(-;\text{bad};\eta)\Big) \le 100\eta, \tag{19.33}$$

and for $\omega \notin \Omega(-;\text{bad};\eta)$,

$$\sup_{\beta\le 0}\left|\text{length}\left\{N_0 \le t \le T: \frac{|\mathcal{Y}_{\omega;t}\cap B| - pN}{\sqrt{pqN}} \le \beta\right\} - \mathcal{N}(\beta)(T-N_0)\right|$$
$$\le \frac{(2+\log_2 T)^{7/2}(2\log_2 N)^2\sqrt{2T}}{\sqrt{\eta}} + \frac{T-N_0}{\sqrt{pqN}} + \frac{2(T-N_0)}{N}. \tag{19.34}$$

Let

$$\Omega(\text{bad};\eta) = \Omega(+;\text{bad};\eta) \cup \Omega(-;\text{bad};\eta), \tag{19.35}$$

then by (19.32)–(19.33),

$$\text{ProdMeas}_{\text{Gauss}}(\Omega(\text{bad};\eta)) \le 200\eta, \tag{19.36}$$

and for $\omega \notin \Omega(\text{bad};\eta)$,

$$\sup_{-\infty<\beta<\infty}\left|\text{length}\left\{N_0 \le t \le T: \frac{|\mathcal{Y}_{\omega;t}\cap B| - pN}{\sqrt{pqN}} \le \beta\right\}\right.$$
$$\left. -\mathcal{N}(\beta)(T-N_0)\right|$$
$$\le \frac{(2+\log_2 T)^{7/2}(2\log_2 N)^2\sqrt{2T}}{\sqrt{\eta}} + \frac{T-N_0}{\sqrt{pqN}} + \frac{2(T-N_0)}{N} \tag{19.37}$$

for all integers $T \ge 2N_0 + 2 = 2\sqrt{\log N} + 2$.

It immediately follows from (19.37) that, for every $\omega \notin \Omega(\text{bad};\eta)$ and every real number $T \ge 2\sqrt{\log N} + 2$,

$$\sup_{-\infty<\beta<\infty}\left|\text{length}\left\{0 \le t \le T: \frac{|\mathcal{Y}_{\omega;t}\cap B| - pN}{\sqrt{pqN}} \le \beta\right\} - \mathcal{N}(\beta)\cdot T\right|$$
$$\le \frac{(2+\log_2 T)^{7/2}(2\log_2 N)^2\sqrt{2T}}{\sqrt{\eta}} + \frac{T-N_0}{\sqrt{pqN}}$$
$$+\frac{2(T-N_0)}{N} + 2\sqrt{\log N} + 3. \tag{19.38}$$

Dividing (19.38) by T, Theorem 19.1 follows. □

Chapter 4

More Results about Randomness and Stability in Equilibrium

20. Simultaneous Square-Root Equilibrium Relative to Nice Sets (I)

As the title of the section suggests, we are going to prove long-term stability of square-root equilibrium in the particle space (= "gas container"). It means uniformity with square-root error simultaneously for a whole family of nice test sets (in the particle space). Of course we have to clarify what *nice sets* mean. Aligned and rotated boxes, tetrahedrons, balls, ellipsoids are certainly *nice*, and they are all convex. Since we study isotropic (= rotation-invariant) velocity distributions, it is natural to consider a rotation-invariant family of test sets. A plausible choice is to consider the family of *all convex sets.*

What we do here is basically a continuation of Sec. 5 working with the 3-dimensional torus model, but the proof also works for the closed orbit models discussed in Secs. 6–7.

We employ a geometric approximation argument, and Sec. 20 is devoted to the technical discussion of this. To make the somewhat cumbersome details as simple as possible, we explain the idea in the 2-dimensional case. It means that we replace the 3-dimensional torus model with the 2-dimensional version, and the family of nice test sets is the family of all convex sets in the unit square $[0, 1]^2$.

We approximate plane convex sets with inscribed polygons. We apply a well-known area approximation theorem of Sas [Sa39] that goes as follows. For every integer $n \geq 3$, every plane convex set of area one contains an inscribed polygon of n vertices (= n-gon) whose area is at least as large as the area of the largest n-gon inscribed in a circle of area one. Since the area of the regular n-gon on the unit circle is exactly $\frac{1}{2}n\sin(2\pi/n)$, Sas's theorem implies that every plane convex set of area one contains an inscribed n-gon such that the area of the set-difference (= area of the convex set minus the n-gon) is

$$\leq 1 - \frac{1}{2\pi}n\sin(2\pi/n) \leq 1 - \frac{1}{2\pi}(2\pi n/n) + \frac{1}{2\pi}n(2\pi/n)^3/6 = \frac{2\pi^2}{3n^2} < \frac{7}{n^2}, \tag{20.1}$$

where we used the well-known inequality $\sin x \geq x - x^3/6$.

Let $B \subset [0,1]^2$ be a convex set in the unit square, and let $2 \leq j < k$ be integers. We define the set $B(k;j)$ as follows. Let $P(B;k)$ denote the largest area inscribed polygon in B with 2^k vertices; let $P(B;k;k-1) = P(P(B;k);k-1)$ denote the largest area inscribed polygon in $P(B;k)$ with 2^{k-1} vertices; let $P(B;k;k-2) = P(P(B;k;k-1);k-2)$ denote the largest area inscribed polygon in $P(B;k;k-1)$ with 2^{k-2} vertices; and so on. That is, $P(B;k;k-\ell) = P(P(B;k;k-\ell+1);k-\ell)$ denotes the largest area inscribed polygon in $P(B;k;k-\ell+1)$ with $2^{k-\ell}$ vertices, where $1 \leq \ell \leq k-2$ (and of course $P(B;k;k) = P(B;k)$). Finally, we define $B(k;j)$ as the set-difference

$$B(k;j) = P(B;k;j+1) \setminus P(B;k;j). \tag{20.2}$$

Thus for every $2 \leq i < k$ we have

$$P(B;k) = P(B;k;i) \cup \bigcup_{i \leq j < k} B(k;j). \tag{20.3}$$

(Note in advance that in the application later we will choose the concrete value $i = 5$.)

By (20.1) we have the area estimations

$$\text{area}\,(B \setminus P(B;k)) < \frac{7}{4^k} \quad \text{and} \quad \text{area}(B(k;j)) < \frac{7}{4^j}. \tag{20.4}$$

It follows from the construction of the sets $P(B;k;j)$, $B(k;j)$ that $B(k;j)$ is the disjoint union of at most 2^{j+1} triangles, whose vertices are the vertices of the two polygons $P(B;k;j)$ and $P(B;k;j+1)$. More precisely, we have exactly 2^{j+1} triangles if the vertices of the two polygons $P(B;k;j)$ and $P(B;k;j+1)$ are different; and in general the number of triangles is $2^{j+1} - c$ if the polygons $P(B;k;j)$ and $P(B;k;j+1)$ have c common vertices. Indeed, every decomposition of a polygon of $2^j + 2^{j+1} - c$ vertices into triangles has $2^j + 2^{j+1} - c - 3$ triangles, and from this number we have to subtract the $2^j - 3$ triangles that come from $P(B; k; j)$.

Let $\Delta = \Delta(B; k; j)$ denote one of these triangles. We approximate triangle Δ with a *special triangle*. Consider the following square lattice in the unit square

$$\left(\frac{a_1}{M}, \frac{a_2}{M}\right), \quad 0 \leq a_1, a_2 \leq M \quad \text{integers.} \tag{20.5}$$

We refer to the square lattice (20.5) as the *M-lattice*, and call its elements *M-lattice-points*.

We need the following almost trivial geometric lemma.

Lemma 20.1. *Let $\Delta^* \subset [0,1]^2$ be a triangle in the unit square with area $\geq 2/M$. Then Δ^* contains an M-lattice-point.*

Proof. We use the well-known fact in elementary geometry

$$\text{area}\,(\Delta^*) = \text{perimeter}\,(\Delta^*) \cdot r\,(\Delta^*)\,/2, \tag{20.6}$$

where $r = r(\Delta^*)$ denotes the radius of the largest inscribed circle in Δ^*. Clearly

$$\text{perimeter}\,(\Delta^*) \leq 4, \tag{20.7}$$

which follows from the well-known inequality that, if $B_1 \subset B_2$ are two bounded plane convex sets then

$$\text{perimeter}(B_1) \leq \text{perimeter}(B_2). \tag{20.8}$$

Combining the hypothesis of the lemma with (20.6) and (20.7), we have

$$r = \frac{2\text{area}(\Delta^*)}{\text{perimeter}(\Delta^*)} \geq \frac{2 \cdot 2/M}{4} = \frac{1}{M},$$

which implies that the largest inscribed circle in Δ^* must contain an aligned square of side length $1/M$, and every such square contains an *M-lattice-point* (see (20.5)). This completes the proof of Lemma 20.1. □

Let us return now to the triangle $\Delta = \Delta(B; k; j)$ defined above. By using its three midpoints, we can decompose triangle Δ into four congruent subtriangles that are "half copies" of Δ. Similarly, for every integer $i \geq 2$ we can divide each side of triangle Δ into i equal parts (by using $i - 1$ points), and joining the corresponding points with line segments, we can divide triangle Δ into i^2 congruent subtriangles of *order* i such that each subtriangle is similar to Δ with the ratio $1 : i$. Let $\Delta(i;\ell)$, $1 \le \ell \le i^2$ denote these congruent subtriangles of order i.

Assume that $\Delta = \Delta(B;k;j)$ satisfies the area condition

$$\text{area}(\Delta) \ge 2^{-2j+5} N^{-1}. \tag{20.9}$$

Under the condition (20.9), let $i_0 = i_0(\Delta(B;k;j)) \ge 4$ be the largest integer such that

$$\frac{\text{area}(\Delta)}{i_0^2} \ge 2^{-2j+1} N^{-1}. \tag{20.10}$$

Let $\Delta(i_0;1)$, $\Delta(i_0;2)$ and $\Delta(i_0;3)$ denote the three congruent subtriangles of Δ of order i_0 which contain the three vertices of Δ. In view of Lemma 20.1 and (20.10), each subtriangle $\Delta(i_0;\ell)$, $1 \le \ell \le 3$ contains an M-lattice-point with $M = 4^j N$; let $\mathbf{v}_\ell = \mathbf{v}_\ell(\Delta(B;k;j))$, $1 \le \ell \le 3$ denote these three $4^j N$-lattice-points (if there are many in a subtriangle, we choose the nearest one to the corresponding vertex of Δ). A simple geometric consideration shows that

$$\text{area}(\mathbf{v}_1\mathbf{v}_2\mathbf{v}_3\text{-triangle}) \ge \left(\frac{i_0 - 3}{i_0}\right)^2 \text{area}\,(\Delta)\,. \tag{20.11}$$

Indeed, the $\mathbf{v}_1\mathbf{v}_2\mathbf{v}_3$-triangle clearly contains a congruent copy of Δ that consists of $(i_0 - 3)^2$ congruent subtriangles of order i_0.

We refer to the $\mathbf{v}_1\mathbf{v}_2\mathbf{v}_3$-triangle as the *inscribed approximation* of triangle $\Delta = \Delta(B;k;j)$ *using* $4^j N$*-lattice-points.*

If the area condition (20.9) is violated, i.e., if $\Delta = \Delta(B;k;j)$ has the property

$$\text{area}(\Delta) < 2^{-2j+5} N^{-1}, \tag{20.12}$$

then we choose $i_0 = i_0(\Delta(B;k;j)) = 3$, and of course (20.11) remains true in a trivial way by choosing $\mathbf{v}_1\mathbf{v}_2\mathbf{v}_3$ to be the empty set ("empty triangle").

We recall that $B(k;j)$ is the disjoint union of at most 2^{j+1} triangles $\Delta = \Delta(B;k;j)$, whose vertices are the vertices of the two polygons $P(B;k;j)$ and $P(B;k;j+1)$. For each of these triangles $\Delta = \Delta(B;k;j)$ we take the corresponding $\mathbf{v}_1\mathbf{v}_2\mathbf{v}_3$-triangle (i.e., inscribed approximation using $4^j N$-lattice-points), and finally, we take the union of these $\mathbf{v}_1\mathbf{v}_2\mathbf{v}_3$-triangles. Let $IN(B(k;j))$ denote this union set. We refer to $IN(B(k;j))$ as the *inscribed approximation* of $B(k;j)$ *using* $4^j N$*-lattice-points.*

Since $IN(B(k;j))$ is the union of at most 2^{j+1} triangles whose vertices are 4^jN-lattice-points, we have the following trivial upper bound.

Fact 1: *The total number of sets $IN(B(k;j))$, where j is fixed, $k > j$, and $B \subset [0,1]^2$ runs over all convex subsets of the unit square, is at most*

$$\binom{(4^jN)^2}{3}^{2^{j+1}}.$$

Exactly the same way as we defined the inscribed approximation $IN(B(k;k-1))$ of $B(k;k-1)$ using $4^{k-1}N$-lattice-points, we can define the inscribed approximation $IN(P(B;k))$ of $P(B;k)$ using $4^{k-1}N$-lattice-points. Indeed, the only minor difference is that $IN(B(k;k-1))$ is the union of *at most* 2^k triangles whose vertices are $4^{k-1}N$-lattice-points, and $IN(P(B;k))$ is the union of *exactly* 2^k triangles whose vertices are $4^{k-1}N$-lattice-points, but this difference is irrelevant in the corresponding arguments. Thus we obtain the perfect analog of Fact 1.

Fact 2: *The total number of sets $IN(P(B;k))$, where k is fixed and $B \subset [0,1]^2$ runs over all convex subsets of the unit square, is at most*

$$\binom{(4^{k-1}N)^2}{3}^{2^k}.$$

Next we describe the *circumscribed approximation* of $B(k;j)$ *using 4^jN-lattice-points* — it is a simpler construction. For every triangle $\Delta = \Delta(B;k;j)$ consider its three vertices, and for each vertex consider the *elementary 4^jN-square in* (20.5) containing this vertex. An *elementary 4^jN-square in* (20.5) has four vertices of the form

$$\left(\frac{a_1+\delta_1}{4^jN}, \frac{a_2+\delta_2}{4^jN}\right), \quad 0 \le a_1, a_2 < 4^jN, \ 0 \le \delta_1, \delta_2 \le 1 \ \text{ integers}, \tag{20.13}$$

and this elementary 4^jN-square is clearly determined by its lower left corner point

$$\left(\frac{a_1}{4^jN}, \frac{a_2}{4^jN}\right), \tag{20.14}$$

which is of course a 4^jN-lattice-point.

The *circumscribed approximation* of triangle $\Delta = \Delta(B; k; j)$ *using* $4^j N$*-lattice-points* is defined as the convex hull of the three elementary $4^j N$-squares in (20.5) containing the three vertices of the triangle.

It follows from the construction that the area of the set-difference (i.e., the area of the circumscribed approximation of $\Delta = \Delta(B; k; j)$ minus the area of Δ) is estimated from above by the sum

$$\sum_{3 \text{ sides of } \Delta} \left(\text{SideLength} + \frac{8}{4^j N}\right) \frac{2}{4^j N} = \frac{2}{4^j N}\text{perimeter}(\Delta) + \frac{3 \cdot 8 \cdot 2}{(4^j N)^2}. \tag{20.15}$$

We know that $B(k; j)$ is the disjoint union of at most 2^{j+1} triangles $\Delta = \Delta(B; k; j)$ (whose vertices are the vertices of the two polygons $P(B; k; j)$ and $P(B; k; j+1)$); for each of these triangles $\Delta = \Delta(B; k; j)$ we take the corresponding *circumscribed approximation* of the triangle, and finally we take the union of these sets. Let $OUT(B(k; j))$ denote this union set. We refer to $OUT(B(k; j))$ as the *circumscribed approximation* of $B(k; j)$ *using* $4^j N$*-lattice-points.*

Note that $OUT(B(k; j))$ is the union of at most 2^{j+1} parts, where each part is the convex hull of three elementary $4^j N$-squares in (20.5). Since every elementary $4^j N$-square in (20.5) is determined by its lower left corner point, repeating the proof of Fact 1 we obtain the following result.

Fact 3: *The total number of set-difference sets* $OUT(B(k; j)) \setminus IN(B(k; j))$ *where* j *is fixed,* $k > j$*, and* $B \subset [0, 1]^2$ *runs over all convex subsets of the unit square, is at most*

$$\binom{(4^j N)^2}{3}^{2^{j+1}} \binom{(4^j N)^2}{3}^{2^{j+1}}.$$

Clearly

$$IN(B(k; j)) \subset B(k; j) \subset OUT(B(k; j)). \tag{20.16}$$

By (20.15),

$$\begin{aligned}
&\text{area}(OUT(B(k; j)) \setminus B(k; j)) \\
&\quad \le \frac{2}{4^j N}(\text{perimeter}(P(B; k; j)) + \text{perimeter}(P(B; k; j+1))) \\
&\quad\quad + 2^{j+1} \frac{3 \cdot 8 \cdot 2}{(4^j N)^2} \le \frac{2}{4^j N}(4 + 4) + 2^{j+1} \frac{3 \cdot 8 \cdot 2}{(4^j N)^2} = \frac{1}{4^{j-2} N} + \frac{3/2}{8^{j-2} N^2},
\end{aligned} \tag{20.17}$$

where the upper bound $4+4$ comes from combining the facts $P(B;k;j) \subset [0,1]^2$ and $P(B;k;j+1) \subset [0,1]^2$ with (20.8).

Next we switch to the inscribed approximation "error". By (20.11) we have,

$$\text{area}(\Delta \setminus \mathbf{v}_1\mathbf{v}_2\mathbf{v}_3\text{-triangle}) \le \left(1 - \left(\frac{i_0 - 3}{i_0}\right)^2\right) \text{area}(\Delta) < \frac{6}{i_0}\text{area}(\Delta). \tag{20.18}$$

Under the condition (20.9), (20.10) implies

$$\frac{\text{area}(\Delta)}{(i_0+1)^2} < 2^{-2j+1}N^{-1},$$

so

$$i_0 + 1 > \sqrt{\text{area}(\Delta)}2^{(2j-1)/2}\sqrt{N}. \tag{20.19}$$

Since $i_0 \ge 3$, by (20.19) we have

$$i_0 > \frac{i_0+1}{\sqrt{2}} > \sqrt{\text{area}(\Delta)}2^{j-1}\sqrt{N},$$

and using it in (20.18), we obtain

$$\text{area}(\Delta \setminus \mathbf{v}_1\mathbf{v}_2\mathbf{v}_3\text{-triangle}) < \frac{6}{i_0}\text{area}(\Delta) < \frac{12}{2^j\sqrt{N}}\sqrt{\text{area}(\Delta)}, \tag{20.20}$$

assuming $\Delta = \Delta(B;k;j)$ satisfies (20.9). Combining (20.12) and (20.20) with the Cauchy–Schwarz inequality, we have

$$\begin{aligned}
&\text{area}\,(B(k;j) \setminus IN(B(k;j))) \\
&\le \frac{12}{2^j\sqrt{N}} \left(\sum_{\substack{\Delta=\Delta(B;k;j) \\ \text{satisfies (20.9)}}} \sqrt{\text{area}\,(\Delta)} \right) + \sum_{\substack{\Delta=\Delta(B;k;j) \\ \text{satisfies (20.12)}}} \text{area}\,(\Delta) \\
&\le \frac{12}{2^j\sqrt{N}} \left(\sum_{\substack{\Delta=\Delta(B;k;j) \\ \text{satisfies (20.9)}}} \text{area}\,(\Delta) \right)^{1/2} \left(2^{j+1}\right)^{1/2} + 2^{j+1} \cdot 2^{-2j+5}N^{-1} \\
&\le \frac{12}{2^j\sqrt{N}} \left(\text{area}\,(B(k;j))\right)^{1/2} 2^{(j+1)/2} + 2^{-j+6}N^{-1}
\end{aligned}$$

$$\leq \frac{12}{2^j\sqrt{N}}\left(7\cdot 4^{-j}\right)^{1/2} 2^{(j+1)/2} + 2^{-j+6}N^{-1}$$

$$\leq \frac{50}{2^{3j/2}\sqrt{N}} + 2^{-j+6}N^{-1}, \tag{20.21}$$

where in the last step we used (20.4).

Repeating the argument of (20.21) with $j = k-1$, we obtain the following analog upper bound:

$$\text{area}\,(P(B;k)\setminus IN(P(B;k)))$$

$$\leq \frac{12}{2^{k-1}\sqrt{N}}\left(\text{area}\,(P(B;k))\right)^{1/2} 2^{k/2} + 2^{-(k-1)+6}N^{-1}$$

$$\leq \frac{24}{2^{k/2}\sqrt{N}} + 2^{-k+7}N^{-1}. \tag{20.22}$$

Note that the whole argument between (20.2) and (20.22) was based on (20.1) and $P(B;k)$ (= largest area inscribed polygon of the plane convex set $B \subset [0,1]^2$ with 2^k vertices). Next we switch to the circumscribed approximation of $B \subset [0,1]^2$. For circumscribed approximation we do not know a sharp result like Sas's theorem; nevertheless, it is easy to prove a "quadratic approximation" type result like (20.1); see (20.25) below. Let S be a plane convex set of area one; by Sas's theorem for every $m \geq 3$ there exists an inscribed m-gon $P_m(S) \subset S$ such that (see (20.1))

$$\text{area}(S \setminus P_m(S)) < \frac{7}{m^2}. \tag{20.23}$$

Let $P_m^+(S) \supset S$ be the circumscribed polynomial such that the sides of $P_m^+(S)$ are parallel to the sides of $P_m(S)$. The sides of $P_m^+(S)$ are tangent lines of S; we also consider the tangent lines of S at the vertices of $P_m(S)$. These $2m$ tangent lines have angles

$$0 < \theta_1 < \theta_2 < \theta_3 < \cdots < \theta_{2m} < 2\pi. \tag{20.24}$$

It might happen that there is an exceptionally large jump among these angles in the precise sense that $\theta_{i+2} - \theta_i > \pi$ for some $1 \leq i \leq 2m$ (where $\theta_{2m+1} = \theta_1 + 2\pi$ and $\theta_{2m+2} = \theta_2 + 2\pi$). We can easily eliminate the "exceptionally large jump" by including an extra angle either between θ_i and θ_{i+1} or between θ_{i+1} and θ_{i+2}. If we also take the corresponding tangent line with this possible extra angle, we obtain at most $2m+1$ tangent lines of S, and they define a circumscribed polynomial $P_{2m+1}^+(*;S) \supset S$ of the convex set S.

Lemma 20.2. *For every integer $m \geq 3$ and plane convex set S of area one.*

$$\text{area}\,(P^{+}_{2m+1}(*;S) \setminus S) \leq \text{area}(S \setminus P_m(S)).$$

Proof. We use the following simple geometric fact. Consider the trapezoid with vertices $v_1 = (0,0), v_2 = (a,0), v_3 = (b,h), v_4 = (c,h)$ such that $a \geq b - c > 0$ and $h > 0$. We refer to the side v_1v_2 of the trapezoid as the *longer base*. Let S' be a convex subset of the $v_1v_2v_3v_4$-trapezoid such that S' contains the whole v_1v_2-line-segment, and S' intersects the v_3v_4-line-segment. Then the area of S' is at least half of the area of the $v_1v_2v_3v_4$-trapezoid.

The proof of this "trapezoid fact" is almost trivial. Indeed, let w be a point of S' on the v_3v_4-line-segment. Then S' contains the v_1v_2w-triangle, which has area $ah/2$. On the other hand, the $v_1v_2v_3v_4$-trapezoid has area

$$= (a + (b-c))h/2 \leq (a+a)h/2 = ah,$$

proving the "trapezoid fact".

The proof of the lemma is a straightforward application of the "trapezoid fact" to m trapezoids such that the *longer base* runs through every side of the inscribed polygon $P_m(S)$. □

Combining (20.23) with Lemma 20.2, we obtain the upper bound

$$\text{area}\left(P^{+}_{2m+1}(*;S) \setminus S\right) < \frac{7}{m^2}. \tag{20.25}$$

For $k \geq 4$, let $P^{+}(B;k)$ denote the smallest area circumscribed polygon of B in the unit square with 2^k vertices. (That is, $P^{+}(B;k) \subset [0,1]^2$, which means that we may include four more tangent lines of B — parallel to the sides of the unit square — to define the circumscribed polygon of B.) By (20.25) we have the area estimation

$$\text{area}\left(P^{+}(B;k) \setminus B\right) < \frac{7}{4^{k-2}}. \tag{20.26}$$

We define the *circumscribed approximation* $OUT(P^{+}(B;k))$ of $P^{+}(B;k)$ *using 4^kN-lattice-points* the same way as we defined the circumscribed approximation $OUT(B(k;j))$ of $B(k;j)$ using 4^jN-lattice-points. That is, for each vertex of $P^{+}(B;k)$ consider the elementary 4^kN-square in (20.5) containing this vertex. We recall that an elementary 4^kN-square in (20.5)) has four vertices of the form (20.13) with $j = k$, and this elementary 4^kN-square is clearly determined by its lower left corner point (see (20.14) with $j = k$), which is of course a 4^kN-lattice-point. $OUT(P^+(B;k))$ is defined as the convex hull of the 2^k elementary 4^kN-squares in (20.5) containing the vertices of $P^+(B;k)$.

We have the analog of (20.17):

$$\begin{aligned}
&\text{area}(OUT(P^+(B;k)) \setminus P^+(B;k)) \\
&\le \frac{2}{4^kN}\text{perimeter}(P^+(B;k)) + 2^k\frac{3\cdot 8\cdot 2}{(4^kN)^2} \\
&\le \frac{2}{4^kN}\cdot 4 + 2^k\frac{3\cdot 8\cdot 2}{(4^kN)^2} \\
&\le \frac{1}{4^{k-2}N} + \frac{1}{8^{k-2}N^2}.
\end{aligned} \tag{20.27}$$

Moreover, combining Fact 2 with the usual counting argument, we have the following.

Fact 4: *The total number of set-difference sets $OUT(P^+(B;k)) \setminus OUT(P(B;k))$, where k is fixed and $B \subset [0,1]^2$ runs over all convex subsets of the unit square, is at most*

$$\binom{(4^kN)^2}{2^k}\binom{(4^{k-1}N)^2}{3}^{2^k}.$$

By (20.3) we have for every $2 \le i < k$,

$$\begin{aligned}
&P(B;k;i) \cup \bigcup_{i\le j<k} IN(B(k;j)) \\
&\subset P(B;k;i) \cup \bigcup_{i\le j<k} B(k;j) = P(B;k) \subset B
\end{aligned} \tag{20.28}$$

and

$$\begin{aligned}
&\left(P(B;k;i) \cup \bigcup_{i\le j<k} IN(B(k;j))\right) \cup \left(\bigcup_{i\le j<k} OUT(B(k;j)) \setminus IN(B(k;j))\right) \\
&\cup\, OUT(P^+(B;k)) \setminus IN(P(B;k)) = OUT(P^+(B;k)) \supset B.
\end{aligned} \tag{20.29}$$

The last step in the approximation process is to define an inscribed and circumscribed approximation of the "beginning" $P(B;k;i)$ in (20.28)–(20.29). We do it in the usual way by using M-lattice-points with the novel choice $M = N^{4/3}$. First we define the circumscribed approximation: for each vertex of $P(B;k;i)$ consider the elementary $N^{4/3}$-square in (20.5) containing this vertex. $OUT(P(B;k;i))$ is defined as the convex hull of the 2^i elementary $N^{4/3}$-squares in (20.5) containing the vertices of $P(B;k;i)$.

To define the inscribed approximation, we start, as usual, with a triangle decomposition: we decompose $P(B;k;i)$ into 2^i triangles. Let $\Delta = \Delta(P(B;k;i))$ denote one of these triangles. We approximate triangle Δ with a special triangle by repeating the arguments of (20.9)–(20.12) with $M = N^{4/3}$ instead of $M = 4^j N$,

Assume that $\Delta = \Delta(P(B;k;i))$ satisfies the area condition

$$\text{area}(\Delta) \geq 32N^{-4/3}. \tag{20.30}$$

Under the condition (20.30), let $i_0 = i_0(\Delta(P(B;k;i))) \geq 4$ be the largest integer such that

$$\frac{\text{area}(\Delta)}{i_0^2} \geq 2N^{-4/3}. \tag{20.31}$$

Let $\Delta(i_0;1)$, $\Delta(i_0;2)$ and $\Delta(i_0;3)$ denote the three congruent subtriangles of Δ of order i_0 which contain the three vertices of Δ. In view of Lemma 20.1 and (20.31), each subtriangle $\Delta(i_0;\ell)$, $1 \leq \ell \leq 3$ contains an M-lattice-point with $M = N^{4/3}$; let $\mathbf{v}_\ell = \mathbf{v}_\ell(\Delta(B;k;j))$, $1 \leq \ell \leq 3$ denote these three $N^{4/3}$-lattice-points (if there are many in a subtriangle, we choose the nearest one to the corresponding vertex of Δ). We have the analog of (20.11):

$$\text{area}\,(\mathbf{v}_1\mathbf{v}_2\mathbf{v}_3\text{-triangle}) \geq \left(\frac{i_0 - 3}{i_0}\right)^2 \text{area}(\Delta). \tag{20.32}$$

We refer to the $\mathbf{v}_1\mathbf{v}_2\mathbf{v}_3$-triangle as the inscribed approximation of triangle $\Delta = \Delta(P(B;k;i))$ using $N^{4/3}$-lattice-points.

If the area condition (20.30) is violated, i.e., if $\Delta = \Delta(P(B;k;i))$ has the property

$$\text{area}(\Delta) < 32N^{-4/3}, \tag{20.33}$$

then we choose $i_0 = i_0(\Delta P(B;k;i)) = 3$, and of course (20.32) remains true in a trivial way by choosing $\mathbf{v}_1\mathbf{v}_2\mathbf{v}_3$ to be the empty set ("empty triangle").

Note that $P(B;k;i)$ is the disjoint union of 2^i triangles $\Delta = \Delta(P(B;k;i))$, and for each of these triangles Δ we take the corresponding

$\mathbf{v}_1\mathbf{v}_2\mathbf{v}_3$-triangle (i.e., inscribed approximation using $N^{4/3}$-lattice-points), and finally, we take the union of these $\mathbf{v}_1\mathbf{v}_2\mathbf{v}_3$-triangles. Let $IN(P(B;k;i))$ denote this union set. We refer to $IN(P(B;k;i))$ as the inscribed approximation of $P(B;k;i)$ using $N^{2/3}$-lattice-points.

We have the following trivial upper bounds.

Fact 5: (a) *The total number of sets $OUT(P(B;k;i))$, where i is fixed, $k \geq i$, and $B \subset [0,1]^2$ runs over all convex subsets of the unit square, is at most*

$$\binom{(N^{4/3})^2}{2^i}.$$

(b) *The total number of sets $IN(P(B;k;i))$, where i is fixed, $k \geq i$, and $B \subset [0,1]^2$ runs over all convex subsets of the unit square, is at most*

$$\binom{(N^{4/3})^2}{3}^{2^i}.$$

We also have the following analogs of (20.17) and (20.21) with $M = N^{4/3}$ instead of $M = 4^j N$:

$$\text{area}(OUT\,(P(B;k;i)) \setminus P(B;k;i)) \leq 16N^{-4/3} + 2^{i+6}N^{-8/3} \tag{20.34}$$

and

$$\begin{aligned} &\text{area}\,(P(B;k;i) \setminus IN(P(B;k;i))) \\ &\quad \leq \frac{12}{N^{2/3}} 2^{i/2} + 2^i \cdot 32N^{-4/3} < 2^{(i+8)/2}N^{-2/3} + 2^{i+5}N^{-4/3}. \end{aligned} \tag{20.35}$$

We conclude the approximation process by involving $IN\,(P(B;k;i))$ and $OUT\,(P(B;k;i))$ in (20.28)–(20.29): for every $2 \leq i < k$, we have

$$IN(P(B;k;i)) \cup \bigcup_{i \leq j < k} IN(B(k;j)) \subset B \tag{20.36}$$

and

$$\begin{aligned} &\left(IN(P(B;k;i)) \cup \bigcup_{i\leq j<k} IN(B(k;j))\right) \cup \left(\bigcup_{i\leq j<k} OUT(B(k;j)) \setminus IN(B(k;j))\right) \\ &\cup\Big(OUT(P^+(B;k)) \setminus IN(P(B;k))\Big) \cup \Big(OUT\,(P(B;k;i)) \setminus IN\,(P(B;k;i))\Big) \supset B. \end{aligned} \tag{20.37}$$

We have the following area estimations: by (20.16) and (20.21),

$$\begin{aligned}
&\text{area}(OUT(B(k;j)) \setminus IN(B(k;j))) \\
&= \text{area}(OUT(B(k;j)) \setminus B(k;j)) + \text{area}(B(k;j) \setminus IN(B(k;j))) \\
&\le \frac{1}{4^{j-2}N} + \frac{3/2}{8^{j-2}N^2} + \frac{50}{2^{3j/2}\sqrt{N}} + \frac{1}{2^{j-6}N} \le \frac{1}{2^{j-5}\sqrt{N}} \text{ for } j \ge 5,
\end{aligned} \tag{20.38}$$

by (20.4), (20.22), (20.26) and (20.27),

$$\begin{aligned}
&\text{area}\left(OUT(P^+(B;k)) \setminus IN(P(B;k))\right) \\
&= \text{area}\left(OUT(P^+(B;k)) \setminus P^+(B;k)\right) + \text{area}\left(P^+(B;k) \setminus B\right) \\
&\quad + \text{area}\left(B \setminus P(B;k)\right) + \text{area}\left(P(B;k) \setminus IN(P(B;k))\right) \\
&\le \frac{1}{4^{k-2}N} + \frac{1}{8^{k-2}N^2} + \frac{7}{4^{k-2}} + \frac{7}{4^k} + \frac{24}{2^{k/2}\sqrt{N}} + 2^{-k+7}N^{-1},
\end{aligned} \tag{20.39}$$

and by (20.34) and (20.35),

$$\begin{aligned}
&\text{area}\left(OUT\left(P(B;k;i)\right) \setminus IN(P(B;k;i))\right) \\
&= \text{area}\left(OUT\left(P(B;k;i)\right) \setminus P(B;k;i)\right) + \text{area}\left(P(B;k;i) \setminus IN(P(B;k;i))\right) \\
&\le 16N^{-4/3} + 2^{i+6}N^{-8/3} + 2^{(i+8)/2}N^{-2/3} + 2^{i+5}N^{-4/3}.
\end{aligned} \tag{20.40}$$

Finally, we choose the integers i and k as follows:

$$i = 5 \quad \text{and } k \text{ is defined by the inequality } N \le 8^k < 8N. \tag{20.41}$$

21. Simultaneous Square-Root Equilibrium Relative to Nice Sets (II)

The family of time evolutions $\mathcal{Y}(\text{Gauss};\omega;t)$, $\omega \in \Omega_{\text{Gauss}}$ of the simplest 2-dimensional Gaussian torus model is represented by the family of torus lines (see (5.2))

$$\vec{\mathcal{Y}}(\text{Gauss};\omega;t) = \vec{\mathcal{Y}} + t\mathbf{v}(\omega) \ \text{ modulo one} \tag{21.1}$$

in the configuration space $I^d = [0,1)^d$ with $d = 2N$, all starting from the same point

$$\vec{\mathcal{Y}} = (\mathbf{y}_1, \ldots, \mathbf{y}_N) \in I^d$$

(where, as usual, $\mathcal{Y} = \{\mathbf{y}_1, \ldots, \mathbf{y}_N\}$ denotes the given N-set of initial point configuration in the unit square $[0,1)^2$).

For an arbitrary measurable test set $A \subset [0,1)^2$ and $\gamma > 0$ define the following (very complicated!) set in the configuration space

$$S(A;\gamma) = \left\{\vec{\mathcal{Z}} \in I^d : ||\mathcal{Z} \cap A| - \text{area}(A)N| > \gamma\sqrt{N}\right\}, \qquad (21.2)$$

where

$$\vec{\mathcal{Z}} = (z_1, \ldots, z_{2N}) \quad \text{and} \quad \mathcal{Z} = \{\mathbf{z}_1, \ldots, \mathbf{z}_N\}$$

with $\mathbf{z}_k = (z_{2k-1}, z_{2k})$, $1 \le k \le N$.

Inserted Remark. Similarly to definition (5.7) in Sec. 5, definition (21.2) becomes useless for "small" test sets A with

$$\text{area}(A) = O\left(N^{-1/2}\right).$$

Replacing (21.2) with the following modified set in the configuration space

$$S'(A;\gamma) = \left\{\vec{\mathcal{Z}} \in I^d \ : \ ||\mathcal{Z} \cap A| - \text{area}(A)N| > \gamma\sqrt{pqN}\right\}, \qquad (21.2')$$

where $p = \text{area}(A) \le 1/2$ and $q = 1 - p \ge 1/2$, we can easily prove useful *relative* square-root discrepancy even for test sets as small as

$$p = \text{area}(A) \ge C \cdot N^{-1},$$

where C is a "large" absolute constant. Note that $\text{area}(A) \ge C \cdot N^{-1}$ describes a much larger family of test sets than $\text{area}(A) \ge C \cdot N^{-1/2}$, and it is best possible in the sense that it is the largest class for which the *relative* square-root discrepancy estimation is still useful.

We need the 2-dimensional analog of Theorem 5.1 (see (5.15); the proof is exactly the same: combining Theorem 4.2 with Bernstein's inequality (5.9)).

Lemma 21.1 (2-dimensional analog of Theorem 5.1) *Let* $\mathcal{Y}(\text{Gauss}; \omega; t)$, $\omega \in \Omega_{\text{Gauss}}$ *be the simplest 2-dimensional Gaussian torus model, and let* $A \subset [0,1)^2$ *be a measurable test with 2-dimensional Lebesgue measure* $\text{area}(A)$. *Assume that*

$$U \ge 1 \quad \textit{and} \quad e^{\pi^2 U^2/2} \ge 3dU = 6NU. \qquad (21.3)$$

Then for every $\gamma > 0$ and every integer $\ell \geq 1$,

$$\mathrm{Var}(U;\ell;A;\gamma) \leq 2\exp\left(-\frac{\gamma^2(2p(1-p))^{-1}}{1+\frac{\gamma}{3p(1-p)\sqrt{N}}}\right) 10\ell((2^\ell-1)U+1),$$

where

$$\mathrm{Var}(U;\ell;A;\gamma) = \mathrm{Var}(2;N;Gauss;\mathcal{Y};U;\ell;A;\gamma)$$
$$= \int_{\omega\in\Omega_{\mathrm{Gauss}}} \Bigg(\mathrm{length}\left\{U \leq t \leq 2^\ell U : \ ||\mathcal{Y}(\mathrm{Gauss};\omega;t)\cap A|\right.$$
$$\left. - \mathrm{area}(A)N| > \gamma\sqrt{N}\right\} - \mathrm{vol}_d(S(A;\gamma))(2^\ell-1)U\Bigg)^2 d\mathrm{ProdMeas}_{\mathrm{Gauss}}(\omega) \tag{21.4}$$

and

$$\mathrm{vol}_d(S(A;\gamma)) < 2\exp\left(-\frac{\gamma^2(2p(1-p))^{-1}}{1+\frac{\gamma}{3p(1-p)\sqrt{N}}}\right) \quad \textit{with} \quad p = \mathrm{area}(A).$$

□

The notation Var in (21.4) indicates that the square integral (21.4) is basically a "variance".

What we actually need is the following average

$$\mathrm{Aver}(U;\ell;A;\gamma) = \mathrm{Aver}(2;N;Gauss;\mathcal{Y};U;\ell;A;\gamma)$$
$$= \int_{\omega\in\Omega_{\mathrm{Ga}}} \mathrm{le}\left\{U \leq t \leq 2^\ell U : \ ||\mathcal{Y}(\mathrm{Ga};\omega;t)\cap A|\right.$$
$$\left. - \mathrm{ar}(A)N| > \gamma\sqrt{N}\right\} d\mathrm{ProdMe}_{\mathrm{Ga}}(\omega), \tag{21.5}$$

where Ga, le, ar and Me stand, respectively, for Gauss, length, area and measure.

By using the well-known inequality between the L_1-norm and the L_2-norm, we have

$$\mathrm{Aver}(2;N;Gauss;\mathcal{Y};U;\ell;A;\gamma)$$
$$\leq \int_{\omega\in\Omega_{\mathrm{Gauss}}} \Bigg|\mathrm{length}\left\{U \leq t \leq 2^\ell U : \ ||\mathcal{Y}(\mathrm{Gauss};\omega;t)\cap A|\right.$$
$$\left. - \mathrm{area}(A)N| > \gamma\sqrt{N}\right\} - \mathrm{vol}_d(S(A;\gamma))(2^\ell-1)U\Bigg| d\mathrm{ProdMeas}_{\mathrm{Gauss}}(\omega)$$

$$+ \mathrm{vol}_d(S(A;\gamma))(2^\ell - 1)U$$

$$\leq (\mathrm{Var}(2; N; Gauss; \mathcal{Y}; U; \ell; A; \gamma))^{1/2} + \mathrm{vol}_d(S(A;\gamma))(2^\ell - 1)U$$

$$\leq 2\,(\mathrm{Var}(2; N; Gauss; \mathcal{Y}; U; \ell; A; \gamma))^{1/2},$$

if

$$(\mathrm{Var}(2; N; Gauss; \mathcal{Y}; U; \ell; A; \gamma))^{1/2} \geq \mathrm{vol}_d(S(A;\gamma))(2^\ell - 1)U.$$

Combining this with Lemma 21.1, we obtain the following lemma.

Lemma 21.2. *Under the condition of Lemma* 21.1, *for every* $\gamma > 0$ *and every integer* $\ell \geq 1$,

$$\mathrm{Aver}(U;\ell;A;\gamma) \leq 2\left(2\exp\left(-\frac{\gamma^2(2p(1-p))^{-1}}{1+\frac{\gamma}{3p(1-p)\sqrt{N}}}\right)10\ell\left((2^\ell-1)U+1\right)\right)^{1/2},$$

if

$$2\exp\left(-\frac{\gamma^2(2p(1-p))^{-1}}{1+\frac{\gamma}{3p(1-p)\sqrt{N}}}\right)10\ell\left((2^\ell-1)U+1\right) \leq 1. \tag{21.6}$$

□

Motivated by (20.36)–(20.37), we define the following families of subsets of the unit square. Let $j \geq 2$ be an integer such that $8^j \leq N$; then $\mathcal{A}_j^{(1)}$ denotes the family of all sets of the form $IN(B(k;j))$, and similarly, let $\mathcal{A}_j^{(2)}$ denote the family of all sets of the form $OUT(B(k;j)) \setminus IN(B(k;j))$, where $j < k$, k is defined in (20.41), and $B \subset [0,1]^2$ runs over all convex subsets of the unit square. Let $\mathcal{A}_k^{(3)}$ denote the family of all sets of the form $OUT(P^+(B;k)) \setminus IN(P(B;k))$, where $B \subset [0,1]^2$ runs over all convex subsets of the unit square. We recall that $i = 5$; see (20.41). Let $\mathcal{A}_5^{(4)}$ denote the family of all sets of the form $OUT(P(B;k;5)) \setminus IN(P(B;k;5))$, and let $\mathcal{A}_5^{(5)}$ denote the family of all sets of the form $IN(P(B;k;5))$, where k is defined in (20.41) and, as usual, $B \subset [0,1]^2$ runs over all convex subsets of the unit square.

We recall (20.4),

$$\mathrm{area}(B(k;j)) < 7 \cdot 4^{-j}. \tag{21.7}$$

Combining Fact 1, Lemma 21.2 and (21.7), we obtain the following inequality: under the condition (21.3), for every $\gamma > 0$ and every integer $\ell \geq 1$,

$$\sum_{A \in \mathcal{A}_j^{(1)}} \text{Aver}(U; \ell; A; \gamma)$$
$$\leq \binom{(4^j N)^2}{3}^{2^{j+1}} \cdot 2 \left(2 \exp\left(- \frac{\gamma^2 \left(2 \cdot 7 \cdot 4^{-j}\right)^{-1}}{1 + \frac{\gamma}{3 \cdot 7 \cdot 4^{-j} \sqrt{N}}} \right) \right.$$
$$\left. \times\ 10\ell \left((2^\ell - 1)U + 1\right) \right)^{1/2}, \tag{21.8}$$

assuming (21.6) holds, i.e.,

$$2 \exp\left(- \frac{\gamma^2 \left(2 \cdot 7 \cdot 4^{-j}\right)^{-1}}{1 + \frac{\gamma}{3 \cdot 7 \cdot 4^{-j} \sqrt{N}}} \right) 10\ell \left((2^\ell - 1)U + 1\right) \leq 1. \tag{21.9}$$

Applying the trivial inequality $\binom{n}{m} \leq n^m$, we have

$$\binom{\left(4^j N\right)^2}{3}^{2^{j+1}} \leq \left(4^j N\right)^{3 \cdot 2^{j+2}} = \exp(12(\log N + j \log 4) 2^j). \tag{21.10}$$

Using (21.10) in (21.8), we obtain the following: under the condition (21.3), for every $\gamma > 0$ and every integer $\ell \geq 1$,

$$\sum_{A \in \mathcal{A}_j^{(1)}} \text{Aver}(U; \ell; A; \gamma)$$
$$\leq \exp\left(12(\log N + j \log 4) 2^j - \frac{\gamma^2 4^j}{28 + \frac{4\gamma}{3 \cdot 4^{-j} \sqrt{N}}} \right)$$
$$\times\ 2(20\ell((2^\ell - 1)U + 1))^{1/2}, \tag{21.11}$$

assuming

$$2 \exp\left(- \frac{\gamma^2 4^j}{14 + \frac{2\gamma}{3 \cdot 4^{-j} \sqrt{N}}} \right) 20\ell \left((2^\ell - 1)U + 1\right) \leq 1. \tag{21.12}$$

Note that (21.11) is a key step in the polygon-approximation technique. To illustrate the power of (21.11), we specify the parameters as follows: let

$$N = 10^{27}, \ U = 4, \ \ell = 100, \tag{21.13}$$

which implies $d = 2 \cdot 10^{27}$, and notice that with the choice (21.13) condition (21.3) is satisfied; moreover, let

$$\gamma = \gamma(1;j) = \frac{100}{(j-3)(j-4)} \quad \text{for } j \geq 5 \text{ with } 8^j \leq N. \tag{21.14}$$

We are going to use the simple fact

$$\sum_{j \geq 5} \frac{1}{(j-3)(j-4)} = 1. \tag{21.15}$$

(Indeed, in view of the identity $\frac{1}{i(i+1)} = \frac{1}{i} - \frac{1}{i+1}$, (21.15) is a telescopic sum.)

To motivate the choice of the parameters in (21.13), we recall that there are roughly 10^{27} gas molecules in a cubic box of volume $1\,\mathrm{m}^3$, and realistic gas molecules have average speed in the range of 10^3 meter per second. For technical reasons in Secs. 20 and 21 we restrict ourselves to the 2-dimensional case, which means that here we replace the cubic box of volume $1\,\mathrm{m}^3$ (= "gas container") with a square of area $1\,\mathrm{m}^2$; nevertheless, we still assume that the average speed of the $N = 10^{27}$ particles is in the range of 10^3 meter per second. In the 2-dimensional Bernoulli gas model the gas molecules are represented by point billiards, and as usual, we apply the geometric trick of *unfolding* (see Sec. 3); thus we reduce the billiards-in-a-square model to the torus-billiard model. In fact, what we have here is the simplest 2-dimensional Gaussian torus model. We will show that the threshold $U = 4$ represents — roughly speaking — the "relaxation distance for simultaneous convex equilibrium in the particle space".

Let us return now to (21.11). Using (21.13)–(21.15) in (21.11), we see that (21.12) is satisfied, and we have the following upper bound via trivial numerical calculations:

$$\begin{aligned}
&\sum_{\substack{j\geq 5:\\ 8^j \leq N}} \sum_{A \in \mathcal{A}_j^{(1)}} \mathrm{Aver}(U=4; \ell = 100; A; \gamma(1;j)) \\
&\quad \leq 2\left(2000\left((2^{100}-1)4+1\right)\right)^{1/2} \\
&\quad \quad \cdot \left(\sum_{\substack{j\geq 5:\\ 8^j \leq N}} \exp\left(12(27\log 10 + j \log 4)2^j - \frac{10^4 \cdot 4^j \cdot (j-3)^{-2}(j-4)^{-2}}{28 + \frac{100 \cdot 4^{j+1}\gamma(1;j)}{3\cdot(j-3)(j-4)\sqrt{N}}} \right) \right) \\
&\quad < e^{-10000}.
\end{aligned} \tag{21.16}$$

It is based on the fact that the infinite sum is dominated by its first few terms (and also $N \geq 8^j$).

Next we switch to the family $\mathcal{A}_5^{(5)}$, i.e., the family of all sets of the form $IN(P(B;k;5))$, where k is defined in (20.41) and $B \subset [0,1]^2$ runs over all convex subsets of the unit square.

Combining Fact 5(b) and Lemma 21.2 with $p(1-p) \leq 1/4$, we obtain the following inequality: under the condition (21.3) and (21.6), for every $\gamma > 0$ and every integer $\ell \geq 1$,

$$\sum_{A \in \mathcal{A}_5^{(5)}} \text{Aver}(U;\ell;A;\gamma)$$

$$\leq \binom{N^{8/3}}{3}^{2^5} 2\left(\exp\left(-\frac{2\gamma^2}{1+\frac{2\gamma}{\sqrt{N}}}\right) 20\ell\left((2^\ell-1)U+1\right)\right)^{1/2}$$

$$\leq \exp\left(2^8 \log N - \frac{\gamma^2}{1+\frac{2\gamma}{\sqrt{N}}}\right) 2\left(20\ell\left((2^\ell-1)U+1\right)\right)^{1/2}. \quad (21.17)$$

Let

$$\gamma = \gamma(5) = 150. \quad (21.18)$$

Applying (21.13) and (21.18) in (21.17), we have the following upper bound via trivial numerical calculation:

$$\sum_{A \in \mathcal{A}_5^{(5)}} \text{Aver}(U=4;\ell=100;A;\gamma(5)=150)$$

$$\leq \exp\left(2^8 \cdot 27 \log 10 - \frac{150^2}{1+\frac{2\gamma}{\sqrt{N}}}\right)$$

$$\times 2\left(2000\left((2^{100}-1)4+1\right)\right)^{1/2} < e^{-3001}, \quad (21.19)$$

where we used that

$$\frac{\gamma}{\sqrt{N}} = \frac{\gamma(5)}{\sqrt{N}} = 150 \cdot 10^{-27/2}$$

is totally negligible.

Next we switch to $\mathcal{A}_j^{(2)}$, i.e., the family of all sets of the form $OUT(B(k;j)) \setminus IN(B(k;j))$, where $B \subset [0,1]^2$ runs over all convex subsets of the unit square. By (20.38),

$$\text{area}(OUT(B(k;j)) \setminus IN(B(k;j))) \leq \frac{1}{2^{j-5}\sqrt{N}} \text{ for all } j \geq 5. \tag{21.20}$$

Combining Fact 3, Lemma 21.2 and (21.20), we obtain the following inequality: under the condition (21.3) and (21.6), for every $\gamma > 0$ and every integer $\ell \geq 1$,

$$\begin{aligned}
&\sum_{A \in \mathcal{A}_j^{(2)}} \text{Aver}(U;\ell;A;\gamma) \\
&\leq \left(\frac{(4^j N)^2}{3}\right)^{2^{j+2}} \cdot 2\left(2\exp\left(-\frac{\gamma^2\left(2 \cdot 2^{5-j} \cdot N^{-1/2}\right)^{-1}}{1+\frac{\gamma}{3 \cdot 2^{5-j} \cdot N^{-1/2}\sqrt{N}}}\right)\right. \\
&\quad \left.\times\, 10\ell((2^\ell - 1)U + 1)\right)^{1/2} \\
&\leq \exp\left(24(\log N + j\log 4)2^j - \frac{\gamma^2 2^{j-7}\sqrt{N}}{1+\frac{\gamma}{3 \cdot 2^{5-j}}}\right) 2(20\ell((2^\ell - 1)U + 1))^{1/2}.
\end{aligned} \tag{21.21}$$

Let

$$\gamma = \gamma(2;j) = \frac{100}{j^{3/2}} \quad \text{for} \quad j \geq 5 \quad \text{with} \quad 8^j \leq N. \tag{21.22}$$

Applying (21.13) and (21.22) in (21.21), we have the following upper bound via trivial numerical calculations:

$$\begin{aligned}
&\sum_{\substack{j \geq 5: \\ 8^j \leq N}} \sum_{A \in \mathcal{A}_j^{(2)}} \text{Aver}(U=4;\ell=100;A;\gamma(2;j)) \leq 2(2000((2^{100}-1)4+1))^{1/2} \\
&\quad \cdot \sum_{\substack{j \geq 5: \\ 8^j \leq N}} \exp\left(24(27\log 10 + j\log 4)2^j - \frac{10^4 j^{-3} 2^{j-7} 10^{27/2}}{1+\frac{100 \cdot 2^{j-5}}{3j^{3/2}}}\right) < e^{-10000},
\end{aligned} \tag{21.23}$$

where we used that $\log N = 27\log 10 < 70$ and $10^{27} = N \geq 8^j$.

Next we study the family $\mathcal{A}_k^{(3)}$, i.e., the family of all sets of the form $OUT(P^+(B;k)) \setminus IN(P(B;k))$, where k is defined in (20.41) and $B \subset [0,1]^2$

runs over all convex subsets of the unit square. By (20.39) and (20.41),

$$\text{area}\left(OUT(P^+(B;k)) \setminus IN(P(B;k))\right)$$
$$\leq \frac{1}{4^{k-2}N} + \frac{1}{8^{k-2}N^2} + \frac{7}{4^{k-2}} + \frac{7}{4^k} + \frac{24}{2^{k/2}\sqrt{N}} + 2^{-k+7}N^{-1} \leq \frac{2^5}{N^{5/8}}. \tag{21.24}$$

Combining Fact 4, Lemma 21.2 and (21.24), we obtain the following inequality: under the condition (21.3) and (21.6), for every $\gamma > 0$ and every integer $\ell \geq 1$,

$$\sum_{A \in \mathcal{A}_k^{(3)}} \text{Aver}(U;\ell;A;\gamma) \leq \left(\binom{(4^kN)^2}{2^k}\binom{(4^{k-1}N)^2}{3}\right)^{2^k}$$
$$\cdot 2\left(\exp\left(-\frac{\gamma^2\left(2\cdot 2^5\cdot N^{-5/8}\right)^{-1}}{1+\frac{\gamma}{3\cdot 2^5\cdot N^{-5/8}\sqrt{N}}}\right)10\ell\left((2^\ell-1)U+1\right)\right)^{1/2}$$
$$\leq \exp\left(2^5N^{1/3}\log N - \frac{\gamma^2 2^{-7}N^{5/8}}{1+\frac{\gamma}{3\cdot 2^5\cdot N^{-5/8}\sqrt{N}}}\right)2\left(20\ell\left((2^\ell-1)U+1\right)\right)^{1/2}. \tag{21.25}$$

Let

$$\gamma = \gamma(3) = 100. \tag{21.26}$$

Using (21.13) and (21.26) in (21.25), we have the following upper bound via trivial numerical calculations:

$$\sum_{A \in \mathcal{A}_k^{(3)}} \text{Aver}(U=4;\ell=100;A;\gamma(3))$$
$$\leq 2\left(2000\left((2^{100}-1)4+1\right)\right)^{1/2}$$
$$\times \exp\left(2^5\cdot 10^{27/3}\cdot 27\log 10 - \frac{10^4 2^{-7}10^{135/8}}{1+\frac{10^2\cdot 10^{27/8}}{3\cdot 2^5}}\right) < e^{-100000}. \tag{21.27}$$

Finally we study the family $\mathcal{A}_i^{(4)}$ with $i=5$, i.e., the family of all sets of the form $OUT(P(B;k;5)) \setminus IN(P(B;k;5))$, where k is defined in (20.41) and, as always, $B \subset [0,1]^2$ runs over all convex subsets of the unit square. By (20.40) with $i = 5$,

$$\begin{aligned}&\text{area}\left(OUT(P(B;k;5)) \setminus IN(P(B;k;5))\right)\\&\quad\le 16N^{-4/3} + 2^{5+6}N^{-8/3} + 2^{(5+8)/2}N^{-2/3} + 2^{5+5}N^{-4/3} < 2^7 N^{-2/3},\end{aligned} \tag{21.28}$$

since $N = 10^{27}$. Combining Fact 5, Lemma 21.2 and (21.28), we obtain the following inequality: under the condition (21.3) and (21.6), for every $\gamma > 0$ and every integer $\ell \ge 1$,

$$\begin{aligned}\sum_{A\in\mathcal{A}_5^{(4)}} \text{Aver}(U;\ell;A;\gamma) &\le \binom{\left(N^{4/3}\right)^2}{2^5}\binom{\left(N^{4/3}\right)^2}{3}^{2^5}\\&\quad\cdot 2\left(\exp\left(-\frac{\gamma^2\left(2\cdot 2^7\cdot N^{-2/3}\right)^{-1}}{1+\frac{\gamma}{3\cdot 2^7\cdot N^{-2/3}\sqrt{N}}}\right)20\ell\left((2^\ell-1)U+1\right)\right)^{1/2}\\&\le \exp\left(2^{12}\log N - \frac{\gamma^2 2^{-9}N^{2/3}}{1+\frac{\gamma N^{1/6}}{3\cdot 2^7}}\right)2\left(20\ell\left((2^\ell-1)U+1\right)\right)^{1/2}.\end{aligned} \tag{21.29}$$

Let

$$\gamma = \gamma(4) = 1. \tag{21.30}$$

Using (21.13) and (21.30) in (21.29), we have the following upper bound via trivial numerical calculations:

$$\begin{aligned}&\sum_{A\in\mathcal{A}_5^{(4)}} \text{Aver}(U=4;\ell=100;A;\gamma(4))\\&\quad\le 2\left(2000\left((2^{100}-1)4+1\right)\right)^{1/2}\\&\qquad\times\exp\left(2^{12}\cdot 27\log 10 - \frac{2^{-9}10^{18}}{1+\frac{10^{27/6}}{3\cdot 2^7}}\right) < e^{-100000}.\end{aligned} \tag{21.31}$$

Combining (21.16), (21.19), (21.23), (21.27) and (21.31), we have

$$\begin{aligned}&\sum_{\substack{j\ge 5:\\ 8^j\le N}}\ \sum_{A\in\mathcal{A}_j^{(1)}} \text{Aver}(U=4;\ell=100;A;\gamma(1;j))\\&\quad+\sum_{A\in\mathcal{A}_5^{(5)}} \text{Aver}(U=4;\ell=100;A;\gamma(5))\end{aligned}$$

$$+ \sum_{\substack{j\ge 5:\\ 8^j\le N}} \sum_{A\in\mathcal{A}_j^{(2)}} \mathrm{Aver}(U=4;\ell=100;A;\gamma(2;j))$$

$$+ \sum_{A\in\mathcal{A}_k^{(3)}} \mathrm{Aver}(U=4;\ell=100;A;\gamma(3))$$

$$+ \sum_{A\in\mathcal{A}_5^{(4)}} \mathrm{Aver}(U=4;\ell=100;A;\gamma(4))$$

$$\le e^{-10000} + e^{-3001} + e^{-10000} + e^{-100000} + e^{-100000} < e^{-3000}. \qquad (21.32)$$

We recall (21.5) with the choice $N = 10^{27}$, $U = 4$, $\ell = 100$ (see (21.13)):

$$\begin{aligned} &\mathrm{Aver}(U=4;\ell=100;A;\gamma) \\ &\quad= \mathrm{Aver}(2;N=10^{27};Gauss;\mathcal{Y};U=4;\ell=100;A;\gamma) \\ &\quad= \int_{\omega\in\Omega_{\mathrm{Ga}}} \mathrm{le}\,\Big\{4\le t\le 2^{102}: \\ &\qquad ||\mathcal{Y}(\mathrm{Ga};\omega;t)\cap A| - \mathrm{ar}(A)N| > \gamma\sqrt{N}\Big\}\, d\mathrm{ProdMe}_{\mathrm{Ga}}(\omega), \end{aligned} \qquad (21.33)$$

where Ga, le, ar and Me stand, respectively, for Gauss, length, area and Measure.

What we do next is a routine application of the simple Markov's inequality (see (21.38) below). Let

$$\mathcal{A}_0 = \bigcup_{5\le j<k} \mathcal{A}_j^{(1)} \cup \mathcal{A}_5^{(5)} \cup \bigcup_{5\le j<k} \mathcal{A}_j^{(2)} \cup \mathcal{A}_k^{(3)} \cup \mathcal{A}_5^{(4)}, \qquad (21.34)$$

where k is defined by (20.41). Let $\Omega_{\mathrm{Gauss}}(\mathrm{bad})$ be the set of those $\omega\in\Omega_{\mathrm{Gauss}}$ such that

$$\sum_{A\in\mathcal{A}_0} \mathrm{length}\,\Big\{4\le t\le 2^{102}:\ ||\mathcal{Y}(\mathrm{Ga};\omega;t)\cap A| - \mathrm{area}(A)N| > \gamma(A)\sqrt{N}\Big\} \ge e^{-1000}, \qquad (21.35)$$

where

$$\gamma(A) = \gamma(1;j) \ \text{ or } \ \gamma(5) \ \text{ or } \ \gamma(2;j) \ \text{ or } \ \gamma(3) \ \text{ or } \ \gamma(4) \qquad (21.36)$$

according as

$$A\in\mathcal{A}_j^{(1)} \ \text{ or } \ \mathcal{A}_5^{(5)} \ \text{ or } \ \mathcal{A}_j^{(2)} \ \text{ or } \ \mathcal{A}_k^{(3)} \ \text{ or } \ \mathcal{A}_5^{(4)}. \qquad (21.37)$$

Combining (21.32)–(21.37), we obtain the upper bound

$$\text{ProdMeas}_{\text{Gauss}}(\Omega_{\text{Gauss}}(\text{bad})) < e^{-2000}. \tag{21.38}$$

Let

$$\Omega_{\text{Gauss}}(\text{good}) = \Omega_{\text{Gauss}} \setminus \Omega_{\text{Gauss}}(\text{bad}). \tag{21.39}$$

By (21.38)

$$\text{ProdMeas}_{\text{Gauss}}(\Omega_{\text{Gauss}}(\text{good})) \geq 1 - e^{-2000}, \tag{21.40}$$

representing a totally overwhelming majority. By (21.35), for every $\omega \in \Omega_{\text{Gauss}}(\text{good})$

$$\sum_{A\in\mathcal{A}_0} \Big|\text{length}\Big\{4 \leq t \leq 2^{102} : \,||\mathcal{Y}(\text{Ga};\omega;t) \cap A| - \text{area}(A)N| > \gamma(A)\sqrt{N}\Big\}\Big| < e^{-1000}. \tag{21.41}$$

Let $\omega \in \Omega_{\text{Gauss}}(\text{good})$ be fixed. Then for every real number $0 < \alpha < e^{-1000}$, let $Z(\alpha) = Z(\omega;\alpha)$ denote the number of pairs (A, m) with $A \in \mathcal{A}_0$ and $0 \leq m < (2^{102} - 4)e^{1000}$ integer such that

$$\Big||\mathcal{Y}(\text{Ga};\omega;t = 4 + \alpha + me^{-1000}) \cap A\big| - \text{area}(A)N\Big| > \gamma(A)\sqrt{N}, \tag{21.42}$$

holds with $\gamma(A)$ defined by (21.36)–(21.37). A standard double counting argument gives the equality

$$\int_0^{e^{-1000}} Z(\alpha)\, d\alpha = \sum_{A\in\mathcal{A}_0} \text{length}\Big\{4 \leq t \leq 2^{102} : \,||\mathcal{Y}(\text{Ga};\omega;t) \cap A| - \text{area}(A)N| > \gamma(A)\sqrt{N}\Big\}. \tag{21.43}$$

By (21.41)–(21.43) we have

$$\int_0^{e^{-1000}} Z(\alpha)\, d\alpha < e^{-1000},$$

which implies that $Z(\alpha) = 0$ for some $0 < \alpha < e^{-1000}$. That is, for every $\omega \in \Omega_{\text{Gauss}}(\text{good})$ there exists a real number $0 < \alpha = \alpha(\omega) < e^{-1000}$ such that

$$||\mathcal{Y}(\text{Gauss};\omega;t = 4 + \alpha + me^{-1000}) \cap A| - \text{area}(A)N| \leq \gamma(A)\sqrt{N} \tag{21.44}$$

holds simultaneously for every set $A \in \mathcal{A}_0$ and every integer $0 \leq m < (2^{102} - 4)e^{1000}$, and $\gamma(A)$ is defined by (21.36)–(21.37).

22. Simultaneous Square-Root Equilibrium Relative to Nice Sets (III)

For every convex set $B \subset [0,1]^2$, let

$$\mathrm{inscr}(B) = IN(P(B;k;5)) \cup \bigcup_{5 \le j < k} IN(B(k;j)) \tag{22.1}$$

denote the *inscribed approximation of* B (see (20.36) with $i = 5$, and k is determined by (20.41)), and let

$$\begin{aligned}\mathrm{cirscr}(B) = \mathrm{inscr}(B) &\cup \left(\bigcup_{5 \le j < k} OUT(B(k;j)) \setminus IN(B(k;j)) \right) \\ &\cup (OUT(P^{+}(B;k)) \setminus IN(P(B;k))) \\ &\cup (OUT(P(B;k;5)) \setminus IN(P(B;k;5)))\end{aligned} \tag{22.2}$$

denote the *circumscribed approximation of* B (see (20.37)).

Let $\omega \in \Omega_{\mathrm{Gauss}}(\mathrm{good})$ be fixed, and let $0 < \alpha = \alpha(\omega) < e^{-1000}$ be the corresponding real number in (21.44). Since (22.1) is a disjoint union, by (21.44) we have for all integers $0 \le m < (2^{102} - 4)e^{1000}$,

$$\begin{aligned}&\left|\mathcal{Y}(\mathrm{Gauss};\omega;t = 4+\alpha+me^{-1000}) \cap \mathrm{inscr}(B)\right| \\ &= \left|\mathcal{Y}(\mathrm{Gauss};\omega;t = 4+\alpha+me^{-1000}) \cap IN(P(B;k;5))\right| \\ &\quad + \sum_{5 \le j < k} \left|\mathcal{Y}(\mathrm{Gauss};\omega;t = 4+\alpha+me^{-1000}) \cap IN(B(k;j))\right| \\ &\ge \mathrm{area}(\mathrm{inscr}(B)) - \left(\gamma(5) + \sum_{5 \le j < k} \gamma(1;j) \right) \sqrt{N} \\ &\ge \mathrm{area}(\mathrm{inscr}(B)) - \left(150 + \sum_{j \ge 5} \frac{100}{(j-3)(j-4)} \right) \sqrt{N} \\ &\ge \mathrm{area}(\mathrm{inscr}(B)) - 250\sqrt{N},\end{aligned} \tag{22.3}$$

and repeating the same argument in the other direction,

$$\begin{aligned}&\left|\mathcal{Y}(\text{Gauss};\omega;t=4+\alpha+me^{-1000})\cap\text{inscr}(B)\right|\\&\le\text{area}(\text{inscr}(B))+250\sqrt{N},\end{aligned}\tag{22.4}$$

where we used (21.14), (21.15) and (21.18).

Similarly, by (22.2), (22.4) and (21.44) we have

$$\begin{aligned}&\left|\mathcal{Y}(\text{Gauss};\omega;t=4+\alpha+me^{-1000})\cap\text{cirscr}(B)\right|\\&\quad\le\left|\mathcal{Y}(\text{Gauss};\omega;t=4+\alpha+me^{-1000})\cap\text{inscr}(B)\right|\\&\qquad+\sum_{5\le j<k}\left|\mathcal{Y}(\text{Gauss};\omega;t=4+\alpha+me^{-1000})\right.\\&\qquad\left.\cap\,(OUT(B(k;j))\setminus IN(B(k;j)))\right|\\&\qquad+\left|\mathcal{Y}(\text{Gauss};\omega;t=4+\alpha+me^{-1000})\right.\\&\qquad\left.\cap\left(OUT(P^+(B;k))\setminus IN(P(B;k))\right)\right|\\&\qquad+\left|\mathcal{Y}(\text{Gauss};\omega;t=4+\alpha+me^{-1000})\right.\\&\qquad\left.\cap\,(OUT(P(B;k;5))\setminus IN(P(B;k;5)))\right|\\&\quad\le\text{area}(\text{inscr}(B))+250\sqrt{N}+\sum_{5\le j<k}\text{area}\,(OUT(B(k;j))\setminus IN(B(k;j)))\\&\qquad+\text{area}\left(OUT(P^+(B;k))\setminus IN(P(B;k))\right)\\&\qquad+\text{area}\,(OUT(P(B;k;5))\setminus IN(P(B;k;5)))\\&\qquad+\left(\sum_{5\le j<k}\gamma(2;j)+\gamma(3)+\gamma(4)\right)\sqrt{N}\\&\quad=\text{AREA}(+;B)+250\sqrt{N}+\left(100\sum_{5\le j<k}j^{-3/2}+100+1\right)\sqrt{N}\\&\quad<\text{AREA}(+;B)+451\sqrt{N},\end{aligned}\tag{22.5}$$

where

$$\text{AREA}(+;B) = \text{area}(\text{inscr}(B)) + \sum_{5\le j<k} \text{area}\left(OUT(B(k;j)) \setminus IN(B(k;j))\right)$$
$$+ \text{area}\left(OUT(P^+(B;k)) \setminus IN(P(B;k))\right)$$
$$+ \text{area}\left(OUT(P(B;k;5)) \setminus IN(P(B;k;5))\right), \tag{22.6}$$

and we used the numerical fact

$$\sum_{5\le j<k} j^{-3/2} < \int_4^\infty x^{-3/2}\,dx = 1.$$

By (21.20), (21.24) and (21.28),

$$\sum_{5\le j<k} \text{area}\left(OUT(B(k;j)) \setminus IN(B(k;j))\right)$$
$$+ \text{area}\left(OUT(P^+(B;k)) \setminus IN(P(B;k))\right)$$
$$+ \text{area}\left(OUT(P(B;k;5)) \setminus IN(P(B;k;5))\right)$$
$$\le \sum_{5\le j<k} 2^{5-j}N^{-1/2} + 2^5N^{-3/4} + 2^7N^{-2/3} < 3N^{-1/2},$$

and using it in (22.6), we have

$$\text{area}(\text{inscr}(B)) \le \text{area}(B) \le \text{AREA}(+;B) \le \text{area}(\text{inscr}(B)) + 3N^{-1/2}. \tag{22.7}$$

By (22.4) and (22.7),

$$\left|\mathcal{Y}(\text{Gauss};\omega;t = 4+\alpha+me^{-1000}) \cap B\right| - \text{area}(B)N$$
$$\ge \left|\mathcal{Y}(\text{Gauss};\omega;t = 4+\alpha+me^{-1000}) \cap \text{inscr}(B)\right| - \text{AREA}(+;B)N$$
$$\ge \text{area}(\text{inscr}(B)) - 250\sqrt{N} - \text{area}(\text{inscr}(B)) - 3\sqrt{N} = -253\sqrt{N}, \tag{22.8}$$

and similarly, by (22.5) and (22.7), and repeating the same argument in the other direction,

$$\left|\mathcal{Y}(\text{Gauss};\omega;t = 4+\alpha+me^{-1000}) \cap B\right| - \text{area}(B)N$$
$$\le \left|\mathcal{Y}(\text{Gauss};\omega;t = 4+\alpha+me^{-1000}) \cap \text{cirscr}(B)\right|$$

$$- \text{area}(\text{inscr}(B))N$$
$$\leq \text{area}(\text{inscr}(B)) + 451\sqrt{N} - \text{area}(\text{inscr}(B)) = 451\sqrt{N}. \quad (22.9)$$

Combining (21.40), (22.8), (22.9) we obtain the following lemma.

Lemma 22.1. *Let* $\mathcal{Y}(\text{Gauss}; \omega; t)$, $\omega \in \Omega_{\text{Gauss}}$ *be the simplest* 2*-dimensional Gaussian torus model with* $N = 10^{27}$ *particles. Then there exists a (measurable) subset* $\Omega_{\text{Gauss}}(\text{good})$ *of* Ω_{Gauss}, *representing an overwhelming majority in the quantitative sense*

$$\text{ProdMeas}_{\text{Gauss}}(\Omega_{\text{Gauss}}(\text{good})) \geq 1 - e^{-2000}, \quad (22.10)$$

which satisfies the following property. For every $\omega \in \Omega_{\text{Gauss}}(\text{good})$ *there is a real number* $0 < \alpha = \alpha(\omega) < e^{-1000}$ *such that*

$$\left| |\mathcal{Y}(\text{Gauss}; \omega; t = 4 + \alpha + me^{-1000}) \cap B| - \text{area}(B)N \right| \leq 451\sqrt{N} \quad (22.11)$$

holds simultaneously for all convex sets $B \subset [0,1]^2$ *and all integers* $0 \leq m < (2^{102} - 4)e^{1000}$. □

We recall that the average speed of a particle in our model is 10^3 m/s, and the particle space is the unit square $[0,1]^2$ of side length 1 m. Then the time interval $4 \leq t \leq 4 \cdot 2^{100}$ is about 10^{27} seconds, which is an incredibly long time: it is roughly billion times the age of the universe. The sequence $4+\alpha+me^{-1000}$, where m runs over the integers in $0 \leq m < (2^{102}-4)e^{1000}$, is an extremely dense arithmetic progression of time points in the very long interval $4 \leq t \leq 4 \cdot 2^{100}$. The message of Lemma 22.1 is that, starting from an arbitrary initial configuration, the overwhelming majority of the time evolutions of the 2-dimensional Gaussian torus model with $N = 10^{27}$ particles (see (22.10)) reaches "simultaneous convex equilibrium in the particle space" very fast. Indeed, the threshold $U = 4$ represents (roughly speaking) the "relaxation distance for simultaneous convex equilibrium". This equilibrium is in fact a square-root equilibrium (see (22.11)), and the system stays in this square-root equilibrium for a very long time (billion times the age of the universe).

The last almost trivial step is to upgrade "every time point of an extremely dense arithmetic progression in the very long time interval $4 \leq t \leq 4 \cdot 2^{100}$" to "every time point in the very long time interval $4 \leq t \leq 4 \cdot 2^{100}$", that is, to extend the discrete sequence to the whole interval. It is a routine approximation argument that goes as follows. Given a compact plane set $A \subset \mathbb{R}^2$ and an arbitrary real number $\delta > 0$, let

$$A(-\delta) = \left\{\mathbf{x} \in A : \text{ distance}(\mathbf{x}; \mathbb{R}^2 \setminus A) \geq \delta\right\}, \tag{22.12}$$

and similarly, let

$$A(\delta) = \left\{\mathbf{x} \in \mathbb{R}^2 : \text{ distance}(\mathbf{x}; A) \leq \delta\right\}. \tag{22.13}$$

It is a well-known fact that if A is convex then both $A(-\delta)$ and $A(\delta)$ are convex. (Indeed, the statements are almost trivial for convex polygons, and we can finish by taking limit.)

Now let $\omega \in \Omega_{\text{Gauss}}(\text{good})$, and also assume the extra requirement that no particle has speed $\geq e^{500}$ meter per second. Since the average speed is 10^3 meter per second, and the tail of the Gaussian velocity distribution is superexponentially small, it means the exclusion of a tiny part of Ω_{Gauss} such that its $\text{ProdMeas}_{\text{Gauss}}$-measure is *much* less than e^{-2000}. Let $A \subset [0,1]^2$ be an arbitrary convex set, and let $t' \in [4, 2^{102}]$ be an arbitrary time instant. There is a member $t_m = 4 + \alpha + me^{-1000}$ (with some $0 < \alpha = \alpha(\omega) < e^{-1000}$ and $0 \leq m < (2^{102} - 4)e^{1000}$ integer) of the arithmetic progression in Lemma 22.1 such that $|t' - t_m| < e^{-1000}$. Choosing $\delta = e^{-500}$, consider the subset $A(-\delta)$ of A. We have the trivial area estimation

$$\text{area}(A \setminus A(-\delta)) \leq \delta \cdot \text{perimeter}(A) \leq 4\delta < e^{-498}. \tag{22.14}$$

By Lemma 22.1 and (22.14),

$$\begin{aligned} |\mathcal{Y}(\text{Gauss}; \omega; t_m) \cap A(-\delta)| &\geq \text{area}(A(-\delta))N - 451\sqrt{N} \\ &> \text{area}(A)N - 451\sqrt{N} - e^{-498}. \end{aligned} \tag{22.15}$$

If a particle is in $A(-\delta)$ at $t = t_m$ but not at $t = t'$, then that particle must have exceptionally large speed. Indeed, since $|t' - t_m| < e^{-1000}$ and $\delta = e^{-500}$, that particle must travel a distance $\geq \delta = e^{-500}$ meter in a time interval of less than e^{-1000} seconds, implying a speed $\geq e^{500}$, which is impossible by hypothesis. Thus we have

$$\begin{aligned} &|\mathcal{Y}(\text{Gauss}; \omega; t') \cap A| \geq |\mathcal{Y}(\text{Gauss}; \omega; t_m) \cap A(-\delta)| \\ &\quad > \text{area}(A)N - 451\sqrt{N} - e^{-498} > \text{area}(A)N - 500\sqrt{N}. \end{aligned} \tag{22.16}$$

Repeating the argument with $A(\delta)$ instead of $A(-\delta)$, we obtain the other direction

$$|\mathcal{Y}(\text{Gauss}; \omega; t') \cap A| < \text{area}(A)N + 500\sqrt{N}. \tag{22.17}$$

This implies the following upgrading of Lemma 22.1.

Lemma 22.2. *Let* $\mathcal{Y}(\text{Gauss}; \omega; t)$, $\omega \in \Omega_{\text{Gauss}}$ *be the simplest* 2*-dimensional Gaussian torus model with* $N = 10^{27}$ *particles. Then there exists a (measurable) subset* $\Omega_{\text{Gauss}}(\text{good})$ *of* Ω_{Gauss}, *representing an overwhelming majority in the quantitative sense*

$$\text{ProdMeas}_{\text{Gauss}}(\Omega_{\text{Gauss}}(\text{good})) \geq 1 - 2e^{-2000},$$

which satisfies the following property. For every $\omega \in \Omega_{\text{Gauss}}(\text{good})$

$$||\mathcal{Y}(\text{Gauss}; \omega; t) \cap B| - \text{area}(B)N| < 500\sqrt{N}$$

holds simultaneously for all convex sets $B \subset [0,1]^2$ *and all times* $4 \leq t \leq 2^{102}$. □

Of course the message is exactly the same as that of Lemma 22.1. Again the threshold $U = 4$ represents the surprisingly short "relaxation distance for simultaneous convex equilibrium", and this equilibrium is a square-root equilibrium. The system stays in this square-root equilibrium for a very long time (billion times the age of the universe). The relatively large constant factor 500 is basically accidental. It comes from the complexity of the approximation technique, and with more work one should be able to substantially reduce its value.

What happens if we increase the number N of particles? If we replace $N = 10^{27}$ with an arbitrary large N, then, repeating the proof above, the upper bound $500\sqrt{N}$ (with $N = 10^{27}$) in Lemma 22.2 is replaced by $c_0\sqrt{\log N}\sqrt{N}$ and the threshold $U = 4$ is replaced by $U = c_1 \sqrt{\log N}$, where $c_0 > 0$ and $c_1 > 0$ are some absolute constants independent of N. This justifies the term simultaneous convex *square-root* equilibrium.

Note that Lemma 22.2 has a straightforward analog for the closed orbit models discussed in Secs. 12–13 (the same proof works).

The 2-dimensional approximation argument above can be generalized for the 3-space, and a key step in the proof is to replace Sas's theorem by the following result of Macbeath in higher dimensions; see [Mac51]. For every dimension $d \geq 3$ and integer $n \geq d+1$, among all d-dimensional convex bodies of volume one, the only extremal sets that minimize the volume of the largest inscribed polytopes with n vertices are the spheres and the ellipsoids. We apply Macbeath theorem in the 3-space. A novelty of the 3-space is that there is no analog of the regular n-gon inscribed in a circle in higher dimensions. Instead we choose a random n-element point set on the unit sphere with uniform distribution (with respect to the surface area), and consider the convex hull of this point set. This defines a "random n-polyhedron", which can succesfully replace the regular n-gon in many cases. It is easy to see that the volume of the set-difference = ball of unit volume minus its inscribed "random n-polyhedron" is $O(1/n)$ (instead of $O(1/n^2)$ on the plane), which is *just*(!) enough for our purposes. The rest is a more-or-less straightforward generalization of the 2-dimensional approximation argument above; we omit the details. The price that we pay for the weaker approximation in the 3-space ($O(1/n)$ instead of $O(1/n^2)$) is an extra factor of $\log N$-power, which is basically negligible. More precisely, in the 3-dimensional case the "simultaneous convex equilibrium" is still a square-root equilibrium apart from a $\log N$-power (like $\sqrt{N}\sqrt{\log N}$ is replaced by $\sqrt{N}(\log N)^2$).

What makes dimensions 2 and 3 special is that in dimension $d \geq 4$ the "simultaneous convex equilibrium" is *not* a square-root equilibrium, due to an inevitable discrepancy of size $N^{(d-1)/d+1)}$ (note that $(d-1)/(d+1) \geq 3/5 > 1/2$ for $d \geq 4$). The "inevitable discrepancy" comes from the following result (see Schmidt [Sch75]).

Schmidt's discrepancy theorem for convex sets. *Given an arbitrary set $\mathcal{X} = \{\mathbf{x}_1, \mathbf{x}_2, \ldots, \mathbf{x}_N\}$ of N points in the d-dimensional unit cube $[0,1]^d$, there exists a convex set $A \subset [0,1]^d$ such that*

$$||\mathcal{X} \cap A| - \text{vol}_d(A)N| \geq c_0(d)N^{(d-1)/(d+1)},$$

where the constant factor $c_0(d) > 0$ depends only on the dimension.

For the sake of completeness we include the simple but ingeneous proof of Schmidt's theorem. We can summarize the proof in one sentence: we start with a ball, consider many disjoint solid spherical caps of certain size, and enforce large discrepancy by a process of removing or keeping a solid spherical cap depending on whether it is "rich" (contains a point from $\mathcal{X}$) or "poor" (does not contain a point from $\mathcal{X}$); what is left at the end of the removal process is clearly a convex set.

Here comes the detailed proof. Let B be the ball of diameter one contained in $[0,1]^d$, and let S denote its surface. Let C_1 be a solid spherical cap of B such that its d-volume is $1/2N$ (a solid spherical cap is a solid bounded by the sphere and a $(d-1)$-dimensional hyperplane in the d-space). We now pick as many pairwise disjoint solid spherical caps of B with d-volume $1/2N$

as possible, say $C_1, C_2, \ldots, C_m$. Elementary calculation gives

$$m \geq c_2(d) N^{(d-1)/(d+1)}. \tag{22.18}$$

Indeed, if the "radius" of a small solid cap of B is r, then its surface area is around r^{d-1} and its "height" is around r^2, so the volume is around r^{d+1}. For C_i we have $r^{d+1} \approx 1/2N$, which implies that r is around $N^{-1/d+1}$, so C_1 has surface area around $N^{-(d-1)/(d+1)}$, and the disjointness of the caps finishes the proof of (22.18).

Given a sequence $\varepsilon_1 \in \{\pm 1\}, \varepsilon_2 \in \{\pm 1\}, \ldots, \varepsilon_m \in \{\pm 1\}$, we define a convex subset of B by removing the solid spherical caps C_i with $\varepsilon_i = -1$:

$$B(\varepsilon_1, \varepsilon_2, \ldots, \varepsilon_m) = B \setminus \bigcup_{1 \leq i \leq m:\ \varepsilon_i = -1} C_i.$$

We clearly have

$$\begin{aligned} &(|\mathcal{X} \cap B(\varepsilon_1, \varepsilon_2, \ldots, \varepsilon_m)| - \mathrm{vol}_d(B(\varepsilon_1, \varepsilon_2, \ldots, \varepsilon_m))N) \\ &\quad - (|\mathcal{X} \cap B(-\varepsilon_1, -\varepsilon_2, \ldots, -\varepsilon_m)| - \mathrm{vol}_d(B(-\varepsilon_1, -\varepsilon_2, \ldots, -\varepsilon_m))N) \\ &= \sum_{i=1}^{m} \varepsilon_i \left(|\mathcal{X} \cap C_i| - \mathrm{vol}_d(C_i)N\right). \end{aligned} \tag{22.19}$$

But

$$|\mathcal{X} \cap C_i| - \mathrm{vol}_d(C_i)N = |\mathcal{X} \cap C_i| - \frac{1}{2},$$

which is either $-1/2$ or $\geq 1/2$. Thus we can choose $\varepsilon_i \in \{\pm 1\}$ such that

$$\varepsilon_i \left(|\mathcal{X} \cap C_i| - \mathrm{vol}_d(C_i)N\right) \geq \frac{1}{2}$$

for every $1 \leq i \leq m$, and using it in (22.19), we have

$$\begin{aligned} &(|\mathcal{X} \cap B(\varepsilon_1, \varepsilon_2, \ldots, \varepsilon_m)| - \mathrm{vol}_d(B(\varepsilon_1, \varepsilon_2, \ldots, \varepsilon_m))N) \\ &\quad - (|\mathcal{X} \cap B(-\varepsilon_1, -\varepsilon_2, \ldots, -\varepsilon_m)| - \mathrm{vol}_d(B(-\varepsilon_1, -\varepsilon_2, \ldots, -\varepsilon_m))N) \geq \frac{m}{2}. \end{aligned}$$

And so either $A = B(\varepsilon_1, \varepsilon_2, \ldots, \varepsilon_m)$ or $A = B(-\varepsilon_1, -\varepsilon_2, \ldots, -\varepsilon_m)$ has the property

$$||\mathcal{X} \cap A| - \mathrm{vol}_d(A)N| \geq \frac{m}{4} \geq \frac{c_1(d)N^{(d-1)/(d+1)}}{4},$$

where in the last step we used (22.18). This completes the proof of Schmidt's theorem. □

We conclude Sec. 22 with the following long-term problem.

What can we say about the alternation of equilibrium and off-equilibrium states? To answer this question, we consider the simplest 3-dimensional Gaussian torus model ($\mathbf{r}_1(t) = \cdots = \mathbf{r}_N(t) = (t, 0, 0)$), and let $\varepsilon > 0$ be a fixed small positive constant (e.g. $\varepsilon = 1/10$ or $1/100$). We modify the concept of square-root equilibrium (in the particle space) by replacing $N^{1/2}$ with $N^{(1+\varepsilon)/2}$. We say that at time t the system is in $(1+\varepsilon)/2$-*power box-equilibrium* if for every rotated box in the unit cube, the number of particles in the box is very close to the expected value (= N times the volume of the box) in the quantitative sense that the difference has absolute value less than $N^{(1+\varepsilon)/2}$. (That is, we replace square-root with $(1+\varepsilon)/2$-power, and restrict the test sets to rotated boxes.) Since rotated boxes are convex sets, the results of Secs. 20–22 imply the following. Starting from an arbitrary initial configuration $\mathcal{Y} = \{\mathbf{y}_1, \mathbf{y}_2, \ldots, \mathbf{y}_N\}$, the typical individual time evolution of the system reaches the state of $(1+\varepsilon)/2$-power box-equilibrium extremely fast in time less than $\sqrt{\log N}$, and the system stays in the state of $(1+\varepsilon)/2$-power box-equilibrium for an exponentially long time in the range of

$$e^{N^\varepsilon/2}. \tag{22.20}$$

In view of Theorem 3.1, the individual time evolution of the system eventually leaves the state of $(1+\varepsilon)/2$-power box-equilibrium, but of course later it goes back to the "majority state" of $(1+\varepsilon)/2$-power box-equilibrium again, and we have this infinite alternation of equilibrium and off-equilibrium states. In a sufficiently long time interval $0 \le t \le T$, the total length of the times t for which the system is *not* in $(1+\varepsilon)/2$-power box-equilibrium is an exponentially small part of T in the range of

$$e^{-N^\varepsilon/2}T. \tag{22.21}$$

Here is a nutshell description of a typical time evolution: if the system is in off-equilibrium, then it goes back very fast to $(1+\varepsilon)/2$-power box-equilibrium, and the system stays in this state for an exponentially long time in the range of (22.20); later in view of Theorem 3.1, the system eventually leaves the state of $(1+\varepsilon)/2$-power box-equilibrium, then it goes back again from off-equilibrium to $(1+\varepsilon)/2$-power box-equilibrium, and the system stays in this state for an exponentially long time in the range of (22.20), and so on. A typical time evolution has infinitely many alternations of long time intervals of $(1+\varepsilon)/2$-power box-equilibrium separated by short off-equilibrium intervals. The equilibrium blocks are exponentially long in the range of (22.20), and the off-equilibrium blocks are much shorter. For example, starting from off-equilibrium, the typical time evolution goes back to equilibrium extremely fast in time less than $\sqrt{\log N}$. We can prove, however, that if we wait long enough, there will be *relatively long* off-equilibrium blocks, much longer than $\sqrt{\log N}$. This follows from a combination of Theorem 3.1 ("superdiscrepancy") and the following result. For the details of the deduction, see the Remarks below.

Theorem 22.1. *Consider the simplest 3-dimensional Gaussian torus model kind* $(\mathcal{Y}; \Omega_{\text{Gauss}})$ *where* $\mathbf{r}_1(t) = \cdots = \mathbf{r}_N(t) = (t, 0, 0)$ *(i.e., each particle moves on a straight line with constant speed; different particles have different speed). Let* $1 > \varepsilon > 0$ *be arbitrary but fixed. Then there is a subset* $\Omega^*_{\text{Gauss}} \subset \Omega_{\text{Gauss}}$ *such that* $\text{ProdMeas}_{\text{Gauss}}(\Omega^*_1) > 9/10$, *and for every*

$$\omega = (\rho_1, \vartheta_1, \rho_2, \vartheta_2, \ldots, \rho_N, \vartheta_N) \in \Omega^*_1$$

there is an initial configuration $\mathcal{Z} = \mathcal{Z}(\varepsilon; \omega) = \{\mathbf{z}_1, \ldots, \mathbf{z}_N\}$ *such that, starting from* $(\mathcal{Z}, \omega)$ *at* $t = 0$, *in the interval*

$$0 \le t \le 2^{-5} N^{(1-\varepsilon)/2} \tag{22.22}$$

the time evolution of the system does not reach any $(1+\varepsilon)/2$*-power box-equilibrium state, assuming* N *is large enough in terms of* ε. *What is more, if* t *is in the interval* (35.18), *then the number of particles in the left half* $[0, 1/2] \times [0, 1] \times [0, 1]$ *of the unit cube is more than* $N/2 + N^{(1+\varepsilon)/2}$ *(assuming* N *is large enough in terms of* ε*).*

Remarks. First note that there is an analog result for the closed orbit models introduced in Secs. 6–7 (the proof is the same).

The existence of *relatively long* off-equilibrium blocks (in the sense of being much longer than $\sqrt{\log N}$) can be deduced from Theorem 3.1 and Theorem 22.1 as follows. We can clearly assume that for $\omega \in \Omega^*_1$ the $3N$ initial velocity coordinates are linearly independent over the rationals (since the complement set has zero ProdMeas_1-measure). Therefore, starting from $(\mathcal{Y}, \omega)$, if we wait long enough, the Kronecker–Weyl equidistribution theorem implies that the system will get arbitrarily close to $(\mathcal{Z}, \omega)$ with $\mathcal{Z} = \mathcal{Z}(\varepsilon; \omega) = \{\mathbf{z}_1, \ldots, \mathbf{z}_N\}$. More precisely, given any $\delta > 0$, it will happen that the image of $\mathbf{y}_i$ gets δ-close to $\mathbf{z}_i$, $1 \le i \le N$. Since $\delta > 0$ can be arbitrarily small, Theorem 22.1 applies, and guarantees the existence of an off-equilibrium block much longer than $\sqrt{\log N}$ (= length of the initial off-equilibrium block); namely, the length is polynomial in N, see (22.22). In fact, the maximum length can be close to $\sqrt{N}$ if $\varepsilon > 0$ is small.

On the other hand, we mention (without proof) that for typical time evolution the maximum length of the off-equilibrium blocks is certainly less than N, independently of the initial configuration.

Next consider the $(1+\varepsilon)/2$-power box-equilibrium blocks. The average length is in the exponential range of (22.20). On the other hand, we can prove (but omit the proof) that, for typical time evolution the maximum length of the equilibrium blocks is less than some exponential function of N, independently of the initial configuration.

Proof of Theorem 22.1. Let

$$\begin{aligned}\Omega^*_{\text{Gauss}} = \{\omega \in \Omega_{\text{Gauss}} :\ &\text{there are } \geq 2N^{(1+\varepsilon)/2} \text{ initial velocities}\\ &\mathbf{v}_i(\omega) = (v_{i,1}, v_{i,2}, v_{i,3}) \text{ such that } |v_{i,1}| < 8N^{-(1-\varepsilon)/2}, \text{ and}\\ &|\mathbf{v}_j(\omega)| < N^{1/4},\ 1 \leq j \leq N\}.\end{aligned} \tag{22.23}$$

It is trivial from the definition of the Gaussian/Maxwellian velocity distribution that

$$\text{ProdMeas}_{\text{Gauss}}(\Omega^*_{\text{Gauss}}) > 9/10,$$

assuming N is large enough in terms of $\varepsilon > 0$.

Let $\omega \in \Omega^*_{\text{Gauss}}$ be arbitrary. By rearranging the indices, we can assume that

$$|v_{i,1}| < 8N^{-(1+\varepsilon)/2} \text{ holds for all } N_1 = N - 2N^{(1+\varepsilon)/2} < i \leq N. \tag{22.24}$$

For every $1 \leq j \leq N_1 = N - 2N^{(1+\varepsilon)/2}$, let ξ_j be a random variable uniformly distributed in the unit interval, and assume that ξ_j, $1 \leq j \leq N_1$ are independent. Since the torus is translation invariant, for notational convenience we can switch from the interval $[0, 1/2)$ to $[1/4, 3/4)$, i.e., we translate the left half of the unit cube to the right by $1/4$. The product form of the test set explains why it suffices to study the first coordinates of the particles.

Let $\eta = N^{-3/4}$, and write

$$I_1 = [1/4 + \eta, 3/4 - \eta] \subset I_2 = [1/4, 3/4).$$

For $1 \leq j \leq N_1$, let

$$\zeta_j(t) = 1 \text{ if } \{\xi_j + tv_{j,i}\} \in I_1, \text{ and zero otherwise,} \tag{22.25}$$

where $\{y\}$ stands for the fractional part of a real number y. Notice that for a fixed t, $\zeta_j(t)$, $1 \le j \le N_1$ are independent 0-1 valued Bernoulli random variables with $p = \text{length}(I_1) = 1/2 - 2\eta$. Using Bernstein's large deviation inequality,

$$\Pr\left[\sum_{j=1}^{N_1} \zeta_j(t) \le \left(\frac{1}{2} - 2\eta\right) N_1 - \lambda\sqrt{N_1}\right] \le e^{-\lambda^2/2} \tag{22.26}$$

holds for every fixed time instant t. Let

$$0 < T < \frac{e^{N^\varepsilon/8}}{N} \tag{22.27}$$

be fixed. We apply (22.26) with $\lambda = \frac{1}{2}N^{\varepsilon/2}$ for every t of the form $0 < t = k/N \le T$, k integer (which represents TN time points). By (22.27)

$$e^{-\lambda^2/2} = e^{-N^\varepsilon/8} < \frac{1}{TN},$$

so by (22.26), for every t of the form $0 < t = k/N \le T$, k integer, there exist real numbers $\xi_j(0) \in [0, 1)$, $1 \le j \le N_1$ such that

$$\left\{\xi_j(0) + \frac{k}{N} v_{j,i}\right\} \in I_1 = [1/4 + \eta, 3/4 - \eta]$$

holds for at least

$$\begin{aligned}\left(\frac{1}{2} - 2\eta\right) N_1 - \lambda\sqrt{N_1} &\ge \left(\frac{1}{2} - 2\eta\right)\left(N - 2N^{(1+\varepsilon)/2}\right) - \lambda\sqrt{N} \\ &\ge \frac{1}{2}N - 2\eta N - \frac{1}{2}N^{(1+\varepsilon)/2} > \frac{1}{2}N - N^{(1+\varepsilon)/2}\end{aligned} \tag{22.28}$$

indices j in the interval $1 \le j \le N_1$ (in the last step of (22.28) we used that N is large enough). Since no particle has speed $\ge N^{1/4}$, if a particle was in the box $B_1 = I_1 \times [0, 1) \times [0, 1)$ at time $t = k/N$, then this particle stays in the larger box $B_2 = I_2 \times [0, 1) \times [0, 1)$ in the whole time interval $k/N \le t < (k + 1)/N$.

Now we are ready to specify the desired point configuration $\mathcal{Z} = \{\mathbf{z}_1, \ldots, \mathbf{z}_N\}$ for the box B_2. Let $\xi_j(0)$ be the first coordinate of $\mathbf{z}_j$ for $1 \le j \le N_1$, and let $1/2$ be the first coordinate of $\mathbf{z}_j$ for $N_0 < j \le N$ (the second and third coordinates can be chosen freely).

Then it follows from (22.28) that the box $B_2 = [1/4, 3/4) \times [0,1) \times [0,1)$ contains more than

$$\frac{1}{2}N - N^{(1+\varepsilon)/2}$$

particles with index j in $1 \leq j \leq N_1$ during $0 \leq t \leq T$, assuming (22.27) holds.

On the other hand, by (22.24),

$$|v_{i,1}| < 8N^{-(1-\varepsilon)/2} \text{ holds for all } N_1 = N - 2N^{(1+\varepsilon)/2} < i \leq N. \tag{22.29}$$

Choosing

$$T_0 = N^{(1-\varepsilon)/2}/32, \tag{22.30}$$

and combining the fact

$$T_0 \cdot 8N^{-(1-\varepsilon)/2} = \frac{1}{4} = \text{radius of the interval } I_2 = [1/4, 3/4].$$

with (22.29), it follows that the box $B_2 = [1/4, 3/4) \times [0,1) \times [0,1)$ contains every particle with index j in $N_1 < j \leq N$ during $0 \leq t \leq T_0$ (see (22.30), and use the fact that $1/2$ is the first coordinate of the starting point $\mathbf{z}_j$ for $N_1 < j \leq N$).

Comparing (22.27) and (22.30), we see that T_0 satisfies (22.27) assuming N is large enough in terms of $\varepsilon > 0$.

Summarizing, the box $B_2 = [1/4, 3/4) \times [0,1) \times [0,1)$ contains more than

$$\frac{1}{2}N - N^{(1+\varepsilon)/2} + (N - N_1) = \frac{1}{2}N - N^{(1+\varepsilon)/2} + 2N^{(1+\varepsilon)/2} = \frac{1}{2}N + N^{(1+\varepsilon)/2}$$

particles during $0 \leq t \leq T_0$, which completes the proof of Theorem 22.1. □

23. On the Square-Root Logarithmic Threshold in the Gaussian Case

The majority of sections so far were all applications of Theorem 4.2. Theorem 4.2 has a somewhat peculiar "square-root logarithmic threshold" in the following sense. The key condition in (4.12)

$$e^{\pi^2 U^2/2} \geq 3dU$$

is equivalent to

$$U \geq \frac{\sqrt{\log d}}{\pi} + o(1). \tag{23.1}$$

The square-root-logarithmic (23.1) is the (shockingly small!) threshold for "conf-space equilibrium". We may say that it is the threshold when the typical time evolution of a system with $N = d/3$ particles and Gaussian initial velocity distribution reaches equilibrium in the configuration space.

We have the same threshold in Theorem 4.1 where the threshold is denoted by $T_0 = T_0(d)$. Theorems 4.1–4.2 are both *dimension-free, complexity-free and start-free* results. The only dependence on the dimension d is in the square-root logarithmic threshold — it represents an extremely weak dependence. For example, choosing $d = 10^{1000}$ in Theorem 4.1, we have

$$T_0 = T_0(d) = \frac{\sqrt{1 + \log 100 + 1000 \log 10}}{\pi} < 18.$$

Perhaps the reader is wondering: why do we need the strange threshold $T_0 = T_0(d)$ in Theorem 4.1? Or equivalently: why do we ignore the initial part

$$0 \le t \le T_0 = \frac{\sqrt{\log d}}{\pi} + o(1) \tag{23.2}$$

of the torus lines? Iss it really necessary? The answer is yes: we have to ignore the initial part, because we certainly cannot expect any kind of uniformity in the slightly shorter interval

$$0 \le t < \frac{\sqrt{\log d}}{2\pi} - o(1). \tag{23.3}$$

Notice that interval (23.2) is basically twice longer than (23.3), so they have the same order $\sqrt{\log d}$.

In this section we explain why we cannot expect any kind of uniformity in the interval (23.3). Consider the simplest 3-dimensional Gaussian torus model starting from the origin (Big Bang). The curve $\Gamma_{BB;\omega}(t)$ of the whole system in the configuration space I^d is a straight line in the d-space (with $d = 3N$) modulo one:

$$\vec{\Gamma}_{BB;\omega}(t) = t(\rho_1\mathbf{e}_1, \rho_2\mathbf{e}_2, \ldots, \rho_N\mathbf{e}_N) \text{ modulo one} = t\mathbf{v}(\omega) \text{ modulo one,} \tag{23.4}$$

where

$$\omega = (\rho_1, \mathbf{e}_1, \rho_2, \mathbf{e}_2, \ldots, \rho_N, \mathbf{e}_N) \in \Omega_{\text{Gauss}} = \left([0,\infty) \times \mathbf{S}^2\right)^N \tag{23.5}$$

and

$$\mathbf{v}(\omega) = (\rho_1\mathbf{e}_1, \rho_2\mathbf{e}_2, \ldots, \rho_N\mathbf{e}_N). \tag{23.6}$$

Note that BB in $\vec{\Gamma}_{BB;\omega}(t)$ refers to Big Bang, and the product space Ω_{Gauss} is equipped with the usual (Gaussian) product measure $\text{ProdMeas}_{\text{Gauss}}$.

Omitting the arrow in $\vec{\Gamma}_{BB;\omega}(t)$, we obtain the N-element *point set* in the particle space $I^3 = [0,1)^3$ (3-dimensional unit torus):

$$\Gamma_{BB;\omega}(t) = \{t\rho_1\mathbf{e}_1, t\rho_2\mathbf{e}_2, \ldots, t\rho_N\mathbf{e}_N\} \text{ modulo one.} \tag{23.7}$$

Consider the following very simple "nice" test function in the particle space

$$f_0 = f_0(x_1, x_2, x_3) = e^{2\pi \mathrm{i} x_1}. \tag{23.8}$$

We claim that, for every t in the time interval

$$0 \le t < T = \frac{\sqrt{\log N}}{2\pi} - o(1) = \frac{\sqrt{\log(d/3)}}{2\pi} - o(1) = \frac{\sqrt{\log d}}{2\pi} - o(1) \tag{23.9}$$

(which is the same as (23.3)), we have the following lower bound for the integral

$$\int_{I^3} \left| \sum_{\mathbf{z} \in \Gamma_{BB;\omega}(t)} f_0(\mathbf{z} + \mathbf{w}) \right|^2 d\mathbf{w} > 60N, \tag{23.10}$$

where (23.10) holds for the majority of $\omega \in \Omega_{\text{Gauss}}$. More precisely, we prove the following lemma.

Lemma 23.1. *For every $t \in [0, T]$, where T is defined in (23.8), there exists a (measurable) subset $\Omega^*(t) \subset \Omega_{\text{Gauss}}$ with (say)*

$$\text{ProdMeas}_{\text{Gauss}}(\Omega^*(t)) \ge 1 - 10^{-4} = 99.99\%$$

such that the integral

$$\mathcal{I}[\omega; f_0; t] = \int_{I^3} \left| \sum_{\mathbf{z} \in \Gamma_{BB;\omega}(t)} f_0(\mathbf{z} + \mathbf{w}) \right|^2 d\mathbf{w}$$

has the lower bound

$$\mathcal{I}[\omega; f_0; t] > 60N,$$

which holds for every $\omega \in \Omega^*(t)$.

Remark. The choices of the constants 60 and 99.99% are accidental. With an appropriate choice of "$-o(1)$" in (23.8) if N is large, we could upgrade

99.99% to 99.9999%, and so on. As the proof below shows, this is just a matter of routine calculation.

Proof of Lemma 23.1. First we evaluate the integral $\mathcal{I}[\omega; f_0; t]$; see (23.8) and (23.10). By using the well-known fact

$$\int_{I^3} e^{-2\pi i \mathbf{n}\cdot\mathbf{w}}\, d\mathbf{w} = 0$$

for all $\mathbf{n} \in \mathbb{Z}^3$ except the trivial case $\mathbf{n} = \mathbf{0}$, we have with $\mathbf{n}_1 = (1, 0, 0)$,

$$\mathcal{I}[\omega; f_0; t] = N + \sum_{1\le k,\ell\le N: k\ne\ell} e^{2\pi i t \mathbf{n}_1\cdot(\rho_k \mathbf{e}_k - \rho_\ell \mathbf{e}_\ell)}. \tag{23.11}$$

Next we integrate (23.11) over the big initial velocity space $\Omega_{\text{Gauss}} = ([0,\infty)\times \mathbf{S}^2)^N$:

$$\begin{aligned}
\mathcal{I}[\Omega_{\text{Gauss}}; f_0; t] &= \sqrt{\frac{2}{\pi}}\int_{\rho_1=0}^{\infty}\int_{\mathbf{e}_1\in\mathbf{S}^2}\cdots\sqrt{\frac{2}{\pi}}\int_{\rho_N=0}^{\infty}\int_{\mathbf{e}_N\in\mathbf{S}^2} \mathcal{I}[\omega; f_0; t]\rho_1^2 \\
&\quad\cdot e^{-\rho_1^2/2}\, d\rho_1\, d\mathrm{SA}^*(\mathbf{e}_1)\cdots\rho_N^2 e^{-\rho_N^2/2}\, d\rho_N\, d\mathrm{SA}^*(\mathbf{e}_N) \\
&= N + \sum_{1\le k,\ell\le N: k\ne\ell}\sqrt{\frac{2}{\pi}}\int_{\rho_1=0}^{\infty}\int_{\mathbf{e}_1\in\mathbf{S}^2}\cdots \\
&\quad\times\sqrt{\frac{2}{\pi}}\int_{\rho_N=0}^{\infty}\int_{\mathbf{e}_N\in\mathbf{S}^2} e^{2\pi i t\mathbf{n}_1\cdot(\rho_k\mathbf{e}_k-\rho_\ell\mathbf{e}_\ell)}\rho_1^2 \\
&\quad\cdot e^{-\rho_1^2/2}\, d\rho_1\, d\mathrm{SA}^*(\mathbf{e}_1)\cdots\rho_N^2 e^{-\rho_N^2/2}\, d\rho_N\, d\mathrm{SA}^*(\mathbf{e}_N) \\
&= N + \sum_{1\le k,\ell\le N: k\ne\ell}\sqrt{\frac{2}{\pi}}\int_0^{\infty}\int_{\mathbf{e}_k\in\mathbf{S}^2} e^{2\pi i t\rho\mathbf{e}_k\cdot\mathbf{n}_1}\rho^2 e^{-\rho^2/2} \\
&\quad\times d\rho\, d\mathrm{SA}^*(\mathbf{e}_k)\cdot\int_{\mathbf{e}_\ell\in\mathbf{S}^2} e^{2\pi i t\rho\mathbf{e}_\ell\cdot\mathbf{n}_1}\rho^2 e^{-\rho^2/2}\, d\rho\, d\mathrm{SA}^*(\mathbf{e}_\ell).
\end{aligned} \tag{23.12}$$

Applying Lemma 9.1 with $d = 3$ in (23.12), we obtain

$$\begin{aligned}
&\mathcal{I}[\Omega_{\text{Gauss}}; f_0; t] \\
&= N + \sum_{1\le k,\ell\le N: k\ne\ell}\exp\left(-(2\pi t|\mathbf{e}_k|\,|\mathbf{n}_1|)^2/2\right)\exp\left(-(2\pi|\mathbf{e}_\ell|\,|\mathbf{n}_1|)^2/2\right) \\
&= N + N(N-1)\exp\left(-(2\pi t)^2\right).
\end{aligned} \tag{23.13}$$

The next step in the proof of the lemma is to estimate the "variance"

$$\begin{aligned}\text{Variance} &= \int_{\Omega_{\text{Gauss}}} \left(\mathcal{I}[\omega; f_0; t] - \mathcal{I}[\Omega_{\text{Gauss}}; f_0; t]\right)^2 \, d\omega \\ &= \int_{\Omega_{\text{Gauss}}} \left(\mathcal{I}[\omega; f_0; t]\right)^2 \, d\omega - \left(\mathcal{I}[\Omega_{\text{Gauss}}; f_0; t]\right)^2, \quad (23.14)\end{aligned}$$

where for simplicity we write $d\omega = d\text{ProdMeas}_{\text{Gauss}}(\omega)$.

By using (23.11) we have

$$\begin{aligned}\left(\mathcal{I}[\omega; f_0; t]\right)^2 &= \left| N + \sum_{1 \le k, \ell \le N : k \ne \ell} e^{2\pi i t \mathbf{n}_1 \cdot (\rho_k \mathbf{e}_k - \rho_\ell \mathbf{e}_\ell)} \right|^2 \\ &= N^2 + 2N \sum_{1 \le k, \ell \le N : k \ne \ell} e^{2\pi i t \mathbf{n}_1 \cdot (\rho_k \mathbf{e}_k - \rho_\ell \mathbf{e}_\ell)} \\ &\quad + \sum_{1 \le k_1, \ell_1 \le N : k_1 \ne \ell_1} \sum_{1 \le k_2, \ell_2 \le N : k_2 \ne \ell_2} \\ &\quad \cdot e^{2\pi i t \mathbf{n}_1 \cdot \left(\rho_{k_1} \mathbf{e}_{k_1} - \rho_{\ell_1} \mathbf{e}_{\ell_1} - \rho_{k_2} \mathbf{e}_{k_2} + \rho_{\ell_2} \mathbf{e}_{\ell_2}\right)}. \quad (23.15)\end{aligned}$$

Integrating (23.15) over Ω_{Gauss}, and using again Lemma 9.1 with $d = 3$, we have

$$\begin{aligned}\int_{\Omega_{\text{Gauss}}} \left(\mathcal{I}[\omega; f_0; t]\right)^2 \, d\omega &= N^2 + 2N \cdot N(N-1) \exp\left(-(2\pi t)^2\right) \\ &\quad + \mathcal{S}_1 + \mathcal{S}_2 + 2\mathcal{S}_3 + 2\mathcal{S}_4 + \mathcal{S}_5, \quad (23.16)\end{aligned}$$

where

$$\mathcal{S}_1 = N(N-1), \quad (23.17)$$

which corresponds to the case $(k_1, \ell_1) = (k_2, \ell_2)$ in (23.15),

$$\mathcal{S}_2 = N(N-1) \exp\left(-(2\pi 4t)^2\right), \quad (23.18)$$

which corresponds to the case $(k_1, \ell_1) = (\ell_2, k_2)$ in (23.15) (where the factor four comes from the fact that $4\mathbf{n}_1 = (4, 0, 0)$),

$$\mathcal{S}_3 = N(N-1)(N-2) \exp\left(-(2\pi t)^2\right), \quad (23.19)$$

which corresponds to the case $k_1 = k_2$, $\ell_1 \ne \ell_2$ in (23.15) (notice that $\mathcal{S}_3$ has coefficient 2 in (23.16) due to the symmetric case $k_1 \ne k_2$, $\ell_1 = \ell_2$ that has exactly the same contribution),

$$\mathcal{S}_4 = N(N-1)(N-2)\exp\left(-(2\pi 2t)^2\right), \tag{23.20}$$

which corresponds to the case $k_1 = \ell_2$, $\ell_1 \neq k_2$ in (23.15) (again $\mathcal{S}_4$ has coefficient 2 in (23.16) due to the symmetric case $k_2 = \ell_1$, $\ell_2 \neq k_1$ that has exactly the same contribution), and finally,

$$\mathcal{S}_5 = N(N-1)(N-2)(N-3)\exp\left(-2(2\pi t)^2\right), \tag{23.21}$$

which corresponds to the case where k_1, ℓ_1, k_2, ℓ_2 are all different.

On the other hand, (by (23.13)

$$\begin{aligned}(\mathcal{I}[\Omega_{\text{Gauss}}; f_0; t])^2 &= \left(N + N(N-1)\exp\left(-(2\pi t)^2\right)\right)^2 \\ &= N^2 + 2N \cdot N(N-1)\exp\left(-(2\pi t)^2\right) \\ &\quad + N^2(N-1)^2\exp\left(-2(2\pi t)^2\right). \end{aligned} \tag{23.22}$$

Combining (23.14)–(23.22), we have

$$\begin{aligned}\text{Variance} &= \int_{\Omega_{\text{Gauss}}} (\mathcal{I}[\omega; f_0; t])^2 \, d\omega - (\mathcal{I}[\Omega_{\text{Gauss}}; f_0; t])^2 \\ &= \mathcal{S}_1 + \mathcal{S}_2 + 2\mathcal{S}_3 + 2\mathcal{S}_4 + \mathcal{S}_5 - N^2(N-1)^2\exp\left(-2(2\pi t)^2\right) \\ &= N(N-1) + N(N-1)\exp\left(-(2\pi 4t)^2\right) \\ &\quad + 2N(N-1)(N-2)\exp\left(-(2\pi t)^2\right) \\ &\quad +2\, N(N-1)(N-2)\exp\left(-(2\pi 2t)^2\right) \\ &\quad + (N(N-1)(N-2)(N-3) - N^2(N-1)^2)\exp(-2(2\pi t)^2) \\ &\le N(N-1) + N(N-1)(\exp\left(-(2\pi 4t)^2\right) \\ &\quad + 2N(N-1)(N-2)\exp\left(-(2\pi t)^2\right) \\ &\quad + 2N(N-1)(N-2)\exp\left(-(2\pi 2t)^2\right) \\ &\le N^2 + 4N^3\exp(-(2\pi t)^2) \le N^2(2N^{1/2}\exp(-2\pi^2 t^2) + 1)^2. \end{aligned} \tag{23.23}$$

Let

$$\Omega_1(t) = \Big\{\omega \in \Omega_{\text{Gauss}} : \ |\mathcal{I}[\omega; f_0; t] - \mathcal{I}[\Omega_{\text{Gauss}}; f_0; t]| \\ \leq 10^2 N \left(2N^{1/2} e^{-2\pi^2 t^2} + 1\right)\Big\}. \tag{23.24}$$

By (23.23), (23.24) and Chebyshev's inequality, we obtain

$$\text{ProdMeas}_{\text{Gauss}}(\Omega_1(t)) \geq 1 - 10^{-4}. \tag{23.25}$$

Let $t = T > 0$ be the solution of the equation

$$N e^{-4\pi^2 t^2} = 10^6, \tag{23.26}$$

that is, with $d = 3N$,

$$T = \frac{\sqrt{\log N - 6 \log 10}}{2\pi} = \frac{\sqrt{\log d - \log 3 - 6 \log 10}}{2\pi}.$$

Clearly

$$T = \frac{\sqrt{\log d - \log 3 - 6 \log 10}}{2\pi} > \frac{\sqrt{\log d}}{2\pi} - \frac{\sqrt{\log 3 + 6 \log 10}}{2\pi \sqrt{\log d}},$$

so, we can write

$$T = \frac{\sqrt{\log d}}{2\pi} - o(1),$$

which justifies the choice of T in (23.9).

By (23.13),

$$\left|\mathcal{I}[\Omega_{\text{Gauss}}; f_0; t] - N^2 e^{-4\pi^2 t^2}\right| \leq N. \tag{23.27}$$

Combining (23.24) and (23.27), for $0 \leq t \leq T$ and $\omega \in \Omega_1(t)$ we have

$$\left|\mathcal{I}[\omega; f_0; t] - N^2 e^{-4\pi^2 t^2}\right| \leq N + 10^2 N \left(2N^{1/2} e^{-2\pi^2 t^2} + 1\right),$$

which implies that

$$\mathcal{I}[\omega; f_0; t] \geq N \left(N e^{-4\pi^2 t^2} - 200 \left(N e^{-4\pi^2 t^2}\right)^{1/2} - 101\right). \tag{23.28}$$

Clearly

$$x - 200\sqrt{x} - 101 > x/2 \text{ for all } x \geq 10^6,$$

and combining this fact with (23.28), we obtain that for every $0 \le t \le T$ and every $\omega \in \Omega_1(t)$

$$\mathcal{I}[\omega; f_0; t] \ge N \cdot \frac{1}{2} N e^{-4\pi^2 t^2} \ge \frac{10^6}{2} N, \tag{23.29}$$

where in the last step we used (23.26). Since $10^6/2 > 60$, by choosing $\Omega^*(t) = \Omega_1(t)$, Lemma 23.1 follows from (23.25) and (23.29). □

Next write

$$\Omega^{**} = \{\omega \in \Omega_{\text{Gauss}} : \ \omega \in \Omega^*(t) \text{ holds for at least } 99\% \text{ of } t \in [0, T]\}, \tag{23.30}$$

where $\Omega^*(t)$ is defined in Lemma 23.1.

Using Lemma 23.1 we can easily derive the lower bound

$$\text{ProdMeas}_{\text{Gauss}}(\Omega^{**}) \ge 1 - 10^{-2}. \tag{23.31}$$

Indeed, let

$$f(\omega, t) = \begin{cases} 1 & \text{if } \omega \notin \Omega^*(t) \\ 0 & \text{if } \omega \in \Omega^*(t). \end{cases}$$

Then by Lemma 23.1, (23.30) and Fubini's theorem ("double-counting argument")

$$\begin{aligned} 10^{-4} T &\ge \int_{t=0}^{T} \int_{\omega \in \Omega_{\text{Gauss}}} f(\omega, t) = \int_{\omega \in \Omega_{\text{Gauss}}} \int_{t=0}^{T} f(\omega, t) \\ &\ge \int_{\omega \in \Omega_{\text{Gauss}} \setminus \Omega^{**}} \int_{t=0}^{T} f(\omega, t) \\ &\ge \text{ProdMeas}_{\text{Gauss}}(\Omega_{\text{Gauss}} \setminus \Omega^{**}) \cdot 10^{-2} T, \end{aligned}$$

which implies (23.31).

Given an N-element set in the 3-dimensional unit cube

$$\mathcal{Z} = \{\mathbf{z}_1, \ldots, \mathbf{z}_N\} \subset [0, 1)^3,$$

let $(d = 3N)$

$$\vec{\mathcal{Z}} = (\mathbf{z}_1, \ldots, \mathbf{z}_N) \in [0, 1)^d$$

denote the corresponding point in the configuration space.

Let $S = S(f_0; 60) \subset [0,1)^d$, $d = 3N$ be the set of those points

$$\vec{\mathcal{Z}} = (\mathbf{z}_1, \ldots, \mathbf{z}_N) \in [0,1)^d$$

in the configuration space which have average square-discrepancy $> 60N$ with respect to the test function f_0 defined in (23.8) (the motivation for the factor 60 comes from Lemma 20.1). Formally,

$$S = S(f_0; 60) = \left\{ \vec{\mathcal{Z}} \in [0,1)^d : \int_{I^3} \left| \sum_{\mathbf{z} \in \mathcal{Z}} f_0(\mathbf{z} + \mathbf{w}) \right|^2 d\mathbf{w} > 60N \right\}. \tag{23.32}$$

We claim that

$$\mathrm{vol}_d(S(f_0; 60)) < 10^{-2}, \tag{23.33}$$

where vol_d stands for the d-dimensional Lebesgue measure.

To prove (23.33), we introduce the following auxiliary subsets of the configuration space: for every $k \geq 1$ let

$$S_k = \left\{ \vec{\mathcal{Z}} \in [0,1)^d : \left| \sum_{\mathbf{z} \in \mathcal{Z}} f_0(\mathbf{z} + \mathbf{w}) \right|^2 > 8(k+3)N \right.$$
$$\left. \text{holds for at least } 2^{-k} \text{part of } \mathbf{w} \in [0,1)^3 \right\}. \tag{23.34}$$

Notice that the union of the sets S_k covers $S(f_0; 60)$:

$$S(f_0; 60) \subset \bigcup_{k=1}^{\infty} S_k. \tag{23.35}$$

Indeed, for

$$\vec{\mathcal{Z}} \in [0,1)^d \setminus \bigcup_{k=1}^{\infty} S_k$$

we have

$$\int_{I^3} \left| \sum_{\mathbf{z} \in \mathcal{Z}} f_0(\mathbf{z} + \mathbf{w}) \right|^2 d\mathbf{w}$$
$$\leq \frac{1}{2} 32N + \frac{1}{4} 40N + \frac{1}{8} 48N + \frac{1}{16} 56N + \cdots < 60N,$$

implying that $\vec{\mathcal{Z}} \notin S(f_0; 60)$.

On the other hand, it is easy to estimate the measure of the set S_k from above. Indeed, for a *fixed* translation $\mathbf{w}$ we have the upper bound

$$\mathrm{vol}_d\left(\left\{\vec{\mathcal{Z}} \in [0,1)^d : \left|\sum_{\mathbf{z}\in\mathcal{Z}} f_0(\mathbf{z}+\mathbf{w})\right|^2 > 8(k+3)N\right\}\right)$$
$$\leq 4e^{-4(k+3)/2}. \tag{23.36}$$

We show that (23.36) follows from Bernstein's large deviation type inequality in probability theory. Indeed, since vol_d is a product measure, the sum in (23.36) (see also (23.8))

$$\sum_{\mathbf{z}\in\mathcal{Z}} f_0(\mathbf{z}+\mathbf{w}) = \sum_{\mathbf{z}\in\mathcal{Z}} f_1(\mathbf{z}+\mathbf{w}) + \mathrm{i}\sum_{\mathbf{z}\in\mathcal{Z}} f_2(\mathbf{z}+\mathbf{w}) \tag{23.37}$$

with

$$f_1 = f_1(x_1,x_2,x_3) = \cos(2\pi x_1) \text{ and } f_2 = f_2(x_1,x_2,x_3) = \sin(2\pi x_1) \tag{23.38}$$

represents two sums of independent random variables, where f_1 and f_2 are real valued.

To prove (23.36), we apply the general form of Bernstein's inequality (see the upper bound before (11.8)) for f_j, $j = 1,2$ with $n = N$, $M = 1$, $\tau = 2\sqrt{(k+3)N}$, $|f_1| \leq 1$ and $|f_2| \leq 1$, and obtain the upper bound

$$\mathrm{vol}_d\left(\left\{\vec{\mathcal{Z}} \in [0,1)^d : \left(\sum_{\mathbf{z}\in\mathcal{Z}} f_j(\mathbf{z}+\mathbf{w})\right)^2 > 4(k+3)N\right\}\right)$$
$$\leq 2\exp\left(-\frac{\tau^2/2}{(\sum_{i=1}^n \mathbf{E}X_i^2) + (\tau M/3)}\right)$$
$$= 2\exp\left(-\frac{4(k+3)N/2}{(N/2)+(\tau/3)}\right) \leq 2\exp\left(-\frac{4(k+3)N/2}{(N/2)+(N/3)}\right)$$
$$\leq 2e^{-4(k+3)/2}. \tag{23.39}$$

In the last step we used $\tau \leq N$, since otherwise $4(k+2)N > N^2$, implying that the volume in (23.36) is trivially zero.

Since (see (23.37)–(23.38))

$$\left|\sum_{\mathbf{z}\in\mathcal{Z}} f_0(\mathbf{z}+\mathbf{w})\right|^2 = \left(\sum_{\mathbf{z}\in\mathcal{Z}} f_1(\mathbf{z}+\mathbf{w})\right)^2 + \left(\sum_{\mathbf{z}\in\mathcal{Z}} f_2(\mathbf{z}+\mathbf{w})\right)^2,$$

(23.39) implies (23.36).

Applying (23.36) in (23.34), we have

$$\mathrm{vol}_d(S_k) \le 2^k \cdot 4e^{-4(k+3)/2} = 2^{k+2}e^{-2(k+3)}. \tag{23.40}$$

Combining (23.35) and (23.40),

$$\begin{aligned}\mathrm{vol}_d(S(f_0;60)) &\le \sum_{k=1}^{\infty} \mathrm{vol}_d(S_k) \le \sum_{k=1}^{\infty} 2^{k+2}e^{-2(k+3)} \\ &= \frac{8}{e^8} + \frac{16}{e^{10}} + \frac{32}{e^{12}} + \cdots < 10^{-2},\end{aligned}$$

which proves (23.33).

Summarizing, by (23.31) the majority of the torus lines in the configuration space $[0,1)^d$ ($d = 3N$) starting from the origin — representing the time evolution of the Gaussian system (see (23.7) and (23.30)) $\Gamma_{BB;\omega}(t)$, $\omega \in \Omega^{**}$ starting from Big Bang — spend at least 99% of the time $0 \le t \le T$ (see (23.30)) in the small subset $S(f_0;60)$, which is less than 1% (see (23.33)) of the configuration space. The huge discrepancy between the 99% actual time and the 1% expected time contradicts uniformity in the initial time interval (23.3). This proves what we claimed at the beginning of the section: in the Gaussian case the square-root logarithmic threshold is unavoidable.

In the next section we show, among others, that, by replacing the Gaussian initial velocity distribution with some other velocity distributions, the already shockingly small square-root logarithmic threshold can be upgraded to an even better absolute constant (independent of the number of particles, or equivalently, independent of the dimension of the configuration space). This means that, independently of the initial point configuration, the typical time evolution of the system reaches conf-space equilibrium at time t_0, where t_0 is less than an absolute constant. We may say, therefore, that the (typical) time evolution of such a system exhibits *instant conf-space equilibrium.*

24. Beyond the Applications of Theorem 4.2

The message of Theorem 4.2 is that the typical time evolution of an off-equilibrium Bernoulli gas model with Gaussian initial velocity distribution reaches equilibrium in the configuration space — state of Configuration Equilibrium — very quickly: superexponentially fast. We briefly recall some of the consequences of Theorem 4.2 that we discussed so far.

In Sec. 5 we studied the classical Bernoulli gas model, where the gas molecules are represented by point billiards, and assumed Gaussian initial velocities (via *unfolding* this Bernoulli billiards-in-a-box model can be reduced to the simplest Gaussian torus-billiard model). We proved that, starting from an arbitrary initial configuration, for the typical time evolution the relaxation distance to reach square-root equilibrium in the particle space — i.e., the necessary number of "jumps" per particle in the torus model, which is half of the number of bounces in the billiard model — is about $\sqrt{\log N}$, that is, square-root logarithmic in terms of the number of particles. Moreover, the system stays in square-root equilibrium (meaning an arbitrary but fixed measurable test set in the unit cube) for a very long time; say, in the range of $N^{\log N}$, with the possible exception of a totally negligible set of times t.

In Secs. 6 and 7 we proved similar long-term stability results for two spherical and two polar models (closed orbit systems). Again the relaxation distance to reach square-root equilibrium was the same surprisingly small $\sqrt{\log N}$.

In Secs. 8 and 15 we proved that in the simplest 3-dimensional Gaussian torus model the relaxation distance to reach snapshot randomness (Poisson and CLT) is the same $\sqrt{\log N}$, and again we have long-term stability. Section 16 is an illustration of how to extend these results to closed orbit systems.

In Secs. 20–22 we proved that in the simplest (at least 2-dimensional) Gaussian torus model the relaxation distance to reach simultaneous square-root equilibrium relative to nice sets is the usual $\sqrt{\log N}$, and again we have long-term stability.

The common feature of these results (all based on Theorem 4.2) is the square-root logarithmic threshold for the relaxation distance in the Gaussian case, and the long-term stability of the equilibrium. The square-root logarithmic threshold represents a superexponentially fast approach to equilibrium, and as we proved in Sec. 24, in the Gaussian case the square-root logarithmic threshold is best possible — it cannot be improved.

Superexponentially fast approach to equilibrium is already very impressive, but as it turns out, we can do even better. Replacing Theorem 4.2 with Theorem 13.4 in the proofs of the results listed above, we can reduce the threshold $\sqrt{\log N}$ to an absolute constant independent of N.

Instant conf-space equilibrium. To give an illustration, we derive an analog of Theorem 5.1 by replacing Theorem 4.2 with Theorem 13.4 (one can obtain exactly the same way the analogs of the rest of the corollaries of Theorem 4.2 mentioned above). According to Sec. 13 (and in particular to Theorem 13.4) we consider now a more general torus model where the particles move on torus lines with (possibly) varying speeds, and every particle has its own (almost arbitrary) smooth initial speed distribution. The curve of the whole system in the configuration space is denoted by

$$\begin{aligned}&\vec{\mathcal{Y}}(\mathcal{R};\rho_1,\mathbf{e}_1,\ldots,\rho_N,\mathbf{e}_N;t)\\&\quad=\vec{\mathcal{Y}}+\vec{\Gamma}(\mathcal{R};\rho_1,\mathbf{e}_1,\ldots,\rho_N,\mathbf{e}_N;t)\text{ modulo one}\\&\quad=(\mathbf{y}_1+\rho_1r_1(t)\mathbf{e}_1,\mathbf{y}_2+\rho_2r_2(t)\mathbf{e}_2,\ldots,\mathbf{y}_N+\rho_Nr_N(t)\mathbf{e}_N)\text{ modulo one,}\end{aligned} \tag{24.1}$$

where t is the time,

$$\vec{\mathcal{Y}}=(\mathbf{y}_1,\mathbf{y}_2,\ldots,\mathbf{y}_N) \tag{24.2}$$

is the vector (= point in the configuration space) formed from the set

$$\mathcal{Y}=\{\mathbf{y}_1,\mathbf{y}_2,\ldots,\mathbf{y}_N\}$$

of the initial point configuration at the start $t=0$,

$$\mathcal{R}=(r_1(t),r_2(t),\ldots,r_N(t)) \tag{24.3}$$

represents a family of functions such that

$$\mathcal{R}^{(\mathrm{inf})}(t)=\min_{1\le k\le N}\inf_{x\ge t}r_k(x)\ge t^{\alpha}\text{ for all }t\ge\alpha_0 \tag{24.4}$$

and

$$\mathcal{R}^{(\text{sup})}(t) = \max_{1\le k\le N} \sup_{x\le t} r_k(x) \le t^{\alpha_1} \text{ for all } t \ge \alpha_0, \tag{24.5}$$

where $\alpha_0 \ge 1$, $0 < \alpha \le \alpha_1$ are some absolute constants.

Moreover, let $\mathcal{I}$ denote an arbitrary finite interval of positive length. Assume that

$$\max_{1\le k\le N} \sup_{\mathcal{I}\subset\mathbb{R}} \frac{\text{length}\{0 \le t \le T:\ r_k(t) \in \mathcal{I}\}}{\text{length}(\mathcal{I})} \le T^{\beta} \tag{24.6}$$

uniformly for all $T \ge \beta_0$, where $\beta_0 \ge 1$ and $0 \le \beta < 1$ are absolute constants. (About the question of how to satisfy conditions (24.4)–(24.6); see the beginning of Sec. 13.)

The trajectory of the kth particle is $\mathbf{y}_k + \rho_k r_k(t)\mathbf{e}_k \in \mathbb{R}^3$, $1 \le k \le N$ modulo one, representing a (possibly) varying speed motion on a torus line in the direction $\mathbf{e}_k$. For the *set* of N particles at time t we use the notation

$$\begin{aligned}\mathcal{Y}(\mathcal{R};\rho_1,\mathbf{e}_1,\ldots,\rho_N,\mathbf{e}_N;t)\\ = \{\mathbf{y}_1 + \rho_1 r_1(t)\mathbf{e}_1,\ldots,\mathbf{y}_N + \rho_N r_N(t)\mathbf{e}_N\} \text{ modulo one}\end{aligned} \tag{24.7}$$

(as usual, the switch from the point set in the particle space to the point in the configuration space is indicated by an extra arrow $\vec{\mathcal{Y}}$ on top of $\mathcal{Y}$).

Let $d = 3N$, let $f \in L_2(I^d)$, $I^d = [0,1)^d$ is the d-dimensional unit torus (i.e., we extend f over the whole d-space $\mathbb{R}^d$ periodically), and, as usual, write

$$\sigma_0^2(f) = \int_{I^d} |f|^2\, d\mathrm{V} - \left|\int_{I^d} f\, d\mathrm{V}\right|^2 = \int_{I^d} \left|f(\mathbf{y}) - \int_{I^d} f\, d\mathrm{V}\right|^2 d\mathbf{y}.$$

Since the torus is translation invariant, it suffices to study the time discrepancy of the $\vec{\mathcal{Y}}$-independent part $\vec{\Gamma}(t)$ in (24.1):

$$D_f(\mathcal{R};\rho_1,\mathbf{e}_1,\ldots,\rho_N,\mathbf{e}_N;T_1,T_2) = \int_{T_1}^{T_2} f(\vec{\Gamma}(t))\, dt - (T_2 - T_1)\int_{I^d} f\, d\mathrm{V}, \tag{24.8}$$

where $\vec{\Gamma}(t) = \vec{\Gamma}(\mathcal{R};\rho_1,\mathbf{e}_1,\ldots,\rho_N,\mathbf{e}_N;t)$ and $0 < T_1 < T_2$.

We turn now to the family

$$\mathcal{G} = (g_1, g_2, \ldots, g_N)$$

of probability density functions, where the kth function g_k defines the initial speed distribution $\Pr[\rho_k \le u] = \int_0^u g_k(y)\,dy$ of the kth particle ($1 \le k \le N$). We assume that

$$\infty > g_k(u) \ge 0 \text{ for } 0 \le u < \infty, \int_0^\infty g_k(u)\,du = 1 \text{ and } \int_0^\infty u g_k(u)\,du < \infty,$$

where the last requirement means that the "expectation = mean-speed is finite". The corresponding initial velocity space is

$$\omega = (\rho_1, \mathbf{e}_1, \rho_2, \mathbf{e}_2, \ldots, \rho_N, \mathbf{e}_N) \in \Omega_{\mathcal{G}} = ([0,\infty) \times \mathbf{S}^2)^N, \tag{24.9}$$

equipped with the product measure $\text{ProdMeas}_{\mathcal{G}}$, where the kth factor $[0,\infty) \times \mathbf{S}^2$ in (24.9) has the normed surface area for the unit sphere $\mathbf{S}^2 \subset \mathbb{R}^3$ and has the probability density function g_k for $[0,\infty)$.

By using (24.9), we can rewrite the time discrepancy (24.8) as follows:

$$\begin{aligned} D_f(\mathcal{R};\omega;T_1,T_2) &= D_f(\mathcal{R};\rho_1,\mathbf{e}_1,\ldots,\rho_N,\mathbf{e}_N;T_1,T_2) \\ &= \int_{T_1}^{T_2} f(\vec{\Gamma}_\omega(t))\,dt - (T_2 - T_1)\int_{I^d} f\,d\mathrm{V}, \end{aligned} \tag{24.10}$$

where

$$\begin{aligned} \vec{\Gamma}_\omega(t) &= \vec{\Gamma}(\mathcal{R};\omega;t) = \vec{\Gamma}(\mathcal{R};\rho_1,\mathbf{e}_1,\ldots,\rho_N,\mathbf{e}_N;t), \\ \omega &= (\rho_1,\mathbf{e}_1,\ldots,\rho_N,\mathbf{e}_N) \in \Omega_{\mathcal{G}} \end{aligned}$$

and $0 < T_1 < T_2$.

We take the square of the absolute value of the time discrepancy

$$D_f(\mathcal{R};\omega;T_1,T_2) = D_f(\mathcal{R};\rho_1,\mathbf{e}_1,\ldots,\rho_N,\mathbf{e}_N;T_1,T_2)$$

and study the following average ($d = 3N$)

$$\begin{aligned} \Delta_f^2(\mathcal{G};\mathcal{R};T_1,T_2) &= \int_{\omega\in\Omega_{\mathcal{G}}} |D_f(\mathcal{R};\omega;T_1,T_2)|^2\,d\mathrm{ProdMeas}_{\mathcal{G}}(\omega) \\ &= \int_{\rho_1=0}^\infty \int_{\mathbf{e}_1\in\mathbf{S}^2} \cdots \int_{\rho_N=0}^\infty \int_{\mathbf{e}_N\in\mathbf{S}^2} \end{aligned}$$

$$\cdot |D_f(\mathcal{R}; \rho_1, \mathbf{e}_1, \ldots, \rho_N, \mathbf{e}_N; T_1, T_2)|^2$$
$$\cdot g_1(\rho_1)\, d\rho_1\, d\mathrm{SA}^\star(\mathbf{e}_1) \ldots g_N(\rho_N)\, d\rho_N\, d\mathrm{SA}^\star(\mathbf{e}_N), \tag{24.11}$$

where $d\mathrm{SA}^\star(\mathbf{e})$ means the normed surface area on the unit sphere $\mathbf{S}^2$, i.e., $\mathrm{SA}^\star(\mathbf{S}^2) = 1$.

We recall the definition of the so-called 3-dimensional "isotropic Fourier transform" of $g = g_k$, $1 \leq k \leq N$:

$$\begin{aligned}\mathbf{F}_g(\mathbf{w}) &= \int_{u=0}^{\infty} \frac{1}{4\pi} \int_{\mathbf{v}\in\mathbf{S}^2} e^{iu\mathbf{v}\cdot\mathbf{w}} g(u)\, du\, d\mathrm{SA}(\mathbf{v}) \\ &= \int_{u=0}^{\infty} \frac{1}{4\pi} \int_{\mathbf{v}\in\mathbf{S}^2} \cos(u\mathbf{v}\cdot\mathbf{w}) g(u)\, du\, d\mathrm{SA}(\mathbf{v}),\end{aligned} \tag{24.12}$$

where $\mathbf{w}$ is an arbitrary 3-dimensional vector, $d\mathrm{SA}(\mathbf{v})$ represents the surface area. $\mathbf{F}_g(\mathbf{w})$ defines a real-valued function depending only on the length $w = |\mathbf{w}|$ of the input vector, and we have

$$\mathbf{F}_g(0) = \mathbf{F}_g(\mathbf{0}) = 1 \quad \text{and} \quad |\mathbf{F}_g(w)| = |\mathbf{F}_g(\mathbf{w})| < 1 \quad \text{for} \quad w = |\mathbf{w}| > 0.$$

By the Riemann–Lebesgue theorem $\mathbf{F}_g(w) \to 0$ as $w \to \infty$.

We recall Theorem 13.4.

Theorem 13.4. *Assume $\mathcal{R}$ satisfies* (24.4)–(24.6), *$\mathcal{G}$ satisfies the "vanishing property"*

$$\max_{1\leq k\leq N} |\mathbf{F}_{g_k}(x)| = 0 \quad \textit{for all} \quad x \geq \gamma_0$$

for the 3-dimensional "isotropic Fourier transforms" $\mathbf{F}_{g_k}$ (for concrete examples of such probability distributions, see (13.46)–(13.49)), *and*

$$W > U \geq \max\left\{\alpha_0, \beta_0, \left(\frac{\gamma_0}{2\pi}\right)^{1/\alpha}\right\}.$$

Then

$$\Delta_f^2(\mathcal{G}; \mathcal{R}; U, W) \leq 2\lceil \log_2(W/U)\rceil \cdot \sigma_0^2(f)(W-U)W^\beta$$
$$\cdot (1+\gamma_0)(1 + (\alpha_1 - \alpha)\log W + \alpha_1 \log 2 + \log 3). \tag{24.13}$$

Let $B \subset I^3 = [0,1)^3$ be an arbitrary but fixed measurable test set in the particle space (= unit cube), where $\mathrm{vol}(B)$ denotes the 3-dimensional Lebesgue measure.

Consider now a 3-dimensional torus model $\mathcal{Y}(\mathcal{R};\omega;t)$ (in the particle space I^3) satisfying the hypothesis of Theorem 13.4. The family of time evolutions $\omega \in \Omega_{\mathcal{G}}$ of this torus model is represented by the family of curves lines (see (24.2))

$$\vec{\mathcal{Y}}(\mathcal{R};\omega;t) = \vec{\mathcal{Y}} + \vec{\Gamma}(\mathcal{R};\omega;t) \text{ modulo one} \tag{24.14}$$

in the configuration space I^d, all starting from the same point $\vec{\mathcal{Y}} = (\mathbf{y}_1,\ldots,\mathbf{y}_N) \in I^d$.

Repeating (5.7), for an arbitrary $\tau > 0$ write

$$S(B;\tau) = \{\vec{\mathcal{Z}} \in I^d : ||\mathcal{Z} \cap B| - \mathrm{vol}(B)N| > \tau\sqrt{N}\}, \tag{24.15}$$

where

$$\vec{\mathcal{Z}} = (z_1,\ldots,z_{3N}) \quad \text{and} \quad \mathcal{Z} = \{\mathbf{z}_1,\ldots,\mathbf{z}_N\}$$

with $\mathbf{z}_k = (z_{3k-2}, z_{3k-1}, z_{3k})$, $1 \le k \le N$.

Since the torus I^d is translation invariant, we apply Theorem 13.4 with $f = \chi_S$ where S is the translated copy $S = S(B;\tau) - \vec{\mathcal{Y}}$ of $S(B;\tau)$ in the torus I^d. Note that in the special case of a characteristic function $f = \chi_S$, $S \subset I^d$,

$$\sigma_0^2(f) = \mathrm{vol}_d(S)(1 - \mathrm{vol}_d(S)) \le \mathrm{vol}_d(S).$$

We recall the upper bound in (11.11):

$$\mathrm{vol}_d(S) = \mathrm{vol}_d(S(B;\tau)) < 2\exp\left(-\frac{2\tau^2}{1+\frac{\tau}{3\sqrt{N}}}\right)$$

and using it in Theorem 13.4 with $W = 2^k U$, (24.13) yields

$$\int_{\omega\in\Omega_{\mathcal{G}}} \left(\int_U^{2^k U} \chi_S(\vec{\Gamma}(\mathcal{R};\omega;t))dt - \mathrm{vol}_d(S)(2^k-1)U\right)^2 d\mathrm{ProdMeas}_{\mathcal{G}}(\omega)$$

$$\le 2\lceil \log_2(2^k)\rceil \cdot \sigma_0^2(f)(2^k-1)U(2^kU)^{\beta}$$

$$\cdot (1+\gamma_0)(1+(\alpha_1-\alpha)\log(2^kU) + \alpha_1\log 2 + \log 3)$$

$$\leq 2k \cdot 2\exp\left(-\frac{2\tau^2}{1+\frac{\tau}{3\sqrt{N}}}\right) \cdot (2^k-1)U(2^kU)^\beta$$

$$\cdot (1+\gamma_0)(1+(\alpha_1-\alpha)(k+\log U)+\alpha_1\log 2+\log 3), \tag{24.16}$$

assuming

$$U \geq \max\left\{\alpha_0, \beta_0, \left(\frac{\gamma_0}{2\pi}\right)^{1/\alpha}\right\}.$$

By (24.14), (24.15), and using $S = S(B;\tau) - \vec{\mathcal{Y}}$, we have

$$\int_{\omega\in\Omega_{\mathcal{G}}} \left(\int_U^{2^kU} \chi_S(\vec{\Gamma}(\mathcal{R};\omega;t))\,dt - \mathrm{vol}_d(S)(2^k-1)U\right)^2 d\mathrm{ProdMeas}_{\mathcal{G}}(\omega)$$

$$= \int_{\omega\in\Omega_{\mathcal{G}}} (\mathrm{length}\{U \leq t \leq 2^kU :\ ||\mathcal{Y}(\mathcal{R};\omega;t)\cap B| - \mathrm{vol}(B)N| > \tau\sqrt{N}\}$$

$$- \mathrm{vol}_d(S(B;\tau))(2^k-1)U)^2\, d\mathrm{ProdMeas}_{\mathcal{G}}(\omega), \tag{24.17}$$

where, as usual, length stands for the 1-dimensional Lebesgue measure.

Combining (24.16) and (24.17), we obtain the following result.

Theorem 24.1. *Let $\mathcal{Y}(\mathcal{R};\omega;t)$, $\omega \in \Omega_{\mathcal{G}}$ be a 3-dimensional torus model satisfying the hypothesis of Theorem 13.4, and let $B \subset [0,1)^3$ be a measurable test with 3-dimensional Lebesgue measure $\mathrm{vol}(B)$. Assume that*

$$U \geq \max\left\{\alpha_0, \beta_0, \left(\frac{\gamma_0}{2\pi}\right)^{1/\alpha}\right\}.$$

Then for every $\tau > 0$ and every integer $k \geq 1$,

$$\int_{\omega\in\Omega_{\mathcal{G}}} (\mathrm{length}\{U \leq t \leq 2^kU :\ ||\mathcal{Y}(\mathcal{R};\omega;t)\cap B| - \mathrm{vol}(B)N| > \tau\sqrt{N}\}$$

$$-\mathrm{vol}_d(S(B;\tau))(2^k-1)U)^2\, d\mathrm{ProdMeas}_{\mathcal{G}}(\omega)$$

$$\leq 2k \cdot 2\exp\left(-\frac{2\tau^2}{1+\frac{\tau}{3\sqrt{N}}}\right) \cdot (2^k-1)U(2^kU)^\beta$$

$$\cdot (1+\gamma_0)(1+(\alpha_1-\alpha)(k+\log U)+\alpha_1\log 2+\log 3), \tag{24.18}$$

where

$$\mathrm{vol}_d(S(B;\tau)) < 2\exp\left(-\frac{2\tau^2}{1+\frac{\tau}{3\sqrt{N}}}\right).$$

□

Similarly to Theorem 5.1, Theorem 24.1 is an extremely powerful result, due to the same reason: the superexponentially small factor

$$\exp\left(-\frac{2\tau^2}{1+\frac{\tau}{3\sqrt{N}}}\right).$$

This guarantees the long-term stability exactly the same way as we demonstrated on examples in Sec. 11.

The crucial difference between Theorems 5.1 and 24.1 is that the threshold $\sqrt{\log N}$ in the former is reduced to a constant

$$\text{threshold} = \max\left\{\alpha_0, \beta_0, \left(\frac{\gamma_0}{2\pi}\right)^{1/\alpha}\right\} \tag{24.19}$$

in the latter, where the constant threshold (24.19) does not depend on N. This is why we say that the (typical) time evolution of a system in Theorem 24.1 exhibits *instant conf-space equilibrium*.

Note that Theorem 24.1 has another advantage over Theorem 5.1: it is about a more general torus model where the particles move on torus lines with (possibly) varying speeds, and every particle has its own, possible different, smooth initial speed distribution. The two crucial requirements for *instant conf-space equilibrium* are: (1) the "vanishing property", and (2) the constant exponent β in (24.6) satisfies the inequality $\beta < 1$.

Beyond Theorems 4.2 and 13.4. Of course in the proof of Theorem 24.1 we can replace Theorem 13.4 with Theorem 13.1 or with Theorem 13.2 or with Theorem 13.3. Thus we obtain three analogs of Theorem 24.1 with the following thresholds:

(1) if $\mathcal{R}$ satisfies (24.4)–(24.6), $\mathcal{G}$ satisfies the "polynomial decay" condition

$$\max_{1\le k\le N}\sup_{x\ge w}|\mathbf{F}_{g_k}(x)| \le w^{-\gamma} \quad \text{for all} \quad w \ge \gamma_0,$$

and $\gamma > 3,\ \alpha\gamma > 1, \beta < 1$,

$$\text{then the threshold is } N^{\frac{2}{\alpha\gamma}+\varepsilon} \text{ with some } \varepsilon > 0; \tag{24.20}$$

(2) if $\mathcal{R}$ satisfies (24.4)–(24.6), $\mathcal{G}$ satisfies the "exponential decay" condition

$$\max_{1\le k\le N} \sup_{x\ge w} |\mathbf{F}_{g_k}(x)| \le e^{-\gamma w} \quad \text{for all} \quad w \ge \gamma_0,$$

and $\beta < 1$,

$$\text{then the threshold is } \left(\frac{\log N}{\pi\gamma}\right)^{1/\alpha}; \tag{24.21}$$

and finally,

(3) if $\mathcal{R}$ satisfies (24.4)–(24.6), $\mathcal{G}$ satisfies the "superexponential decay" condition

$$\max_{1\le k\le N} \sup_{x\ge w} |\mathbf{F}_{g_k}(x)| \le e^{-\gamma w^2} \quad \text{for all} \quad w \ge \gamma_0,$$

and $\beta < 1$,

$$\text{then the threshold is } \left(\frac{\log N}{2\pi^2\gamma}\right)^{1/(2\alpha)}. \tag{24.22}$$

In view of Lemma 13.1 we may say that the threshold depends mainly on the smoothness of the probability density functions g_k, $1 \le k \le N$.

Two-dimensional case where the particles move on general non-linear curves. In Theorems 4.2 and 13.1–13.4 all particles stay on torus lines, and the only difference is whether the particles move with constant speeds or with varying speeds. In Sec. 14 we explained how to extend Theorems 13.1–13.4 from torus lines to arbitrary non-linear curves in the 2-dimensional case (the case of dimension $d \ge 3$ is an open problem).

We recall the notation of Sec. 14: in the 2-dimensional case the particle space is the unit torus $[0,1)^2 = I^2$, and the curve of the whole system in the configuration space has the form

$$\begin{aligned} &\vec{\mathcal{Y}}(\vec{\mathcal{R}}; \rho_1, \vartheta_1, \ldots, \rho_N, \vartheta_N) \\ &\quad = \vec{\mathcal{Y}} + \vec{\Gamma}(\vec{\mathcal{R}}; \rho_1, \vartheta_1, \ldots, \rho_N, \vartheta_N) \text{ modulo one} \\ &\quad = (\mathbf{y}_1 + \rho_1\vartheta_1\mathbf{r}_1(t), \mathbf{y}_2 + \rho_2\vartheta_2\mathbf{r}_2(t), \ldots, \mathbf{y}_N + \rho_N\vartheta_N\mathbf{r}_N(t) \text{ modulo one}, \end{aligned} \tag{24.23}$$

where t is the time,

$$\vec{\mathcal{Y}} = (\mathbf{y}_1, \mathbf{y}_2, \ldots, \mathbf{y}_N) \tag{24.24}$$

is the vector (= point in the configuration space) formed from the set

$$\mathcal{Y} = \{\mathbf{y}_1, \mathbf{y}_2, \ldots, \mathbf{y}_N\}$$

of the initial point configuration at the start $t = 0$, $\vartheta_k \in \mathrm{SO}(2)$ are 2-dimensional rotations (we could simply write $\vartheta_k \in [0, 2\pi)$), and finally

$$\vec{\mathcal{R}} = (\mathbf{r}_1(t), \mathbf{r}_2(t), \ldots, \mathbf{r}_N(t)) \tag{24.25}$$

represents a family of parametrized curves on the plane. Note that the 1-dimensional $r_k(t)$ in (24.3) is replaced by the 2-dimensional $\mathbf{r}_k(t)$ in (24.25); the dimension increase is indicated by putting an arrow on top of $\mathcal{R}$.

We assume that

$$\mathcal{R}^{(\inf)}(t) = \min_{1 \le k \le N} \inf_{x \ge t} |\mathbf{r}_k(x)| \ge t^{\alpha} \quad \text{for all} \quad t \ge \alpha_0, \tag{24.26}$$

$$\sup_{x \le t} \max_{1 \le k \le N} |\mathbf{r}'_k(x)| \le t^{\kappa} \quad \text{for all} \quad t \ge \alpha_0 \quad \text{and}$$

$$\max_{1 \le k \le N} |\mathbf{r}'_k(x)| \le \kappa_0 \quad \text{for} \quad 0 < t < \alpha_0, \tag{24.27}$$

where $1 + \kappa > \alpha > 0$, $\alpha_0 \ge 1$, $\kappa \ge 0$, and $\mathbf{r}'_k(x)$ is the derivative of the parametrized curve $\mathbf{r}_k(x)$.

Moreover, let $C(\mathbf{c}; \varrho)$ denote an arbitrary circle centered at $\mathbf{c} \in \mathbb{R}^2$ with radius ϱ, and assume that

$$\sup_{\mathbf{c} \in \mathbb{R}^2, \varrho > 0} \max_{1 \le k \le N} \frac{\text{length}\{0 \le t \le T : \ \mathbf{r}_k(t) \in C(\mathbf{c}; \varrho)\}}{\varrho} \le T^{\beta} \tag{24.28}$$

uniformly for all $T \ge \beta_0$ (note that $1 + \kappa > \alpha > 0$, $\alpha_0 \ge 1$, $\kappa \ge 0$, κ_0, $\beta_0 \ge 1$ and $0 \le \beta < 1$ are absolute constants).

We also need a family of probability density functions $\mathcal{G} = (g_1, g_2, \ldots, g_N)$, where g_k is the distribution of the "magnification factor" ρ_k, i.e., $\Pr[\rho_k \le u] = \int_0^u g_k(y)\, dy$. An interesting special case is the simplest 2-dimensional Gaussian distribution

$$g_1(u) = \cdots = g_N(u) = ue^{-u^2/2}, \ 0 \le u < \infty. \tag{24.29}$$

We indicate condition (24.29) by replacing $\mathcal{G}$ with "Gauss".

Let us return now to Theorem 24.1. If in the proof of Theorem 24.1 we replace Theorem 13.4 with Theorem 14.1, we obtain an analog of Theorem 24.1 with the following threshold:

(4) assuming $\vec{\mathcal{R}}$ satisfies (24.26)–(24.28), $\mathcal{G}$ satisfies (24.29), and $\alpha > \beta + \kappa$,

$$\text{then the threshold is } \left(\frac{2\log N}{\pi^2}\right)^{1/(2\alpha)}. \tag{24.30}$$

Next we switch from the 2-dimensional Gaussian distribution (24.29) to the (still 2-dimensional) case of general probability density functions $\mathcal{G} = (g_1, g_2, \ldots, g_N)$.

Condition of polynomial decay. Assume that there exist real numbers $\gamma > 2$ and $\gamma_0 \geq 1$ such that

$$\max_{1\leq k\leq N} \sup_{x\geq w} |\mathbf{F}_{g_k}(x)| \leq w^{-\gamma} \quad \text{for all} \quad w \geq \gamma_0. \tag{24.31}$$

Note that Lemma 14.2 provides a sufficient condition for a probability density function g to guarantee that its 2-dimensional "isotropic Fourier transform" $\mathbf{F}_g$ satisfies the polynomial decay condition (24.31).

If in the proof of Theorem 24.1 we replace Theorem 13.4 with Theorem 14.2, we obtain an analog of Theorem 24.1 with the following threshold:

(5) assuming $\vec{\mathcal{R}}$ satisfies (24.26)–(24.28), $\mathcal{G}$ satisfies (24.31), $\gamma > 2$, $\alpha\gamma > 1$, $\alpha > \beta + \kappa$,

$$\text{then the threshold is } N^{\frac{2}{\alpha\gamma}+\varepsilon} \text{ with some } \varepsilon > 0. \tag{24.32}$$

Finally, note that Theorem 24.1 and its analogs guarantee the long-term stability of equilibrium exactly the same way as Theorem 5.1 did in the illustrations/examples in Sec. 5.

25. The Case of Singular Underlying Measure

The simplest example of singular underlying measure is the class of models that we call *compound systems with circles*, or simply *circles system*. These closed orbit models can be considered as a far-reaching generalization of the spherical Great-Circle system introduced in Sec. 6. The *singularity* of the underlying measure (relative to the Lebesgue measure) comes from the fact that the closed orbits are 1-dimensional curves, i.e., they have lower dimension than the particle space. (Another source of singularity where each particle is restricted to a surface, and the particle space is 3-dimensional.)

Let $C_1, C_2, C_3, \ldots, C_N$ be N not necessarily different circles (we mean the curves, not the disks) in the 3-space $\mathbb{R}^3$, and suppose that there are N particles such that the kth particle orbits on C_k, $1 \le k \le N$. For simplicity assume that kth particle orbits on C_k with constant speed, where the speed depends only on k ($1 \le k \le N$). More precisely, assume that the *relative* speed of the kth particle is v_k, that is, in a time interval $T \le t \le T + \ell$ of length ℓ the kth particle travels arclength distance $2\pi R_k v_k \ell$, where R_k is the radius of C_k.

Let $\mathbf{y}_k \in C_k$ denote the starting point of the kth particle at $t = 0$. For every circle we choose an orientation, and assume that the relative constant speeds v_k, $1 \le k \le N$ are chosen *independently* by the 1-dimensional normal (= Gaussian) distribution

$$\Pr[v_k \le u] = \frac{1}{\sqrt{2\pi}} \int_{-\infty}^{u} e^{-y^2/2}\, dy. \tag{25.1}$$

(Negative speed means that the particle goes backward following the negative orientation.)

As usual, let

$$\mathcal{Y} = \{\mathbf{y}_1, \mathbf{y}_2, \mathbf{y}_3, \ldots, \mathbf{y}_N\} \quad \text{with} \quad \mathbf{y}_k \in C_k \tag{25.2}$$

denote the initial configuration at $t = 0$. We refer to the system $(\mathcal{C}; \mathcal{Y}; \text{Gauss})$ as a *circles system.*

For simplicity assume that the N circles C_k, $1 \le k \le N$ are all in the unit cube $[0,1]^3$, and we define the following measure on the family $\mathcal{B}$ of Borel sets in the unit cube $[0,1]^3$:

$$\text{Meas}_{\mathcal{C}}(A) = \frac{1}{N} \sum_{k=1}^{N} \frac{1}{2\pi R_k} \text{length}(A \cap C_k), \quad A \in \mathcal{B}, \tag{25.3}$$

where R_k is the radius of C_k. We refer to $\text{Meas}_{\mathcal{C}}$ as the *underlying measure* of the circles system $(\mathcal{C}; \mathcal{Y}; \text{Gauss})$.

We study the usual questions: given an arbitrary initial configuration

$$\mathcal{Y} = \{\mathbf{y}_1, \mathbf{y}_2, \mathbf{y}_3, \ldots, \mathbf{y}_N\}$$

with $\mathbf{y}_k \in C_k$, what can we say about the typical time evolution of the circles system $(\mathcal{C}; \mathcal{Y}; \text{Gauss})$? How long does it take to reach equilibrium? What can we say about the stability of equilibrium?

The novelty of the circles system is that the underlying measure — the key concept to define equilibrium — is completely different from the usual (homogeneous) Lebesgue measure: the underlying measure is *singular*. In other words, it is not absolutely continuous with respect to Lebesgue measure; so, we do not have a corresponding density function (= Radon–Nikodym derivative). Note that the inhomogeneous underlying measures of the closed orbit systems in Secs. 12–13 were absolutely continuous with respect to the corresponding Lebesgue measure (= surface area for spherical systems).

To be very precise, we explain why $\text{Meas}_{\mathcal{C}}$ is a well-defined measure (i.e., σ-additive) on the Borel σ-algebra $\mathcal{B}$. The proof is a straightforward application of a basic Extension Theorem from measure theory. First we need some standard definitions. An *elementary set* in the 3-space $\mathbb{R}^3$ means a union of a finite number of axis-parallel (bounded) boxes (including all possibilities of open, closed, half-open, etc.). Let $\mathcal{E}$ denote the family of elementary sets in $\mathbb{R}^3$.

A finite, nonnegative, additive set function ϕ defined on $\mathcal{E}$ is said to be *regular* if for every $A \in \mathcal{E}$ and every $\varepsilon > 0$ there exist sets $F, G \in \mathcal{E}$ such that F is closed, G is open, $F \subset A \subset G$, and

$$\phi(G) - \varepsilon \leq \phi(A) \leq \phi(F) + \varepsilon.$$

Let ϕ be a finite, regular, nonnegative additive set function defined on $\mathcal{E}$. Then we can define the ϕ-outer measure ϕ^* on the family of all subset $E \subset \mathbb{R}^3$ as follows:

$$\phi^*(E) = \inf_{E \subset \bigcup_{n=1}^{\infty} A_n} \sum_{n=1}^{\infty} \phi(A_n),$$

where every $A_n \in \mathcal{E}$ is open.

Extension Theorem. *Let ϕ be a finite, regular, nonnegative additive set function defined on $\mathcal{E}$. Then $\phi^*(A) = \phi(A)$ for every $A \in \mathcal{E}$; moreover, ϕ^* is a measure (= σ-additive) on a σ-algebra that contains the Borel sets.*

More precisely, ϕ^* is a measure on the smallest σ-algebra that contains the Borel sets and the sets $E \subset \mathbb{R}^3$ with $\phi^*(E) = 0$.

It is easy to check that $\text{Meas}_{\mathcal{C}}$ is a finite, regular, nonnegative, additive set function on the family of elementary sets $\mathcal{E}$. So, the Extension Theorem applies, and gives that $\text{Meas}_{\mathcal{C}}$ is a measure on the Borel σ-algebra $\mathcal{B}$.

We explain why we can trivially repeat the argument of Sec. 5 for the circles system $(\mathcal{C}; \mathcal{Y}; \text{Gauss})$, and obtain long-term stability of square-root equilibrium in the particle space $[0, 1]^3$ with respect to an arbitrary but fixed Borel test set. The only change that we need to make is that in the definition of square-root equilibrium we have to replace the 3-dimensional Lebesgue measure with the singular measure $\text{Meas}_{\mathcal{C}}$. (We talk about Borel sets instead of Lebesgue measurable sets, because $\text{Meas}_{\mathcal{C}}$ is singular: singularity implies that the σ-algebra of $\text{Meas}_{\mathcal{C}}$-zero sets is different from the σ-algebra of Lebesgue-zero sets.)

Of course the 3-dimensional torus system of Sec. 5 is completely different from the circles system $(\mathcal{C}; \mathcal{Y}; \text{Gauss})$ here, *but* the curve in the configuration space $I^d = [0, 1)^d$ with $d = N$, representing the time evolution of the whole circles system, is the *same*. Indeed, it is again a *straight line* in $\mathbb{R}^d$ modulo one (in fact, here the lines can start from the origin)

$$t\mathbf{v} \text{ modulo one,} \tag{25.4}$$

where

$$\omega = \mathbf{v} = (v_1, v_2, \ldots, v_N) \in \Omega_{\text{Gauss}} = (-\infty, \infty)^N. \tag{25.5}$$

The product space Ω_{Gauss} is equipped with the product measure $\text{ProdMeas}_{\text{Gauss}}$, where the real line $(-\infty, \infty)$ has the probability density function

$$g(u) = \frac{1}{\sqrt{2\pi}} e^{-u^2/2}$$

("1-dimensional normal distribution"). Notice that (25.4)–(25.5) is the perfect analog of (11.2)–(11.3) with the minor change that $d = 3N$ is reduced to $d = N$ (which just makes the case of "circles systems" slightly simpler).

Since the curve of the system in the configuration space is the same, we can repeat the rest of Sec. 5 with the natural modifications mentioned above. That is, again let $B \subset I^3 = [0, 1)^3$ be an arbitrary but fixed test set in the particle space $[0, 1]^3$, but now B has to be Borel measurable (instead of Lebesgue measurable), and the 3-dimensional Lebesgue measure $\text{vol}(B)$ is replaced by the singular measure $\text{Meas}_{\mathcal{C}}(B)$.

Again the application of Theorem 4.2 in the configuration space yields a perfect analog of Theorem 5.1, and the same holds for the rest of Sec. 5.

What happens if we want to prove long-term stability of square-root equilibrium with respect to a whole family of nice sets (instead of a fixed Borel test set)? Since the curve of the system in the configuration space is the same, a straightforward repetition of the arguments in Secs. 20–22 give the analog results as follows. Starting from an arbitrary but fixed initial configuration $\mathcal{Y}$, after the circles system reaches conf-space equilibrium, the typical time evolution of the circles system stays in square-root equilibrium in the particle space with respect to *all nice* test sets (say, all boxes) simultaneously for a very, very long time, without any violator time instant t.

Of course long-term stability does not mean that the circles system stays in equilibrium forever. Indeed, the "superdiscrepancy" result Theorem 3.1 implies that, given any kind of equilibrium state, the system will leave this equilibrium for an infinite sequence of times tending to infinity.

Moreover, since the curve of the system in the configuration space is the same, we also have snapshot randomness results for the circles systems. Thus we have an analog of Poisson randomness in Sec. 8, have an analog of CLT randomness in Sec. 15, and so on. The same proof works.

Also, we can extend the class of *compund systems with circles* much further by relaxing the restriction of constant speed motions, and allowing general motions of the particles. For example, we can assume that the motion of the kth particle on C_k is general in the sense of (13.4)–(13.6). More precisely, again let $\mathbf{c}_k$ denote the center, let R_k denote the radius, let $\mathbf{n}_k$ denote a unit normal vector of C_j, and finally, let $\mathbf{s}_k \in C_k$ denote the starting point of the kth particle. The "standard circle of radius R_k" is given in the xy-plane

$$x = R_k \cos(2\pi\theta), \; y = R_k \sin(2\pi\theta), \; z = 0, \text{ where } 0 \leq \theta < 1. \tag{25.6}$$

Let $\vartheta_k \in \mathrm{SO}(3)$ denote the rotation that maps $(R_k, 0, 0)$ to $\mathbf{s}_k - \mathbf{c}_k$ and maps $(0, 0, 1)$ to $\mathbf{n}_k$, then by (25.6) we have

$$C_k = \mathbf{c}_k + \vartheta_k \left(R_k \cos(2\pi\theta), R_k \sin(2\pi\theta), 0\right), \text{ where } 0 \leq \theta < 1. \tag{25.7}$$

Constant speed motion on the circle C_j means that $\theta = \theta(t) = ct$ in (25.7) with some fixed constant c. We study the general case where the kth particle moves with (possibly) varying speed on the circle C_k, and the limitations are similar to (13.4)–(13.6) as follows.

Assume that the motion of the kth particle is detemined by a function $\theta = \rho_k \theta_k(t)$ in (25.7), where ρ_k is a constant (depending on k) to be specified later, and

$$\inf_{x \geq t} \theta_k(x) \geq t^{\alpha} \quad \text{for all} \quad t \geq \alpha_0 \tag{25.8}$$

and

$$\sup_{x \leq t} \theta_k(x) \leq t^{\alpha_1} \quad \text{for all} \quad t \geq \alpha_0, \tag{25.9}$$

where $\alpha_0 \geq 1$, $0 < \alpha \leq \alpha_1$ are some absolute constants.

Moreover, let $\mathcal{I}$ denote an arbitrary finite interval of positive length. We also assume that

$$\sup_{\mathcal{I} \subset \mathbb{R}} \frac{\text{length}\{0 \leq t \leq T : \ r_k(t) \in \mathcal{I}\}}{\text{length}(\mathcal{I})} \leq T^{\beta} \tag{25.10}$$

uniformly for all $T \geq \beta_0$, z where $\beta_0 \geq 1$ and $0 \leq \beta < 1$ are absolute constants.

Note that it is easy to satisfy conditions (25.8)–(25.10). For example, consider the forward motions where the speed $r_k'(t)$ of the kth particle satisfies the inequalities

$$\min\left\{\kappa_1 t^{-\zeta_1}, \kappa_1\right\} \leq r_k'(t) \leq \max\left\{\kappa_2 t^{\zeta_2}, \kappa_2\right\}, \ t > 0,$$

which hold for some absolute constants $0 \leq \zeta_1 < 1$ and $\zeta_2 \geq 0$.

Also there are many other ways, where we mix forward and backward motion of the kth particle, such that (25.8)–(25.10) are still satisfied.

Since we already defined the starting point $\mathbf{s}_k \in C_k$ of the kth particle, and we assume that (25.8)–(25.10) hold for all particles $1 \leq k \leq N$, the last step is to specify the constant parameter ρ_k (depending on k) in $\theta = \rho_k \theta_k(t)$; see (25.7). We assume that the constant parameters ρ_k, $1 \leq k \leq N$ are chosen *independently* by the 1-dimensional normal (= Gaussian) distribution

$$\Pr[\rho_k \le u] = \frac{1}{\sqrt{2\pi}} \int_{-\infty}^{u} e^{-y^2/2}\,dy. \tag{25.11}$$

Let $(\mathcal{C}; \vec{\theta}; \mathcal{Y}; \text{Gauss})$ denote the system defined by the family of N circles

$$\mathcal{C} = \{C_1, C_2, C_3, \ldots, C_N\},$$

by the family of functions

$$\vec{\theta} = \{\theta_1(t), \theta_2(t), \theta_3(t), \ldots, \theta_N(t)\} \tag{25.12}$$

satisfying (25.8)–(25.10), by the initial configuration

$$\mathcal{Y} = \{\mathbf{s}_1, \mathbf{s}_2, \mathbf{s}_3, \ldots, \mathbf{s}_N\}$$

with $\mathbf{s}_k \in C_k$, and the constant parameters ρ_k (where $\theta = \rho_k\theta_k(t)$ in (25.7)) chosen independently by the probability distribution (25.11).

We study the usual questions: given an arbitrary initial configuration

$$\mathcal{Y} = \{\mathbf{s}_1, \mathbf{s}_2, \mathbf{s}_3, \ldots, \mathbf{s}_N\}$$

with $\mathbf{s}_k \in C_k$, what can we say about the typical time evolution of the system $(\mathcal{C}; \vec{\theta}; \mathcal{Y}; \text{Gauss})$? How long does it take to reach equilibrium? What can we say about the stability of equilibrium?

Well, we can easily prove results similar to the above-mentioned results in the special case of constant speed motions. The only difference is that we apply Theorem 13.3 instead of Theorem 4.2.

Another generalization is to replace the circles with other closed orbits like ellipse, or any other "reasonable closed curve".

So far we have been studying closed orbit models, where the orbits are closed curves (= 1-dimensional torus). Another possible generalization is to switch from closed curves to closed surfaces (for example, 2-dimensional torus). For motivation note that in some physical problems the perturbation of a periodic closed orbit motion of a particle becomes a quasi-periodic motion on a geometric torus surface (e.g., a donut surface). This means upgrading the 1-dimensional torus to a 2-dimensional torus. We can also go beyond the donut surface. For example, we can study the geodesic flow on a regular tetrahedron surface or in general on any equifacial tetrahedron surface, i.e., a particle moves along a (piecewise linear) geodesic of the tetrahedron surface (see the beginning of Sec. 7). And so on.

Again we can prove similar results. Since the closed curve orbits are replaced by closed surface orbits, we have to apply the results of Sec. 14 (replacing Theorem 4.2 and the results of Sec. 13).

Finally, we can combine these constructions to obtain even more general compound models. Assume that some particles orbit on circles and other closed curves, some other particles move on closed surfaces (like the donut surface, the tetrahedron surface, and so on). Finally, assume that some particles move inside a few disjoint boxes like point billiards. The union system is what we call a *general compound model*.

We study the usual questions: given an arbitrary initial configuration $\mathcal{Y}$, what can we say about the typical time evolution of the *general compound model*? How long does it take to reach equilibrium? What can we say about the stability of equilibrium?

The class of *general compound models* is the largest class for which we can answer these questions by proving results similar to the results up to this point, employing the machinery of short-time ergodic theorems.

Chapter 5

More Proofs

26. Proof of Theorem 4.1

This section is devoted to the deduction of Theorem 4.1 from Theorem 9.1. The somewhat strange definition of the threshold $T_0 = T_0(d)$ in Theorem 4.1 becomes well motivated by the key condition (4.12) in Theorem 4.2 (which is repeated in Theorem 9.1). Indeed, the equation

$$100dTe^{-\pi^2T^2/2} = 1, \tag{26.1}$$

guarantees that we can apply Theorem 9.1 for every $U \geq T_0$, where $T_0 = T_0(d)$ is the solution of (26.1). Note that (26.1) implies that

$$T_0 = T_0(d) = \frac{\sqrt{2}}{\pi}\sqrt{\log d} + o(1), \tag{26.2}$$

where $o(1) = o_d(1) \to 0$ as $d \to \infty$.

For every $\rho > 0$ we have the equality (trivial via substitution)

$$D_f(\sqrt{d}\mathbf{e}; T_0, T_1) = \frac{\rho}{\sqrt{d}} D_f(\rho\mathbf{e}; T_0\sqrt{d}/\rho, T_1\sqrt{d}/\rho). \tag{26.3}$$

At a later stage of the proof we specify the value of $\rho = \rho_0$ in the interval $\sqrt{d-2} - 1 \leq \rho_0 \leq \sqrt{d-2} + 1$ as the solution of an optimization problem (see (26.20)). For the application of Theorem 9.1, it is convenient to decompose the interval $[T_0\sqrt{d}/\rho, T_1\sqrt{d}/\rho)$ into "simple" subintervals, where we call an interval $[U, V)$ *simple* if $0 < U < V \leq 2U$. Since we do not know in advance which value of $\rho \in [\sqrt{d-2} - 1, \leq \sqrt{d-2} + 1]$ will turn out to be the optimal value, we have to define the decomposition in such a way that the simple subintervals that we use are from a "universal" family. This "universal" family should work equally well for every choice of ρ in the interval $\sqrt{d-2} - 1 \leq \rho \leq \sqrt{d-2} + 1$.

Clearly

$$\frac{\sqrt{d}T_i}{\sqrt{d-2}+1} \le \frac{\sqrt{d}T_i}{\rho} \le \frac{\sqrt{d}T_i}{\sqrt{d-2}-1}, i = 0, 1, \tag{26.4}$$

and motivated by this fact, write

$$W_0 = \frac{\sqrt{d}T_0}{\sqrt{d-2}-1} \quad \text{and} \quad W_k = 2^k W_0 \text{ for } k \ge 1, \tag{26.5}$$

where T_0 is defined in (26.1)–(26.2).

Thus we have the decomposition into subintervals

$$[T_0\sqrt{d}/\rho, T_1\sqrt{d}/\rho) = [T_0\sqrt{d}/\rho, W_0) \cup \bigcup_{k=0}^{\ell} [W_k, W_{k+1}) \cup [W_{\ell+1}, \sqrt{d}T_1/\rho), \tag{26.6}$$

where

$$W_{\ell+1} < \frac{\sqrt{d}T_1}{\sqrt{d-2}+1} \le W_{\ell+2} = 2W_{\ell+1}, \tag{26.7}$$

The intervals $[W_k, W_{k+1})$, $0 \le k \le \ell$ are all simple, and they are also independent of the choice of $\rho \in [\sqrt{d-2}-1, \sqrt{d-2}+1]$.

The first interval in (26.6) $[\sqrt{d}T_0/\rho, W_0)$ has length (see (26.1)–(26.2))

$$\le \frac{\sqrt{d}T_0}{\sqrt{d-2}-1} - \frac{\sqrt{d}T_0}{\sqrt{d-2}+1} \le \frac{2\sqrt{d}T_0}{d-3}, \tag{26.8}$$

which is clearly ≤ 1 if $d \ge 10^3$.

The last interval in (26.6) $[W_{\ell+1}, \sqrt{d}T_1/\rho)$ is not necessarily simple, but it is "almost simple"; see (26.7) (we will explain this later).

To handle the last interval in (26.6), we consider the following basically binary decomposition. Let $\mathbf{k} = (k_1, \ldots, k_r) \in \mathbb{Z}^r$, $r \ge 2$ be an arbitrary at least 2-dimensional integer vector such that the coordinates are strictly decreasing $k_1 > \cdots > k_r$, and let $I(\mathbf{k})$ denote the interval (note that $W_0 = \sqrt{d}T_0/(\sqrt{d-2}-1)$ is defined in (26.5))

$$I(\mathbf{k}) = [(2^{k_1} + \cdots + 2^{k_{r-1}})W_0, (2^{k_1} + \cdots + 2^{k_r})W_0) \text{ for } r \ge 3 \quad \text{and}$$

$$I(\mathbf{k}) = [2^{k_1}W_0, (2^{k_1} + 2^{k_2})W_0) \quad \text{for } r \ge 2. \tag{26.9}$$

Note that $I(\mathbf{k})$ has length $2^{k_r} W_0$, where k_r is the last coordinate of the vector $\mathbf{k}$.

For notational coherence, write

$$J(k) = [W_k, W_{k+1}) = \left[2^k W_0, 2^{k+1} W_0\right). \tag{26.10}$$

Note that (26.9) and (26.10) are all simple intervals.

Write (see (26.2))

$$c_0(d) = \log_2 W_0 = \log_2 \frac{\sqrt{d} T_0}{(\sqrt{d-2} - 1)}. \tag{26.11}$$

Note that

$$c_0(d) \approx \frac{1}{2} \log_2 \log d \text{ if } d \text{ is large.} \tag{26.12}$$

We are now ready to apply Theorem 9.1: we have

$$\begin{aligned}
&\sum_{k=0}^{\infty} \Delta_f^2(\text{Gauss}; J(k))(k+1)^{-2} 2^{-k} W_0^{-1} \\
&\quad + \sum_{\substack{\mathbf{k}=(k_1,\ldots,k_r)\in\mathbb{Z}^r, r\geq 2 \\ k_1 > \cdots > k_r \geq -c_0(d), k_1 \geq 0}} \Delta_f^2(\text{Gauss}; I(\mathbf{k}))(k_1+1)^{-2} \\
&\quad \times (k_1 + 1 + c_0(d))^{-1} 2^{-k_1} W_0^{-1} \\
&\leq \sigma_0^2(f) \cdot \left(\sum_{k=0}^{\infty} \left(9 \cdot 2^k W_0 + 1\right) \cdot (k+1)^{-2} 2^{-k} W_0^{-1} \right. \\
&\quad + \sum_{\substack{\mathbf{k}=(k_1,\ldots,k_r)\in\mathbb{Z}^r, r\geq 2 \\ k_1 > \cdots > k_r \geq -c_0(d), k_1 \geq 0}} \left(9 \cdot 2^{k_r} W_0 + 1\right) \cdot (k_1+1)^{-2} \\
&\quad \left. \times (k_1 + 1 + c_0(d))^{-1} 2^{-k_1} W_0^{-1} \right).
\end{aligned} \tag{26.13}$$

Fixing the first coordinate $k_1 \geq 0$, we have

$$\begin{aligned}
&\sum_{\substack{\mathbf{k}=(k_1,\ldots,k_r)\in\mathbb{Z}^r, r\geq 2 \\ k_1 > \cdots > k_r \geq -c_0(d), k_1 \geq 0}}^{k_1 \text{ fixed}} 2^{k_r} W_0 (k_1+1)^{-2} (k_1 + 1 + c_0(d))^{-1} 2^{-k_1} W_0^{-1} \\
&= \sum_{k_r:\ -c_0(d) \leq k_r < k_1 = \text{fixed}} 2^{k_1 - k_r} \cdot 2^{k_r} (k_1+1)^{-2} (k_1 + 1 + c_0(d))^{-1} 2^{-k_1}
\end{aligned}$$

$$\le (k_1+1+c_0(d))\cdot(k_1+1)^{-2}(k_1+1+c_0(d))^{-1}$$
$$= (k_1+1)^{-2}. \tag{26.14}$$

Using (26.14) in (26.13), we have

$$\begin{aligned}
&\sum_{k=0}^{\infty}\Delta_f^2(\text{Gauss}; J(k))(k+1)^{-2}2^{-k}W_0^{-1}\\
&\quad+\sum_{\substack{\mathbf{k}=(k_1,\ldots,k_r)\in\mathbb{Z}^r, r\ge 2\\ k_1>\cdots>k_r\ge -c_0(d), k_1\ge 0}}\Delta_f^2(\text{Gauss}; I(\mathbf{k}))\\
&\quad\times(k_1+1)^{-2}(k_1+1+c_0(d))^{-1}2^{-k_1}W_0^{-1}\\
&\le \sigma_0^2(f)\left(10\sum_{k=0}^{\infty}(k+1)^{-2}+10\sum_{k_1=0}^{\infty}(k_1+1)^{-2}\right)\\
&\le 40\sigma_0^2(f).
\end{aligned}\tag{26.15}$$

We recall the definition of the Gaussian square-integral (since at the end we choose $f=\chi_S$, we can assume that f is real-valued; $0\le W'<W''$ are arbitrary)

$$\Delta_f^2(\text{Gauss}; W', W'') = \int_0^{\infty} D_f^2(\rho; W', W'')\frac{\rho^{d-1}e^{-\rho^2/2}}{C_d}\,d\rho, \tag{26.16}$$

where

$$D_f^2(\rho; W', W'') = \int_{\mathbf{e}\in\mathbf{S}^{d-1}} D_f^2(\rho\mathbf{e}; W', W'')\,d\text{SA}^\star(\mathbf{e}), \tag{26.17}$$

and $d\text{SA}^\star$ stands for the normalized surface area.

Lemma 26.1. *For $d\ge 10^3$ and $0\le W'<W''$ we have*

$$\Delta_f^2(\text{Gauss}; W', W'') \ge \frac{1}{6}\int_{\sqrt{d-2}-1}^{\sqrt{d-2}+1} D_f^2(\rho; W', W'')\,d\rho. \tag{26.18}$$

Proof. Notice that inequality (26.18) immediately follows from the lower bound

$$g(\rho) = \frac{\rho^{d-1}e^{-\rho^2/2}}{C_d} \ge \frac{1}{6} \text{ if } |\rho-\sqrt{d-2}|\le 1 \quad\text{and}\quad d\ge 10^3. \tag{26.19}$$

To prove (26.19), we use Stirling's formula (for the definition of C_d, see (6.30)–(6.31))

$$\begin{aligned} C_d &= (1+o(1))2^{\frac{d}{2}-1}\left(\frac{d}{2}-1\right)! \\ &= (1+o(1))2^{\frac{d}{2}-1}\left(\frac{\frac{d}{2}-1}{e}\right)^{\frac{d}{2}-1}\sqrt{2\pi\left(\frac{d}{2}-1\right)} \\ &= (1+o(1))\left(\frac{d-2}{e}\right)^{(d-2)/2}\sqrt{\pi(d-2)}, \end{aligned}$$

which gives

$$\begin{aligned} g(\rho) &= \frac{\rho^{d-1}e^{-\rho^2/2}}{C_d} = (1+o(1))\frac{\rho^{d-1}e^{-\rho^2/2}}{\left(\frac{d-2}{e}\right)^{(d-2)/2}\sqrt{\pi(d-2)}} \\ &= \frac{1+o(1)}{\sqrt{\pi}}\frac{\rho}{\sqrt{d-2}}\cdot\left(\frac{\rho}{\sqrt{d-2}}\right)^{d-2}e^{-\rho^2/2+(d-2)/2}, \end{aligned}$$

and using $\rho=\sqrt{d-2}+c$, $-1\le c\le 1$ and $\exp(x)=e^x$, we obtain

$$\begin{aligned} &\left(1+\frac{c}{\sqrt{d-2}}\right)^{d-2}e^{-\rho^2/2+(d-2)/2} \\ &= \exp\left((d-2)\log(1+c/\sqrt{d-2})\right)\cdot e^{(\sqrt{d-2}-r)(\sqrt{d-2}+\rho)/2} \\ &= \exp\left((d-2)\left(\frac{c}{\sqrt{d-2}}-\frac{c^2}{2(d-2)}\right.\right. \\ &\qquad \left.\left.+\frac{c^3}{3(d-2)^{3/2}}-\frac{c^4}{4(d-2)^2}\pm\cdots\right)\right) \\ &\qquad \cdot e^{-c(2\sqrt{d-2}+c)/2} = \exp\left(-c^2+\frac{c^3}{3\sqrt{d-2}}-\frac{c^4}{4(d-2)}\pm\cdots\right) \\ &\ge \frac{1}{3} \end{aligned}$$

if $|c|\le 1$ and $d\ge 10^3$, and (26.19) immediately follows via simple calculation. This completes the proof of Lemma 26.1. □

Combining (26.15)–(26.17) and Lemma 26.1, we have

$$
\begin{aligned}
40\sigma_0^2(f) &\geq \sum_{k=0}^{\infty} \Delta_f^2(\text{Gauss}; J(k))(k+1)^{-2}2^{-k}W_0^{-1} \\
&\quad + \sum_{\substack{\mathbf{k}=(k_1,\ldots,k_r)\in\mathbb{Z}^r, r\geq 2 \\ k_1>\cdots>k_r\geq -c_0(d), k_1\geq 0}} \Delta_f^2(\text{Gauss}; I(\mathbf{k})) \\
&\quad \times (k_1+1)^{-2}(k_1+1+c_0(d))^{-1}2^{-k_1}W_0^{-1} \\
&\geq \frac{1}{6}\int_{\sqrt{d-2}-1}^{\sqrt{d-2}+1} \left(\sum_{k=0}^{\infty} D_f^2(\rho; J(k))(k+1)^{-2}2^{-k}W_0^{-1} \right. \\
&\quad + \sum_{\substack{\mathbf{k}=(k_1,\ldots,k_r)\in\mathbb{Z}^r, r\geq 2 \\ k_1>\cdots>k_r\geq -c_0(d), k_1\geq 0}} D_f^2(\rho; I(\mathbf{k}))(k_1+1)^{-2} \\
&\quad \left. \times (k_1+1+c_0(d))^{-1}2^{-k_1}W_0^{-1} \right) d\rho \\
&\geq \frac{1}{3} \min_{\sqrt{d-2}-1\leq\rho\leq\sqrt{d-2}+1} \left(\sum_{k=0}^{\infty} D_f^2(\rho; J(k))(k+1)^{-2}2^{-k}W_0^{-1} \right. \\
&\quad + \sum_{\substack{\mathbf{k}=(k_1,\ldots,k_r)\in\mathbb{Z}^r, r\geq 2 \\ k_1>\cdots>k_r\geq -c_0(d), k_1\geq 0}} D_f^2(\rho; I(\mathbf{k}))(k_1+1)^{-2} \\
&\quad \left. \times (k_1+1+c_0(d))^{-1}2^{-k_1}W_0^{-1} \right). \qquad (26.20)
\end{aligned}
$$

Assume that the minimum for $\rho \in [\sqrt{d-2}-1, \sqrt{d-2}+1]$ in the last line of (26.20) is attained at $\rho = \rho_0$. Then (26.20) gives

$$
\begin{aligned}
&\sum_{k=0}^{\infty} D_f^2(\rho_0; J(k))(k+1)^{-2}2^{-k}W_0^{-1} + \sum_{\substack{\mathbf{k}=(k_1,\ldots,k_r)\in\mathbb{Z}^r, r\geq 2 \\ k_1>\cdots>k_r\geq -c_0(d), k_1\geq 0}} \\
&\quad \times D_f^2(\rho_0; I(\mathbf{k}))(k_1+1)^{-2}(k_1+1+c_0(d))^{-1}2^{-k_1}W_0^{-1} \\
&\leq 120\sigma_0^2(f). \qquad (26.21)
\end{aligned}
$$

By (26.17) and (26.21),

$$\int_{\mathbf{e}\in\mathbf{S}^{d-1}} D_f^2(\rho_0\mathbf{e}; J(k))\, d\mathrm{SA}^\star(\mathbf{e}) \le (k+1)^2 2^k W_0 \cdot 120\sigma_0^2(f) \tag{26.22}$$

holds for every $k \ge 0$. Similarly,

$$\int_{\mathbf{e}\in\mathbf{S}^{d-1}} \sum_{\substack{\mathbf{k}=(k_1,\ldots,k_r)\in\mathbb{Z}^r, r\ge 2\\ k_1>\cdots>k_r\ge -c_0(d), k_1\ge 0}}^{k_1 \text{ fixed}} D_f^2(\rho_0\mathbf{e}; I(\mathbf{k}))\, d\mathrm{SA}^\star(\mathbf{e})$$

$$\le (k_1+1)^2(k_1+1+c_0(d))2^{k_1} W_0 \cdot 120\sigma_0^2(f) \tag{26.23}$$

holds for every $k_1 \ge 0$.

It immediately follows from (26.22) and the definition of integral ("Markov's inequality") that for every $k \ge 0$ there exists a measurable subset of the (hyper)sphere $\mathcal{A}_{1,k} \subset \mathbf{S}^{d-1}$ such that the normalized surface area

$$\mathrm{SA}^\star(\mathcal{A}_{1,k}) \ge 1 - \frac{\varepsilon}{4(k+1)^2} \tag{26.24}$$

(i.e., $\mathrm{SA}^\star(\mathbf{S}^{d-1}) = 1$) and

$$D_f^2(\rho_0\mathbf{e}; J(k)) \le \frac{4(k+1)^2}{\varepsilon} \cdot (k+1)^2 2^k W_0 \cdot 120\sigma_0^2(f) \text{ for every } \mathbf{e} \in \mathcal{A}_{1,k}. \tag{26.25}$$

Similarly, by (26.23) for every $k_1 \ge 0$ there exists a measurable subset of the (hyper)sphere $\mathcal{A}_{2,k_1} \subset \mathbf{S}^{d-1}$ such that

$$\mathrm{SA}^\star(\mathcal{A}_{2,k_1}) \ge 1 - \frac{\varepsilon}{4(k_1+1)^2} \tag{26.26}$$

and

$$\sum_{\substack{\mathbf{k}=(k_1,\ldots,k_r)\in\mathbb{Z}^r, r\ge 2\\ k_1>\cdots>k_r\ge -c_0(d), k_1\ge 0}}^{k_1 \text{ fixed}} D_f^2(\rho_0\mathbf{e}; I(\mathbf{k}))$$

$$\le \frac{4(k_1+1)^2}{\varepsilon} \cdot (k_1+1)^2(k_1+1+c_0(d))2^{k_1} W_0$$

$$\cdot 120\sigma_0^2(f) \text{ for every } \mathbf{e} \in \mathcal{A}_{2,k_1}. \tag{26.27}$$

Let

$$\mathcal{A} = \bigcap_{k=0}^{\infty} \mathcal{A}_{1,k} \cap \bigcap_{k_1=0}^{\infty} \mathcal{A}_{2,k_1}. \tag{26.28}$$

By (26.24), (26.26) and (26.28), the normalized surface area

$$\mathrm{SA}^\star(\mathcal{A}) \geq 1 - \sum_{k=0}^{\infty} \frac{\varepsilon}{4(k+1)^2} - \sum_{k_1=0}^{\infty} \frac{\varepsilon}{4(k_1+1)^2} > 1 - \varepsilon. \tag{26.29}$$

Let $\mathbf{e} \in \mathcal{A}$ be arbitrary.

We recall (26.3) with $\rho = \rho_0 \in [\sqrt{d-2} - 1, \sqrt{d-2} + 1]$

$$D_f(\sqrt{d}\mathbf{e}; T_0, T_1) = \frac{\rho_0}{\sqrt{d}} D_f(\rho_0 \mathbf{e}; T_0\sqrt{d}/\rho_0, T_1\sqrt{d}/\rho_0). \tag{26.30}$$

By (26.6)–(26.10),

$$[T_0\sqrt{d}/\rho_0, T_1\sqrt{d}/\rho_0) = [T_0\sqrt{d}/\rho_0, W_0) \cup \bigcup_{k=0}^{\ell} J(k) \cup [W_{\ell+1}, \sqrt{d}T_1/\rho_0), \tag{26.31}$$

where

$$W_{\ell+1} < \frac{\sqrt{d}T_1}{\sqrt{d-2}+1} \leq W_{\ell+2} = 2W_{\ell+1}. \tag{26.32}$$

Since

$$\frac{\sqrt{d}T_1}{\rho_0} \leq \frac{\sqrt{d}T_1}{\sqrt{d-2}-1},$$

by (26.32) we have two cases for $d \geq 10^3$:

$$\text{either } 1 \leq \frac{\sqrt{d}T_1/\rho_0}{W_{\ell+1}} < 2 \quad \text{or} \quad 2 \leq \frac{\sqrt{d}T_1/\rho_0}{W_{\ell+1}} < 4. \tag{26.33}$$

We just discuss the first case; the second case goes similarly.

Case 1:

$$1 \leq \frac{\sqrt{d}T_1/\rho_0}{W_{\ell+1}} < 2.$$

Consider the binary expansion

$$\frac{1}{W_0}\left(\frac{\sqrt{d}T_1}{\rho_0} - W_{\ell+1}\right) = 2^{\ell_2} + 2^{\ell_3} + 2^{\ell_4} + \cdots, \tag{26.34}$$

where $\ell_2 > \ell_3 > \ell_4 > \cdots$ is a strictly decreasing sequence of integers. In Case 1 we have $\ell + 1 > \ell_2$.

Using (26.34) in (26.31), we have

$$\begin{aligned}
&[T_0\sqrt{d}/\rho_0, T_1\sqrt{d}/\rho_0) = [T_0\sqrt{d}/\rho_0, W_0) \\
&\cup \bigcup_{k=0}^{\ell} J(k) \cup I((\ell+1,\ell_2)) \cup I((\ell+1,\ell_2,\ell_3)) \cup I((\ell+1,\ell_2,\ell_3,\ell_4)) \cup \dots \\
&= [T_0\sqrt{d}/\rho_0, W_0) \cup \bigcup_{k=0}^{\ell} J(k) \cup \bigcup_{r\geq 2} I((\ell+1,\dots,\ell_r)) \\
&= [T_0\sqrt{d}/\rho_0, W_0) \cup \bigcup_{k=0}^{\ell} J(k) \cup \bigcup_{\substack{r\geq 2\\ \ell_r\geq -c_0(d)}} I((\ell+1,\dots,\ell_r)) \cup I_{tail}.
\end{aligned} \tag{26.35}$$

It follows from the definition of $c_0(d)$ in (26.11) that I_{tail} is an interval of length less than 2.

Next we use the Cauchy–Schwarz inequality: combining (26.25), (26.27) and (26.35), we have

$$\begin{aligned}
&\left(\sum_{k=0}^{\ell} D_f(\rho_0\mathbf{e}; J(k)) + \sum_{r\geq 2:\ \ell_r\geq -c_0(d)} D_f(\rho_0\mathbf{e}; I((\ell+1,\dots,\ell_r)))\right)^2 \\
&\leq \left(\sum_{k=0}^{\ell} D_f^2(\rho_0\mathbf{e}; J(k)) + \sum_{r\geq 2:\ \ell_r\geq -c_0(d)} D_f^2(\rho_0\mathbf{e}; I((\ell+1,\dots,\ell_r)))\right) \\
&\quad \cdot ((\ell+1) + (\ell+1+c_0(d))) \\
&\leq \left(\sum_{k=0}^{\ell} 480\frac{(k+1)^4}{\varepsilon}2^k W_0\sigma_0^2(f) + 480\frac{(\ell+2)^4}{\varepsilon}\right. \\
&\quad \left.\times (\ell+2+c_0(d))2^{\ell+1}W_0\sigma_0^2(f)\right) \cdot (2\ell+2+c_0(d)).
\end{aligned} \tag{26.36}$$

Using the facts ($d \geq 10^3$)

$$W_{\ell+1} = 2^{\ell+1}W_0 < \frac{\sqrt{d}T_1}{\sqrt{d-2}+1} \leq T_1$$

(see (26.32)) and $c_0(d) = \log_2 W_0$ (see (26.11)) in (26.36), we obtain

$$\left(\sum_{k=0}^{\ell} D_f(\rho_0\mathbf{e}; J(k)) + \sum_{r\geq 2:\ \ell_r\geq -c_0(d)} D_f(\rho_0\mathbf{e}; I((\ell+1,\ldots,\ell_r)))\right)^2$$

$$\leq \frac{480}{\varepsilon}\sigma_0^2(f)2^{\ell+3}W_0(\ell+2)^4(\ell+2+c_0(d))^2$$

$$\leq \frac{480}{\varepsilon}\sigma_0^2(f)5T_1(2+\log_2 T_1)^6 = \frac{2400}{\varepsilon}\sigma_0^2(f)T_1(2+\log_2 T_1)^6,$$

which implies

$$\left|\sum_{k=0}^{\ell} D_f(\rho_0\mathbf{e}; J(k)) + \sum_{r\geq 2:\ \ell_r\geq -c_0(d)} D_f(\rho_0\mathbf{e}; I((\ell+1,\ldots,\ell_r)))\right|$$

$$\leq \frac{49}{\sqrt{\varepsilon}}\sigma_0(f)\sqrt{T_1}(2+\log_2 T_1)^3. \tag{26.37}$$

By (26.30), (26.35) and (26.37), for every $\mathbf{e} \in \mathcal{A}$,

$$D_f(\sqrt{d}\mathbf{e}; T_0, T_1) = \frac{\rho_0}{\sqrt{d}} D_f(\rho_0\mathbf{e}; T_0\sqrt{d}/\rho_0, T_1\sqrt{d}/\rho_0)$$

$$= \frac{\rho_0}{\sqrt{d}}\Bigg(D_f(\rho_0\mathbf{e}; T_0\sqrt{d}/\rho_0, W_0) + D_f(\rho_0\mathbf{e}; I_{tail})$$

$$+\sum_{k=0}^{\ell} D_f(\rho_0\mathbf{e}; J(k)) + \sum_{r\geq 2:\ \ell_r\geq -c_0(d)} D_f(\rho_0\mathbf{e}; I((\ell+1,\ldots,\ell_r)))\Bigg), \tag{26.38}$$

where I_{tail} is an interval of length less than 2 and $[T_0\sqrt{d}/\rho_0, W_0)$ is an interval of length ≤ 1 (see (26.8)).

In the special case $f = \chi_S$ where $S \subset [0,1)^d$ is a measurable test set we have the trivial upper bound

$$\left|D_f(\rho_0\mathbf{e}; T_0\sqrt{d}/\rho_0, W_0)\right| \leq \text{length} \leq 1,$$

and similarly, $|D_f(\rho_0\mathbf{e}; I_{tail})| \leq 2$. Combining these facts with (26.37)–(26.38), we have

$$
\begin{aligned}
&|\mathrm{TimeDiscrep}_S(\sqrt{d}\mathbf{e}; T_0, T_1)| = |D_f(\sqrt{d}\mathbf{e}; T_0, T_1)| \\
&\leq \frac{\rho_0}{\sqrt{d}}\left(1 + 2 + \frac{49}{\sqrt{\varepsilon}}\sigma_0(f)\sqrt{T_1}(2 + \log_2 T_1)^3\right) \\
&\leq \frac{\sqrt{d-2}+1}{\sqrt{d}}\left(3 + \frac{49}{\sqrt{\varepsilon}}\sigma_0(f)\sqrt{T_1}(2 + \log_2 T_1)^3\right) \\
&\leq \left(5 + \frac{50}{\sqrt{\varepsilon}}\sigma_0(f)\sqrt{T_1}(2 + \log_2 T_1)^3\right), \qquad (26.39)
\end{aligned}
$$

which holds for every $\mathbf{e} \in \mathcal{A}$, where the normalized surface area of $\mathcal{A} \subset \mathbf{S}^{d-1}$ is $> 1 - \varepsilon$. In the special case $f = \chi_S$ we have $\sigma_0^2(f) = p(1-p)$ where $p = \mathrm{vol}_d(S)$, and thus (26.39) completes the deduction of Theorem 4.1 from Theorem 9.1. □

27. Starting the Proofs of Theorems 13.1–13.4

The proofs of Theorems 13.1–13.4 have the same beginning, which is very similar to the argument in Sec. 9. We recall the notation that the curve $\Gamma(t) = \vec{\Gamma}(t)$ of the whole system in the configuration space is

$$
\begin{aligned}
\Gamma(t) &= \vec{\Gamma}(t) = \vec{\Gamma}(\mathcal{R}; \rho_1, \mathbf{e}_1, \ldots, \rho_N, \mathbf{e}_N; t) \\
&= (\rho_1 r_1(t)\mathbf{e}_1, \rho_2 r_2(t)\mathbf{e}_2, \ldots, \rho_N r_N(t)\mathbf{e}_N) \text{ modulo one}, \qquad (27.1)
\end{aligned}
$$

where t is the time,

$$\mathcal{R} = (r_1(t), r_2(t), \ldots, r_N(t))$$

is a (vector) family of functions such that

$$\mathcal{R}^{(\inf)}(t) = \min_{1\leq k\leq N} \inf_{x\geq t} r_k(x) \geq t^{\alpha} \text{ for all } t \geq \alpha_0 \qquad (27.2)$$

and

$$\mathcal{R}^{(\sup)}(t) = \max_{1\leq k\leq N} \sup_{x\leq t} r_k(x) \leq t^{\alpha_1} \text{ for all } t \geq \alpha_0, \qquad (27.3)$$

where $\alpha_0 \geq 1$, $0 < \alpha \leq \alpha_1$ are some absolute constants.

Moreover, let I denote an arbitrary finite interval of positive length. Assume that

$$\max_{1\le k\le N}\sup_{I\subset\mathbb{R}}\frac{\text{length}\{0\le t\le T:\ r_k(t)\in I\}}{\text{length}(I)}\le T^{\beta} \tag{27.4}$$

uniformly for all $T\ge\beta_0$, where $\beta_0\ge 1$ and $0\le\beta<1$ are absolute constants.

Since $\rho_k r_k(t)\mathbf{e}_k\in\mathbb{R}^3$, $1\le k\le N$, (27.1) is a curve in the $3N$-dimensional unit torus $I^d=[0,1)^d$, which is the configuration space.

Let $d=3N$. Let $f\in L^2(I^d)$ be a complex-valued Lebesgue square-integrable function in the d-dimensional unit torus (i.e., we extend f over the whole d-space $\mathbb{R}^d$ periodically), and consider the Fourier expansion of f:

$$f(\mathbf{u})=\sum_{\mathbf{n}\in\mathbb{Z}^d}a_{\mathbf{n}}e^{2\pi i\mathbf{n}\cdot\mathbf{u}},$$

where

$$a_{\mathbf{n}}=\int_{I^d}f(\mathbf{y})e^{-2\pi i\mathbf{n}\cdot\mathbf{y}}\,d\mathbf{y},$$

and $\mathbf{v}\cdot\mathbf{w}=v^{(1)}w^{(1)}+\cdots+v^{(d)}w^{(d)}$ denotes the usual dot product.

Write

$$\sigma_0^2(f)=\int_{I^d}|f|^2\,d\mathrm{V}-\left|\int_{I^d}f\,d\mathrm{V}\right|^2=\int_{I^d}\left|f(\mathbf{y})-\int_{I^d}f\,d\mathrm{V}\right|^2d\mathbf{y}.$$

Clearly

$$a_{\mathbf{0}}=\int_{I^d}f\,d\mathrm{V},\ \text{and so}\ \sum_{\mathbf{n}\in\mathbb{Z}^d\setminus\mathbf{0}}|a_{\mathbf{n}}|^2=\sigma_0^2(f),$$

where we used Parseval's formula.

Combining the Fourier series of f and the curve

$$\Gamma(t)=\vec{\Gamma}(t)=\vec{\Gamma}(\mathcal{R};\rho_1,\mathbf{e}_1,\ldots,\rho_N,\mathbf{e}_N;t)$$

of the whole system, we have

$$f(\vec{\Gamma}(t))-\int_{I^d}f\,d\mathrm{V}=\sum_{\mathbf{n}\in\mathbb{Z}^d\setminus\mathbf{0}}a_{\mathbf{n}}e^{2\pi i\mathbf{n}\cdot\vec{\Gamma}(t)}.$$

Consider the integral

$$D_f(\mathcal{R}; \rho_1, \mathbf{e}_1, \ldots, \rho_N, \mathbf{e}_N; T_1, T_2)$$
$$= \int_{T_1}^{T_2} f(\vec{\Gamma}(t))\, dt - (T_2 - T_1) \int_{I^d} f\, d\mathrm{V} = \int_{T_1}^{T_2} \sum_{\mathbf{n} \in \mathbb{Z}^d \setminus \mathbf{0}} a_{\mathbf{n}} e^{2\pi i \mathbf{n} \cdot \vec{\Gamma}(t)}\, dt, \tag{27.5}$$

where $\vec{\Gamma}(t) = \vec{\Gamma}(\mathcal{R}; \rho_1, \mathbf{e}_1, \ldots, \rho_N, \mathbf{e}_N; t)$ and $0 < T_1 < T_2$.

The corresponding initial velocity space is

$$\omega = (\rho_1, \vartheta_1, \rho_2, \vartheta_2, \ldots, \rho_N, \vartheta_N) \in \Omega_{\mathcal{G}} = \left([0, \infty) \times \mathbf{S}^2\right)^N, \tag{27.6}$$

where

$$\mathcal{G} = (g_1, g_2, \ldots, g_N)$$

is the vector of probability density functions g_k that define the initial speed distribution $\Pr[\rho_k \le u] = \int_0^u g_k(y)\, dy$ of the kth particle ($1 \le k \le N$). $\Omega_{\mathcal{G}}$ is equipped with the product measure such that for the kth factor $[0, \infty) \times \mathbf{S}^2$ in (27.6) we have the probability density function g_k for $[0, \infty)$ and the normalized surface area for the sphere $\mathbf{S}^2$.

Write

$$\mathbf{F}^{(\sup)}(w) = \mathbf{F}_{\mathcal{G}}^{(\sup)}(w) = \max_{1 \le k \le N} \sup_{x \ge w} |\mathbf{F}_{g_k}(x)|, \tag{27.7}$$

where

$$\mathbf{F}_g(\mathbf{w}) = \int_0^\infty \frac{1}{4\pi} \int_{\mathbf{v} \in \mathbf{S}^2} e^{iu\mathbf{v} \cdot \mathbf{w}} g(u)\, du\, d\mathbf{v}$$
$$= \int_0^\infty \frac{1}{4\pi} \int_{\mathbf{v} \in \mathbf{S}^2} \cos(u\mathbf{v} \cdot \mathbf{w}) g(u)\, du\, d\mathbf{v}, \tag{27.8}$$

is the "isotropic Fourier transform". $\mathbf{F}_g(\mathbf{w})$ defines a real-valued function depending only on the length $w = |\mathbf{w}|$ of the input vector, and we repeatedly make use of the facts that

$$\mathbf{F}_g(0) = \mathbf{F}_g(\mathbf{0}) = 1, |\mathbf{F}_g(w)| = |\mathbf{F}_g(\mathbf{w})| < 1 \quad \text{for } w = |\mathbf{w}| > 0, \tag{27.9}$$

and $\mathbf{F}_g^{(\text{sup})}(w)$ is a monotone decreasing positive function tending to zero as $w \to \infty$.

We take the square of the absolute value of

$$D_f(\mathcal{R}; \rho_1, \mathbf{e}_1, \ldots, \rho_N, \mathbf{e}_N; T_1, T_2),$$

and study the following average

$$\begin{aligned}\Delta_f^2(\mathcal{G}; \mathcal{R}; T_1, T_2) = &\int_{\rho_1=0}^{\infty} \int_{\mathbf{e}_1 \in \mathbf{S}^2} \cdots \int_{\rho_N=0}^{\infty} \int_{\mathbf{e}_N \in \mathbf{S}^2} \\ &\cdot |D_f(\mathcal{R}; \rho_1, \mathbf{e}_1, \ldots, \rho_N, \mathbf{e}_N; T_1, T_2)|^2 \\ &\cdot g_1(\rho_1)\, d\rho_1\, d\text{SA}^\star(\mathbf{e}_1) \ldots g_N(\rho_N)\, d\rho_N\, d\text{SA}^\star(\mathbf{e}_N),\end{aligned} \tag{27.10}$$

where, as usual, $d\text{SA}^\star(\mathbf{e})$ denotes the normed surface area on the unit sphere $\mathbf{S}^2$, i.e.,

$$\int_{\mathbf{S}^2} 1\, d\text{SA}^\star(\mathbf{e}) = \frac{1}{4\pi} \int_{\mathbf{S}^2} 1\, d\text{SA}(\mathbf{e}) = 1.$$

We need the following simple lemma.

Lemma 27.1. *For $\mathbf{n}_j = (n_{j,1}, n_{j,2}, \ldots, n_{j,d})$, $j = 1, 2$, let $\mathbf{n}_j(k) = (n_{j,3k-2}, n_{j,3k-1}, n_{j,3k})$ denote the kth block of triples of the coordinates, $1 \le k \le d/3 = N$. We have*

$$\begin{aligned}\Delta_f^2(\mathcal{G}; \mathcal{R}; T_1, T_2) = &\sum_{\mathbf{n}_1 \in \mathbb{Z}^d \setminus \mathbf{0}} \sum_{\mathbf{n}_2 \in \mathbb{Z}^d \setminus \mathbf{0}} a_{\mathbf{n}_1} \overline{a_{\mathbf{n}_2}} \\ &\cdot \int_{T_1}^{T_2} \int_{T_1}^{T_2} \prod_{k=1}^{N} \mathbf{F}_{g_k}(2\pi |r_k(t_1)\mathbf{n}_1(k) - r_k(t_2)\mathbf{n}_2(k)|)\, dt_1\, dt_2.\end{aligned} \tag{27.11}$$

Proof of Lemma 27.1. Using the trivial fact

$$\int_{a_1}^{b_1} h(x)\, dx \int_{a_2}^{b_2} g(y)\, dy = \int_{a_1}^{b_1} \int_{a_2}^{b_2} h(x) g(y)\, dx\, dy$$

in (27.5), we have

$$|D_f(\mathcal{R};\rho_1,\mathbf{e}_1,\ldots,\rho_N,\mathbf{e}_N;T_1,T_2)|^2 = \int_{T_1}^{T_2}\int_{T_1}^{T_2} \sum_{\mathbf{n}_1\in\mathbb{Z}^d\setminus\mathbf{0}} \sum_{\mathbf{n}_2\in\mathbb{Z}^d\setminus\mathbf{0}} a_{\mathbf{n}_1} \cdot \overline{a_{\mathbf{n}_2}} e^{2\pi\mathrm{i}(\vec{\Gamma}(t_1)\cdot\mathbf{n}_1-\vec{\Gamma}(t_2)\cdot\mathbf{n}_2)}\,dt_1\,dt_2, \tag{27.12}$$

where, as usual, $\overline{z} = x - \mathrm{i}y$ denotes the complex conjugate of a complex number $z = x + \mathrm{i}y$.

Applying (27.12) in (27.10), we obtain

$$\begin{aligned}
&\Delta_f^2(\mathcal{G};\mathcal{R};T_1,T_2)\\
&= \int_{\rho_1=0}^{\infty}\int_{\mathbf{e}_1\in\mathbf{S}^2}\cdots\int_{\rho_N=0}^{\infty}\int_{\mathbf{e}_N\in\mathbf{S}^2}\\
&\quad\cdot\left(\int_{T_1}^{T_2}\int_{T_1}^{T_2}\sum_{\mathbf{n}_1\in\mathbb{Z}^d\setminus\mathbf{0}}\sum_{\mathbf{n}_2\in\mathbb{Z}^d\setminus\mathbf{0}} a_{\mathbf{n}_1}\overline{a_{\mathbf{n}_2}} e^{2\pi\mathrm{i}(\vec{\Gamma}(t_1)\cdot\mathbf{n}_1-\vec{\Gamma}(t_2)\cdot\mathbf{n}_2)}\,dt_1\,dt_2\right)\\
&\quad\cdot g_1(\rho_1)\,d\rho_1\,d\mathrm{SA}^\star(\mathbf{e}_1)\ldots g_N(\rho_N)\,d\rho_N\,d\mathrm{SA}^\star(\mathbf{e}_N)\\
&= \sum_{\mathbf{n}_1\in\mathbb{Z}^d\setminus\mathbf{0}}\sum_{\mathbf{n}_2\in\mathbb{Z}^d\setminus\mathbf{0}} a_{\mathbf{n}_1}\overline{a_{\mathbf{n}_2}}\int_{T_1}^{T_2}\int_{T_1}^{T_2}\left(\int_{\rho_1=0}^{\infty}\int_{\mathbf{e}_1\in\mathbf{S}^2}\cdots\int_{\rho_N=0}^{\infty}\int_{\mathbf{e}_N\in\mathbf{S}^2}\right.\\
&\quad\cdot e^{2\pi\mathrm{i}(\vec{\Gamma}(t_1)\cdot\mathbf{n}_1-\vec{\Gamma}(t_2)\cdot\mathbf{n}_2)} g_1(\rho_1)\,d\rho_1\,d\mathrm{SA}^\star(\mathbf{e}_1)\ldots\\
&\quad\left. g_N(\rho_N)\,d\rho_N\,d\mathrm{SA}^\star(\mathbf{e}_N)\right)\,dt_1\,dt_2.
\end{aligned} \tag{27.13}$$

Since

$$e^{2\pi\mathrm{i}(\vec{\Gamma}(t_1)\cdot\mathbf{n}_1-\vec{\Gamma}(t_2)\cdot\mathbf{n}_2)} = \prod_{k=1}^{N} e^{2\pi\mathrm{i}\rho_k(r_k(t_1)\mathbf{n}_1(k)-r_k(t_2)\mathbf{n}_2(k))\cdot\mathbf{e}_k},$$

applying (27.8) we have

$$\int_{\rho_1=0}^{\infty}\int_{\mathbf{e}_1\in\mathbf{S}^2}\cdots\int_{\rho_N=0}^{\infty}\int_{\mathbf{e}_N\in\mathbf{S}^2} \times e^{2\pi\mathrm{i}(\vec{\Gamma}(t_1)\cdot\mathbf{n}_1-\vec{\Gamma}(t_2)\cdot\mathbf{n}_2)} g_1(\rho_1)\,d\rho_1\,d\mathrm{SA}^\star(\mathbf{e}_1)\ldots g_N(\rho_N)\,d\rho_N\,d\mathrm{SA}^\star(\mathbf{e}_N)$$

$$= \prod_{k=1}^{N} \int_{\rho_k=0}^{\infty} \int_{\mathbf{e}_k \in \mathbf{S}^2} e^{2\pi i \rho_k (r_k(t_1)\mathbf{n}_1(k) - r_k(t_2)\mathbf{n}_2(k))\cdot \mathbf{e}_k} g_k(\rho_k)\, d\rho_k\, d\mathrm{SA}^\star(\mathbf{e}_k)$$

$$= \prod_{k=1}^{N} \mathbf{F}_{g_k}(2\pi |r_k(t_1)\mathbf{n}_1(k) - r_k(t_2)\mathbf{n}_2(k)|), \tag{27.14}$$

where $N = d/3$. Combining (27.13) and (27.14), Lemma 27.1 follows. $\square$

By Lemma 27.1

$$\Delta_f^2(\mathcal{G}; \mathcal{R}; U, V;) = \sum_{\mathbf{n}_1 \in \mathbb{Z}^d \setminus \mathbf{0}} \sum_{\mathbf{n}_2 \in \mathbb{Z}^d \setminus \mathbf{0}} a_{\mathbf{n}_1} \overline{a_{\mathbf{n}_2}}$$

$$\cdot \int_U^V \int_U^V \prod_{j=1}^{N} \mathbf{F}_{g_j}(2\pi |r_j(t_1)\mathbf{n}_1(j) - r_j(t_2)\mathbf{n}_2(j)|)\, dt_1\, dt_2, \tag{27.15}$$

and first we estimate it from above under the condition that $1 \le U < V \le 2U$. The next lemma is brute force combinatorics.

Lemma 27.2. *Let* $1 \le U < V \le 2U$, *and write*

$$\mathcal{R}^{(\inf)}(U) = \min_{1 \le k \le N} \inf_{x \ge U} r_k(x) = U_1.$$

Under the condition of (27.2)–(27.4) *and*

$$N \sum_{j=1}^{\infty} 2^{3j+2} \mathbf{F}^{(\sup)}(2^{j-1}\pi U_1) < 1/10,$$

we have

$$\left|\Delta_f^2(\mathcal{G}; \mathcal{R}; U, V)\right| \le \sigma_0^2(f)(V-U)U$$

$$\cdot \left(25N^2 \sum_{j=1}^{\infty} 2^{3j+2} \mathbf{F}^{(\sup)}(2^j \pi U_1) + 5N \left(N \sum_{j=1}^{\infty} 2^{3j+2} \mathbf{F}^{(\sup)}(2^j \pi U_1) \right)^2 \right)$$

$$+ 2\sigma_0^2(f)(V-U)(2U)^\beta \left(1 + (\alpha_1 - \alpha)\log U + \alpha_1 \log 2 + \log 3\right)$$

$$\times \sum_{\ell=1}^{\infty} \ell \cdot \mathbf{F}^{(\sup)}(2\pi(\ell - 1)).$$

Proof. For every $\mathbf{n} = (n_1, \ldots, n_d) = (\mathbf{n}(1), \ldots, \mathbf{n}(N)) \in \mathbb{Z}^d \setminus \mathbf{0}$ ($d = 3N$) write $L(\mathbf{n}) = \{1 \le i \le N : \ \mathbf{n}(i) = (n_{3i-2}, n_{3i-1}, n_{3i}) \neq \mathbf{0}\}$. Applying the simple inequality $|a_{\mathbf{n}_1}\overline{a_{\mathbf{n}_2}}| \le (|a_{\mathbf{n}_1}|^2 + |a_{\mathbf{n}_2}|^2)/2$ in (27.15), and using (27.7) we have

$$\left|\Delta_f^2(\mathcal{G};\mathcal{R};U,V)\right| \le \sum_{\mathbf{n}_1\in\mathbb{Z}^d\setminus\mathbf{0}} \sum_{\mathbf{n}_2\in\mathbb{Z}^d\setminus\mathbf{0}} \frac{1}{2}\left(|a_{\mathbf{n}_1}|^2 + |a_{\mathbf{n}_2}|^2\right) \cdot \int_U^V \int_U^V$$

$$\times \prod_{j=1}^N \mathbf{F}^{(\mathrm{sup})}(2\pi|r_j(t_1)\mathbf{n}_1(j) - r_j(t_2)\mathbf{n}_2(j)|)\,dt_1\,dt_2 = \sum_{\mathbf{n}_1\in\mathbb{Z}^d\setminus\mathbf{0}} |a_{\mathbf{n}_1}|^2$$

$$\cdot \sum_{\mathbf{n}_2\in\mathbb{Z}^d\setminus\mathbf{0}} \int_U^V \int_U^V \prod_{j=1}^N \mathbf{F}^{(\mathrm{sup})}(2\pi|r_j(t_1)\mathbf{n}_1(j) - r_j(t_2)\mathbf{n}_2(j)|)\,dt_1\,dt_2$$

$$= \int_{t_1=U}^V \sum_{\mathbf{n}_1\in\mathbb{Z}^d\setminus\mathbf{0}} |a_{\mathbf{n}_1}|^2 \left(\int_{t_2=U}^V \sum_{\mathbf{n}_2\in\mathbb{Z}^d\setminus\mathbf{0}}\right.$$

$$\left.\cdot \prod_{j=1}^N \mathbf{F}^{(\mathrm{sup})}(2\pi|r_j(t_1)\mathbf{n}_1(j) - r_j(t_2)\mathbf{n}_2(j)|)\,dt_2\right) dt_1$$

$$= \sum_{\lambda_1=1}^N \int_{t_1=U}^V \sum_{\substack{\mathbf{n}_1\in\mathbb{Z}^d\setminus\mathbf{0}:\\ |L(\mathbf{n}_1)|=\lambda_1}} |a_{\mathbf{n}_1}|^2 \left(\sum_{L_{1,2}\subseteq L(\mathbf{n}_1)} \sum_{\lambda_2=\max\{|L_{1,2}|,1\}}^N \right.$$

$$\cdot \int_{t_2=U}^V \sum_{\substack{\mathbf{n}_2\in\mathbb{Z}^d\setminus\mathbf{0}:\\ |L(\mathbf{n}_2)|=\lambda_2 \text{ and } L(\mathbf{n}_2)\cap L(\mathbf{n}_1)=L_{1,2}}} \prod_{j=1}^N \mathbf{F}^{(\mathrm{sup})}(2\pi|r_j(t_1)\mathbf{n}_1(j)$$

$$\left. -r_j(t_2)\mathbf{n}_2(j)|)\,dt_2\right) dt_1. \tag{27.16}$$

We fix $t_1 \in [U,V]$, $\mathbf{n}_1 \in \mathbb{Z}^d \setminus \mathbf{0}$, $L_{1,2} \subseteq L(\mathbf{n}_1)$ and λ_2, and focus on the inner integral at the end of (27.16).

Write

$$\lambda_{1,2} = |L_{1,2}| = |\{1 \le i \le N : \ \mathbf{n}_1(i) \neq \mathbf{0} \quad \text{and} \quad \mathbf{n}_2(i) \neq \mathbf{0}\}|. \tag{27.17}$$

Let $k_1(\mathbf{n}_2)$ denote the number of indices i such that the coordinate-triple $\mathbf{n}_2(i)$ of $\mathbf{n}_2$ satisfies $1 \le |\mathbf{n}_2(i)| < 2$ and $\mathbf{n}_1(i) = \mathbf{0}$; let $k_2(\mathbf{n}_2)$ denote the number of indices i such that the coordinate-triple $\mathbf{n}_2(i)$ of $\mathbf{n}_2$ satisfies $2 \le |\mathbf{n}_2(i)| < 4$ and $\mathbf{n}_1(i) = \mathbf{0}$; let $k_3(\mathbf{n}_2)$ denote the number of indices i such that the coordinate-triple $\mathbf{n}_2(i)$ of $\mathbf{n}_2$ satisfies $4 \le |\mathbf{n}_2(i)| < 8$ and $\mathbf{n}_1(i) = \mathbf{0}$; and so on, keeping up the doubling. Note that

$$k_1(\mathbf{n}_2) + k_2(\mathbf{n}_2) + k_3(\mathbf{n}_2) + \cdots = \lambda_2 - \lambda_{1,2}. \tag{27.18}$$

Let $h_0(t_2; \mathbf{n}_2)$ denote the number of indices $j \in L_{1,2}$ such that $|r_j(t_1)\mathbf{n}_1(j) - r_j(t_2)\mathbf{n}_2(j)| < U_1/2$; let $h_1(t_2; \mathbf{n}_2)$ denote the number of indices $j \in L_{1,2}$ such that $U_1/2 \le |r_j(t_1)\mathbf{n}_1(j) - r_j(t_2)\mathbf{n}_2(j)| < U_1$; let $h_2(t_2; \mathbf{n}_2)$ denote the number of indices $j \in L_{1,2}$ such that $U_1 \le |r_j(t_1)\mathbf{n}_1(j) - r_j(t_2)\mathbf{n}_2(j)| < 2U_1$; let $h_3(t_2; \mathbf{n}_2)$ denote the number of indices $j \in L_{1,2}$ such that $2U_1 \le |r_j(t_1)\mathbf{n}_1(j) - r_j(t_2)\mathbf{n}_2(j)| < 4U_1$; and so on, keeping up the doubling. Note that

$$h_0(t_2; \mathbf{n}_2) + h_1(t_2; \mathbf{n}_2) + h_2(t_2; \mathbf{n}_2) + h_3(t_2; \mathbf{n}_2) + \cdots = \lambda_{1,2}. \tag{27.19}$$

By using the definitions of $k_i(\mathbf{n}_2)$ and $h_i(t_2; \mathbf{n}_2)$, we have

$$\begin{aligned}
&\int_{t_2=U}^{V} \sum_{\substack{\mathbf{n}_2 \in \mathbb{Z}^d \setminus \mathbf{0}: \\ |L(\mathbf{n}_2)| = \lambda_2 \text{ and } L(\mathbf{n}_2) \cap L(\mathbf{n}_1) = L_{1,2}}} \prod_{j=1}^{N} \mathbf{F}^{(\sup)}(2\pi |r_j(t_1)\mathbf{n}_1(j) \\
&\quad - r_j(t_2)\mathbf{n}_2(j)|)\, dt_2 \le \prod_{j \in L(\mathbf{n}_1) \setminus L_{1,2}} \mathbf{F}^{(\sup)}(2\pi |r_j(t_1)\mathbf{n}_1(j)|) \\
&\quad \times \int_{t_2=U}^{V} \sum_{\substack{\mathbf{n}_2 \in \mathbb{Z}^d \setminus \mathbf{0}: \\ |L(\mathbf{n}_2)| = \lambda_2 \text{ and } L(\mathbf{n}_2) \cap L(\mathbf{n}_1) = L_{1,2}}} \\
&\quad \cdot \left(\mathbf{F}^{(\sup)}(2\pi \mathcal{R}^{(\inf)}(t_2))\right)^{k_1(\mathbf{n}_2)} (\mathbf{F}^{(\sup)}(2\pi \cdot 2\mathcal{R}^{(\inf)}(t_2))^{k_2(\mathbf{n}_2)} \\
&\quad \cdot \left(\mathbf{F}^{(\sup)}(2\pi \cdot 2^2 \mathcal{R}^{(\inf)}(t_2))\right)^{k_3(\mathbf{n}_2)} \cdots \\
&\quad \cdot \left(\mathbf{F}^{(\sup)}(2\pi U_1/2)\right)^{h_0(t_2;\mathbf{n}_2)} \left(\mathbf{F}^{(\sup)}(2\pi U_1)\right)^{h_1(t_2;\mathbf{n}_2)} \\
&\quad \times \left(\mathbf{F}^{(\sup)}(2\pi \cdot 2U_1)\right)^{h_2(t_2;\mathbf{n}_2)} \\
&\quad \cdot \left(\mathbf{F}^{(\sup)}(2\pi \cdot 2^2 U_1)\right)^{h_3(t_2;\mathbf{n}_2)} \cdots \, dt_2.
\end{aligned} \tag{27.20}$$

We estimate the long sum at the end of (27.20). By using the definitions of $k_i(\mathbf{n}_2)$, $h_i(t_2;\mathbf{n}_2)$, (27.18) and (27.19), we obtain the upper bound

$$\begin{aligned}
&\text{EndSum of (27.20)}\\
&= \sum_{\substack{\mathbf{n}_2\in\mathbb{Z}^d\setminus\mathbf{0}:\\ |L(\mathbf{n}_2)|=\lambda_2 \text{ and } L(\mathbf{n}_2)\cap L(\mathbf{n}_1)=L_{1,2}}} (\mathbf{F}^{(\sup)}(2\pi\mathcal{R}^{(\inf)}(t_2)))^{k_1(\mathbf{n}_2)}\\
&\quad\times(\mathbf{F}^{(\sup)}(2\pi\cdot 2\mathcal{R}^{(\inf)}(t_2)))^{k_2(\mathbf{n}_2)}\cdot(\mathbf{F}^{(\sup)}(2\pi\cdot 2^2\mathcal{R}^{(\inf)}(t_2)))^{k_3(\mathbf{n}_2)}\cdots\\
&\quad\cdot(\mathbf{F}^{(\sup)}(2\pi U_1/2))^{h_1(t_2;\mathbf{n}_2)}(\mathbf{F}^{(\sup)}(2\pi U_1))^{h_2(t_2;\mathbf{n}_2)}\cdot\\
&\quad\cdot(\mathbf{F}^{(\sup)}(2\pi\cdot 2U_1))^{h_3(t_2;\mathbf{n}_2)}(\mathbf{F}^{(\sup)}(2\pi\cdot 2^2U_1))^{h_4(t_2;\mathbf{n}_2)}\cdots\\
&\le \sum_{\substack{(k_1,\dots,k_r)\text{ with } r\ge 1:\\ k_1+\cdots+k_r=\lambda_2-\lambda_{1,2},k_j\ge 0,1\le j\le r-1 \text{ and } k_r\ge 1}} \binom{N-\lambda_1}{k_1}((2+1)^3)^{k_1}\\
&\quad\times(\mathbf{F}^{(\sup)}(2\pi\mathcal{R}^{(\inf)}(t_2)))^{k_1}\cdot\\
&\quad\cdot\binom{N-\lambda_1-k_1}{k_2}((2\cdot 2+1)^3)^{k_2}(\mathbf{F}^{(\sup)}(2\pi\cdot 2\mathcal{R}^{(\inf)}(t_2)))^{k_2}\cdot\\
&\quad\cdot\binom{N-\lambda_1-k_1-k_2}{k_3}((2\cdot 2^2+1)^3)^{k_3}(\mathbf{F}^{(\sup)}(2\pi\cdot 2^2\mathcal{R}^{(\inf)}(t_2)))^{k_3}\cdots\\
&\quad\cdot\sum_{\substack{(h_0,h_1,\dots,h_r)\text{ with } r\ge 0:\\ h_0+h_1+\cdots+h_r=\lambda_{1,2},h_j\ge 0,0\le j\le r-1 \text{ and } h_r\ge 1}} \binom{\lambda_{1,2}}{h_0}\cdot\\
&\quad\cdot\binom{\lambda_{1,2}-h_0}{h_1}((2+1)^3)^{h_1}(\mathbf{F}^{(\sup)}(2\pi U_1/2))^{h_1}\cdot\\
&\quad\cdot\binom{\lambda_{1,2}-h_0-h_1}{h_2}((2\cdot 2+1)^3)^{h_2}(\mathbf{F}^{(\sup)}(2\pi U_1))^{h_2}\cdot\\
&\quad\cdot\binom{\lambda_{1,2}-h_0-h_1-h_2}{h_3}((2\cdot 2^2+1)^3)^{h_3}(\mathbf{F}^{(\sup)}(2\pi\cdot 2U_1))^{h_3}\cdot\\
&\quad\cdot\binom{\lambda_{1,2}-h_0-h_1-h_2-h_3}{h_4}((2\cdot 2^3+1)^3)^{h_4}(\mathbf{F}^{(\sup)}(2\pi\cdot 2^2U_1))^{h_4}\cdots.
\end{aligned} \tag{27.21}$$

Note that (27.21) includes the pathological case $\lambda_2 - \lambda_{1,2} = 0$ with the natural convention that the summation means the single term $(k_1, \ldots, k_r) = (0)$, and similarly, if $\lambda_{1,2} = 0$ then $(h_0, h_1, \ldots, h_r)$ is just the single term (0). Moreover, to explain the binomial coefficient factor $\binom{\lambda_{1,2}}{h_0}$ in (27.21), we note that for every given half-open axis-parallel cube Q of side length U_1, there is at most one integer lattice point $\mathbf{m}_2 \in \mathbb{Z}^3$ such that $r_k(t_2)\mathbf{m}_2 \in Q$ (see the definition of U_1 in Lemma 27.2). The same argument explains the next binomial coefficient factor $\binom{\lambda_{1,2}-h_0}{h_1}$ in (27.21), and so on. The extra factor $((2{+}1)^3)^{h_1}$ is a generous upper bound: it comes from the contribution of the first neighborhood of size $(2+1)^3 - 1$ of a lattice point $\mathbf{m}_2 \in \mathbb{Z}^3$; and so on. The same argument explains the factors

$$\binom{N-\lambda_1}{k_1}((2+1)^3)^{k_1}, \quad \binom{N-\lambda_1-k_1}{k_2}((2\cdot 2+1)^3)^{k_2}, \text{ and so on}$$

in (27.21).

Applying the multinomial theorem twice in (27.21) the same way as we did in (13.15)–(13.16), we have the very generous upper bound (where the extra factors of N come from the trivial upper bound $\binom{N}{k} \le N^k$)

$$\begin{aligned}
&\text{EndSum of (27.20)}\\
&\le \big(N\cdot(2+1)^3\mathbf{F}^{(\mathrm{sup})}(2\pi\mathcal{R}^{(\mathrm{inf})}(t_2))\\
&\quad + N\cdot(2\cdot 2+1)^3\mathbf{F}^{(\mathrm{sup})}(2\pi\cdot 2\mathcal{R}^{(\mathrm{inf})}(t_2))\\
&\quad + N\cdot(2\cdot 2^2+1)^3\mathbf{F}^{(\mathrm{sup})}(2\pi\cdot 2^2\mathcal{R}^{(\mathrm{inf})}(t_2)) + \cdots\big)^{\lambda_2-\lambda_{1,2}}\\
&\quad \cdot\big(1+(2+1)^3\mathbf{F}^{(\mathrm{sup})}(2\pi U_1/2)\\
&\quad + (2\cdot 2+1)^3\mathbf{F}^{(\mathrm{sup})}(2\pi U_1) + (2\cdot 2^2+1)^3\mathbf{F}^{(\mathrm{sup})}(2\pi\cdot 2U_1)\\
&\quad + (2\cdot 2^3+1)^3\mathbf{F}^{(\mathrm{sup})}(2\pi\cdot 2^2U_1) + \cdots\big)^{\lambda_{1,2}}.
\end{aligned} \tag{27.22}$$

Next we use (27.21)–(27.22) in (27.20):

$$\begin{aligned}
&\int_{t_2=U}^{V} \sum_{\substack{\mathbf{n}_2\in\mathbb{Z}^d\setminus\mathbf{0}:\\ |L(\mathbf{n}_2)|=\lambda_2 \text{ and } L(\mathbf{n}_2)\cap L(\mathbf{n}_1)=L_{1,2}}} \prod_{j=1}^{N} \mathbf{F}^{(\mathrm{sup})}(2\pi|r_j(t_1)\mathbf{n}_1(j)\\
&\quad -r_j(t_2)\mathbf{n}_2(j)|)\,dt_2 \le (V-U) \prod_{j\in L(\mathbf{n}_1)\setminus L_{1,2}} \mathbf{F}^{(\mathrm{sup})}(2\pi|r_j(t_1)\mathbf{n}_1(j)|)
\end{aligned}$$

$$\cdot \left(N \sum_{j=1}^{\infty} 2^{3j+2} \mathbf{F}^{(\text{sup})}(2^j \pi U_1) \right)^{\lambda_2 - \lambda_{1,2}}$$

$$\times \left(1 + \sum_{j=1}^{\infty} 2^{3j+2} \mathbf{F}^{(\text{sup})}(2^{j-1} \pi U_1) \right)^{\lambda_{1,2}} . \tag{27.23}$$

Let us return now to (27.16); we have the following decomposition into four parts:

$$\left| \Delta_f^2(\mathcal{G}; \mathcal{R}; U, V) \right| \leq \text{Part One} + \text{Part Two} + \text{Part Three} + \text{Part Four}, \tag{27.24}$$

where

$$\text{Part One} = \sum_{\lambda_1=1}^{N-1} \int_{t_1=U}^{V} \sum_{\substack{\mathbf{n}_1 \in \mathbb{Z}^d \setminus \mathbf{0}: \\ |L(\mathbf{n}_1)|=\lambda_1}} |a_{\mathbf{n}_1}|^2 \left(\sum_{L_{1,2} \subseteq L(\mathbf{n}_1)} \sum_{\lambda_2=\lambda_1+1}^{N} \right.$$

$$\cdot \int_{t_2=U}^{V} \sum_{\substack{\mathbf{n}_2 \in \mathbb{Z}^d \setminus \mathbf{0}: \\ |L(\mathbf{n}_2)|=\lambda_2 \text{ and } L(\mathbf{n}_2) \cap L(\mathbf{n}_1) = L_{1,2}}} \prod_{j=1}^{N} \mathbf{F}^{(\text{sup})}(2\pi |r_j(t_1)\mathbf{n}_1(j)$$

$$\left. - r_j(t_2)\mathbf{n}_2(j)|) \, dt_2 \right) dt_1 \tag{27.25}$$

is characterized by the property $\lambda_1 < \lambda_2$,

$$\text{Part Two} = \sum_{\lambda_1=1}^{N} \int_{t_1=U}^{V} \sum_{\substack{\mathbf{n}_1 \in \mathbb{Z}^d \setminus \mathbf{0}: \\ |L(\mathbf{n}_1)|=\lambda_1}} |a_{\mathbf{n}_1}|^2 \left(\sum_{L_{1,2} \subset L(\mathbf{n}_1)} \sum_{\lambda_2=\max\{|L_{1,2}|,1\}}^{\lambda_1-1} \right.$$

$$\cdot \int_{t_2=U}^{V} \sum_{\substack{\mathbf{n}_2 \in \mathbb{Z}^d \setminus \mathbf{0}: \\ |L(\mathbf{n}_2)|=\lambda_2 \text{ and } L(\mathbf{n}_2) \cap L(\mathbf{n}_1) = L_{1,2}}} \prod_{j=1}^{N} \mathbf{F}^{(\text{sup})}(2\pi |r_j(t_1)\mathbf{n}_1(j)$$

$$\left. - r_j(t_2)\mathbf{n}_2(j)|) \, dt_2 \right) dt_1 \tag{27.26}$$

is characterized by the property $\lambda_1 > \lambda_2$, and finally, we split the case $\lambda_1 = \lambda_2$ into two subcases according as $L(\mathbf{n}_1) \neq L(\mathbf{n}_2)$ or $L(\mathbf{n}_1) = L(\mathbf{n}_2)$:

$$\text{Part Three} = \sum_{\lambda_1=1}^{d} \int_{t_1=U}^{V} \sum_{\substack{\mathbf{n}_1 \in \mathbb{Z}^d \setminus \mathbf{0}: \\ |L(\mathbf{n}_1)|=\lambda_1}} |a_{\mathbf{n}_1}|^2 \left(\sum_{\substack{L_{1,2} \subset L(\mathbf{n}_1): \\ L_{1,2} \neq L(\mathbf{n}_1)}} \right.$$

$$\cdot \int_{t_2=U}^{V} \sum_{\substack{\mathbf{n}_2 \in \mathbb{Z}^d \setminus \mathbf{0}: \\ |L(\mathbf{n}_2)|=\lambda_1 \text{ and } L(\mathbf{n}_2) \cap L(\mathbf{n}_1) = L_{1,2}}} \prod_{j=1}^{N} \mathbf{F}^{(\text{sup})}(2\pi |r_j(t_1)\mathbf{n}_1(j)$$

$$\left. - r_j(t_2)\mathbf{n}_2(j)|)\, dt_2 \right) dt_1 \tag{27.27}$$

and

$$\text{Part Four} = \int_{t_1=U}^{V} \sum_{\mathbf{n}_1 \in \mathbb{Z}^d \setminus \mathbf{0}} |a_{\mathbf{n}_1}|^2$$

$$\cdot \left(\int_{t_2=U}^{V} \sum_{\substack{\mathbf{n}_2 \in \mathbb{Z}^d \setminus \mathbf{0}: \\ L(\mathbf{n}_2)=L(\mathbf{n}_1)}} \prod_{j=1}^{N} \mathbf{F}^{(\text{sup})}(2\pi |r_j(t_1)\mathbf{n}_1(j) \right.$$

$$\left. - r_j(t_2)\mathbf{n}_2(j)|)\, dt_2 \right) dt_1. \tag{27.28}$$

To estimate (27.28), we are going to use a simple but important lemma, which is an analog of Lemma 9.3. First a definition: given real numbers C, C' and a positive function $r(x)$, consider the set

$$\mathcal{B}_U(r; C; C') = \{t \in [U, 2U] : \text{there exists } n \in \mathbb{Z} \setminus \{0\}$$
$$\text{such that } |C - r(t)n| \leq C'\}.$$

We give an upper bound on the 1-dimensional Lebesgue measure (i.e., the length) of the set $\mathcal{B}_U(r; C; C')$.

Lemma 27.3. *Assume that*

$$\inf_{x\geq U} r(x) \geq U_1, \quad 1 \leq C' < U_1/2, \quad \sup_{U\leq x\leq 2U} r(x) \leq U_2, \tag{27.29}$$

and

$$\sup_{I\subset\mathbb{R}} \frac{\text{length}\{U \leq t \leq 2U: \ r(t) \in I\}}{\text{length}(I)} \leq (2U)^{\beta} \tag{27.30}$$

uniformly for all $U \geq \beta_0$, *where the supremum in* (27.30) *is taken over all finite intervals* $I \subset \mathbb{R}$ *of positive length, and* $\beta_0 \geq 1, 0 \leq \beta < 1$ *are absolute constants. Then*

$$\text{length}(\mathcal{B}_U(r; C; C')) \leq 2C' \cdot (2U)^{\beta} \left(1 + \log \frac{3U_2}{U_1}\right).$$

Proof. We can assume without loss of generality that $C \geq 0$. Clearly

$$|C - r(t)n| \leq C' \iff \frac{C - C'}{n} \leq r\left(t \leq \frac{C + C'}{n}\right),$$

so by (27.30)

$$\text{length}(\mathcal{B}_U(r; C; C')) \leq (2U)^{\beta} \sum\nolimits_n^{\star} \frac{2C'}{n}, \tag{27.31}$$

where the summation $\sum_n^{\star}$ is extended over all positive integers n such that

$$\max\left\{\frac{C - C'}{r(t)}, 1\right\} \leq n \leq \frac{C + C'}{r(t)}. \tag{27.32}$$

By (27.29)

$$\frac{C + C'}{r(t)} \leq \frac{C + (U_1/2)}{U_1} = \xi_2 \tag{27.33}$$

and

$$\max\left\{\frac{C - C'}{r(t)}, 1\right\} \geq \max\left\{\frac{C - (U_1/2)}{U_2}, 1\right\} = \xi_1. \tag{27.34}$$

To estimate the sum

$$\sum_{\xi_1\leq n\leq \xi_2} \frac{1}{n},$$

we distinguish two cases. If $0 \le C < U_1$ then $\xi_2 \le 3/2$, and so we have the trivial upper bound

$$\sum_{\xi_1 \le n \le \xi_2} \frac{1}{n} \le \frac{1}{1} = 1. \tag{27.35}$$

If $C \ge U_1$, then

$$\frac{C/2}{U_2} \le \xi_1 \quad \text{and} \quad \xi_2 \le \frac{3C/2}{U_1},$$

and so

$$\begin{aligned}\sum_{\xi_1 \le n \le \xi_2} \frac{1}{n} &\le \sum_{C/(2U_2) \le n \le 3C/(2U_1)} \\ &\le 1 + \log \frac{3C/(2U_1)}{C/(2U_2)} = 1 + \log \frac{3U_2}{U_1}. \end{aligned} \tag{27.36}$$

Combining (27.31)–(27.36), we conclude

$$\text{length}(\mathcal{B}_U(r;C;C')) \le (2U)^{\beta} {\sum_n}^{\star} \frac{2C'}{n} \le 2C' \cdot (2U)^{\beta} \left(1 + \log \frac{3U_2}{U_1}\right),$$

which completes the proof of Lemma 27.3. □

28. Completing the Proof of Lemma 27.2

Applying (27.23) in (27.25), we have

$$\begin{aligned}\text{Part One} \le \sum_{\lambda_1=1}^{N-1} \int_{t_1=U}^{V} \sum_{\substack{\mathbf{n}_1 \in \mathbb{Z}^d \setminus \mathbf{0}: \\ |L(\mathbf{n}_1)|=\lambda_1}} |a_{\mathbf{n}_1}|^2 \Bigg(\sum_{L_{1,2} \subseteq L(\mathbf{n}_1)} \sum_{\lambda_2=\lambda_1+1}^{N} \\ \times (V-U) \cdot \mathbf{F}^{(\text{sup})}(2\pi U_1)^{\lambda_1 - \lambda_{1,2}} \left(N \sum_{j=1}^{\infty} 2^{3j+2} \mathbf{F}^{(\text{sup})}(2^j \pi U_1) \right)^{\lambda_2 - \lambda_{1,2}} \\ \cdot \left(1 + \sum_{j=1}^{\infty} 2^{3j+2} \mathbf{F}^{(\text{sup})}(2^{j-1} \pi U_1) \right)^{\lambda_{1,2}} \Bigg) \, dt_1\end{aligned}$$

$$\leq (V-U)^2 \max_{\lambda_{1,2}\leq N} \left(1+\sum_{j=1}^{\infty} 2^{3j+2}\mathbf{F}^{(\mathrm{sup})}(2^{j-1}\pi U_1)\right)^{\lambda_{1,2}}$$

$$\cdot \sum_{\lambda_2=1}^{N} \left(N\sum_{j=1}^{\infty} 2^{3j+2}\mathbf{F}^{(\mathrm{sup})}(2^{j}\pi U_1)\right)^{\lambda_2} \sum_{\lambda_1=1}^{\lambda_2-1} \sum_{\substack{\mathbf{n}_1\in\mathbb{Z}^d\setminus\mathbf{0}:\\ |L(\mathbf{n}_1)|=\lambda_1}} |a_{\mathbf{n}_1}|^2$$

$$\cdot \left(\sum_{\lambda_{1,2}=0}^{\lambda_1} \binom{\lambda_1}{\lambda_{1,2}} \left(N\sum_{j=1}^{\infty} 2^{3j+2}\mathbf{F}^{(\mathrm{sup})}(2^{j}\pi U_1)\right)^{-\lambda_{1,2}}\right.$$

$$\left.\times \mathbf{F}^{(\mathrm{sup})}(2\pi U_1)^{\lambda_1-\lambda_{1,2}}\right)$$

$$= (V-U)^2 \left(1+\sum_{j=1}^{\infty} 2^{3j+2}\mathbf{F}^{(\mathrm{sup})}(2^{j-1}\pi U_1)\right)^{N}$$

$$\times \sum_{\lambda_2=1}^{N} \left(N\sum_{j=1}^{\infty} 2^{3j+2}\mathbf{F}^{(\mathrm{sup})}(2^{j}\pi U_1)\right)^{\lambda_2} \cdot \sum_{\lambda_1=1}^{\lambda_2-1}$$

$$\times \sum_{\substack{\mathbf{n}_1\in\mathbb{Z}^d\setminus\mathbf{0}:\\ |L(\mathbf{n}_1)|=\lambda_1}} |a_{\mathbf{n}_1}|^2 \left(\left(N\sum_{j=1}^{\infty} 2^{3j+2}\mathbf{F}^{(\mathrm{sup})}(2^{j}\pi U_1)\right)^{-1} + \mathbf{F}^{(\mathrm{sup})}(2\pi U_1)\right)^{\lambda_1}. \tag{28.1}$$

By hypothesis

$$N\sum_{j=1}^{\infty} 2^{3j+2}\mathbf{F}^{(\mathrm{sup})}(2^{j-1}\pi U_1) < 1/10, \tag{28.2}$$

and using it we have

$$\left(\left(N\sum_{j=1}^{\infty} 2^{3j+2}\mathbf{F}^{(\mathrm{sup})}(2^{j}\pi U_1)\right)^{-1} + \mathbf{F}^{(\mathrm{sup})}(2\pi U_1)\right)^{\lambda_1}$$

$$= \left(N\sum_{j=1}^{\infty} 2^{3j+2}\mathbf{F}^{(\mathrm{sup})}(2^{j}\pi U_1)\right)^{-\lambda_1}$$

$$\times \left(1 + \mathbf{F}^{(\text{sup})}(2\pi U_1) \cdot N \sum_{j=1}^{\infty} 2^{3j+2} \mathbf{F}^{(\text{sup})}(2^j \pi U_1)\right)^{\lambda_1}$$

$$\leq \left(N \sum_{j=1}^{\infty} 2^{3j+2} \mathbf{F}^{(\text{sup})}(2^j \pi U_1)\right)^{-\lambda_1}$$

$$\times \left(1 + N \left(\sum_{j=1}^{\infty} 2^{3j+2} \mathbf{F}^{(\text{sup})}(2^j \pi U_1)\right)^2\right)^{\lambda_1}$$

$$\leq \left(N \sum_{j=1}^{\infty} 2^{3j+2} \mathbf{F}^{(\text{sup})}(2^j \pi U_1)\right)^{-\lambda_1} \left(1 + \frac{1}{N}\right)^{\lambda_1}$$

$$\leq 3 \left(N \sum_{j=1}^{\infty} 2^{3j+2} \mathbf{F}^{(\text{sup})}(2^j \pi U_1)\right)^{-\lambda_1}, \tag{28.3}$$

since

$$\left(1 + \frac{1}{N}\right)^{\lambda_1} \leq \left(1 + \frac{1}{N}\right)^{N} < 3.$$

By using (28.2)–(28.3) in (28.1), we obtain

$$\text{Part One} \leq (V - U)^2 \left(1 + \frac{1}{N}\right)^{N}$$

$$\times \sum_{\lambda_2=1}^{N} \sum_{\lambda_1=1}^{\lambda_2-1} \left(N \sum_{j=1}^{\infty} 2^{3j+2} \mathbf{F}^{(\text{sup})}(2^j \pi U_1)\right)^{\lambda_2}$$

$$\cdot 3 \left(N \sum_{j=1}^{\infty} 2^{3j+2} \mathbf{F}^{(\text{sup})}(2^j \pi U_1)\right)^{-\lambda_1} \sum_{\mathbf{n} \in \mathbb{Z}^d \setminus \mathbf{0}} |a_{\mathbf{n}}|^2$$

$$\leq 3 \cdot 3(V - U)^2 \sum_{\lambda_2=1}^{N} \sum_{\lambda_1=1}^{\lambda_2-1} \left(N \sum_{j=1}^{\infty} 2^{3j+2} \mathbf{F}^{(\text{sup})}(2^j \pi U_1)\right)^{\lambda_2-\lambda_1} \sum_{\mathbf{n} \in \mathbb{Z}^d \setminus \mathbf{0}} |a_{\mathbf{n}}|^2. \tag{28.4}$$

By using the substitution $i = \lambda_2 - \lambda_1 \geq 1$ in (28.4), we obtain the upper bound

$$\text{Part One} \leq 9(V-U)^2 \sum_{\mathbf{n} \in \mathbb{Z}^d \setminus \mathbf{0}} |a_{\mathbf{n}}|^2$$

$$\cdot N \sum_{i=1}^{\infty} \left(N \sum_{j=1}^{\infty} 2^{3j+2} \mathbf{F}^{(\sup)}(2^j \pi U_1) \right)^i$$

$$= 9(V-U)^2 \frac{N^2 \sum_{j=1}^{\infty} 2^{3j+2} \mathbf{F}^{(\sup)}(2^j \pi U_1)}{1 - N \sum_{j=1}^{\infty} 2^{3j+2} \mathbf{F}^{(\sup)}(2^j \pi U_1)} \sigma_0^2(f), \qquad (28.5)$$

where in the last step we used Parseval's formula.

Next we apply (27.23) in (27.26):

Part Two

$$\leq \sum_{\lambda_1=1}^{N} \int_{t_1=U}^{V} \sum_{\substack{\mathbf{n}_1 \in \mathbb{Z}^d \setminus \mathbf{0}: \\ |L(\mathbf{n}_1)| = \lambda_1}} |a_{\mathbf{n}_1}|^2 \left(\sum_{L_{1,2} \subset L(\mathbf{n}_1)} \sum_{\lambda_2 = \max\{|L_{1,2}|, 1\}}^{\lambda_1 - 1} \right.$$

$$\times (V-U) \cdot \mathbf{F}^{(\sup)}(2\pi U_1)^{\lambda_1 - \lambda_{1,2}} \left(N \sum_{j=1}^{\infty} 2^{3j+2} \mathbf{F}^{(\sup)}(2^j \pi U_1) \right)^{\lambda_2 - \lambda_{1,2}}$$

$$\left. \cdot \left(1 + \sum_{j=1}^{\infty} 2^{3j+2} \mathbf{F}^{(\sup)}(2^{j-1} \pi U_1) \right)^{\lambda_{1,2}} \right) dt_1$$

$$\leq (V-U)^2 \left(1 + \sum_{j=1}^{\infty} 2^{3j+2} \mathbf{F}^{(\sup)}(2^{j-1} \pi U_1) \right)^N \sum_{\lambda_1=1}^{N} \sum_{\substack{\mathbf{n}_1 \in \mathbb{Z}^d \setminus \mathbf{0}: \\ |L(\mathbf{n}_1)| = \lambda_1}} |a_{\mathbf{n}_1}|^2$$

$$\cdot \sum_{\lambda_2=1}^{\lambda_1 - 1} \sum_{\lambda_{1,2}=0}^{\lambda_2} \binom{\lambda_1}{\lambda_{1,2}} \left(N \sum_{j=1}^{\infty} 2^{3j+2} \mathbf{F}^{(\sup)}(2^j \pi U_1) \right)^{\lambda_2 - \lambda_{1,2}}$$

$$\times \left(\mathbf{F}^{(\sup)}(2\pi U_1) \right)^{\lambda_1 - \lambda_{1,2}}$$

$$\leq (V-U)^2 \left(1+\sum_{j=1}^{\infty} 2^{3j+2}\mathbf{F}^{(\mathrm{sup})}(2^{j-1}\pi U_1)\right)^N \sum_{\lambda_1=1}^{N} \sum_{\substack{\mathbf{n}_1\in\mathbb{Z}^d\setminus\mathbf{0}:\\ |L(\mathbf{n}_1)|=\lambda_1}} |a_{\mathbf{n}_1}|^2$$

$$\cdot \sum_{\lambda_2=1}^{\lambda_1-1} \sum_{\lambda_{1,2}=0}^{\lambda_2} \left(N\sum_{j=1}^{\infty} 2^{3j+2}\mathbf{F}^{(\mathrm{sup})}(2^{j}\pi U_1)\right)^{\lambda_2-\lambda_{1,2}}$$

$$\times \left(N\cdot \mathbf{F}^{(\mathrm{sup})}(2\pi U_1)\right)^{\lambda_1-\lambda_{1,2}} \leq (V-U)^2 \left(1+\frac{1}{N}\right)^N \sum_{\mathbf{n}\in\mathbb{Z}^d\setminus\mathbf{0}} |a_{\mathbf{n}}|^2$$

$$\cdot \sum_{\lambda_1=1}^{N} \sum_{\lambda_2=1}^{\lambda_1-1} \sum_{\lambda_{1,2}=0}^{\lambda_2} \left(N\sum_{j=1}^{\infty} 2^{3j+2}\mathbf{F}^{(\mathrm{sup})}(2^{j}\pi U_1)\right)^{\lambda_1+\lambda_2-2\lambda_{1,2}}, \qquad (28.6)$$

where in the middle of the argument we used the trivial inequality

$$\binom{\lambda_1}{\lambda_{1,2}} = \binom{\lambda_1}{\lambda_1-\lambda_{1,2}} \leq \lambda_1^{\lambda_1-\lambda_{1,2}} \leq N^{\lambda_1-\lambda_{1,2}}, \qquad (28.7)$$

and at the end we used (28.2).

Using the substitution $i=\lambda_1+\lambda_2-2\lambda_{1,2}$ in (28.6), we have

$$\text{Part Two} \leq (V-U)^2\cdot 3\cdot \sum_{\mathbf{n}\in\mathbb{Z}^d\setminus\mathbf{0}} |a_{\mathbf{n}}|^2$$

$$\cdot \sum_{i=1}^{\infty} N(i+1)i \left(N\sum_{j=1}^{\infty} 2^{3j+2}\mathbf{F}^{(\mathrm{sup})}(2^{j}\pi U_1)\right)^{i}$$

$$= 3(V-U)^2 \sum_{\mathbf{n}\in\mathbb{Z}^d\setminus\mathbf{0}} |a_{\mathbf{n}}|^2$$

$$\cdot N\frac{3N\sum_{j=1}^{\infty} 2^{3j+2}\mathbf{F}^{(\mathrm{sup})}(2^{j}\pi U_1)}{\left(1-N\sum_{j=1}^{\infty} 2^{3j+2}\mathbf{F}^{(\mathrm{sup})}(2^{j}\pi U_1)\right)^3}$$

$$= 9(V-U)^2\frac{N^2\sum_{j=1}^{\infty} 2^{3j+2}\mathbf{F}^{(\mathrm{sup})}(2^{j}\pi U_1)}{\left(1-N\sum_{j=1}^{\infty} 2^{3j+2}\mathbf{F}^{(\mathrm{sup})}(2^{j}\pi U_1)\right)^3}\sigma_0^2(f), \qquad (28.8)$$

where in the last step we used Parseval's formula.

Next we apply (27.23) in (27.27):

Part Three

$$\le \sum_{\lambda_1=1}^{N} \int_{t_1=U}^{V} \sum_{\substack{\mathbf{n}_1\in\mathbb{Z}^d\setminus\mathbf{0}:\\ |L(\mathbf{n}_1)|=\lambda_1}} |a_{\mathbf{n}_1}|^2 \left(\sum_{\substack{L_{1,2}\subset L(\mathbf{n}_1):\\ L_{1,2}\ne L(\mathbf{n}_1)}} \right.$$

$$\times (V-U)\cdot \mathbf{F}^{(\mathrm{sup})}(2\pi U_1)^{\lambda_1-\lambda_{1,2}} \left(N \sum_{j=1}^{\infty} 2^{3j+2}\mathbf{F}^{(\mathrm{sup})}(2^j\pi U_1)\right)^{\lambda_2-\lambda_{1,2}}$$

$$\left. \cdot \left(1+\sum_{j=1}^{\infty} 2^{3j+2}\mathbf{F}^{(\mathrm{sup})}(2^{j-1}\pi U_1)\right)^{\lambda_{1,2}} \right) dt_1$$

$$\le (V-U)^2 \left(1+\sum_{j=1}^{\infty} 2^{3j+2}\mathbf{F}^{(\mathrm{sup})}(2^{j-1}\pi U_1)\right)^{N} \sum_{\lambda_1=1}^{N} \sum_{\substack{\mathbf{n}_1\in\mathbb{Z}^d\setminus\mathbf{0}:\\ |L(\mathbf{n}_1)|=\lambda_1}} |a_{\mathbf{n}_1}|^2$$

$$\cdot \sum_{\lambda_{1,2}=0}^{\lambda_1-1} \binom{\lambda_1}{\lambda_{1,2}} \left(\mathbf{F}^{(\mathrm{sup})}(2\pi U_1)\cdot N \sum_{j=1}^{\infty} 2^{3j+2}\mathbf{F}^{(\mathrm{sup})}(2^j\pi U_1)\right)^{\lambda_1-\lambda_{1,2}}. \tag{28.9}$$

Using (28.7), (28.2) and the substitution $i=\lambda_1-\lambda_{1,2}$ in (28.9), we have

Part Three

$$\le (V-U)^2\left(1+\frac{1}{N}\right)^{N} \sum_{\mathbf{n}\in\mathbb{Z}^d\setminus\mathbf{0}} |a_{\mathbf{n}}|^2 \cdot$$

$$\cdot \sum_{\lambda_1=1}^{N} \sum_{\lambda_{1,2}=0}^{\lambda_1-1} \left(N\cdot \mathbf{F}^{(\mathrm{sup})}(2\pi U_1)\cdot N \sum_{j=1}^{\infty} 2^{3j+2}\mathbf{F}^{(\mathrm{sup})}(2^j\pi U_1)\right)^{\lambda_1-\lambda_{1,2}}$$

$$\le (V-U)^2\cdot 3\cdot \sum_{\mathbf{n}\in\mathbb{Z}^d\setminus\mathbf{0}} |a_{\mathbf{n}}|^2 \sum_{\lambda_1=1}^{N}$$

$$\cdot \sum_{\lambda_{1,2}=0}^{\lambda_1-1} \left(N \sum_{j=1}^{\infty} 2^{3j+2}\mathbf{F}^{(\mathrm{sup})}(2^j\pi U_1)\cdot N \sum_{j=1}^{\infty} 2^{3j+2}\mathbf{F}^{(\mathrm{sup})}(2^j\pi U_1)\right)^{\lambda_1-\lambda_{1,2}}$$

$$\le 3(V-U)^2 \sum_{\mathbf{n}\in\mathbb{Z}^d\setminus\mathbf{0}} |a_{\mathbf{n}}|^2 N \sum_{i=1}^{\infty} \left(N \sum_{j=1}^{\infty} 2^{3j+2}\mathbf{F}^{(\mathrm{sup})}(2^j\pi U_1) \right)^{2i}$$

$$= 3(V-U)^2 \frac{N\left(N\sum_{j=1}^{\infty} 2^{3j+2}\mathbf{F}^{(\mathrm{sup})}(2^j\pi U_1)\right)^2}{1-\left(N\sum_{j=1}^{\infty} 2^{3j+2}\mathbf{F}^{(\mathrm{sup})}(2^j\pi U_1)\right)^2}\sigma_0^2(f), \tag{28.10}$$

where in the last step we used Parseval's formula.

Finally we estimate (27.28). We have

$$\text{Part Four} = \text{Part Four A} + \text{Part Four B}, \tag{28.11}$$

where

$$\text{Part Four A}$$
$$= \int_{t_1=U}^{V} \sum_{\mathbf{n}_1\in\mathbb{Z}^d\setminus\mathbf{0}} |a_{\mathbf{n}_1}|^2 \cdot \Bigg(\int_{t_2=U}^{V}$$
$$\times \sum_{\substack{\mathbf{n}_2\in\mathbb{Z}^d\setminus\mathbf{0}: \\ L(\mathbf{n}_2)=L(\mathbf{n}_1),\ h_0(t_2;\mathbf{n}_2)<|L(\mathbf{n}_1)|}} \prod_{j=1}^{N} \mathbf{F}^{(\mathrm{sup})}(2\pi|r_j(t_1)\mathbf{n}_1(j)$$
$$-r_j(t_2)\mathbf{n}_2(j)|)\,dt_2 \Bigg) dt_1, \tag{28.12}$$

and

$$\text{Part Four B}$$
$$= \int_{t_1=U}^{V} \sum_{\mathbf{n}_1\in\mathbb{Z}^d\setminus\mathbf{0}} |a_{\mathbf{n}_1}|^2 \cdot \Bigg(\int_{t_2=U}^{V}$$
$$\times \sum_{\substack{\mathbf{n}_2\in\mathbb{Z}^d\setminus\mathbf{0}: \\ L(\mathbf{n}_2)=L(\mathbf{n}_1),\ h_0(t_2;\mathbf{n}_2)=|L(\mathbf{n}_1)|}} \prod_{j=1}^{N} \mathbf{F}^{(\mathrm{sup})}(2\pi|r_j(t_1)\mathbf{n}_1(j)$$
$$-r_j(t_2)\mathbf{n}_2(j)|)\,dt_2 \Bigg) dt_1. \tag{28.13}$$

We start with Part Four A, and use the notation $\lambda_1 = |L(\mathbf{n}_1)|$. In this special case the argument of (27.20)–(27.23) simplifies to the following upper bound (note that here $\lambda_1 = \lambda_2 = \lambda_{1,2}$)

$$\begin{aligned}
\text{Part Four A} \le \int_{t_1=U}^{V} \sum_{\mathbf{n}_1 \in \mathbb{Z}^d \setminus \mathbf{0}} |a_{\mathbf{n}_1}|^2 (V-U) \cdot \\
\cdot \Bigg(\sum_{\substack{(h_0,h_1,\ldots,h_r) \text{ with } r\ge 0:\ h_0<\lambda_1 \\ h_0+h_1+\cdots+h_r=\lambda_1, h_j\ge 0, 0\le j\le r-1 \text{ and } h_r\ge 1}} \binom{\lambda_1}{h_0} \\
\cdot \binom{\lambda_1 - h_0}{h_1} ((2+1)^3)^{h_1} \left(\mathbf{F}^{(\sup)}(2\pi U_1/2)\right)^{h_1} \\
\cdot \binom{\lambda_1 - h_0 - h_1}{h_2} ((2\cdot 2+1)^3)^{h_2} \left(\mathbf{F}^{(\sup)}(2\pi U_1)\right)^{h_2} \\
\cdot \binom{\lambda_1 - h_0 - h_1 - h_2}{h_3} ((2\cdot 2^2+1)^3)^{h_3} \left(\mathbf{F}^{(\sup)}(2\pi \cdot 2U_1)\right)^{h_3} \\
\cdot \binom{\lambda_1 - h_0 - h_1 - h_2 - h_3}{h_4} ((2\cdot 2^3+1)^3)^{h_4} \\
\times \left(\mathbf{F}^{(\sup)}(2\pi \cdot 2^2 U_1)\right)^{h_4} \cdots \Bigg) \, dt_1 \\
= \int_{t_1=U}^{V} \sum_{\mathbf{n}_1 \in \mathbb{Z}^d \setminus \mathbf{0}} |a_{\mathbf{n}_1}|^2 (V-U) \\
\cdot \left(\left(1 + \sum_{j=1}^{\infty} 2^{3j+2} \mathbf{F}^{(\sup)}(2^{j-1}\pi U_1) \right)^{\lambda_1} - 1 \right) dt_1,
\end{aligned} \tag{28.14}$$

where the term -1 at the end is explained by the restriction $h_0 < \lambda_1$, i.e., the term -1 is due to the fact that in the application of the multinomial theorem we have an *almost* complete sum, where the single missing case is $h_0 = \lambda_1$.

By (28.14)

Part Four A

$$\leq \sum_{\mathbf{n}_1\in\mathbb{Z}^d\setminus\mathbf{0}} |a_{\mathbf{n}_1}|^2 (V-U)^2 \times \left(\left(1+\sum_{j=1}^{\infty} 2^{3j+2}\mathbf{F}^{(\mathrm{sup})}(2^{j-1}\pi U_1)\right)^N - 1\right)$$

$$\leq \sigma_0^2(f)(V-U)^2\left(\exp\left(N\sum_{j=1}^{\infty} 2^{3j+2}\mathbf{F}^{(\mathrm{sup})}(2^{j-1}\pi U_1)\right)-1\right)$$

$$\leq \sigma_0^2(f)(V-U)^2 2N\sum_{j=1}^{\infty} 2^{3j+2}\mathbf{F}^{(\mathrm{sup})}(2^{j-1}\pi U_1), \tag{28.15}$$

if

$$N\sum_{j=1}^{\infty} 2^{3j+2}\mathbf{F}^{(\mathrm{sup})}(2^{j-1}\pi U_1) \leq 1. \tag{28.16}$$

Note that in (28.15) we used Parseval's formula and the simple inequalities $1+x \leq e^x \leq 1+2x$ for $0 \leq x \leq 1$; and of course (28.16) follows from the stronger condition (28.2).

To estimate (28.13)

$$\text{Part Four B} = \int_{t_1=U}^{V} \sum_{\mathbf{n}_1\in\mathbb{Z}^d\setminus\mathbf{0}} |a_{\mathbf{n}_1}|^2 \cdot \left(\int_{t_2=U}^{V} \times \sum_{\substack{\mathbf{n}_2\in\mathbb{Z}^d\setminus\mathbf{0}:\\ L(\mathbf{n}_2)=L(\mathbf{n}_1),\ h_0(t_2;\mathbf{n}_2)=|L(\mathbf{n}_1)|}} \prod_{j=1}^{N} \mathbf{F}^{(\mathrm{sup})}(2\pi|r_j(t_1)\mathbf{n}_1(j) - r_j(t_2)\mathbf{n}_2(j)|)\, dt_2\right) dt_1, \tag{28.17}$$

we make use of the fact that for every given triple $(t_1, \mathbf{n}_1, t_2)$ with $t_1 \in [U,V]$, $\mathbf{n}_1 \in \mathbb{Z}^d\setminus\mathbf{0}$ and $t_2 \in [U,V]$, there is *at most one* term in the endsum

$$\sum_{\substack{\mathbf{n}_2\in\mathbb{Z}^d\setminus\mathbf{0}:\\ L(\mathbf{n}_2)=L(\mathbf{n}_1),\ h_0(t_2;\mathbf{n}_2)=|L(\mathbf{n}_1)|}} \prod_{j=1}^{N} \mathbf{F}^{(\mathrm{sup})}(2\pi|r_j(t_1)\mathbf{n}_1(j) - r_j(t_2)\mathbf{n}_2(j)|)$$

of (28.17). Indeed, it follows from the definition of $h_0(t_2;\mathbf{n}_2)$ that for every given triple $(t_1,\mathbf{n}_1,t_2)$ the inequality $|r_j(t_1)\mathbf{n}_1(j)-r_j(t_2)\mathbf{n}_2(j)|<U_1/2$ has at most one solution $\mathbf{n}_2(j)\in\mathbb{Z}^3$, and because $h_0(t_2;\mathbf{n}_2)=|L(\mathbf{n}_1)|$ and $L(\mathbf{n}_2)=L(\mathbf{n}_1)$, there is at most one $\mathbf{n}_2\in\mathbb{Z}^d\setminus\mathbf{0}$ satisfying the requirements. Let $\mathbf{n}_2^\star\in\mathbb{Z}^d\setminus\mathbf{0}$, $\mathbf{n}_2^\star=\mathbf{n}_2^\star(t_1,\mathbf{n}_1,t_2)$ denote this single integer lattice point, *if it exists*. Thus we can rewrite (28.17) as follows:

$$\text{Part Four B} = \int_{t_1=U}^{V}\sum_{\mathbf{n}_1\in\mathbb{Z}^d\setminus\mathbf{0}}|a_{\mathbf{n}_1}|^2\cdot\left(\int_{t_2=U}^{V}\left\{\prod_{j=1}^{N}\mathbf{F}^{(\text{sup})}(2\pi|r_j(t_1)\mathbf{n}_1(j)\right.\right.$$
$$\left.\left.-\,r_j(t_2)\mathbf{n}_2^\star(j)|)\text{ or }0\right\}dt_2\right)dt_1, \tag{28.18}$$

where $\mathbf{n}_2^\star=\mathbf{n}_2^\star(t_1,\mathbf{n}_1,t_2)$, *if it exists.*

Let $j_0=j_0(\mathbf{n}_1)$ be an arbitrary but fixed element of the set $L(\mathbf{n}_1)$; then we have the following trivial upper bound for (28.18):

$$\text{Part Four B} \le \int_{t_1=U}^{V}\sum_{\mathbf{n}_1\in\mathbb{Z}^d\setminus\mathbf{0}}|a_{\mathbf{n}_1}|^2\cdot\left(\int_{t_2=U}^{V}\left\{\mathbf{F}^{(\text{sup})}(2\pi|r_{j_0}(t_1)\mathbf{n}_1(j_0)\right.\right.$$
$$\left.\left.-\,r_{j_0}(t_2)\mathbf{n}_2^\star(j_0)|)\text{ or }0\right\}dt_2\right)dt_1. \tag{28.19}$$

Given $(t_1,\mathbf{n}_1)$, we consider the following decomposition of the interval $U\le t_2\le V(\le 2U)$:

$$I_\ell(t_1,\mathbf{n}_1)=\{t_2\in[U,V]:\ \ell-1\le|r_{j_0}(t_1)\mathbf{n}_1(j_0)$$
$$-\,r_{j_0}(t_2)\mathbf{n}_2^\star(j_0)|<\ell\},\ \ell=1,2,3,\ldots, \tag{28.20}$$

where $j_0=j_0(\mathbf{n}_1)$ and $\mathbf{n}_2^\star=\mathbf{n}_2^\star(t_1,\mathbf{n}_1,t_2)$.

Using decomposition (28.20) in (28.19), we obtain the upper bound

$$\text{Part Four B} \le \int_{t_1=U}^{V}\sum_{\mathbf{n}_1\in\mathbb{Z}^d\setminus\mathbf{0}}|a_{\mathbf{n}_1}|^2\left(\sum_{\ell=1}^{\infty}\text{length}\,(I_\ell)\mathbf{F}^{(\text{sup})}(2\pi(\ell-1))\right)dt_1, \tag{28.21}$$

where $I_\ell=I_\ell(t_1,\mathbf{n}_1)$. To estimate the length of $I_\ell=I_\ell(t_1,\mathbf{n}_1)$, we use Lemma 27.3. First note that $\mathbf{n}_2^\star(j_0)\in\mathbb{Z}^3\setminus\mathbf{0}$, so it has a nonzero coordinate $n^\star_{2,3j_0-i_0}$ with some $i_0\in\{0,1,2\}$. Applying Lemma 27.3 with

$C = r_{j_0}(t_1)n_{1,3j_0-i_0}$ and $C' = \ell$, we obtain the upper bound

$$\text{length}\,(I_\ell) \le 2\ell \cdot (2U)^\beta \left(1 + \log \frac{3U_2}{U_1}\right), \text{ where } U_2 \le (2U)^{\alpha_1}, \tag{28.22}$$

see (27.3). On the other hand, by (27.2)

$$U_1 = \mathcal{R}^{(\text{inf})}(U) = \min_{1\le k\le N} \inf_{x\ge U} r_k(x) \ge U^\alpha. \tag{28.23}$$

Combining (28.22) and (28.23), we have

$$\begin{aligned} \text{length}\,(I_\ell) &\le 2\ell \cdot (2U)^\beta \left(1 + \log \frac{3(2U)^{\alpha_1}}{U^\alpha}\right) \\ &= 2\ell \cdot (2U)^\beta \left(1 + (\alpha_1 - \alpha)\log U + \alpha_1 \log 2 + \log 3\right). \end{aligned} \tag{28.24}$$

Using (28.24) and Parseval's formula in (28.21), we obtain

$$\begin{aligned} \text{Part Four B} &\le (V-U)\sigma_0^2(f)2\cdot(2U)^\beta \cdot (1 + (\alpha_1 - \alpha)\log U \\ &\quad + \alpha_1 \log 2 + \log 3) \sum_{\ell=1}^{\infty} \ell \cdot \mathbf{F}^{(\text{sup})}(2\pi(\ell-1)). \end{aligned} \tag{28.25}$$

Combining (27.24), (28.5), (28.8), (28.10), (28.11), (28.15) and (28.25), we have

$$\begin{aligned} &\left|\Delta_f^2(\mathcal{G};\mathcal{R};U,V)\right| \\ &\quad \le \text{Part One} + \text{Part Two} + \text{Part Three} + \text{Part Four} \\ &\quad \le 9\sigma_0^2(f)(V-U)^2 \frac{N^2 \sum_{j=1}^\infty 2^{3j+2}\mathbf{F}^{(\text{sup})}(2^j\pi U_1)}{1 - N\sum_{j=1}^\infty 2^{3j+2}\mathbf{F}^{(\text{sup})}(2^j\pi U_1)} \\ &\qquad + 9\sigma_0^2(f)(V-U)^2 \frac{N^2 \sum_{j=1}^\infty 2^{3j+2}\mathbf{F}^{(\text{sup})}(2^j\pi U_1)}{\left(1 - N\sum_{j=1}^\infty 2^{3j+2}\mathbf{F}^{(\text{sup})}(2^j\pi U_1)\right)^3} \\ &\qquad + 3\sigma_0^2(f)(V-U)^2 \frac{N\left(N\sum_{j=1}^\infty 2^{3j+2}\mathbf{F}^{(\text{sup})}(2^j\pi U_1)\right)^2}{1 - \left(N\sum_{j=1}^\infty 2^{3j+2}\mathbf{F}^{(\text{sup})}(2^j\pi U_1)\right)^2} \end{aligned}$$

$$+ 2\sigma_0^2(f)(V-U)^2 N \sum_{j=1}^{\infty} 2^{3j+2}\mathbf{F}^{(\mathrm{sup})}(2^{j-1}\pi U_1)$$

$$+ 2\sigma_0^2(f)(V-U)\cdot(2U)^{\beta}\cdot(1+(\alpha_1-\alpha)\log U$$

$$+ \alpha_1\log 2+\log 3)\sum_{\ell=1}^{\infty}\ell\cdot\mathbf{F}^{(\mathrm{sup})}(2\pi(\ell-1)). \qquad (28.26)$$

By using $V - U \le U$ and (28.2) in (28.26), we obtain the shorter upper bound

$$\left|\Delta_f^2(\mathcal{G};\mathcal{R};U,V)\right| \le \sigma_0^2(f)(V-U)U\cdot\left(25N^2\sum_{j=1}^{\infty}2^{3j+2}\mathbf{F}^{(\mathrm{sup})}(2^j\pi U_1)\right.$$

$$\left.+ 5N\left(N\sum_{j=1}^{\infty}2^{3j+2}\mathbf{F}^{(\mathrm{sup})}(2^j\pi U_1)\right)^2\right)$$

$$+ 2\sigma_0^2(f)(V-U)(2U)^{\beta}\,(1+(\alpha_1-\alpha)\log U$$

$$+ \alpha_1\log 2+\log 3)\sum_{\ell=1}^{\infty}\ell\cdot\mathbf{F}^{(\mathrm{sup})}(2\pi(\ell-1)),$$

which completes the proof of Lemma 27.2. □

29. Finishing the Proofs of Theorems 13.1–13.4

We distinguish four cases according to Theorems 13.1–2-3-4.

Case 1. $\mathbf{F}^{(\mathrm{sup})}(w) \le w^{-\gamma}$ for all $w \ge \gamma_0 \ge 1$

First we estimate the critical sum in Lemma 27.2: for $2\pi U_1 \ge \gamma_0$ and $\gamma > 3$

$$\sum_{j=1}^{\infty}2^{3j+2}\mathbf{F}^{(\mathrm{sup})}(2^j\pi U_1) \le \sum_{j=1}^{\infty}2^{3j+2}(2^j\pi U_1)^{-\gamma}$$

$$= (2\pi U_1)^{-\gamma}\sum_{j=1}^{\infty}2^{(3-\gamma)j+2} = (\pi U_1)^{-\gamma}\frac{2^{5-\gamma}}{1-2^{3-\gamma}}. \qquad (29.1)$$

Moreover, for $U \geq \beta_0$ we have

$$\begin{aligned}\sum_{\ell=1}^{\infty} \ell \cdot \mathbf{F}^{(\sup)}(2\pi(\ell-1)) &\leq \sum_{0\leq k\leq \gamma_0} 1 + \sum_{k>\gamma_0} (2\pi k)^{-\gamma} \\ &\leq 1+\gamma_0+(2\pi)^{-\gamma}\int_{\gamma_0}^{\infty} x^{-\gamma}\,dx \\ &= 1+\gamma_0+(2\pi)^{-\gamma}\frac{\gamma_0^{1-\gamma}}{\gamma-1}. \end{aligned} \tag{29.2}$$

Using (29.1) and (29.2) in Lemma 27.2, under the condition of Case 1, we have

$$\begin{aligned}\left|\Delta_f^2(\mathcal{G};\mathcal{R};U,V)\right| &\leq \sigma_0^2(f)(V-U)U\cdot\left(25N^2(\pi U_1)^{-\gamma}\frac{2^{5-\gamma}}{1-2^{3-\gamma}}\right.\\ &\quad \left.+5N^3(\pi U_1)^{-2\gamma}\left(\frac{2^{5-\gamma}}{1-2^{3-\gamma}}\right)^2\right)+2\sigma_0^2(f)(V-U)(2U)^{\beta}\\ &\quad \times(1+(\alpha_1-\alpha)\log U+\alpha_1\log 2+\log 3)\\ &\quad \cdot\left(1+\gamma_0+(2\pi)^{-\gamma}\frac{\gamma_0^{1-\gamma}}{\gamma-1}\right), \end{aligned} \tag{29.3}$$

assuming

$$\min_{1\leq k\leq N}\inf_{x\geq U} r_k(x) = \mathcal{R}^{(\inf)}(U) = U_1 \geq \gamma_0/2\pi, \quad U\geq\beta_0 \text{ and } \gamma>3. \tag{29.4}$$

By hypothesis (27.2)

$$U_1 \geq U^{\alpha} \quad \text{for all } U\geq \alpha_0. \tag{29.5}$$

Using (29.5) in (29.3)–(29.4), under the condition of Case 1 we have

$$\begin{aligned}\left|\Delta_f^2(\mathcal{G};\mathcal{R};U,V)\right| &\leq \sigma_0^2(f)(V-U)U\left(25N^2\left(\pi U^{\alpha}\right)^{-\gamma}\frac{2^{5-\gamma}}{1-2^{3-\gamma}}\right.\\ &\quad \left.+5N^3\left(\pi U^{\alpha}\right)^{-2\gamma}\left(\frac{2^{5-\gamma}}{1-2^{3-\gamma}}\right)^2\right)+2\sigma_0^2(f)(V-U)(2U)^{\beta}\\ &\quad \times(1+(\alpha_1-\alpha)\log U+\alpha_1\log 2+\log 3)\\ &\quad \cdot\left(1+\gamma_0+(2\pi)^{-\gamma}\frac{\gamma_0^{1-\gamma}}{\gamma-1}\right), \end{aligned} \tag{29.6}$$

assuming

$$U \geq \max\left\{\alpha_0,\beta_0,\left(\frac{\gamma_0}{2\pi}\right)^{1/\alpha}\right\} \quad \text{and} \quad \gamma>3. \tag{29.7}$$

Now we are ready to complete the proof of Theorem 13.1. We repeat the Cauchy–Schwarz argument of the deduction of Theorem 4.2 from Theorem 9.1 (see the end of Sec. 9). Let $2^{k-1}U < W \le 2^k U$. It is easy to find a sequence $W_0 = U < W_1 < \cdots < W_k = W$ such that $W_i < W_{i+1} \le 2W_i$ for all $0 \le i < k$. Assume that $\alpha\gamma > 1$. Combining the Cauchy–Schwarz inequality and (29.6) with the choice $U = W_i$ and $V = W_{i+1}$, $0 \le i < k$, and using the simple fact that $x^{1-\alpha\gamma}$ is a monotone decreasing function of x (since $\alpha\gamma > 1$), we have

$$\begin{aligned}
&\Delta_f^2(\mathcal{G};\mathcal{R};U,W)\\
&\quad= \int_{\omega\in\Omega_{\mathcal{G}}} |D_f(\mathcal{R};\omega;U,W)|^2 \; d\text{ProdMeas}_{\mathcal{G}}(\omega)\\
&\quad= \int_{\omega\in\Omega_{\mathcal{G}}} \left|\sum_{i=0}^{k-1} D_f(\mathcal{R};\omega;W_i,W_{i+1})\right|^2 \; d\text{ProdMeas}_{\mathcal{G}}(\omega)\\
&\quad\le \int_{\omega\in\Omega_{\mathcal{G}}} \left(\sum_{i=0}^{k-1} |D_f(\mathcal{R};\omega;W_i,W_{i+1})|^2\right)\left(\sum_{i=0}^{k-1} 1\right) \; d\text{ProdMeas}_{\mathcal{G}}(\omega)\\
&\quad= \left(\sum_{i=0}^{k-1} \int_{\omega\in\Omega_{\mathcal{G}}} |D_f(\mathcal{R};\omega;W_i,W_{i+1})|^2 \; d\text{ProdMeas}_{\mathcal{G}}(\omega)\right)\left(\sum_{i=0}^{k-1} 1\right)\\
&\quad= \left(\sum_{i=0}^{k-1} \Delta_f^2(\mathcal{G};\mathcal{R};W_i,W_{i+1})\right)\left(\sum_{i=0}^{k-1} 1\right) \le k\cdot\sigma_0^2(f)\left(\sum_{i=0}^{k-1}\left((W_{i+1}-W_i)\right.\right.\\
&\qquad\cdot\left(25N^2\pi^{-\gamma}W_i^{1-\alpha\gamma}\frac{2^{5-\gamma}}{1-2^{3-\gamma}} + 5N^3\pi^{-2\gamma}W_i^{1-2\alpha\gamma}\left(\frac{2^{5-\gamma}}{1-2^{3-\gamma}}\right)^2\right)\Bigg)\\
&\qquad+ 2(W_{i+1}-W_i)(2W_i)^{\beta}\left(1+(\alpha_1-\alpha)\log W_i + \alpha_1\log 2 + \log 3\right)\\
&\qquad\cdot\left(1+\gamma_0+(2\pi)^{-\gamma}\frac{\gamma_0^{1-\gamma}}{\gamma-1}\right)\Bigg)\Bigg) \le k\cdot\sigma_0^2(f)\Bigg((W-U)\\
&\qquad\times\left(25N^2\pi^{-\gamma}U^{1-\alpha\gamma}\frac{2^{5-\gamma}}{1-2^{3-\gamma}} + 5N^3\pi^{-2\gamma}U^{1-2\alpha\gamma}\left(\frac{2^{5-\gamma}}{1-2^{3-\gamma}}\right)^2\right)\\
&\qquad+ 2(W-U)W^{\beta}\left(1+(\alpha_1-\alpha)\log W + \alpha_1\log 2 + \log 3\right)\\
&\qquad\cdot\left(1+\gamma_0+(2\pi)^{-\gamma}\frac{\gamma_0^{1-\gamma}}{\gamma-1}\right)\Bigg),
\end{aligned} \tag{29.8}$$

assuming

$$U \geq \max\left\{\alpha_0, \beta_0, \left(\frac{\gamma_0}{2\pi}\right)^{1/\alpha}\right\}, \ \gamma > 3 \quad \text{and} \quad \alpha\gamma > 1. \tag{29.9}$$

Since $k = \lceil \log_2(W/U) \rceil$ (upper integral part), (29.8)–(29.9) complete the proof of Theorem 13.1. □

Next we replace the polynomial decay in Case 1 with the exponential decay.

Case 2. $\mathbf{F}^{(\text{sup})}(w) \leq e^{-\gamma w}$ **for all** $w \geq \gamma_0$

Again we estimate the critical sum in Lemma 27.2: for $2\pi U_1 \geq \gamma_0$

$$\begin{aligned}\sum_{j=1}^{\infty} 2^{3j+2}\mathbf{F}^{(\text{sup})}(2^j\pi U_1) &\leq \sum_{j=1}^{\infty} e^{-\gamma 2^j \pi U_1} \\ &\leq \sum_{j=1}^{\infty} e^{-\gamma 2j\pi U_1} = \frac{e^{-\gamma 2\pi U_1}}{1-e^{-\gamma 2\pi U_1}},\end{aligned} \tag{29.10}$$

and similarly,

$$\begin{aligned}\sum_{k=0}^{\infty} \mathbf{F}^{(\text{sup})}(2\pi k) &\leq \sum_{0\leq k\leq \gamma_0} 1 + \sum_{k>\gamma_0} e^{-\gamma 2\pi k} \\ &\leq 1 + \gamma_0 + \int_{\gamma_0}^{\infty} e^{-\gamma 2\pi x}\,dx = 1 + \gamma_0 + \frac{e^{-2\pi\gamma\gamma_0}}{2\pi\gamma}.\end{aligned} \tag{29.11}$$

Using (29.10)–(29.11) in Lemma 27.2, under the condition of Case 2 we have

$$\begin{aligned}&\left|\Delta_f^2(\mathcal{G};\mathcal{R};U,V)\right| \leq \sigma_0^2(f)(V-U)U \\ &\quad\cdot\left(25N^2\frac{e^{-\gamma 2\pi U_1}}{1-e^{-\gamma 2\pi U_1}} + 5N^3\left(\frac{e^{-\gamma 2\pi U_1}}{1-e^{-\gamma 2\pi U_1}}\right)^2\right) \\ &\quad + 2\sigma_0^2(f)(V-U)(2U)^{\beta}\left(1+(\alpha_1-\alpha)\log U + \alpha_1\log 2 + \log 3\right) \\ &\quad\cdot\left(1+\gamma_0+\frac{e^{-2\pi\gamma\gamma_0}}{2\pi\gamma}\right),\end{aligned} \tag{29.12}$$

assuming

$$\min_{1\leq k\leq N}\inf_{x\geq U} r_k(x) = \mathcal{R}^{(\text{inf})}(U) = U_1 \geq \frac{\gamma_0}{2\pi} \quad \text{and} \quad U \geq \beta_0. \tag{29.13}$$

By using hypothesis (27.3), we can replace condition (29.13) with

$$U \geq \max\left\{\alpha_0, \beta_0, \left(\frac{\gamma_0}{2\pi}\right)^{1/\alpha}\right\}. \tag{29.14}$$

By using (29.12) we complete the proof of Theorem 13.2 the same way as we completed the proof of Theorem 13.1 in Case 1 above. First we need an elementary fact from calculus. It is a routine derivative calculation that the function

$$g_0(x) = xe^{-2\pi\gamma x^\alpha} \text{ attains its maximum at } x = x_0 = (2\pi\gamma\alpha)^{-1/\alpha}, \tag{29.15}$$

and $g_0(x)$ is monotone decreasing for $x > x_0$. Now let $2^{k-1}U < W \leq 2^k U$. It is easy to find a sequence $W_0 = U < W_1 < \cdots < W_k = W$ such that $W_i < W_{i+1} \leq 2W_i$ for all $0 \leq i < k$. Combining the Cauchy–Schwarz inequality and (29.12) with the choice $U = W_i$ and $V = W_{i+1}$, $0 \leq i < k$, using the fact that $g_0(x)$ is monotone decreasing for $x > x_0$ (see (29.15)), and finally repeating the argument of (29.8), we have

$$\begin{aligned}
\Delta_f^2(\mathcal{G};\mathcal{R};U,W) &\leq \left(\sum_{i=0}^{k-1}\Delta_f^2(\mathcal{G};\mathcal{R};W_i,W_{i+1})\right)\left(\sum_{i=0}^{k-1}1\right)\\
&\leq k\cdot\sigma_0^2(f)\sum_{i=0}^{k-1}\left(2(W_{i+1}-W_i)W^\beta\left(1+\gamma_0+\frac{e^{-2\pi\gamma\gamma_0}}{2\pi\gamma}\right)\right.\\
&\quad\cdot(1+(\alpha_1-\alpha)\log W_i+\alpha_1\log 2+\log 3)\\
&\quad\left.+(W_{i+1}-W_i)\left(25N^2\frac{W_ie^{-2\pi\gamma W_i^\alpha}}{1-e^{-2\pi\gamma W_i^\alpha}}+5N^3\frac{W_ie^{-4\pi\gamma W_i^\alpha}}{(1-e^{-2\pi\gamma W_i^\alpha})^2}\right)\right)\\
&\leq k\cdot\sigma_0^2(f)\left(2(W-U)W^\beta\left(1+\gamma_0+\frac{e^{-2\pi\gamma\gamma_0}}{2\pi\gamma}\right)\right.\\
&\quad\cdot(1+(\alpha_1-\alpha)\log W+\alpha_1\log 2+\log 3)\\
&\quad\left.+(W-U)\left(25N^2\frac{Ue^{-2\pi\gamma U^\alpha}}{1-e^{-2\pi\gamma U^\alpha}}+5N^3\frac{Ue^{-4\pi\gamma U^\alpha}}{(1-e^{-2\pi\gamma U^\alpha})^2}\right)\right),
\end{aligned} \tag{29.16}$$

assuming

$$U \geq \max\left\{\alpha_0, \beta_0, \left(\frac{\gamma_0}{2\pi}\right)^{1/\alpha}\right\} \quad \text{and} \quad U \geq (2\pi\gamma\alpha)^{-1/\alpha}. \tag{29.17}$$

Since $k = \lceil \log_2(W/U) \rceil$, (29.16)–(29.17) complete the proof of Theorem 13.2. □

Next we replace the exponential decay in Case 2 with the superexponential decay of the normal distribution.

Case 3. $\mathbf{F}^{(\text{sup})}(w) \le e^{-\gamma w^2}$ for all $w \ge \gamma_0$

As usual, first we estimate the critical sum in Lemma 27.2: for $2\pi U_1 \ge \gamma_0$

$$\sum_{j=1}^{\infty} 2^{3j+2} \mathbf{F}^{(\text{sup})}(2^j \pi U_1) \le \sum_{j=1}^{\infty} \exp\left(-\gamma 4^j \pi^2 U_1^2\right)$$
$$\le \sum_{j=1}^{\infty} \exp\left(-\gamma 4 j \pi^2 U_1^2\right) = \frac{\exp\left(-\gamma 4\pi^2 U_1^2\right)}{1 - \exp\left(-\gamma 4\pi^2 U_1^2\right)} \tag{29.18}$$

and

$$\sum_{k=0}^{\infty} \mathbf{F}^{(\text{sup})}(2\pi k) \le \sum_{0 \le k \le \gamma_0} 1 + \sum_{k > \gamma_0} e^{-\gamma 4\pi^2 k^2}$$
$$\le 1 + \gamma_0 + e^{-\gamma 4\pi^2 \gamma_0^2}. \tag{29.19}$$

Using (29.18)–(29.19) in Lemma 27.2, under the condition of Case 3 we obtain

$$\left|\Delta_f^2(\mathcal{G}; \mathcal{R}; U, V)\right| \le \sigma_0^2(f)(V-U)U$$
$$\cdot \left(25N^2 \frac{\exp\left(-\gamma 4\pi^2 U_1^2\right)}{1 - \exp\left(-\gamma 4\pi^2 U_1^2\right)} + 5N^3 \left(\frac{\exp\left(-\gamma 4\pi^2 U_1^2\right)}{1 - \exp\left(-\gamma 4\pi^2 U_1^2\right)}\right)^2\right)$$
$$+ 2\sigma_0^2(f)(V-U)(2U)^{\beta} \left(1 + \gamma_0 + e^{-\gamma 4\pi^2 \gamma_0^2}\right)$$
$$\cdot \left(1 + (\alpha_1 - \alpha)\log U + \alpha_1 \log 2 + \log 3\right), \tag{29.20}$$

assuming

$$U \ge \max\left\{\alpha_0, \beta_0, \left(\frac{\gamma_0}{2\pi}\right)^{1/\alpha}\right\}. \tag{29.21}$$

We complete the proof of Theorem 13.3 the same way as we completed the proofs of Theorems 13.1 and 13.2 in Case 1 and Case 2 above. We use the elementary fact from calculus that the function

$$g_1(x) = xe^{-4\pi^2\gamma x^{2\alpha}} \text{ attains its maximum at } x = x_1 = (8\pi^2\gamma\alpha)^{-1/(2\alpha)}, \tag{29.22}$$

and $g_1(x)$ is monotone decreasing for $x > x_1$. Let $2^{k-1}U < W \leq 2^k U$; it is easy to find a sequence $W_0 = U < W_1 < \cdots < W_k = W$ such that $W_i < W_{i+1} \leq 2W_i$ for all $0 \leq i < k$. Combining the Cauchy–Schwarz inequality and (29.20) with the choice $U = W_i$ and $V = W_{i+1}$, $0 \leq i < k$, using the fact that $g_1(x)$ is monotone decreasing for $x > x_1$ (see (29.22)), and finally repeating the argument of (29.8), we have

$$\begin{aligned}
&\Delta_f^2(\mathcal{G}; \mathcal{R}; U, W) \\
&\leq \left(\sum_{i=0}^{k-1} \Delta_f^2(\mathcal{G}; \mathcal{R}; W_i, W_{i+1})\right)\left(\sum_{i=0}^{k-1} 1\right) \\
&\leq k \cdot \sigma_0^2(f) \sum_{i=0}^{k-1} \left(2(W_{i+1} - W_i)W^\beta \left(1 + \gamma_0 + e^{-4\pi^2\gamma\gamma_0^2}\right)\right. \\
&\quad \cdot (1 + (\alpha_1 - \alpha)\log W_i + \alpha_1 \log 2 + \log 3) \\
&\quad \left. + \ (W_{i+1} - W_i)\left(25N^2 \frac{W_i e^{-4\pi^2\gamma W_i^{2\alpha}}}{1 - e^{-4\pi^2\gamma W_i^{2\alpha}}} + 5N^3 \frac{W_i e^{-8\pi^2\gamma W_i^{2\alpha}}}{\left(1 - e^{-4\pi^2\gamma W_i^{2\alpha}}\right)^2}\right)\right) \\
&\leq k \cdot \sigma_0^2(f) \left(2(W - U)W^\beta \left(1 + \gamma_0 + e^{-4\pi^2\gamma\gamma_0^2}\right)\right. \\
&\quad \cdot (1 + (\alpha_1 - \alpha)\log W + \alpha_1 \log 2 + \log 3) \\
&\quad \left. + \ (W - U)\left(25N^2 \frac{U e^{-4\pi^2\gamma U^{2\alpha}}}{1 - e^{-4\pi^2\gamma U^{2\alpha}}} + 5N^3 \frac{U e^{-8\pi^2\gamma U^{2\alpha}}}{\left(1 - e^{-4\pi^2\gamma U^{2\alpha}}\right)^2}\right)\right),
\end{aligned} \tag{29.23}$$

assuming

$$U \geq \max\left\{\alpha_0, \beta_0, \left(\frac{\gamma_0}{2\pi}\right)^{1/\alpha}\right\} \quad \text{and} \quad U \geq (8\pi^2\gamma\alpha)^{-1/(2\alpha)}. \tag{29.24}$$

Since $k = \lceil \log_2(W/U) \rceil$, (29.23)–(29.24) complete the proof of Theorem 13.3. □

Finally we replace the superexponential decay in Case 3 with the "vanishing property".

Case 4. $\mathbf{F}^{(\mathrm{sup})}(w) = 0$ for $w \geq \gamma_0$

For $2\pi U_1 \geq \gamma_0$ the critical sum in Lemma 27.2 vanishes:

$$\sum_{j=1}^{\infty} 2^{3j+2}\mathbf{F}^{(\mathrm{sup})}(2^j \pi U_1) = 0,$$

and similarly

$$\sum_{k=0}^{\infty} \mathbf{F}^{(\mathrm{sup})}(2\pi k) \leq \sum_{0\leq k\leq \gamma_0} 1 \leq 1 + \gamma_0.$$

Using these facts in Lemma 27.2, under the condition of Case 4 we have

$$\begin{aligned} \left|\Delta_f^2(\mathcal{G};\mathcal{R};U,V)\right| &\leq 2\sigma_0^2(f)(V-U)V^{\beta}\,(1+\gamma_0) \\ &\quad \cdot (1 + (\alpha_1 - \alpha)\log U + \alpha_1 \log 2 + \log 3)\,, \qquad (29.25) \end{aligned}$$

assuming

$$U \geq \max\left\{\alpha_0, \beta_0, \left(\frac{\gamma_0}{2\pi}\right)^{1/\alpha}\right\}. \qquad (29.26)$$

We complete the proof of Theorem 13.4 the usual way. Let $2^{k-1}U < W \leq 2^k U$; it is easy to find a sequence $W_0 = U < W_1 < \cdots < W_k = W$ such that $W_i < W_{i+1} \leq 2W_i$ for all $0 \leq i < k$. Combining the Cauchy–Schwarz inequality and (29.25) with the choice $U = W_i$ and $V = W_{i+1}$, $0 \leq i < k$, and again repeating the argument of (29.8), we have

$$\begin{aligned} \Delta_f^2(\mathcal{G};\mathcal{R};U,W) &\leq \left(\sum_{i=0}^{k-1} \Delta_f^2(\mathcal{G};\mathcal{R};W_i,W_{i+1})\right)\left(\sum_{i=0}^{k-1} 1\right) \\ &\leq k\cdot\sigma_0^2(f)\sum_{i=0}^{k-1} 2(W_{i+1}-W_i)W^{\beta}\,(1+\gamma_0) \\ &\quad \times (1 + (\alpha_1-\alpha)\log W + \alpha_1\log 2 + \log 3) \\ &= 2k\cdot\sigma_0^2(f)(W-U)W^{\beta}\,(1+\gamma_0)\,(1+(\alpha_1-\alpha)\log W \\ &\quad + \alpha_1 \log 2 + \log 3)\,, \qquad (29.27) \end{aligned}$$

assuming

$$U \geq \max\left\{\alpha_0, \beta_0, \left(\frac{\gamma_0}{2\pi}\right)^{1/\alpha}\right\}. \qquad (29.28)$$

Since $k = \lceil \log_2(W/U) \rceil$, (29.27)–(29.28) complete the proof of Theorem 13.4. □

30. Starting the Proof of Theorem 14.1

It is convenient to prove first a special case with an upper bound on the ratio W/U.

Theorem 30.1. *Let $f \in L_2(I^d)$ be a test function, where $I^d = [0,1)^d$ with $d = 2N$. Under the condition of (14.4) – (14.6) for 2-dimensional curves*

$$\vec{\mathcal{R}} = (\mathbf{r}_1(t), \mathbf{r}_2(t), \ldots, \mathbf{r}_N(t)),$$

and also assuming $e^{\pi^2 U_1^2/2} \geq 10UN$ with

$$U_1 = \mathcal{R}^{(\inf)}(U) = \min_{1\leq k\leq N} \inf_{x\geq U} |\mathbf{r}_k(x)|,$$

for $2U \geq V > U \geq \max\{\alpha_0, \beta_0, 1\}$ we have

$$\Delta_f^2(\text{Gauss}; \vec{\mathcal{R}}; U, V) \leq 2\sigma_0^2(f) + \sigma_0^2(f)(V-U)\left(2 + U\exp\left(-\pi^2 U_1^2 2^{-8}\right)\right.$$
$$\left. +9\left(3 + 2\log_2\left(1 + \frac{(2U)^{1+\kappa} + \alpha_0\kappa_0}{U_1}\right)\right)\frac{(2U)^{1+\kappa+\beta}}{U_1}\right).$$

Remark. The technical restriction $U < V \leq 2U$ in Theorem 30.1 can be easily eliminated by a routine application of the Cauchy–Schwarz inequality — for the details, see the end of Sec. 31.

Proof of Theorem 30.1. As usual, we apply Fourier analysis in the configuration space $I^d = [0,1)^d$ with $d = 2N$. We use the Fourier expansion of the given Lebesgue square-integrable test function $f \in L_2\left(I^d\right)$ (defined in the d-dimensional unit torus, i.e., we extend f over the whole d-space $\mathbb{R}^d$ periodically):

$$f(\mathbf{u}) = \sum_{\mathbf{n}\in\mathbb{Z}^d} a_{\mathbf{n}} e^{2\pi i \mathbf{n}\cdot\mathbf{u}}. \tag{30.1}$$

Clearly

$$a_{\mathbf{0}} = \int_{I^d} f(\mathbf{y})\, d\mathbf{y}, \text{ and so } \sum_{\mathbf{n}\in\mathbb{Z}^d\setminus\mathbf{0}} |a_{\mathbf{n}}|^2 = \sigma_0^2(f), \tag{30.2}$$

where we used Parseval's formula.

We recall some notation from Sec. 14. For every $\mathbf{n} = (n_1, \ldots, n_d) \in \mathbb{Z}^d \setminus \mathbf{0}$, write $\mathbf{n}(k) = (n_{2k-1}, n_{2k})$, $1 \leq k \leq N = d/2$.

Given t_1 and t_2, write

$$\mathbf{e}_1 = \frac{\mathbf{r}_k(t_1)}{|\mathbf{r}_k(t_1)|} \quad \text{and} \quad \mathbf{e}_2 = \frac{\mathbf{r}_k(t_2)}{|\mathbf{r}_k(t_2)|},$$

where $\mathbf{e}_1, \mathbf{e}_2$ are 2-dimensional unit vectors. Thus we have $\mathbf{r}_k(t_1) = r_k(t_1)\mathbf{e}_1$ and $\mathbf{r}_k(t_2) = r_k(t_2)\mathbf{e}_2$, where $r_k(t_1) = |\mathbf{r}_k(t_1)|$ and $r_k(t_2) = |\mathbf{r}_k(t_2)|$. We use $\theta_{k;t_1\to t_2} \in \mathrm{SO}(2)$ to denote the rotation of the plane that maps $\mathbf{e}_1$ to $\mathbf{e}_2$, i.e., $\mathbf{e}_2 = \theta_{k;t_1\to t_2}\mathbf{e}_1$; and similarly, $\theta_{k;t_2\to t_1} \in \mathrm{SO}(2)$ denotes the inverse of $\theta_{k;t_1\to t_2}$, i.e., $\mathbf{e}_1 = \theta_{k;t_2\to t_1}\mathbf{e}_2$.

We recall Lemma 14.1 from Sec. 14.

Lemma 14.1 *For 2-dimensional curves*

$$\vec{\mathcal{R}} = (\mathbf{r}_1(t), \mathbf{r}_2(t), \ldots, \mathbf{r}_N(t))$$

we have

$$\Delta_f^2(\text{Gauss}; \vec{\mathcal{R}}; T_1, T_2) = \int_{T_1}^{T_2}\int_{T_1}^{T_2} \sum_{\mathbf{n}_1\in\mathbb{Z}^d\setminus\mathbf{0}} \sum_{\mathbf{n}_2\in\mathbb{Z}^d\setminus\mathbf{0}} a_{\mathbf{n}_1}\overline{a_{\mathbf{n}_2}} \cdot \exp\Bigg(-\sum_{k=1}^{N} 2\pi^2 \times \left|r_k(t_1)\mathbf{n}_1(k) - r_k(t_2)\theta_{k;t_2\to t_1}\mathbf{n}_2(k)\right|^2\Bigg)\, dt_1\, dt_2.$$

□

Lemma 14.1 is a curve analog of Lemma 9.2 in the 2-dimensional case. In the rest of the proof we closely follow the brute force combinatorial arguments in Sec. 9, and the similar arguments in Sec. 27–28.

For every $\mathbf{n} = (n_1, \ldots, n_d) \in \mathbb{Z}^d \setminus \mathbf{0}$ write $L(\mathbf{n}) = \{1 \le i \le N = d/2 : \mathbf{n}_i \neq \mathbf{0}\}$. Applying the simple inequality $|a_{\mathbf{n}_1}\overline{a_{\mathbf{n}_2}}| \le (|a_{\mathbf{n}_1}|^2 + |a_{\mathbf{n}_2}|^2)/2$ in Lemma 14.1, we have

$$|\Delta_f^2(\text{Gauss}; \vec{\mathcal{R}}; U, V)| = \Bigg| \sum_{\mathbf{n}_1\in\mathbb{Z}^d\setminus\mathbf{0}} \sum_{\mathbf{n}_2\in\mathbb{Z}^d\setminus\mathbf{0}} a_{\mathbf{n}_1}\overline{a_{\mathbf{n}_2}} \cdot \int_U^V \int_U^V \times \exp\Bigg(-\sum_{j=1}^{N} 2\pi^2 \left|r_j(t_1)\mathbf{n}_1(j) - r_j(t_2)\theta_{j;t_2\to t_1}\mathbf{n}_2(j)\right|^2\Bigg)\, dt_1\, dt_2 \Bigg|$$

$$\le \sum_{\mathbf{n}_1\in\mathbb{Z}^d\setminus\mathbf{0}} \sum_{\mathbf{n}_2\in\mathbb{Z}^d\setminus\mathbf{0}} \frac{1}{2}\left(|a_{\mathbf{n}_1}|^2 + |a_{\mathbf{n}_2}|^2\right).$$

$$\cdot \int_U^V \int_U^V \exp\left(-\sum_{j=1}^{N} 2\pi^2 \left| r_j(t_1)\mathbf{n}_1(j) - r_j(t_2)\theta_{j;t_2\to t_1}\mathbf{n}_2(j) \right|^2 \right) dt_1\, dt_2$$

$$= \sum_{\mathbf{n}_1\in\mathbb{Z}^d\setminus\mathbf{0}} |a_{\mathbf{n}_1}|^2 \cdot \sum_{\mathbf{n}_2\in\mathbb{Z}^d\setminus\mathbf{0}} \int_U^V \int_U^V \exp\left(-\sum_{j=1}^{N} 2\pi^2 \left| r_j(t_1)\mathbf{n}_1(j) \right.\right.$$

$$\left.\left. - r_j(t_2)\theta_{j;t_2\to t_1}\mathbf{n}_2(j) \right|^2 \right) dt_1\, dt_2$$

$$= \int_{t_1=U}^V \sum_{\mathbf{n}_1\in\mathbb{Z}^d\setminus\mathbf{0}} |a_{\mathbf{n}_1}|^2 \cdot \left(\int_{t_2=U}^V \sum_{\mathbf{n}_2\in\mathbb{Z}^d\setminus\mathbf{0}} \exp\left(-\sum_{j=1}^{N} 2\pi^2 \left| r_j(t_1)\mathbf{n}_1(j) \right.\right.\right.$$

$$\left.\left.\left. - r_j(t_2)\theta_{j;t_2\to t_1}\mathbf{n}_2(j) \right|^2 \right) dt_2 \right) dt_1$$

$$= \sum_{\lambda_1=1}^{N} \int_{t_1=U}^V \sum_{\substack{\mathbf{n}_1\in\mathbb{Z}^d\setminus\mathbf{0}:\\ |L(\mathbf{n}_1)|=\lambda_1}} |a_{\mathbf{n}_1}|^2 \left(\sum_{L_{1,2}\subseteq L(\mathbf{n}_1)} \sum_{\lambda_2=\max\{|L_{1,2}|,1\}}^{N} \int_{t_2=U}^V \right.$$

$$\cdot \sum_{\substack{\mathbf{n}_2\in\mathbb{Z}^d\setminus\mathbf{0}:\ |L(\mathbf{n}_2)|=\lambda_2\\ \text{and } L(\mathbf{n}_2)\cap L(\mathbf{n}_1)=L_{1,2}}} \exp\left(-\sum_{j=1}^{N} 2\pi^2 \left| r_j(t_1)\mathbf{n}_1(j) \right.\right.$$

$$\left.\left.\left. -\ r_j(t_2)\theta_{j;t_2\to t_1}\mathbf{n}_2(j) \right|^2 \right) dt_2 \right) dt_1. \tag{30.3}$$

We fix $t_1 \in [U,V]$, $\mathbf{n}_1 \in \mathbb{Z}^d\setminus\mathbf{0}$, $L_{1,2}\subseteq L(\mathbf{n}_1)$ and λ_2, and focus on the inner integral at the end of (30.3)

$$\int_{t_2=U}^V \sum_{\substack{\mathbf{n}_2\in\mathbb{Z}^d\setminus\mathbf{0}:\ |L(\mathbf{n}_2)|=\lambda_2\\ \text{and } L(\mathbf{n}_2)\cap L(\mathbf{n}_1)=L_{1,2}}}$$

$$\times \exp\left(-\sum_{j=1}^{N} 2\pi^2 \left| r_j(t_1)\mathbf{n}_1(j) - r_j(t_2)\theta_{j;t_2\to t_1}\mathbf{n}_2(j) \right|^2 \right) dt_2.$$

Let $k_1(\mathbf{n}_2)$ denote the number of indices i such that the coordinate-pair $\mathbf{n}_2(i)$ of $\mathbf{n}_2$ satisfies $1 \le |\mathbf{n}_2(i)| < 2$ and $\mathbf{n}_1(i) = \mathbf{0}$; let $k_2(\mathbf{n}_2)$ denote the number of indices i such that the coordinate-pair $\mathbf{n}_2(i)$ of $\mathbf{n}_2$ satisfies $2 \le |\mathbf{n}_2(i)| < 4$ and $\mathbf{n}_1(i) = \mathbf{0}$; let $k_3(\mathbf{n}_2)$ denote the number of indices i such that the coordinate-pair $\mathbf{n}_2(i)$ of $\mathbf{n}_2$ satisfies $4 \le |\mathbf{n}_2(i)| < 8$ and $\mathbf{n}_1(i) = \mathbf{0}$; and so on, keeping up the doubling. Note that

$$k_1(\mathbf{n}_2) + k_2(\mathbf{n}_2) + k_3(\mathbf{n}_2) + \cdots = \lambda_2 - \lambda_{1,2}. \tag{30.4}$$

Let $h_0(t_2; \mathbf{n}_2)$ denote the number of indices $j \in L_{1,2}$ such that

$$|r_j(t_1)\mathbf{n}_1(j) - r_j(t_2)\theta_{j;t_2 \to t_1}\mathbf{n}_2(j)| < U_1/2;$$

let $h_1(t_2; \mathbf{n}_2)$ denote the number of indices $j \in L_{1,2}$ such that

$$U_1/2 \le |r_j(t_1)\mathbf{n}_1(j) - r_j(t_2)\theta_{j;t_2 \to t_1}\mathbf{n}_2(j)| < U_1;$$

let $h_2(t_2; \mathbf{n}_2)$ denote the number of indices $j \in L_{1,2}$ such that

$$U_1 \le |r_j(t_1)\mathbf{n}_1(j) - r_j(t_2)\theta_{j;t_2 \to t_1}\mathbf{n}_2(j)| < 2U_1;$$

let $h_3(t_2; \mathbf{n}_2)$ denote the number of indices $j \in L_{1,2}$ such that

$$2U_1 \le |r_j(t_1)\mathbf{n}_1(j) - r_j(t_2)\theta_{j;t_2 \to t_1}\mathbf{n}_2(j)| < 4U_1;$$

and so on, keeping up the doubling. Note that

$$h_0(t_2; \mathbf{n}_2) + h_1(t_2; \mathbf{n}_2) + h_2(t_2; \mathbf{n}_2) + h_3(t_2; \mathbf{n}_2) + \cdots = \lambda_{1,2}. \tag{30.5}$$

By using the definitions of $k_i(\mathbf{n}_2)$ and $h_i(t_2; \mathbf{n}_2)$, we have

$$\int_{t_2=U}^{V} \sum_{\substack{\mathbf{n}_2 \in \mathbb{Z}^d \setminus \mathbf{0}:\ |L(\mathbf{n}_2)|=\lambda_2 \\ \text{and } L(\mathbf{n}_2)\cap L(\mathbf{n}_1)=L_{1,2}}}$$

$$\times \exp\left(-\sum_{j=1}^{N} 2\pi^2 \left| r_j(t_1)\mathbf{n}_1(j) - r_j(t_2)\theta_{j;t_2 \to t_1}\mathbf{n}_2(j) \right|^2 \right) dt_2$$

$$\le \prod_{j \in L(\mathbf{n}_1)\setminus L_{1,2}} \exp\left(-2\pi^2 \left| r_j(t_1)\mathbf{n}_1(j) \right|^2 \right) \int_{t_2=U}^{V} \sum_{\substack{\mathbf{n}_2 \in \mathbb{Z}^d \setminus \mathbf{0}:\ |L(\mathbf{n}_2)|=\lambda_2 \\ \text{and } L(\mathbf{n}_2)\cap L(\mathbf{n}_1)=L_{1,2}}}$$

$$\cdot \exp\left(-k_1(\mathbf{n}_2) \cdot 2(\pi \mathcal{R}^{(\mathrm{inf})}(t_2))^2\right) \exp\left(-k_2(\mathbf{n}_2) \cdot 2(2\pi \mathcal{R}^{(\mathrm{inf})}(t_2))^2\right)$$

$$\cdot \exp(-k_3(\mathbf{n}_2)\cdot 2(2^2\pi\mathcal{R}^{(\inf)}(t_2))^2)\exp(-k_4(\mathbf{n}_2)\cdot 2(2^3\pi\mathcal{R}^{(\inf)}(t_2))^2)\ldots$$
$$\cdot e^{-2\pi^2 h_1(t_2;\mathbf{n}_2)(U_1/2)^2} e^{-2\pi^2 h_2(t_2;\mathbf{n}_2)U_1^2}$$
$$\times e^{-2\pi^2 h_3(t_2;\mathbf{n}_2)2^2U_1^2} e^{-2\pi^2 h_4(t_2;\mathbf{n}_2)2^4U_1^2}\cdots dt_2. \tag{30.6}$$

We estimate the long sum at the end of (30.6). By using the definitions of $k_i(\mathbf{n}_2)$, $h_i(t_2;\mathbf{n}_2)$, (30.4) and (30.5), we obtain the upper bound

EndSum of (30.6)

$$= \sum_{\substack{\mathbf{n}_2\in\mathbb{Z}^d\setminus\mathbf{0}:\ |L(\mathbf{n}_2)|=\lambda_2 \\ \text{and } L(\mathbf{n}_2)\cap L(\mathbf{n}_1)=L_{1,2}}}$$
$$\cdot \exp(-k_1(\mathbf{n}_2)\cdot 2(\pi\mathcal{R}^{(\inf)}(t_2))^2)\exp(-k_2(\mathbf{n}_2)\cdot 2(2\pi\mathcal{R}^{(\inf)}(t_2))^2)$$
$$\cdot \exp(-k_3(\mathbf{n}_2)\cdot 2(2^2\pi\mathcal{R}^{(\inf)}(t_2))^2)\exp(-k_4(\mathbf{n}_2)\cdot 2(2^3\pi\mathcal{R}^{(\inf)}(t_2))^2)\ldots$$
$$\cdot e^{-2h_1(t_2;\mathbf{n}_2)\pi^2(U_1/2)^2} e^{-2h_2(t_2;\mathbf{n}_2)\pi^2(3U_1/2)^2} e^{-2h_3(t_2;\mathbf{n}_2)\pi^2(5U_1/2)^2}\cdots$$
$$\le \sum_{\substack{(k_1,\ldots,k_r)\text{ with } r\ge 1: \\ k_1+\cdots+k_r=\lambda_2-\lambda_{1,2},k_j\ge 0,1\le j\le r-1 \text{ and } k_r\ge 1}} \binom{N-\lambda_1}{k_1}((2+1)^2)^{k_1}$$
$$\cdot \exp(-k_1\cdot 2(\pi\mathcal{R}^{(\inf)}(t_2))^2)$$
$$\cdot \binom{N-\lambda_1-k_1}{k_2}((2\cdot 2+1)^2)^{k_2}\exp(-k_2\cdot 2(2\pi\mathcal{R}^{(\inf)}(t_2))^2)$$
$$\cdot \binom{N-\lambda_1-k_1-k_2}{k_3}((2\cdot 2^2+1)^2)^{k_3}\exp(-k_3\cdot 2(2^2\pi\mathcal{R}^{(\inf)}(t_2))^2)$$
$$\cdot \binom{N-\lambda_1-k_1-k_2-k_3}{k_4}((2\cdot 2^2+1)^2)^{k_4}$$
$$\times \exp(-k_4\cdot 2(2^3\pi\mathcal{R}^{(\inf)}(t_2))^2)\ldots$$
$$\cdot \sum_{\substack{(h_0,h_1,\ldots,h_r)\text{ with } r\ge 0: \\ h_0+h_1+\cdots+h_r=\lambda_{1,2},h_j\ge 0,0\le j\le r-1 \text{ and } h_r\ge 1}} \binom{\lambda_{1,2}}{h_0}$$

$$\cdot \binom{\lambda_{1,2} - h_0}{h_1} ((2+1)^2)^{h_1} e^{-2h_1\pi^2(U_1/2)^2} \cdot$$

$$\cdot \binom{\lambda_{1,2} - h_0 - h_1}{h_2} ((2 \cdot 2+1)^2)^{h_2} e^{-2h_2\pi^2 U_1^2} \cdot$$

$$\cdot \binom{\lambda_{1,2} - h_0 - h_1 - h_2}{h_3} ((2 \cdot 2^2+1)^2)^{h_3} e^{-2h_3\pi^2 2^2 U_1^2} \cdots . \tag{30.7}$$

Note that (30.7) includes the pathological case $\lambda_2 - \lambda_{1,2} = 0$ with the natural convention that the summation means the single term $(k_1, \ldots, k_r) = (0)$, and similarly, if $\lambda_{1,2} = 0$ then $(h_0, h_1, \ldots, h_r)$ is just the single term (0). Moreover, to explain the binomial coefficient factor $\binom{\lambda_{1,2}}{h_0}$ in (30.7), we note that for every given half-open axis-parallel square Q of side length U_1, there is at most one integer lattice point $\mathbf{m}_2 \in \mathbb{Z}^2$ such that $r_k(t_2)\mathbf{m}_2 \in Q$ (see the definition of U_1 in Theorem 30.1). The same argument explains the next binomial coefficient factor $\binom{\lambda_{1,2} - h_0}{h_1}$ in (30.7), and so on. The extra factor $((2+1)^2)^{h_1}$ is a generous upper bound: it comes from the contribution of the first neighborhood of size $(2+1)^2 - 1$ of a lattice point $\mathbf{m}_2 \in \mathbb{Z}^2$; and so on. The same argument explains the factors

$$\binom{N - \lambda_1}{k_1} ((2+1)^2)^{k_1}, \quad \binom{N - \lambda_1 - k_1}{k_2} ((2 \cdot 2+1)^2)^{k_2}, \text{ and so on}$$

in (30.7).

Applying the multinomial theorem twice in (30.7) the same way as we did it in (13.15)–(13.16) (or in (27.22)), we have the very generous upper bound (the extra factor of N comes from the trivial upper bound $\binom{N}{k} \le N^k$)

$$\begin{aligned}
&\text{EndSum of (30.6)} \\
&\quad \le \big(N \cdot (2+1)^2 \exp(-2(\pi \mathcal{R}^{(\text{inf})}(t_2))^2) \\
&\qquad + N \cdot (2 \cdot 2+1)^2 \exp(-2(2\pi \mathcal{R}^{(\text{inf})}(t_2))^2) \\
&\qquad + N \cdot (2 \cdot 2^2+1)^2 \exp(-2(2^2\pi \mathcal{R}^{(\text{inf})}(t_2))^2) \\
&\qquad + N \cdot (2 \cdot 2^2+1)^2 \exp(-2(2^3\pi \mathcal{R}^{(\text{inf})}(t_2))^2) + \cdots \big)^{\lambda_2 - \lambda_{1,2}}
\end{aligned}$$

$$\cdot (1 + (2+1)^2 e^{-2\pi^2 (U_1/2)^2} + (2 \cdot 2 + 1)^2 e^{-2\pi^2 U_1^2}$$
$$+ (2 \cdot 2^2 + 1)^2 e^{-2\pi^2 2^2 U_1^2} + \cdots)^{\lambda_{1,2}}. \tag{30.8}$$

Next we use (30.7)–(30.8) in (30.6):

$$\int_{t_2=U}^{V} \sum_{\substack{\mathbf{n}_2 \in \mathbb{Z}^d \setminus \mathbf{0}:\ |L(\mathbf{n}_2)| = \lambda_2 \\ \text{and } L(\mathbf{n}_2) \cap L(\mathbf{n}_1) = L_{1,2}}} \exp\left(- \sum_{k=1}^{N} 2\pi^2 \left| r_k(t_1)\mathbf{n}_1(k) \right.\right.$$
$$\left.\left. - r_k(t_2)\theta_{k;t_2 \to t_1} \mathbf{n}_2(k) \right|^2 \right) dt_2$$
$$\leq (V-U) \prod_{j \in L(\mathbf{n}_1) \setminus L_{1,2}} \exp(-2\pi^2 |r_j(t_1)\mathbf{n}_1(j)|^2)$$
$$\cdot \left(N \sum_{j=1}^{\infty} 2^{2j+1} \exp(-2(2^{j-1}\pi \mathcal{R}^{(\text{inf})}(U))^2) \right)^{\lambda_2 - \lambda_{1,2}}$$
$$\times \left(1 + \sum_{j=1}^{\infty} 2^{2j+1} e^{-\pi^2 2^{2j-3} U_1^2} \right)^{\lambda_{1,2}}$$
$$\leq (V-U)(e^{-2\pi^2 U_1^2})^{\lambda_1 - \lambda_{1,2}}$$
$$\cdot \left(N \sum_{j=1}^{\infty} 2^{2j+1} e^{-\pi^2 2^{2j-1} U_1^2} \right)^{\lambda_2 - \lambda_{1,2}} \left(1 + \sum_{j=1}^{\infty} 2^{2j+1} e^{-\pi^2 2^{2j-3} U_1^2} \right)^{\lambda_{1,2}}$$
$$\leq (V-U)(e^{-2\pi^2 U_1^2})^{\lambda_1 - \lambda_{1,2}} (10 N e^{-2\pi^2 U_1^2})^{\lambda_2 - \lambda_{1,2}} (1 + 10 e^{-\pi^2 U_1^2/2})^{\lambda_{1,2}}, \tag{30.9}$$

since $\mathcal{R}^{(\text{inf})}(U) = U_1 \geq 1$ and $|L(\mathbf{n}_1) \setminus L_{1,2}| = \lambda_1 - \lambda_{1,2}$.

Let us return now to (30.3); we have the following decomposition into four parts:

$$|\Delta_f^2(\text{Gauss}; \vec{\mathcal{R}}; U, V)| \leq \text{Part One} + \text{Part Two} + \text{Part Three} + \text{Part Four}, \tag{30.10}$$

where

$$\text{Part One} = \sum_{\lambda_1=1}^{N-1} \int_{t_1=U}^{V} \sum_{\substack{\mathbf{n}_1 \in \mathbb{Z}^d \setminus \mathbf{0}: \\ |L(\mathbf{n}_1)|=\lambda_1}} |a_{\mathbf{n}_1}|^2 \Bigg(\sum_{L_{1,2} \subseteq L(\mathbf{n}_1)} \sum_{\lambda_2=\lambda_1+1}^{N} \int_{t_2=U}^{V}$$

$$\cdot \sum_{\substack{\mathbf{n}_2 \in \mathbb{Z}^d \setminus \mathbf{0}:\ |L(\mathbf{n}_2)|=\lambda_2 \\ \text{and } L(\mathbf{n}_2) \cap L(\mathbf{n}_1) = L_{1,2}}} \exp\Bigg(-\sum_{k=1}^{N} 2\pi^2 \,|r_k(t_1)\mathbf{n}_1(k)$$

$$-r_k(t_2)\theta_{k;t_2\to t_1}\mathbf{n}_2(k)|^2 \Bigg)\, dt_2 \Bigg)\, dt_1 \tag{30.11}$$

is characterized by the property $\lambda_1 < \lambda_2$,

$$\text{Part Two} = \sum_{\lambda_1=1}^{N} \int_{t_1=U}^{V} \sum_{\substack{\mathbf{n}_1 \in \mathbb{Z}^d \setminus \mathbf{0}: \\ |L(\mathbf{n}_1)|=\lambda_1}} |a_{\mathbf{n}_1}|^2 \Bigg(\sum_{L_{1,2} \subset L(\mathbf{n}_1)} \sum_{\lambda_2=\max\{|L_{1,2}|,1\}}^{\lambda_1-1} \int_{t_2=U}^{V}$$

$$\cdot \sum_{\substack{\mathbf{n}_2 \in \mathbb{Z}^d \setminus \mathbf{0}:\ |L(\mathbf{n}_2)|=\lambda_2 \\ \text{and } L(\mathbf{n}_2) \cap L(\mathbf{n}_1) = L_{1,2}}} \exp\Bigg(-\sum_{k=1}^{N} 2\pi^2 \,|r_k(t_1)\mathbf{n}_1(k)$$

$$-r_k(t_2)\theta_{k;t_2\to t_1}\mathbf{n}_2(k)|^2 \Bigg)\, dt_2 \Bigg)\, dt_1 \tag{30.12}$$

is characterized by the property $\lambda_1 > \lambda_2$, and finally, we split the case $\lambda_1 = \lambda_2$ into two subcases according as $L(\mathbf{n}_1) \neq L(\mathbf{n}_2)$ or $L(\mathbf{n}_1) = L(\mathbf{n}_2)$:

$$\text{Part Three} = \sum_{\lambda_1=1}^{N} \int_{t_1=U}^{V} \sum_{\substack{\mathbf{n}_1 \in \mathbb{Z}^d \setminus \mathbf{0}: \\ |L(\mathbf{n}_1)|=\lambda_1}} |a_{\mathbf{n}_1}|^2 \Bigg(\sum_{\substack{L_{1,2} \subset L(\mathbf{n}_1): \\ L_{1,2} \neq L(\mathbf{n}_1)}} \int_{t_2=U}^{V}$$

$$\cdot \sum_{\substack{\mathbf{n}_2 \in \mathbb{Z}^d \setminus \mathbf{0}:\ |L(\mathbf{n}_2)|=\lambda_1 \\ \text{and } L(\mathbf{n}_2) \cap L(\mathbf{n}_1) = L_{1,2}}} \exp\Bigg(-\sum_{k=1}^{N} 2\pi^2 \,|r_k(t_1)\mathbf{n}_1(k)$$

$$- r_k(t_2)\theta_{k;t_2\to t_1}\mathbf{n}_2(k)|^2 \Bigg)\, dt_2 \Bigg)\, dt_1 \tag{30.13}$$

and

$$\text{Part Four} = \int_{t_1=U}^{V} \sum_{\mathbf{n}_1 \in \mathbb{Z}^d \setminus \mathbf{0}} |a_{\mathbf{n}_1}|^2 \left(\int_{t_2=U}^{V} \right.$$

$$\cdot \sum_{\substack{\mathbf{n}_2 \in \mathbb{Z}^d \setminus \mathbf{0}: \\ L(\mathbf{n}_2)=L(\mathbf{n}_1)}} \exp\left(- \sum_{k=1}^{N} 2\pi^2 \left| r_k(t_1)\mathbf{n}_1(k) \right.\right.$$

$$\left.\left.\left. - r_k(t_2)\theta_{k;t_2\to t_1}\mathbf{n}_2(k) \right|^2 \right) dt_2 \right) dt_1. \tag{30.14}$$

To estimate (30.14), we are going to use a lemma, which is somewhat similar to Lemmas 9.3 and 27.3.

Lemma 30.1. *Assume that*

$$\inf_{x \geq U} |\mathbf{r}(x)| \geq U_1, \tag{30.15}$$

$$\sup_{x \leq t} |\mathbf{r}'(x)| \leq t^{\kappa} \text{ for all } t \geq \alpha_0 \text{ and } |\mathbf{r}'(x)| \leq \kappa_0 \text{ for } 0 < t < \alpha_0, \tag{30.16}$$

where $\alpha_0 \geq 1$, $\kappa \geq 0$, and $\mathbf{r}'(x)$ is the derivative of the parametrized curve $\mathbf{r}(x)$.

Moreover, let $C(\mathbf{c};\varrho)$ denote an arbitrary circle centered at $\mathbf{c} \in \mathbb{R}^2$ with radius ϱ, and assume that

$$\sup_{\mathbf{c} \in \mathbb{R}^2, \varrho > 0} \frac{\text{length}\{0 \leq t \leq T: \ \mathbf{r}(t) \in C(\mathbf{c};\varrho)\}}{\varrho} \leq T^{\beta} \tag{30.17}$$

uniformly for all $T \geq \beta_0$ (note that $\alpha_0 \geq 1$, $\kappa \geq 0$, κ_0, $\beta_0 \geq 1$ and $0 \leq \beta < 1$ are absolute constants).

Let $U \geq \{\alpha_0, \beta_0, 1\}$; then for every fixed $t_1 \in [U, 2U]$, $\mathbf{m}_1 \in \mathbb{Z}^2 \setminus \mathbf{0}$ and $0 < C' \leq U_1/16$,

$$\text{length}\ \Big\{ t_2 \in [U, 2U]: \ \text{there exists } \mathbf{m}_2 \in \mathbb{Z}^2 \setminus \mathbf{0} \text{ with } |r(t_1)\mathbf{m}_1$$

$$-r(t_2)\theta_{t_2\to t_1}\mathbf{m}_2| \leq C' \Big\}$$

$$\leq 8C' \left(3 + 2\log_2\left(1 + \frac{(2U)^{1+\kappa} + \alpha_0\kappa_0}{U_1} \right) \right) \frac{(2U)^{1+\kappa+\beta}}{U_1}. \tag{30.18}$$

If $U_1/16 < C' < U_1/2$ then the upper bound in (30.18) can be replaced by the trivial upper bound U.

Proof of Lemma 30.1. We study the inequality

$$|r(t_1)\mathbf{m}_1 - r(t_2)\theta_{t_2\to t_1}\mathbf{m}_2| \le C. \tag{30.19}$$

Applying the inverse $\theta_{t_1\to t_2}$ of $\theta_{t_2\to t_1}$, and dividing by $r(t_2)$, we rewrite (30.19) in the equivalent form

$$\left|\frac{r(t_1)\theta_{t_1\to t_2}\mathbf{m}_1}{r(t_2)} - \mathbf{m}_2\right| \le \frac{C'}{r(t_2)}. \tag{30.20}$$

Let $m_i = |\mathbf{m}_i| \ge 1$, $i = 1, 2$, and let $\tilde{\theta} \in \mathrm{SO}(2)$ denote the rotation such that

$$\frac{\mathbf{m}_1}{m_1} = \tilde{\theta}\frac{\mathbf{r}(t_1)}{|r(t_1)|}. \tag{30.21}$$

Using (30.21) in (30.20), we can rewrite (30.20) in the equivalent form

$$\left|\frac{m_1 r(t_1)\tilde{\theta}\mathbf{r}(t_2)}{(r(t_2))^2} - \mathbf{m}_2\right| \le \frac{C'}{r(t_2)}, \tag{30.22}$$

since

$$\frac{\theta_{t_1\to t_2}\mathbf{r}(t_1)}{r(t_1)} = \frac{\mathbf{r}(t_2)}{r(t_2)}.$$

Applying the inverse $\tilde{\theta}^{-1}$ of $\tilde{\theta}$ and dividing (30.22) with $m_1 r(t_1)$, we have

$$\left|\frac{\mathbf{r}(t_2)}{(r(t_2))^2} - \frac{\tilde{\theta}^{-1}\mathbf{m}_2}{m_1 r(t_1)}\right| \le \frac{C'}{m_1 r(t_1) r(t_2)}. \tag{30.23}$$

By hypothesis (30.15), $r(t_2) \ge U_1$. It follows that inequality (30.24)–(30.25) below implies (30.23), where

$$\left|\frac{\mathbf{r}(t_2)}{(r(t_2))^2} - \mathbf{w}\right| \le \varepsilon \tag{30.24}$$

with

$$\mathbf{w} = \mathbf{w}(\mathbf{m}_2) = \frac{\tilde{\theta}^{-1}\mathbf{m}_2}{m_1 r(t_1)} \text{ and } \varepsilon = \frac{C'}{m_1 r(t_1) U_1}. \tag{30.25}$$

Inequality (30.23)–(30.25) motivates the use of the following simple geometric lemma.

Lemma 30.2. *Let* $\mathbf{w} \in \mathbb{R}^2$ *and* $0 < \varepsilon < |\mathbf{w}|/2$, *then*

$$\left\{\mathbf{v} \in \mathbb{R}^2 : \left|\frac{\mathbf{v}}{|\mathbf{v}|^2} - \mathbf{w}\right| \le \varepsilon\right\} \Longleftrightarrow \{\mathbf{v} \in \mathbb{R}^2 : |\mathbf{v} - \mathbf{c}| \le \varrho\}$$

with

$$\mathbf{c} = \frac{\mathbf{w}}{|\mathbf{w}|^2 - \varepsilon^2} \quad \textit{and} \quad \varrho = \frac{\varepsilon}{|\mathbf{w}|^2 - \varepsilon^2}.$$

Proof of Lemma 30.2. Let T_{inv} denote the transformation of inversion on the plane

$$T_{\mathrm{inv}} : (x, y) \longrightarrow \left(\frac{x}{x^2 + y^2}, \frac{y}{x^2 + y^2}\right), \tag{30.26}$$

where the radius of inversion is 1, and the center of inversion is the origin $\mathbf{0} \in \mathbb{R}^2$. Note that an inversion transformation is its own inverse (which provides a good reason for the name of the transformation).

Given $\mathbf{w} \in \mathbb{R}^2 \setminus \mathbf{0}$ and $0 < \varepsilon < |\mathbf{w}|/2$, consider the circle centered at $\mathbf{w}$ and having radius ε:

$$C(\mathbf{w}; \varepsilon) = \left\{\mathbf{u} \in \mathbb{R}^2 : |\mathbf{u} - \mathbf{w}| \le \varepsilon\right\}.$$

It is well known that the image under inversion of a circle not passing through the center of inversion is a circle. Let $C(\mathbf{c}; \varrho)$ denote the T_{inv}-image of $C(\mathbf{w}; \varepsilon)$, where of course $\mathbf{c}$ is the center and ϱ is the radius. By definition (30.26)

$$(|\mathbf{w}| - \varepsilon)(|\mathbf{c}| + \varrho) = 1 = (|\mathbf{w}| + \varepsilon)(|\mathbf{c}| - \varrho),$$

which implies

$$\varrho = \frac{\varepsilon}{|\mathbf{w}|^2 - \varepsilon^2} \quad \text{and} \quad |\mathbf{c}| = \frac{|\mathbf{w}|}{|\mathbf{w}|^2 - \varepsilon^2}. \tag{30.27}$$

Since $\mathbf{0}, \mathbf{w}, \mathbf{c}$ are collinear, we obtain

$$\mathbf{c} = \frac{\mathbf{w}}{|\mathbf{w}|^2 - \varepsilon^2}. \tag{30.28}$$

Combining (30.26), (30.27) and (30.28), Lemma 30.2 follows. □

We also need the following simple fact. Given two lattice points $\mathbf{n} = (n_1, n_2), \mathbf{m} = (m_1, m_2) \in \mathbb{Z}^2$, we say that $\mathbf{n}$ and $\mathbf{m}$ are *neighbors*, if $|n_1 - m_1| \le 1$ and $|n_2 - m_2| \le 1$. So, a lattice point on the plane has eight neighbors.

Let $\mathbf{n}, \mathbf{m} \in \mathbb{Z}^2 \setminus \mathbf{0}$ be two nonzero lattice points that are neighbors, and let $0 < \varepsilon < \delta/2$. Trivial computation shows that

$$\left| \frac{\delta \mathbf{n}}{\delta^2 |\mathbf{n}|^2 - \varepsilon^2} - \frac{\delta \mathbf{m}}{\delta^2 |\mathbf{m}|^2 - \varepsilon^2} \right| \geq \frac{\delta}{\delta^2 \max\{|\mathbf{n}|^2, |\mathbf{m}|^2\} - \varepsilon^2} \geq \frac{1}{\delta \max\{|\mathbf{n}|^2, |\mathbf{m}|^2\}}. \tag{30.29}$$

After these preparations we are ready to complete the proof of Lemma 30.1.

Let $k \geq 0$ be a fixed integer, and let

$$S_k = \big\{ t_2 \in [U, 2U] : \text{(30.24) and (30.25) hold for some } \mathbf{m}_2 \in \mathbb{Z}^2 \text{ with } 2^k \leq m_2 < 2^{k+1} \big\}, \tag{30.30}$$

where $m_2 = |\mathbf{m}_2|$.

Applying Lemma 30.2 in (30.30), we have

$$S_k = \big\{ t_2 \in [U, 2U] : \ \mathbf{r}(t_2) \in C(\mathbf{c}(\mathbf{m}_2); \varrho(\mathbf{m}_2)) \text{ for some } \mathbf{m}_2 \in \mathbb{Z}^2 \text{ with } 2^k \leq m_2 < 2^{k+1} \big\}, \tag{30.31}$$

where

$$\mathbf{c}(\mathbf{m}_2) = \frac{\delta \tilde{\theta}^{-1} \mathbf{m}_2}{(\delta m_2)^2 - \varepsilon^2} \quad \text{and} \quad \varrho(\mathbf{m}_2) = \frac{\varepsilon}{(\delta m_2)^2 - \varepsilon^2} \tag{30.32}$$

$$\text{with} \quad \varepsilon = \frac{\delta C'}{U_1} \text{ and } \delta = \frac{1}{m_1 r(t_1)}. \tag{30.33}$$

By using $0 < \varepsilon < \delta/2$ in (30.32),

$$\varrho(\mathbf{m}_2) = \frac{\varepsilon}{(\delta m_2)^2 - \varepsilon^2} \leq \frac{\varepsilon}{(\delta m_2)^2 - (\delta/2)^2} \leq \frac{\varepsilon}{(\delta m_2)^2 (1 - 1/4)} = \frac{4\varepsilon}{3(\delta m_2)^2} = \frac{4\delta C'/U_1}{3\delta^2 m_2^2} \leq \frac{C'/U_1}{3\delta 4^{k-1}}, \tag{30.34}$$

where in the last step we applied (30.33) and $2^k \leq m_2$.

Using (30.34) in (30.31), we obtain

$$S_k \subset S_k^\star, \tag{30.35}$$

where

$$S_k^\star = \big\{ t_2 \in [U, 2U] : \ \mathbf{r}(t_2) \in C(\mathbf{c}(\mathbf{m}_2); \varrho^\star(k)) \text{ for some } \mathbf{m}_2 \in \mathbb{Z}^2 \text{ with } 2^k \leq m_2 < 2^{k+1} \big\} \tag{30.36}$$

with

$$\varrho^\star(k) = \frac{C'/U_1}{3\delta 4^{k-1}}. \tag{30.37}$$

Combining (30.29) and (30.32), we see that the centers $\mathbf{c}(\mathbf{m}_2)$ of the circles $C(\mathbf{c}(\mathbf{m}_2); \varrho^\star(k))$ listed in (30.36) are separated from each other by a distance

$$\geq \frac{1}{\delta 4^{k+1}}, \tag{30.38}$$

since $m_2 < 2^{k+1}$.

If $C' \leq U_1/16$, then by (30.37) and (30.38) the radius $\varrho^\star(k)$ is at most one-third of the distance between the centers of the circles $C(\mathbf{c}(\mathbf{m}_2); \varrho^\star(k))$ listed in (30.36). This implies that these circles are disjoint, and they are separated from each other by a distance

$$\geq \frac{1}{3\delta 4^{k+1}}. \tag{30.39}$$

It follows that the part of the curve $\mathbf{r}(t)$, $U \leq t \leq 2U$ intersects at most

$$\frac{U(2U)^\kappa}{\frac{1}{3\delta 4^{k+1}}} = 3U(2U)^\kappa \delta 4^{k+1} \tag{30.40}$$

circles $C(\mathbf{c}(\mathbf{m}_2); \varrho^\star(k))$ listed in (30.36). Indeed, the upper bound (30.16) on the speed of the parametrized curve $\mathbf{r}(t)$ implies that the arclength of the part of the curve $\mathbf{r}(t)$, $U \leq t \leq 2U$ is at most (using $U \geq \alpha_0$)

$$U(2U)^\kappa, \tag{30.41}$$

and thus we just divide (30.41) with the lower bound (30.39) of the distance between the disjoint circles $C(\mathbf{c}(\mathbf{m}_2); \varrho^\star(k))$ listed in (30.36).

By hypothesis (30.17), the curve $\mathbf{r}(t)$, $U \leq t \leq 2U$ spends at most $\varrho^\star(k)(2U)^\beta$ time in a circle of radius $\varrho^\star(k)$ (here we use $U \geq \beta_0$). By (30.37)

$$\varrho^\star(k) = \frac{C'/U_1}{3\delta 4^{k-1}},$$

and combining these facts with (30.40), we obtain the upper bound (see (30.35) and (30.36))

$$\begin{aligned} \text{length}(S_k) \leq \text{length}(S_k^\star) &\leq 3U(2U)^\kappa \delta 4^{k+1} \cdot \frac{C'/U_1}{3\delta 4^{k-1}} (2U)^\beta \\ &= 8C'(2U)^{1+\kappa+\beta}/U_1. \end{aligned} \tag{30.42}$$

To finish the proof of Lemma 30.1, we apply (30.42) in (30.23) as follows. Since (30.24)–(30.25) implies (30.23), and (30.23) is equivalent to (30.19), we have

$$\begin{aligned}&\text{length}\left\{t_2 \in [U, 2U]: \text{ there exists } \mathbf{m}_2 \in \mathbb{Z}^2 \setminus \mathbf{0} \text{ such that (30.19) holds}\right\}\\ &\qquad \le \text{length } \left\{t_2 \in [U, 2U]: \text{ (30.24) and (30.25) hold for some}\right.\\ &\qquad\quad \left.\mathbf{m}_2 \in \mathbb{Z}^2 \setminus \mathbf{0}\right\}\\ &\qquad \le \sum_{k_1 \le k < k_2} \text{length}(S_k), \end{aligned} \tag{30.43}$$

where $0 \le k_1 < k_2$ are some appropriate integers (see (30.30)).

To estimate the difference $k_2 - k_1$ (see (30.43)), we recall hypothesis (30.16), which implies

$$\max_{U \le t \le 2U} r(t) \le (2U)^{1+\kappa} + \alpha_0 \kappa_0. \tag{30.44}$$

Moreover, by (30.15),

$$\min_{U \le t \le 2U} r(t) \ge U_1. \tag{30.45}$$

Assume that $\mathbf{m}_2 \in \mathbb{Z}^2 \setminus \mathbf{0}$ satisfies (30.19); then by (30.44) and (30.45),

$$\begin{aligned} m_2 \le \frac{r(t_1)m_1 + (U_1/2)}{r(t_2)} &\le \frac{\left((2U)^{1+\kappa} + \alpha_0\kappa_0\right) m_1 + (U_1/2)}{U_1}\\ &< \frac{\left((2U)^{1+\kappa} + \alpha_0\kappa_0 + U_1\right) m_1}{U_1}, \end{aligned}$$

and similarly,

$$\begin{aligned} m_2 \ge \frac{r(t_1)m_1 - (U_1/2)}{r(t_2)} &\ge \frac{U_1 m_1 - (U_1/2)}{(2U)^{1+\kappa} + \alpha_0\kappa_0}\\ &> \frac{U_1 m_1/2}{(2U)^{1+\kappa} + \alpha_0\kappa_0}. \end{aligned}$$

Thus we have

$$2^{k_1} \le m_2 < 2^{k_2}, \quad \text{where} \tag{30.46}$$

k_1 is the largest integer such that

$$2^{k_1} \le \frac{U_1 m_1/2}{(2U)^{1+\kappa} + \alpha_0\kappa_0},$$

and k_2 is the smallest integer such that

$$\frac{\left((2U)^{1+\kappa}+\alpha_0\kappa_0+(U_1/2)\right)m_1}{U_1}\le 2^{k_2}.$$

It follows that ($\log_2$ denotes the binary logarithm)

$$\begin{aligned}
k_2-k_1 &\le 2+\log_2\frac{\frac{((2U)^{1+\kappa}+\alpha_0\kappa_0+U_1)m_1}{U_1}}{\frac{U_1m_1/2}{(2U)^{1+\kappa}+\alpha_0\kappa_0}}\\
&\le 2+\log_2\frac{\left((2U)^{1+\kappa}+\alpha_0\kappa_0+U_1\right)^2}{(U_1)^2/2}\\
&= 3+2\log_2\frac{(2U)^{1+\kappa}+\alpha_0\kappa_0+U_1}{U_1}\\
&= 3+2\log_2\left(1+\frac{(2U)^{1+\kappa}+\alpha_0\kappa_0}{U_1}\right). \qquad (30.47)
\end{aligned}$$

Combining (30.42), (30.43) and (30.47), we conclude

$$\begin{aligned}
&\text{length}\left\{t_2\in[U,2U]:\ \text{there exists } \mathbf{m}_2\in\mathbb{Z}^2\setminus\mathbf{0} \text{ such that (30.19) holds}\right\}\\
&\le \sum_{k_1\le k<k_2}\text{length}(S_k)\\
&\le \left(3+2\log_2\left(1+\frac{(2U)^{1+\kappa}+\alpha_0\kappa_0}{U_1}\right)\right)\cdot 8C'(2U)^{1+\kappa+\beta}/U_1,
\end{aligned}$$

which completes the proof of Lemma 30.1. □

31. Finishing the Proof of Theorem 14.1

Applying (30.9) in (30.11), we have

$$\begin{aligned}
\text{Part One} \le \sum_{\lambda_1=1}^{N-1}\int_{t_1=U}^{V}\sum_{\substack{\mathbf{n}_1\in\mathbb{Z}^d\setminus\mathbf{0}:\\ |L(\mathbf{n}_1)|=\lambda_1}}|a_{\mathbf{n}_1}|^2\Bigg(\sum_{L_{1,2}\subseteq L(\mathbf{n}_1)}\sum_{\lambda_2=\lambda_1+1}^{N}(V-U)\\
\cdot\,(e^{-2\pi^2U_1^2})^{\lambda_1-\lambda_{1,2}}(10Ne^{-2\pi^2U_1^2})^{\lambda_2-\lambda_{1,2}}(1+10e^{-\pi^2U_1^2/2})^{\lambda_{1,2}}\Bigg)dt_1\\
\le (V-U)^2\max_{\lambda_{1,2}\le N}(1+10e^{-\pi^2U_1^2/2})^{\lambda_{1,2}}.
\end{aligned}$$

$$\cdot \sum_{\lambda_2=1}^{N} (10Ne^{-2\pi^2 U_1^2})^{\lambda_2} \sum_{\lambda_1=1}^{\lambda_2-1} \sum_{\substack{\mathbf{n}_1\in\mathbb{Z}^d\setminus\mathbf{0}:\\ |L(\mathbf{n}_1)|=\lambda_1}} |a_{\mathbf{n}_1}|^2$$

$$\cdot \left(\sum_{\lambda_{1,2}=0}^{\lambda_1} \binom{\lambda_1}{\lambda_{1,2}} (10Ne^{-2\pi^2 U_1^2})^{-\lambda_{1,2}} (e^{-2\pi^2 U_1^2})^{\lambda_1-\lambda_{1,2}} \right)$$

$$= (V-U)^2(1+10e^{-\pi^2 U_1^2/2})^N \sum_{\lambda_2=1}^{N} (10Ne^{-2\pi^2 U_1^2})^{\lambda_2}$$

$$\cdot \sum_{\lambda_1=1}^{\lambda_2-1} \sum_{\substack{\mathbf{n}_1\in\mathbb{Z}^d\setminus\mathbf{0}:\\ |L(\mathbf{n}_1)|=\lambda_1}} |a_{\mathbf{n}_1}|^2((10Ne^{-2\pi^2 U_1^2})^{-1} + e^{-2\pi^2 U_1^2})^{\lambda_1}. \qquad (31.1)$$

By hypothesis

$$UNe^{-\pi^2 U_1^2/2} < 1/10, \qquad (31.2)$$

and using it we have

$$\begin{aligned} &((10Ne^{-2\pi^2 U_1^2})^{-1} + e^{-2\pi^2 U_1^2})^{\lambda_1} \\ &\quad = (10Ne^{-2\pi^2 U_1^2})^{-\lambda_1}(1 + e^{-2\pi^2 U_1^2} \cdot 10Ne^{-2\pi^2 U_1^2})^{\lambda_1} \\ &\quad \le (10Ne^{-2\pi^2 U_1^2})^{-\lambda_1} \left(1+\frac{1}{N}\right)^{\lambda_1} \\ &\quad \le 3(10Ne^{-2\pi^2 U_1^2})^{-\lambda_1}, \end{aligned} \qquad (31.3)$$

since

$$\left(1+\frac{1}{N}\right)^{\lambda_1} \le \left(1+\frac{1}{N}\right)^{N} < 3.$$

By using (31.2)–(31.3) in (31.1), we obtain

$$\text{Part One} \le (V-U)^2 \left(1+\frac{1}{N}\right)^{N} \sum_{\lambda_2=1}^{N} \sum_{\lambda_1=1}^{\lambda_2-1} (10Ne^{-2\pi^2 U_1^2})^{\lambda_2}$$

$$\cdot 3(10Ne^{-2\pi^2 U_1^2})^{-\lambda_1} \sum_{\mathbf{n}\in\mathbb{Z}^d\setminus\mathbf{0}} |a_{\mathbf{n}}|^2$$

$$\le 3\cdot 3(V-U)^2 \sum_{\lambda_2=1}^{N} \sum_{\lambda_1=1}^{\lambda_2-1} (10Ne^{-2\pi^2 U_1^2})^{\lambda_2-\lambda_1} \sum_{\mathbf{n}\in\mathbb{Z}^d\setminus\mathbf{0}} |a_{\mathbf{n}}|^2. \qquad (31.4)$$

By using the substitution $i = \lambda_2 - \lambda_1 \geq 1$ in (31.4), we obtain the upper bound

$$\begin{aligned}\text{Part One} &\leq 9(V-U)^2 \sum_{\mathbf{n}\in\mathbb{Z}^d\setminus\mathbf{0}} |a_{\mathbf{n}}|^2 N \sum_{i=1}^{\infty}\left(10Ne^{-2\pi^2 U_1^2}\right)^i \\ &= 9(V-U)^2\sigma_0^2(f)\frac{10N^2e^{-2\pi^2U_1^2}}{1-10Ne^{-2\pi^2U_1^2}} \\ &\leq 9(V-U)^2\sigma_0^2(f)\frac{10N^2(10UN)^{-4}}{1-\frac{1}{10}} \leq \sigma_0^2(f), \end{aligned} \tag{31.5}$$

where in the last step we used Parseval's formula, (31.2) and $V-U\leq U$.

Next we apply (30.9) in (30.12):

$$\begin{aligned}\text{Part Two} \leq \sum_{\lambda_1=1}^{N}\int_{t_1=U}^{V} &\sum_{\substack{\mathbf{n}_1\in\mathbb{Z}^d\setminus\mathbf{0}:\\ |L(\mathbf{n}_1)|=\lambda_1}} |a_{\mathbf{n}_1}|^2 \Bigg(\sum_{L_{1,2}\subset L(\mathbf{n}_1)}\ \sum_{\lambda_2=\max\{|L_{1,2}|,1\}}^{\lambda_1-1} \\ &\cdot e^{-2\pi^2(\lambda_1-\lambda_{1,2})U_1^2}(V-U)(10Ne^{-2\pi^2U_1^2})^{\lambda_2-\lambda_{1,2}} \\ &\times (1+10e^{-\pi^2U_1^2/2})^{\lambda_{1,2}}\Bigg)\, dt_1 \\ \leq (V-U)^2 &\max_{\lambda_{1,2}\leq N}(1+10e^{-\pi^2U_1^2/2})^{\lambda_{1,2}} \sum_{\lambda_1=1}^{N}\ \sum_{\substack{\mathbf{n}_1\in\mathbb{Z}^d\setminus\mathbf{0}:\\ |L(\mathbf{n}_1)|=\lambda_1}} |a_{\mathbf{n}_1}|^2 \\ &\cdot\sum_{\lambda_2=1}^{\lambda_1-1}\sum_{\lambda_{1,2}=0}^{\lambda_2}\binom{\lambda_1}{\lambda_{1,2}}(10Ne^{-2\pi^2U_1^2})^{\lambda_2-\lambda_{1,2}}e^{-2\pi^2(\lambda_1-\lambda_{1,2})U_1^2} \\ \leq (V-U)^2 &\left(1+\frac{1}{2N}\right)^N \sum_{\lambda_1=1}^{N}\sum_{\lambda_2=1}^{\lambda_1-1}\sum_{\lambda_{1,2}=0}^{\lambda_2} \\ &\cdot(10Ne^{-2\pi^2U_1^2})^{\lambda_2-\lambda_{1,2}}(N\cdot e^{-2\pi^2U_1^2})^{\lambda_1-\lambda_{1,2}} \sum_{\mathbf{n}\in\mathbb{Z}^d\setminus\mathbf{0}}|a_{\mathbf{n}}|^2 \\ \leq (V-U)^2 &\cdot 2\cdot \sum_{\lambda_1=1}^{N}\sum_{\lambda_2=1}^{\lambda_1-1}\sum_{\lambda_{1,2}=0}^{\lambda_2}(10Ne^{-2\pi^2U_1^2})^{\lambda_1+\lambda_2-2\lambda_{1,2}} \\ &\times \sum_{\mathbf{n}\in\mathbb{Z}^d\setminus\mathbf{0}}|a_{\mathbf{n}}|^2, \end{aligned} \tag{31.6}$$

where in the last steps we used (31.2) and the trivial upper bound

$$\binom{\lambda_1}{\lambda_{1,2}} = \binom{\lambda_1}{\lambda_1 - \lambda_{1,2}} \le \lambda_1^{\lambda_1-\lambda_{1,2}} \le N^{\lambda_1-\lambda_{1,2}}. \tag{31.7}$$

Applying (31.2) in (31.6), we obtain

$$\begin{aligned}
\text{Part Two} &\le 2\left(\sum_{\mathbf{n}\in\mathbb{Z}^d\setminus\mathbf{0}} |a_{\mathbf{n}}|^2\right)(V-U)^2 \\
&\quad \cdot \sum_{\lambda_1=1}^{N}\sum_{\lambda_2=1}^{\lambda_1-1}\sum_{\lambda_{1,2}=0}^{\lambda_2} \left(10\cdot 10^{-4}N^{-1}U^{-2}\right)^{\lambda_1+\lambda_2-2\lambda_{1,2}} \\
&\le 2\left(\sum_{\mathbf{n}\in\mathbb{Z}^d\setminus\mathbf{0}} |a_{\mathbf{n}}|^2\right)(V-U)^2 \sum_{j=1}^{\infty} N(j+1)j\left(10^{-3}N^{-1}U^{-2}\right)^j \\
&= 2\left(\sum_{\mathbf{n}\in\mathbb{Z}^d\setminus\mathbf{0}} |a_{\mathbf{n}}|^2\right)(V-U)^2\cdot N\cdot 2\frac{10^{-3}N^{-1}U^{-2}}{(1-10^{-3}N^{-1}U^{-2})^3} \\
&= \frac{4}{27}\left(\sum_{\mathbf{n}\in\mathbb{Z}^d\setminus\mathbf{0}} |a_{\mathbf{n}}|^2\right)(V-U)^2U^{-2}\left(1-10^{-3}N^{-1}U^{-2}\right)^{-3} \\
&\le \frac{1}{4}\sum_{\mathbf{n}\in\mathbb{Z}^d\setminus\mathbf{0}} |a_{\mathbf{n}}|^2 = \frac{1}{4}\sigma_0^2(f),
\end{aligned} \tag{31.8}$$

where we used the substitution $j = \lambda_1+\lambda_2-2\lambda_{1,2}$, the assumption $V-U \le U$, the simple fact

$$\sum_{j=1}^{\infty} j(j+1)x^j = \frac{2x}{(1-x)^3} \quad \text{for all } |x|<1$$

and Parseval's formula.

Next we apply (30.9) in (30.13):

$$\text{Part Three} \leq \sum_{\lambda_1=1}^{N} \int_{t_1=U}^{V} \sum_{\substack{\mathbf{n}_1\in\mathbb{Z}^d\setminus\mathbf{0}: \\ |L(\mathbf{n}_1)|=\lambda_1}} |a_{\mathbf{n}_1}|^2 \left(\sum_{\substack{L_{1,2}\subset L(\mathbf{n}_1): \\ L_{1,2}\neq L(\mathbf{n}_1)}} \cdot e^{-2\pi^2(\lambda_1-\lambda_{1,2})U_1^2} \right.$$

$$\left. \times (V-U)(10Ne^{-2\pi^2U_1^2})^{\lambda_1-\lambda_{1,2}}(1+10e^{-\pi^2U_1^2/2})^{\lambda_{1,2}} \right) dt_1$$

$$\leq (V-U)^2 \max_{\lambda_{1,2}\leq N}(1+10e^{-\pi^2U_1^2/2})^{\lambda_{1,2}} \sum_{\lambda_1=1}^{N} \sum_{\substack{\mathbf{n}_1\in\mathbb{Z}^d\setminus\mathbf{0}: \\ |L(\mathbf{n}_1)|=\lambda_1}} |a_{\mathbf{n}_1}|^2$$

$$\cdot \sum_{\lambda_{1,2}=0}^{\lambda_1-1} \binom{\lambda_1}{\lambda_{1,2}} (10Ne^{-(2+2)\pi^2U_1^2})^{\lambda_1-\lambda_{1,2}}. \tag{31.9}$$

Using (31.7) in (31.9), we have

$$\text{Part Three} \leq (V-U)^2 \max_{\lambda_{1,2}\leq N}(1+10e^{-\pi^2U_1^2/2})^{\lambda_{1,2}}$$

$$\times \sum_{\lambda_1=1}^{N} \sum_{\substack{\mathbf{n}_1\in\mathbb{Z}^d\setminus\mathbf{0}: \\ |L(\mathbf{n}_1)|=\lambda_1}} |a_{\mathbf{n}_1}|^2 \cdot \sum_{\lambda_{1,2}=0}^{\lambda_1-1} (10N^2e^{-4\pi^2U_1^2})^{\lambda_1-\lambda_{1,2}}. \tag{31.10}$$

Using (31.2) in (31.10), we obtain

$$\text{Part Three} \leq (V-U)^2(1+10(NU)^{-1})^N$$

$$\times \sum_{\lambda_1=1}^{N} \sum_{\lambda_{1,2}=0}^{\lambda_1-1} (10(10^4NU)^{-2})^{\lambda_1-\lambda_{1,2}} \cdot \sum_{\mathbf{n}\in\mathbb{Z}^d\setminus\mathbf{0}} |a_{\mathbf{n}}|^2$$

$$\leq (V-U)^2 \left(1+\frac{1}{2N}\right)^N \cdot N \left(\sum_{j=1}^{\infty} (10^7N^2U^2)^{-j} \right) \sum_{\mathbf{n}\in\mathbb{Z}^d\setminus\mathbf{0}} |a_{\mathbf{n}}|^2$$

$$\leq (V-U)^2 \cdot 2 \cdot N \frac{(10^7 N^2 U^2)^{-1}}{1-(10^7 N^2 U^2)^{-1}} \sum_{\mathbf{n} \in \mathbb{Z}^d \setminus \mathbf{0}} |a_{\mathbf{n}}|^2$$

$$= (V-U)^2 U^{-2} \frac{2(10^7 N)^{-1}}{1-(10^7 N^2 U^2)^{-1}} \sum_{\mathbf{n} \in \mathbb{Z}^d \setminus \mathbf{0}} |a_{\mathbf{n}}|^2$$

$$\leq \frac{1}{100N} \sum_{\mathbf{n} \in \mathbb{Z}^d \setminus \mathbf{0}} |a_{\mathbf{n}}|^2 = \frac{\sigma_0^2(f)}{100N}, \tag{31.11}$$

where we used the substitution $j = \lambda_1 - \lambda_{1,2}$, the hypothesis $V - U \leq U$ and Parseval's formula.

Finally we estimate (30.14). We have

$$\text{Part Four} = \text{Part Four A} + \text{Part Four B}, \tag{31.12}$$

where

$$\text{Part Four A} = \int_{t_1=U}^{V} \sum_{\mathbf{n}_1 \in \mathbb{Z}^d \setminus \mathbf{0}} |a_{\mathbf{n}_1}|^2 \left(\int_{t_2=U}^{V} \cdot \sum_{\substack{\mathbf{n}_2 \in \mathbb{Z}^d \setminus \mathbf{0}:\ L(\mathbf{n}_2)=L(\mathbf{n}_1) \\ h_0(t_2;\mathbf{n}_2)<|L(\mathbf{n}_1)|}} \right.$$

$$\times \exp\left(-\sum_{k=1}^{N} 2\pi^2 |r_k(t_1)\mathbf{n}_1(k)\right.$$

$$\left.\left. - r_k(t_2)\theta_{k;t_2\to t_1}\mathbf{n}_2(k)|^2 \right) dt_2 \right) dt_1, \tag{31.13}$$

and

$$\text{Part Four B} = \int_{t_1=U}^{V} \sum_{\mathbf{n}_1 \in \mathbb{Z}^d \setminus \mathbf{0}} |a_{\mathbf{n}_1}|^2 \left(\int_{t_2=U}^{V} \cdot \sum_{\substack{\mathbf{n}_2 \in \mathbb{Z}^d \setminus \mathbf{0}:\ L(\mathbf{n}_2)=L(\mathbf{n}_1) \\ h_0(t_2;\mathbf{n}_2)=|L(\mathbf{n}_1)|}} \right.$$

$$\times \exp\left(-\sum_{k=1}^{N} 2\pi^2 |r_k(t_1)\mathbf{n}_1(k)\right.$$

$$\left.\left. - r_k(t_2)\theta_{k;t_2\to t_1}\mathbf{n}_2(k)|^2 \right) dt_2 \right) dt_1. \tag{31.14}$$

We start with Part Four A, and use the notation $\lambda_1 = |L(\mathbf{n}_1)|$. In this special case we obtain the following analog of (14.15) (note that here $\lambda_1 = \lambda_2 = \lambda_{1,2}$)

$$\text{Part Four A} \le \int_{t_1=U}^{V} \sum_{\mathbf{n}_1 \in \mathbb{Z}^d \setminus \mathbf{0}} |a_{\mathbf{n}_1}|^2 (V-U)$$

$$\cdot \left(\sum_{\substack{(h_0,h_1,\ldots,h_r) \text{ with } r\ge 0:\ h_0<\lambda_1 \\ h_0+h_1+\cdots+h_r=\lambda_1,\, h_j\ge 0,\, 0\le j\le r-1 \text{ and } h_r\ge 1}} \binom{\lambda_1}{h_0} \right.$$

$$\cdot \binom{\lambda_1 - h_0}{h_1} ((2+1)^2)^{h_1} e^{-2\pi^2 h_1 (U_1/2)^2}$$

$$\times \binom{\lambda_1 - h_0 - h_1}{h_2} ((2\cdot 2+1)^2)^{h_2} e^{-2\pi^2 h_2 U_1^2}$$

$$\left. \cdot \binom{\lambda_1 - h_0 - h_1 - h_2}{h_3} ((2\cdot 2^2+1)^2)^{h_3} e^{-2\pi^2 h_3 (2U_1)^2} \cdots \right) dt_1$$

$$= \int_{t_1=U}^{V} \sum_{\mathbf{n}_1 \in \mathbb{Z}^d \setminus \mathbf{0}} |a_{\mathbf{n}_1}|^2 (V-U) \cdot ((1 + (2+1)^2 e^{-\pi^2 U_1^2/2}$$

$$+ (2\cdot 2+1)^2 e^{-2\pi^2 U_1^2} + (2\cdot 2^2+1)^2 e^{-2^3\pi^2 U_1^2} + \cdots)^{\lambda_1} - 1)\, dt_1, \tag{31.15}$$

where the term -1 at the end is explained by the restriction $h_0 < \lambda_1$, i.e., the term -1 is due to the fact that in the application of the multinomial theorem we have an *almost* complete sum, where the single missing case is $h_0 = \lambda_1$.

By (31.15)

$$\text{Part Four A} \le \sum_{\mathbf{n}_1 \in \mathbb{Z}^d \setminus \mathbf{0}} |a_{\mathbf{n}_1}|^2 (V-U)^2 ((1 + 10 e^{-\pi^2 U_1^2/2})^N - 1)$$

$$\le \sigma_0^2(f)(V-U)^2(\exp(10 N e^{-\pi^2 U_1^2/2}) - 1)$$

$$\le \sigma_0^2(f)(V-U)^2 20 N e^{-\pi^2 U_1^2/2}, \tag{31.16}$$

since

$$10Ne^{-\pi^2 U_1^2/2} \leq 1. \tag{31.17}$$

Note that in (31.16) we used Parseval's formula and the simple inequalities $1+x \leq e^x \leq 1+2x$ for $0 \leq x \leq 1$; and of course (31.17) follows from (31.2).

Applying (31.2) in (31.16), we have

$$\text{Part Four A} \leq \sigma_0^2(f)(V-U)^2 \frac{20}{10U} \leq 2\sigma_0^2(f)(V-U), \tag{31.18}$$

since $V - U \leq U$.

To estimate (31.14)

$$\text{Part Four B} = \int_{t_1=U}^{V} \sum_{\mathbf{n}_1 \in \mathbb{Z}^d \setminus \mathbf{0}} |a_{\mathbf{n}_1}|^2 \left(\int_{t_2=U}^{V} \cdot \sum_{\substack{\mathbf{n}_2 \in \mathbb{Z}^d \setminus \mathbf{0}:\ L(\mathbf{n}_2)=L(\mathbf{n}_1), \\ h_0(t_2;\mathbf{n}_2)=|L(\mathbf{n}_1)|}} \right.$$
$$\times \exp\left(-\sum_{k=1}^{N} 2\pi^2 |r_k(t_1)\mathbf{n}_1(k) \right.$$
$$\left.\left. - r_k(t_2)\theta_{k;t_2\to t_1}\mathbf{n}_2(k)|^2 \right) dt_2 \right) dt_1, \tag{31.19}$$

we make use of the fact that for every given triple $(t_1, \mathbf{n}_1, t_2)$ with $t_1 \in [U, V]$, $\mathbf{n}_1 \in \mathbb{Z}^d \setminus \mathbf{0}$ and $t_2 \in [U, V]$, there is *at most one* term in the endsum

$$\sum_{\substack{\mathbf{n}_2 \in \mathbb{Z}^d \setminus \mathbf{0}:\ L(\mathbf{n}_2)=L(\mathbf{n}_1), \\ h_0(t_2;\mathbf{n}_2)=|L(\mathbf{n}_1)|}} \exp\left(-\sum_{k=1}^{N} 2\pi^2 |r_k(t_1)\mathbf{n}_1(k) - r_k(t_2)\theta_{k;t_2\to t_1}\mathbf{n}_2(k)|^2 \right)$$

of (31.19). Indeed, it follows from the definition of $h_0(t_2;\mathbf{n}_2)$ that for every given triple $(t_1, \mathbf{n}_1, t_2)$ the inequality

$$|r_k(t_1)\mathbf{n}_1(k) - r_k(t_2)\theta_{k;t_2\to t_1}\mathbf{n}_2(k)| < U_1/2$$

has at most one integer lattice point solution $\mathbf{n}_2(k) \in \mathbb{Z}^2$, and because $h_0(t_2;\mathbf{n}_2) = |L(\mathbf{n}_1)|$ and $L(\mathbf{n}_2) = L(\mathbf{n}_1)$, there is at most one $\mathbf{n}_2 \in \mathbb{Z}^d \setminus \mathbf{0}$ satisfying the requirements. Let $\mathbf{n}_2^* \in \mathbb{Z}^d \setminus \mathbf{0}$, $\mathbf{n}_2^* = \mathbf{n}_2^*(t_1, \mathbf{n}_1, t_2)$ denote this single integer lattice point, *if it exists*. Thus we can rewrite (31.19) as follows:

$$\text{Part Four B} = \int_{t_1=U}^{V} \sum_{\mathbf{n}_1 \in \mathbb{Z}^d \setminus \mathbf{0}} |a_{\mathbf{n}_1}|^2 \cdot \left(\int_{t_2=U}^{V} \left\{ \exp\left(-\sum_{k=1}^{N} 2\pi^2 |r_k(t_1)\mathbf{n}_1(k) \right.\right.\right.$$
$$\left.\left.\left. - r_k(t_2)\theta_{k;t_2\to t_1}\mathbf{n}_2^*(k)|^2 \right) \text{ or } 0 \right\} dt_2 \right) dt_1, \tag{31.20}$$

where $\mathbf{n}_2^* = \mathbf{n}_2^*(t_1, \mathbf{n}_1, t_2)$, *if it exists.*

Let $j_0 = j_0(\mathbf{n}_1)$ be an arbitrary but fixed element of the set $L(\mathbf{n}_1)$; then we have the following trivial upper bound for (31.20):

$$\text{Part Four B} \le \int_{t_1=U}^{V} \sum_{\mathbf{n}_1 \in \mathbb{Z}^d \setminus \mathbf{0}} |a_{\mathbf{n}_1}|^2 \left(\int_{t_2=U}^{V} \cdot \{\exp(-2\pi^2 |r_{j_0}(t_1)\mathbf{n}_1(j_0) \right.$$
$$\left. - r_{j_0}(t_2)\theta_{j_0;t_2\to t_1}\mathbf{n}_2^*(j_0)|^2) \text{ or } 0\} \, dt_2 \right) dt_1. \tag{31.21}$$

Given $(t_1, \mathbf{n}_1)$, we consider the following decomposition of the interval $U \le t_2 \le V (\le 2U)$: for $\ell = 1, 2, 3, \ldots$ let

$$I_\ell(t_1, \mathbf{n}_1) = \{t_2 \in [U, V] : \ \ell - 1 \le |r_{j_0}(t_1)\mathbf{n}_1(j_0)$$
$$- r_{j_0}(t_2)\theta_{j_0;t_2\to t_1}\mathbf{n}_2^*(j_0)| < \ell\}, \tag{31.22}$$

where $j_0 = j_0(\mathbf{n}_1)$ and $\mathbf{n}_2^* = \mathbf{n}_2^*(t_1, \mathbf{n}_1, t_2)$.

Using decomposition (31.22) in (31.21), we obtain the upper bound

$$\text{Part Four B} \le \int_{t_1=U}^{V} \sum_{\mathbf{n}_1 \in \mathbb{Z}^d \setminus \mathbf{0}} |a_{\mathbf{n}_1}|^2 \left(\sum_{\ell=1}^{\infty} \text{length}(I_\ell) e^{-2\pi^2(\ell-1)^2} \right) dt_1, \tag{31.23}$$

where $I_\ell = I_\ell(t_1, \mathbf{n}_1)$. Applying Lemma 30.1 with $C' = \ell$, we obtain the upper bound

$$\text{length}(I_\ell) \le 8\ell \left(3 + 2\log_2 \left(1 + \frac{(2U)^{1+\kappa} + \alpha_0\kappa_0}{U_1} \right) \right) \frac{(2U)^{1+\kappa+\beta}}{U_1} \tag{31.24}$$

if $\ell \le U_1/16$ and

$$\text{length}(I_\ell) < U \tag{31.25}$$

if $U_1/16 < \ell < U_1/2$. Using (31.24), (31.25) and Parseval's formula in (31.23), we have

$$\begin{aligned}
\text{Part Four B} &\le (V-U)\sigma_0^2(f)\left(\left(\sum_{\ell=1}^{\infty} 8\ell e^{-2\pi^2(\ell-1)^2}\right)\right.\\
&\quad \cdot \left(3 + 2\log_2\left(1 + \frac{(2U)^{1+\kappa} + \alpha_0\kappa_0}{U_1}\right)\right)\\
&\quad \left.\times \frac{(2U)^{1+\kappa+\beta}}{U_1} + U \sum_{\ell > U_1/16} e^{-2\pi^2(\ell-1)^2}\right)\\
&\le \sigma_0^2(f)(V-U)\left(U\exp(-\pi^2 U_1^2 2^{-8}) + 9\left(3 + 2\log_2\right.\right.\\
&\quad \left.\left.\times\left(1 + \frac{(2U)^{1+\kappa} + \alpha_0\kappa_0}{U_1}\right)\right)\frac{(2U)^{1+\kappa+\beta}}{U_1}\right).
\end{aligned} \tag{31.26}$$

Combining (30.10), (31.12), (31.5), (31.8), (31.11), (31.18) and (31.26),

$$\begin{aligned}
|\Delta_f^2(\text{Gauss}; \vec{\mathcal{R}}; U, V)| &\le \text{Part One} + \text{Part Two}\\
&\quad + \text{Part Three} + \text{Part Four A} + \text{Part Four B}\\
&\le \sigma_0^2(f) + \frac{1}{4}\sigma_0^2(f) + \frac{1}{100N}\sigma_0^2(f) + 2\sigma_0^2(f)(V-U)\\
&\quad + \sigma_0^2(f)(V-U)\left(U\exp(-\pi^2 U_1^2 2^{-8}) + 9\left(3 + 2\log_2\right.\right.\\
&\quad \left.\left.\times\left(1 + \frac{(2U)^{1+\kappa} + \alpha_0\kappa_0}{U_1}\right)\right)\frac{(2U)^{1+\kappa+\beta}}{U_1}\right)\\
&\le 2\sigma_0^2(f) + \sigma_0^2(f)(V-U)\left(2 + U\exp(-\pi^2 U_1^2 2^{-8}) + 9\right.\\
&\quad \left.\times\left(3 + 2\log_2\left(1 + \frac{(2U)^{1+\kappa} + \alpha_0\kappa_0}{U_1}\right)\right)\frac{(2U)^{1+\kappa+\beta}}{U_1}\right),
\end{aligned} \tag{31.27}$$

since by hypothesis $U \ge \max\{\alpha_0, \beta_0, 1\}$. (31.27) completes the proof of Theorem 30.1. □

Now we are ready to complete the proof of Theorem 14.1. By hypothesis

$$U_1 = \mathcal{R}^{(\inf)}(U) = \min_{1\le k\le N} \inf_{x\ge U} r_k(x) \ge U^\alpha,$$

and using this fact in (31.27), we have

$$\begin{aligned}|\Delta_f^2(\text{Gauss}; \vec{\mathcal{R}}; U, V)| \le 2\sigma_0^2(f) + \sigma_0^2(f)(V-U) \\ \times \Bigg(2 + U\exp(-\pi^2 U^{2\alpha}2^{-8}) + 9\Bigg(3 + 2\log_2 \\ \times \left(1 + \frac{(2U)^{1+\kappa} + \alpha_0\kappa_0}{U^\alpha}\right)\Bigg)\frac{(2U)^{1+\kappa+\beta}}{U^\alpha}\Bigg).\end{aligned} \tag{31.28}$$

We repeat the usual Cauchy–Schwarz argument (like the deduction of Theorem 4.2 from Theorem 9.1; see the end of Sec. 9). Let $2^{k-1}U < W \le 2^k U$. It is easy to find a sequence $W_0 = U < W_1 < \cdots < W_k = W$ such that $W_i < W_{i+1} \le 2W_i$ for all $0 \le i < k$. Combining the Cauchy–Schwarz inequality and (31.28) with the choice $U = W_i$ and $V = W_{i+1}$, $0 \le i < k$, and using $0 < \alpha \le 1+\kappa$, we have

$$\begin{aligned}&\Delta_f^2(\text{Gauss}; \vec{\mathcal{R}}; U, W) \\ &= \int_{\omega\in\Omega_{\text{Gauss}}} |D_f(\vec{\mathcal{R}}; \omega; U, W)|^2 \, d\text{ProdMeas}_{\text{Gauss}}(\omega) \\ &= \int_{\omega\in\Omega_{\text{Gauss}}} \left|\sum_{i=0}^{k-1} D_f(\vec{\mathcal{R}}; \omega; W_i, W_{i+1})\right|^2 d\text{ProdMeas}_{\text{Gauss}}(\omega) \\ &\le \int_{\omega\in\Omega_{\text{Gauss}}} \left(\sum_{i=0}^{k-1} |D_f(\vec{\mathcal{R}}; \omega; W_i, W_{i+1})|^2\right) \\ &\quad \times \left(\sum_{i=0}^{k-1} 1\right) d\text{ProdMeas}_{\text{Gauss}}(\omega) \\ &= \Bigg(\sum_{i=0}^{k-1} \int_{\omega\in\Omega_{\text{Gauss}}} \left|D_f(\vec{\mathcal{R}}; \omega; W_i, W_{i+1})\right|^2 \\ &\quad \times d\text{ProdMeas}_{\text{Gauss}}(\omega)\Bigg)\left(\sum_{i=0}^{k-1} 1\right)\end{aligned}$$

$$
\begin{aligned}
&= \left(\sum_{i=0}^{k-1} \Delta_f^2(\text{Gauss}; \vec{\mathcal{R}}; W_i, W_{i+1})\right)\left(\sum_{i=0}^{k-1} 1\right) \\
&\le k \cdot \sigma_0^2(f) \sum_{i=0}^{k-1} \Bigg(2 + (W_{i+1} - W_i)\Bigg(2 + W_i \exp(-\pi^2 W_i^{2\alpha} 2^{-8}) \\
&\quad + 9\left(3 + 2\log_2\left(1 + \frac{(2W_i)^{1+\kappa} + \alpha_0\kappa_0}{W_i^\alpha}\right)\right)\frac{(2W_i)^{1+\kappa+\beta}}{W_i^\alpha}\Bigg)\Bigg) \\
&\le k \cdot \sigma_0^2(f)\Bigg(2k + (W - U)\Bigg(2 + \max_{x \ge U} x \exp(-\pi^2 x^{2\alpha} 2^{-8}) \\
&\quad + 9\left(3 + 2\log_2\left(1 + \frac{(2W)^{1+\kappa} + \alpha_0\kappa_0}{W^\alpha}\right)\right)(2W)^{1+\kappa+\beta-\alpha}\Bigg)\Bigg),
\end{aligned}
\tag{31.29}
$$

under the condition

$$U \ge \max\{\alpha_0, \beta_0, 1\}. \tag{31.30}$$

Since $k = \lceil \log_2(W/U) \rceil$ (upper integral part), (31.29)–(31.30) complete the proof of Theorem 14.1. □

32. Proof of Theorem 14.2

It is a straightforward adaptation of the arguments in Secs. 27–31. For the sake of completeness we include — or rather outline — the details. The following lemma is an analog of Lemma 27.2.

Lemma 32.1. *Let* $1 \le U < V \le 2U$, *and write*

$$\mathcal{R}^{(\inf)}(U) = \min_{1 \le k \le N} \inf_{x \ge U} r_k(x) = U_1.$$

Under the condition of (30.15)–(30.17) *and*

$$N \sum_{j=1}^{\infty} 2^{2j+2} \mathbf{F}^{(\sup)}(2^{j-1}\pi U_1) < 1/10,$$

we have

$$
\begin{aligned}
|\Delta_f^2(\mathcal{G};\vec{\mathcal{R}};U,V)| &\le \sigma_0^2(f)(V-U)U \\
&\quad \cdot \left(25N^2 \sum_{j=1}^{\infty} 2^{2j+2}\mathbf{F}^{(\mathrm{sup})}(2^j\pi U_1) \right. \\
&\quad \left. + 5N \left(N \sum_{j=1}^{\infty} 2^{2j+2}\mathbf{F}^{(\mathrm{sup})}(2^j\pi U_1)\right)^2 \right) \\
&\quad + 8\sigma_0^2(f)(V-U) \left(\sum_{\ell=1}^{\infty} \ell\cdot \mathbf{F}^{(\mathrm{sup})}(2\pi(\ell-1)) \right) \\
&\quad \cdot \left(3 + 2\log_2\left(1+\frac{(2U)^{1+\kappa}+\alpha_0\kappa_0}{U_1}\right)\right)\frac{(2U)^{1+\kappa+\beta}}{U_1}.
\end{aligned}
$$

Proof. For every $\mathbf{n}=(n_1,\ldots,n_d)=(\mathbf{n}(1),\ldots,\mathbf{n}(N))\in\mathbb{Z}^d\setminus\mathbf{0}$ $(d=2N)$ write $L(\mathbf{n})=\{1\le i\le N:\ \mathbf{n}(i)=(n_{2i-1},n_{2i})\ne\mathbf{0}\}$. Applying the simple inequality $|a_{\mathbf{n}_1}\overline{a_{\mathbf{n}_2}}|\le(|a_{\mathbf{n}_1}|^2+|a_{\mathbf{n}_2}|^2)/2$, we have the following analog of (27.16) (and (30.3)):

$$
\begin{aligned}
&|\Delta_f^2(\mathcal{G};\vec{\mathcal{R}};U,V)| \\
&\le \sum_{\lambda_1=1}^{N}\int_{t_1=U}^{V} \sum_{\substack{\mathbf{n}_1\in\mathbb{Z}^d\setminus\mathbf{0}:\\ |L(\mathbf{n}_1)|=\lambda_1}} |a_{\mathbf{n}_1}|^2\cdot \left(\sum_{L_{1,2}\subseteq L(\mathbf{n}_1)} \right. \\
&\quad \times \sum_{\lambda_2=\max\{|L_{1,2}|,1\}}^{N} \int_{t_2=U}^{V} \sum_{\substack{\mathbf{n}_2\in\mathbb{Z}^d\setminus\mathbf{0}:\\ |L(\mathbf{n}_2)|=\lambda_2 \text{ and } L(\mathbf{n}_2)\cap L(\mathbf{n}_1)=L_{1,2}}} \\
&\quad \left. \cdot \prod_{k=1}^{N}\mathbf{F}^{(\mathrm{sup})}(2\pi|r_k(t_1)\mathbf{n}_1(k)-r_k(t_2)\theta_{k;t_2\to t_1}\mathbf{n}_2(k)|)dt_2 \right) dt_1. \qquad (32.1)
\end{aligned}
$$

We fix $t_1\in[U,V]$, $\mathbf{n}_1\in\mathbb{Z}^d\setminus\mathbf{0}$, $L_{1,2}\subseteq L(\mathbf{n}_1)$ and λ_2, and focus on the inner integral at the end of (32.1).

Write

$$
\lambda_{1,2}=|L_{1,2}|=|\{1\le i\le N:\ \mathbf{n}_1(i)\ne\mathbf{0}\quad\text{and}\quad \mathbf{n}_2(i)\ne\mathbf{0}\}|. \qquad (32.2)
$$

We define $k_i(\mathbf{n}_2)$ and $h_i(t_2;\mathbf{n}_2)$ the same way as in Sec. 27 with the straightforward modification that $|r_j(t_1)\mathbf{n}_1(j)-r_j(t_2)\mathbf{n}_2(j)|$ is replaced with $|r_k(t_1)\mathbf{n}_1(k)-r_k(t_2)\theta_{k;t_2\to t_1}\mathbf{n}_2(k)|$, and of course dimension 3 is reduced to dimension 2; see (30.4) and (30.5). Then we obtain the following analog of (27.20) (and (30.6)):

$$\begin{aligned}
&\int_{t_2=U}^{V} \sum_{\substack{\mathbf{n}_2\in\mathbb{Z}^d\setminus\mathbf{0}:\\ |L(\mathbf{n}_2)|=\lambda_2 \text{ and } L(\mathbf{n}_2)\cap L(\mathbf{n}_1)=L_{1,2}}} \\
&\quad\times \prod_{j=1}^{N} \mathbf{F}^{(\sup)}(2\pi|r_j(t_1)\mathbf{n}_1(j)-r_j(t_2)\theta_{j;t_2\to t_1}\mathbf{n}_2(j)|)\,dt_2 \\
&\leq \prod_{j\in L(\mathbf{n}_1)\setminus L_{1,2}} \mathbf{F}^{(\sup)}(2\pi|r_j(t_1)\mathbf{n}_1(j)|)\int_{t_2=U}^{V} \sum_{\substack{\mathbf{n}_2\in\mathbb{Z}^d\setminus\mathbf{0}:\\ |L(\mathbf{n}_2)|=\lambda_2 \text{ and } L(\mathbf{n}_2)\cap L(\mathbf{n}_1)=L_{1,2}}} \\
&\quad\cdot(\mathbf{F}^{(\sup)}(2\pi\mathcal{R}^{(\inf)}(t_2)))^{k_1(\mathbf{n}_2)}(\mathbf{F}^{(\sup)}(2\pi\cdot 2\mathcal{R}^{(\inf)}(t_2))^{k_2(\mathbf{n}_2)} \\
&\quad\cdot(\mathbf{F}^{(\sup)}(2\pi\cdot 2^2\mathcal{R}^{(\inf)}(t_2)))^{k_3(\mathbf{n}_2)}\cdots(\mathbf{F}^{(\sup)}(2\pi U_1/2))^{h_0(t_2;\mathbf{n}_2)} \\
&\quad\times(\mathbf{F}^{(\sup)}(2\pi U_1))^{h_1(t_2;\mathbf{n}_2)}(\mathbf{F}^{(\sup)}(2\pi\cdot 2U_1))^{h_2(t_2;\mathbf{n}_2)} \\
&\quad\cdot(\mathbf{F}^{(\sup)}(2\pi\cdot 2^2U_1))^{h_3(t_2;\mathbf{n}_2)}\cdots\,dt_2.
\end{aligned}\tag{32.3}$$

Next we estimate the long sum at the end of (32.3) the same way as we estimated the long sum at the end of (27.20); we obtain the following analog of (27.21) (and (30.7)):

EndSum of (32.3)

$$\begin{aligned}
&= \sum_{\substack{\mathbf{n}_2\in\mathbb{Z}^d\setminus\mathbf{0}:\\ |L(\mathbf{n}_2)|=\lambda_2 \text{ and } L(\mathbf{n}_2)\cap L(\mathbf{n}_1)=L_{1,2}}} \cdot(\mathbf{F}^{(\sup)}(2\pi\mathcal{R}^{(\inf)}(t_2)))^{k_1(\mathbf{n}_2)} \\
&\quad\times(\mathbf{F}^{(\sup)}(2\pi\cdot 2\mathcal{R}^{(\inf)}(t_2)))^{k_2(\mathbf{n}_2)} \\
&\quad\cdot(\mathbf{F}^{(\sup)}(2\pi\cdot 2^2\mathcal{R}^{(\inf)}(t_2)))^{k_3(\mathbf{n}_2)}\cdots \\
&\quad\cdot(\mathbf{F}^{(\sup)}(2\pi U_1/2))^{h_1(t_2;\mathbf{n}_2)}(\mathbf{F}^{(\sup)}(2\pi U_1))^{h_2(t_2;\mathbf{n}_2)} \\
&\quad\cdot(\mathbf{F}^{(\sup)}(2\pi\cdot 2U_1))^{h_3(t_2;\mathbf{n}_2)}(\mathbf{F}^{(\sup)}(2\pi\cdot 2^2U_1))^{h_4(t_2;\mathbf{n}_2)}\cdots \\
&\leq \sum_{\substack{(k_1,\ldots,k_r) \text{ with } r\geq 1:\\ k_1+\cdots+k_r=\lambda_2-\lambda_{1,2},k_j\geq 0,1\leq j\leq r-1 \text{ and } k_r\geq 1}} \binom{N-\lambda_1}{k_1}
\end{aligned}$$

$$
\begin{aligned}
&\times ((2+1)^2)^{k_1}(\mathbf{F}^{(\sup)}(2\pi\mathcal{R}^{(\inf)}(t_2)))^{k_1}\\
&\cdot \binom{N-\lambda_1-k_1}{k_2}((2\cdot 2+1)^2)^{k_2}(\mathbf{F}^{(\sup)}(2\pi\cdot 2\mathcal{R}^{(\inf)}(t_2)))^{k_2}\\
&\cdot \binom{N-\lambda_1-k_1-k_2}{k_3}((2\cdot 2^2+1)^2)^{k_3}(\mathbf{F}^{(\sup)}(2\pi\cdot 2^2\mathcal{R}^{(\inf)}(t_2)))^{k_3}\cdots\\
&\cdot \sum_{\substack{(h_0,h_1,\dots,h_r)\text{ with } r\ge 0:\\ h_0+h_1+\cdots+h_r=\lambda_{1,2},h_j\ge 0,0\le j\le r-1 \text{ and } h_r\ge 1}} \binom{\lambda_{1,2}}{h_0}\\
&\cdot \binom{\lambda_{1,2}-h_0}{h_1}((2+1)^2)^{h_1}(\mathbf{F}^{(\sup)}(2\pi U_1/2))^{h_1}\\
&\cdot \binom{\lambda_{1,2}-h_0-h_1}{h_2}((2\cdot 2+1)^2)^{h_2}(\mathbf{F}^{(\sup)}(2\pi U_1))^{h_2}\\
&\cdot \binom{\lambda_{1,2}-h_0-h_1-h_2}{h_3}((2\cdot 2^2+1)^2)^{h_3}(\mathbf{F}^{(\sup)}(2\pi\cdot 2U_1))^{h_3}\\
&\cdot \binom{\lambda_{1,2}-h_0-h_1-h_2-h_3}{h_4}((2\cdot 2^3+1)^2)^{h_4}(\mathbf{F}^{(\sup)}(2\pi\cdot 2^2U_1))^{h_4}\cdots .
\end{aligned}
\tag{32.4}
$$

Applying the multinomial theorem twice in (32.4) the same way as we did it in (27.22), we have the very generous upper bound (where the extra factors of N come from the trivial upper bound $\binom{N}{k}\le N^k$)

$$
\begin{aligned}
&\text{EndSum of (32.3)}\\
&\quad \le (N\cdot(2+1)^2\mathbf{F}^{(\sup)}(2\pi\mathcal{R}^{(\inf)}(t_2))\\
&\qquad + N\cdot(2\cdot 2+1)^2\mathbf{F}^{(\sup)}(2\pi\cdot 2\mathcal{R}^{(\inf)}(t_2))\\
&\qquad + N\cdot(2\cdot 2^2+1)^2\mathbf{F}^{(\sup)}(2\pi\cdot 2^2\mathcal{R}^{(\inf)}(t_2))+\cdots)^{\lambda_2-\lambda_{1,2}}\\
&\qquad \cdot(1+(2+1)^2\mathbf{F}^{(\sup)}(2\pi U_1/2)+(2\cdot 2+1)^2\mathbf{F}^{(\sup)}(2\pi U_1)\\
&\qquad + (2\cdot 2^2+1)^2\mathbf{F}^{(\sup)}(2\pi\cdot 2U_1)\\
&\qquad + (2\cdot 2^3+1)^2\mathbf{F}^{(\sup)}(2\pi\cdot 2^2U_1)+\cdots)^{\lambda_{1,2}},
\end{aligned}
\tag{32.5}
$$

which is an analog of (27.22) (and (30.8)). Next we use (32.4)–(32.5) in (32.3), and obtain the following analog of (27.23) (and (30.9)):

$$\int_{t_2=U}^{V} \sum_{\substack{\mathbf{n}_2\in\mathbb{Z}^d\setminus\mathbf{0}: \\ |L(\mathbf{n}_2)|=\lambda_2 \text{ and } L(\mathbf{n}_2)\cap L(\mathbf{n}_1)=L_{1,2}}} \cdot \prod_{j=1}^{N} \mathbf{F}^{(\text{sup})}(2\pi|r_j(t_1)\mathbf{n}_1(j) - r_j(t_2)\theta_{j;t_2\to t_1}\mathbf{n}_2(j)|)\, dt_2$$

$$\le (V-U) \prod_{j\in L(\mathbf{n}_1)\setminus L_{1,2}} \mathbf{F}^{(\text{sup})}(2\pi|r_j(t_1)\mathbf{n}_1(j)|)$$

$$\cdot \left(N\sum_{j=1}^{\infty} 2^{2j+2}\mathbf{F}^{(\text{sup})}(2^j\pi U_1)\right)^{\lambda_2-\lambda_{1,2}}$$

$$\times \left(1+\sum_{j=1}^{\infty} 2^{2j+2}\mathbf{F}^{(\text{sup})}(2^{j-1}\pi U_1)\right)^{\lambda_{1,2}}. \qquad (32.6)$$

Let us return now to (32.1); we have the following decomposition into four parts (an analog of (27.24)–(27.28); see also (30.10)–(30.14)):

$$|\Delta_f^2(\mathcal{G};\vec{\mathcal{R}};U,V)| \le \text{Part One} + \text{Part Two} + \text{Part Three} + \text{Part Four}, \qquad (32.7)$$

where

$$\text{Part One} = \sum_{\lambda_1=1}^{N-1} \int_{t_1=U}^{V} \sum_{\substack{\mathbf{n}_1\in\mathbb{Z}^d\setminus\mathbf{0}: \\ |L(\mathbf{n}_1)|=\lambda_1}} |a_{\mathbf{n}_1}|^2$$

$$\cdot \left(\sum_{L_{1,2}\subseteq L(\mathbf{n}_1)} \sum_{\lambda_2=\lambda_1+1}^{N} \int_{t_2=U}^{V} \sum_{\substack{\mathbf{n}_2\in\mathbb{Z}^d\setminus\mathbf{0}: \\ |L(\mathbf{n}_2)|=\lambda_2 \text{ and } L(\mathbf{n}_2)\cap L(\mathbf{n}_1)=L_{1,2}}} \right.$$

$$\left. \cdot \prod_{k=1}^{N} \mathbf{F}^{(\text{sup})}(2\pi|r_k(t_1)\mathbf{n}_1(k) - r_k(t_2)\theta_{k;t_2\to t_1}\mathbf{n}_2(k)|)dt_2 \right) dt_1 \qquad (32.8)$$

is characterized by the property $\lambda_1 < \lambda_2$,

$$\text{Part Two} = \sum_{\lambda_1=1}^{N} \int_{t_1=U}^{V} \sum_{\substack{\mathbf{n}_1 \in \mathbb{Z}^d \setminus \mathbf{0}: \\ |L(\mathbf{n}_1)|=\lambda_1}} |a_{\mathbf{n}_1}|^2 \cdot \left(\sum_{L_{1,2} \subset L(\mathbf{n}_1)} \right.$$

$$\times \sum_{\lambda_2=\max\{|L_{1,2}|,1\}}^{\lambda_1 - 1} \int_{t_2=U}^{V} \sum_{\substack{\mathbf{n}_2 \in \mathbb{Z}^d \setminus \mathbf{0}: \\ |L(\mathbf{n}_2)|=\lambda_2 \text{ and } L(\mathbf{n}_2) \cap L(\mathbf{n}_1) = L_{1,2}}}$$

$$\left. \cdot \prod_{k=1}^{N} \mathbf{F}^{(\sup)}(2\pi |r_k(t_1)\mathbf{n}_1(k) - r_k(t_2)\theta_{k;t_2\to t_1}\mathbf{n}_2(k)|)dt_2 \right) dt_1 \tag{32.9}$$

is characterized by the property $\lambda_1 > \lambda_2$, and finally, we split the case $\lambda_1 = \lambda_2$ into two subcases according as $L(\mathbf{n}_1) \neq L(\mathbf{n}_2)$ or $L(\mathbf{n}_1) = L(\mathbf{n}_2)$:

$$\text{Part Three} = \sum_{\lambda_1=1}^{d} \int_{t_1=U}^{V} \sum_{\substack{\mathbf{n}_1 \in \mathbb{Z}^d \setminus \mathbf{0}: \\ |L(\mathbf{n}_1)|=\lambda_1}} |a_{\mathbf{n}_1}|^2$$

$$\cdot \left(\sum_{\substack{L_{1,2} \subset L(\mathbf{n}_1): \\ L_{1,2} \neq L(\mathbf{n}_1)}} \int_{t_2=U}^{V} \sum_{\substack{\mathbf{n}_2 \in \mathbb{Z}^d \setminus \mathbf{0}: \\ |L(\mathbf{n}_2)|=\lambda_1 \text{ and } L(\mathbf{n}_2) \cap L(\mathbf{n}_1) = L_{1,2}}} \right.$$

$$\left. \cdot \prod_{k=1}^{N} \mathbf{F}^{(\sup)}(2\pi |r_k(t_1)\mathbf{n}_1(k) - r_k(t_2)\theta_{k;t_2\to t_1}\mathbf{n}_2(k)|)dt_2 \right) dt_1, \tag{32.10}$$

and

$$\text{Part Four} = \int_{t_1=U}^{V} \sum_{\mathbf{n}_1 \in \mathbb{Z}^d \setminus \mathbf{0}} |a_{\mathbf{n}_1}|^2 \left(\int_{t_2=U}^{V} \sum_{\substack{\mathbf{n}_2 \in \mathbb{Z}^d \setminus \mathbf{0}: \\ L(\mathbf{n}_2)=L(\mathbf{n}_1)}} \right.$$

$$\left. \cdot \prod_{k=1}^{N} \mathbf{F}^{(\sup)}(2\pi |r_k(t_1)\mathbf{n}_1(k) - r_k(t_2)\theta_{k;t_2\to t_1}\mathbf{n}_2(k)|)dt_2 \right) dt_1. \tag{32.11}$$

Applying (32.6) in (32.8), and repeating the argument of (28.1)–(28.4), we have the following analog of (28.5) (and (31.5)):

$$\text{Part One} \le 9(V-U)^2 \frac{N^2 \sum_{j=1}^{\infty} 2^{2j+2}\mathbf{F}^{(\mathrm{sup})}(2^j\pi U_1)}{1-N\sum_{j=1}^{\infty} 2^{2j+2}\mathbf{F}^{(\mathrm{sup})}(2^j\pi U_1)}\sigma_0^2(f). \tag{32.12}$$

Next we apply (32.6) in (32.9), and repeat the argument of (28.6)–(28.8):

$$\text{Part Two} \le 9(V-U)^2 \frac{N^2 \sum_{j=1}^{\infty} 2^{2j+2}\mathbf{F}^{(\mathrm{sup})}(2^j\pi U_1)}{\left(1-N\sum_{j=1}^{\infty} 2^{2j+2}\mathbf{F}^{(\mathrm{sup})}(2^j\pi U_1)\right)^3}\sigma_0^2(f), \tag{32.13}$$

which is an analog of (28.8) (and (31.8)).

Similarly, we apply (32.6) in (32.10), and repeat the argument of (32.9)–(32.10):

$$\text{Part Three} \le 3(V-U)^2 \frac{N\left(N \sum_{j=1}^{\infty} 2^{2j+2}\mathbf{F}^{(\mathrm{sup})}(2^j\pi U_1)\right)^2}{1-\left(N\sum_{j=1}^{\infty} 2^{2j+2}\mathbf{F}^{(\mathrm{sup})}(2^j\pi U_1)\right)^2}\sigma_0^2(f), \tag{32.14}$$

which is an analog of (28.10) (and (31.11)).

Finally we estimate (32.11). Similarly to (28.11)–(28.13) (and (31.12)–(31.14)), we have

$$\text{Part Four} = \text{Part Four A} + \text{Part Four B}, \tag{32.15}$$

where

$$\text{Part Four A} = \int_{t_1=U}^{V} \sum_{\mathbf{n}_1\in\mathbb{Z}^d\setminus\mathbf{0}} |a_{\mathbf{n}_1}|^2 \left(\int_{t_2=U}^{V} \sum_{\substack{\mathbf{n}_2\in\mathbb{Z}^d\setminus\mathbf{0}:\\ L(\mathbf{n}_2)=L(\mathbf{n}_1),\ h_0(t_2;\mathbf{n}_2)<|L(\mathbf{n}_1)|}} \cdot \prod_{k=1}^{N} \mathbf{F}^{(\mathrm{sup})}(2\pi|r_k(t_1)\mathbf{n}_1(k) - r_k(t_2)\theta_{k;t_2\to t_1}\mathbf{n}_2(k)|)dt_2 \right) dt_1 \tag{32.16}$$

and

$$\text{Part Four B} = \int_{t_1=U}^{V} \sum_{\mathbf{n}_1 \in \mathbb{Z}^d \setminus \mathbf{0}} |a_{\mathbf{n}_1}|^2 \left(\int_{t_2=U}^{V} \sum_{\substack{\mathbf{n}_2 \in \mathbb{Z}^d \setminus \mathbf{0}: \\ L(\mathbf{n}_2)=L(\mathbf{n}_1),\ h_0(t_2;\mathbf{n}_2)=|L(\mathbf{n}_1)|}} \right.$$
$$\left. \cdot \prod_{k=1}^{N} \mathbf{F}^{(\mathrm{sup})}(2\pi |r_k(t_1)\mathbf{n}_1(k) - r_k(t_2)\theta_{k;t_2\to t_1}\mathbf{n}_2(k)|)dt_2 \right) dt_1. \tag{32.17}$$

By repeating the argument of (28.14)–(28.16) (and (31.15)–(31.18)), we have

$$\text{Part Four A} \le \sigma_0^2(f)(V-U)^2 2N \sum_{j=1}^{\infty} 2^{2j+2}\mathbf{F}^{(\mathrm{sup})}(2^{j-1}\pi U_1). \tag{32.18}$$

To estimate (32.17)

$$\text{Part Four B} = \int_{t_1=U}^{V} \sum_{\mathbf{n}_1 \in \mathbb{Z}^d \setminus \mathbf{0}} |a_{\mathbf{n}_1}|^2 \left(\int_{t_2=U}^{V} \sum_{\substack{\mathbf{n}_2 \in \mathbb{Z}^d \setminus \mathbf{0}: \\ L(\mathbf{n}_2)=L(\mathbf{n}_1),\ h_0(t_2;\mathbf{n}_2)=|L(\mathbf{n}_1)|}} \right.$$
$$\left. \cdot \prod_{k=1}^{N} \mathbf{F}^{(\mathrm{sup})}(2\pi |r_k(t_1)\mathbf{n}_1(k) - r_k(t_2)\theta_{k;t_2\to t_1}\mathbf{n}_2(k)|)dt_2 \right) dt_1, \tag{32.19}$$

we make use of the fact that for every given triple $(t_1, \mathbf{n}_1, t_2)$ with $t_1 \in [U, V]$, $\mathbf{n}_1 \in \mathbb{Z}^d \setminus \mathbf{0}$ and $t_2 \in [U, V]$, there is *at most one* term in the endsum

$$\sum_{\substack{\mathbf{n}_2 \in \mathbb{Z}^d \setminus \mathbf{0}: \\ L(\mathbf{n}_2)=L(\mathbf{n}_1),\ h_0(t_2;\mathbf{n}_2)=|L(\mathbf{n}_1)|}} \prod_{k=1}^{N} \mathbf{F}^{(\mathrm{sup})}(2\pi |r_k(t_1)\mathbf{n}_1(k) - r_k(t_2)\theta_{k;t_2\to t_1}\mathbf{n}_2(k)|)$$

of (32.19). Indeed, it follows from the definition of $h_0(t_2; \mathbf{n}_2)$ that for every given triple $(t_1, \mathbf{n}_1, t_2)$ the inequality

$$|r_k(t_1)\mathbf{n}_1(k) - r_k(t_2)\theta_{k;t_2\to t_1}\mathbf{n}_2(k)| < U_1/2$$

has at most one solution $\mathbf{n}_2(j) \in \mathbb{Z}^2$, and because $h_0(t_2;\mathbf{n}_2) = |L(\mathbf{n}_1)|$ and $L(\mathbf{n}_2) = L(\mathbf{n}_1)$, there is at most one $\mathbf{n}_2 \in \mathbb{Z}^d \setminus \mathbf{0}$ satisfying the requirements. Let $\mathbf{n}_2^* \in \mathbb{Z}^d \setminus \mathbf{0}$, $\mathbf{n}_2^* = \mathbf{n}_2^*(t_1, \mathbf{n}_1, t_2)$ denote this single integer lattice point, *if it exists*. Thus we can rewrite (32.19) as follows:

$$\text{Part Four B} = \int_{t_1=U}^{V} \sum_{\mathbf{n}_1 \in \mathbb{Z}^d \setminus \mathbf{0}} |a_{\mathbf{n}_1}|^2 \left(\int_{t_2=U}^{V} \left\{ \prod_{k=1}^{N} \mathbf{F}^{(\text{sup})}(2\pi |r_k(t_1)\mathbf{n}_1(k) - r_k(t_2)\theta_{k;t_2\to t_1}\mathbf{n}_2^*(k)|) \text{ or } 0 \right\} dt_2 \right) dt_1, \tag{32.20}$$

where $\mathbf{n}_2^* = \mathbf{n}_2^*(t_1, \mathbf{n}_1, t_2)$, *if it exists.*

Let $k_0 = k_0(\mathbf{n}_1)$ be an arbitrary but fixed element of the set $L(\mathbf{n}_1)$; then we have the following trivial upper bound for (32.20):

$$\text{Part Four B} \le \int_{t_1=U}^{V} \sum_{\mathbf{n}_1 \in \mathbb{Z}^d \setminus \mathbf{0}} |a_{\mathbf{n}_1}|^2 \cdot \left(\int_{t_2=U}^{V} \{ \mathbf{F}^{(\text{sup})}(2\pi |r_{k_0}(t_1)\mathbf{n}_1(k_0) - r_{k_0}(t_2)\theta_{k_0;t_2\to t_1}\mathbf{n}_2^*(k_0)|) \text{ or } 0 \} dt_2 \right) dt_1. \tag{32.21}$$

Note that (32.21) is an analog of (28.19) (and (31.20)).

Given $(t_1, \mathbf{n}_1)$, we consider the following decomposition of the interval $U \le t_2 \le V (\le 2U)$:

$$I_\ell(t_1, \mathbf{n}_1) = \{ t_2 \in [U, V] : \ \ell - 1 \le |r_{k_0}(t_1)\mathbf{n}_1(k_0) - r_{k_0}(t_2)\theta_{k_0;t_2\to t_1}\mathbf{n}_2^*(k_0)| < \ell \} \tag{32.22}$$

with $\ell = 1, 2, 3, \ldots$, where $k_0 = k_0(\mathbf{n}_1)$ and $\mathbf{n}_2^* = \mathbf{n}_2^*(t_1, \mathbf{n}_1, t_2)$.

Using decomposition (32.22) in (32.21), we obtain the upper bound

$$\text{Part Four B} \le \int_{t_1=U}^{V} \sum_{\mathbf{n}_1 \in \mathbb{Z}^d \setminus \mathbf{0}} |a_{\mathbf{n}_1}|^2 \left(\sum_{\ell=1}^{\infty} \text{length}t(I_\ell) \mathbf{F}^{(\text{sup})}(2\pi(\ell - 1)) \right) dt_1, \tag{32.23}$$

where $I_\ell = I_\ell(t_1, \mathbf{n}_1)$. To estimate the length of $I_\ell = I_\ell(t_1, \mathbf{n}_1)$, we apply Lemma 30.1 with $C' = \ell$, and obtain the upper bound

$$\text{length}(I_\ell) \le 8\ell \left(3 + 2\log_2 \left(1 + \frac{(2U)^{1+\kappa} + \alpha_0 \kappa_0}{U_1} \right) \right) \frac{(2U)^{1+\kappa+\beta}}{U_1} \tag{32.24}$$

if $\ell \le U_1/16$ and

$$\text{length}(I_\ell) < U \tag{32.25}$$

if $U_1/16 < \ell < U_1/2$. Using (32.24), (32.25) and Parseval's formula in (32.23), we have

$$\begin{aligned}\text{Part Four B} &\le (V-U)\sigma_0^2(f)\left(\sum_{\ell=1}^{\infty} 8\ell\cdot \mathbf{F}^{(\text{sup})}(2\pi(\ell-1))\right)\\ &\quad\cdot\left(3+2\log_2\left(1+\frac{(2U)^{1+\kappa}+\alpha_0\kappa_0}{U_1}\right)\right)\frac{(2U)^{1+\kappa+\beta}}{U_1},\end{aligned} \tag{32.26}$$

which is an analog of (31.26).

Combining (32.7), (32.12), (32.13), (32.14), (32.15), (32.18) and (32.26), we have

$$\begin{aligned}|\Delta_f^2(\mathcal{G};\vec{\mathcal{R}};U,V)| &\le \text{Part One} + \text{Part Two} + \text{Part Three} + \text{Part Four}\\ &\le 9\sigma_0^2(f)(V-U)^2\frac{N^2\sum_{j=1}^{\infty}2^{2j+2}\mathbf{F}^{(\text{sup})}(2^j\pi U_1)}{1-N\sum_{j=1}^{\infty}2^{2j+2}\mathbf{F}^{(\text{sup})}(2^j\pi U_1)}\\ &\quad+9\sigma_0^2(f)(V-U)^2\frac{N^2\sum_{j=1}^{\infty}2^{2j+2}\mathbf{F}^{(\text{sup})}(2^j\pi U_1)}{\left(1-N\sum_{j=1}^{\infty}2^{2j+2}\mathbf{F}^{(\text{sup})}(2^j\pi U_1)\right)^3}\\ &\quad+3\sigma_0^2(f)(V-U)^2\frac{N\left(N\sum_{j=1}^{\infty}2^{2j+2}\mathbf{F}^{(\text{sup})}(2^j\pi U_1)\right)^2}{1-\left(N\sum_{j=1}^{\infty}2^{2j+2}\mathbf{F}^{(\text{sup})}(2^j\pi U_1)\right)^2}\\ &\quad+2\sigma_0^2(f)(V-U)^2N\sum_{j=1}^{\infty}2^{2j+2}\mathbf{F}^{(\text{sup})}(2^{j-1}\pi U_1)\\ &\quad+8\sigma_0^2(f)(V-U)\left(\sum_{\ell=1}^{\infty}\ell\cdot\mathbf{F}^{(\text{sup})}(2\pi(\ell-1))\right)\\ &\quad\cdot\left(3+2\log_2\left(1+\frac{(2U)^{1+\kappa}+\alpha_0\kappa_0}{U_1}\right)\right)\frac{(2U)^{1+\kappa+\beta}}{U_1}.\end{aligned} \tag{32.27}$$

By using $V - U \le U$, we obtain the shorter upper bound

$$|\Delta_f^2(\mathcal{G};\vec{\mathcal{R}};U,V)| \le \sigma_0^2(f)(V-U)U$$
$$\cdot \left(25N^2 \sum_{j=1}^{\infty} 2^{2j+2}\mathbf{F}^{(\mathrm{sup})}(2^j\pi U_1) \right.$$
$$\left. + 5N\left(N\sum_{j=1}^{\infty} 2^{2j+2}\mathbf{F}^{(\mathrm{sup})}(2^j\pi U_1)\right)^2\right)$$
$$+ 8\sigma_0^2(f)(V-U)\left(\sum_{\ell=1}^{\infty}\ell\cdot\mathbf{F}^{(\mathrm{sup})}(2\pi(\ell-1))\right)$$
$$\cdot\left(3 + 2\log_2\left(1 + \frac{(2U)^{1+\kappa}+\alpha_0\kappa_0}{U_1}\right)\right)\frac{(2U)^{1+\kappa+\beta}}{U_1},$$

which completes the proof of Lemma 32.1. □

We recall the hypothesis of Theorem 14.2:

$$\mathbf{F}^{(\mathrm{sup})}(w) \le w^{-\gamma} \quad \text{for all } w \ge \gamma_0 \ge 1. \tag{32.28}$$

First we estimate the critical sum in Lemma 32.1: by using (32.28), for $2\pi U_1 \ge \gamma_0$ and $\gamma > 2$ we have

$$\sum_{j=1}^{\infty} 2^{2j+2}\mathbf{F}^{(\mathrm{sup})}(2^j\pi U_1) \le \sum_{j=1}^{\infty} 2^{2j+2}(2^j\pi U_1)^{-\gamma}$$
$$= (2\pi U_1)^{-\gamma}\sum_{j=1}^{\infty} 2^{(2-\gamma)j+2} = (\pi U_1)^{-\gamma}\frac{2^{4-\gamma}}{1-2^{2-\gamma}}. \tag{32.29}$$

Moreover, for $U \ge \beta_0$ we have

$$\sum_{\ell=1}^{\infty}\ell\cdot\mathbf{F}^{(\mathrm{sup})}(2\pi(\ell-1)) \le \sum_{0\le k\le\gamma_0} 1 + \sum_{k>\gamma_0}(2\pi k)^{-\gamma}$$
$$\le 1 + \gamma_0 + (2\pi)^{-\gamma}\int_{\gamma_0}^{\infty} x^{-\gamma}\,dx$$
$$= 1 + \gamma_0 + (2\pi)^{-\gamma}\frac{\gamma_0^{1-\gamma}}{\gamma-1}. \tag{32.30}$$

Using (32.29) and (32.30) in Lemma 32.1, under the condition of (32.28) we have

$$|\Delta_f^2(\mathcal{G};\vec{\mathcal{R}};U,V)| \le \sigma_0^2(f)(V-U)U\cdot\left(25N^2(\pi U_1)^{-\gamma}\frac{2^{4-\gamma}}{1-2^{2-\gamma}}\right.$$
$$\left.+\,5N^3(\pi U_1)^{-2\gamma}\left(\frac{2^{4-\gamma}}{1-2^{2-\gamma}}\right)^2\right)$$
$$+\,8\sigma_0^2(f)(V-U)\left(1+\gamma_0+(2\pi)^{-\gamma}\frac{\gamma_0^{1-\gamma}}{\gamma-1}\right)$$
$$\cdot\left(3+2\log_2\left(1+\frac{(2U)^{1+\kappa}+\alpha_0\kappa_0}{U_1}\right)\right)\frac{(2U)^{1+\kappa+\beta}}{U_1}, \tag{32.31}$$

assuming

$$\min_{1\le k\le N}\inf_{x\ge U} r_k(x) = \mathcal{R}^{(\inf)}(U) = U_1 \ge \gamma_0/2\pi,\ U\ge\beta_0 \quad\text{and}\quad \gamma>2. \tag{32.32}$$

By hypothesis

$$U_1 \ge U^\alpha \quad \text{for all } U\ge\alpha_0. \tag{32.33}$$

Using (32.33) in (32.31)–(32.32), under the condition of (32.28) we have

$$|\Delta_f^2(\mathcal{G};\vec{\mathcal{R}};U,V)| \le \sigma_0^2(f)(V-U)U\left(25N^2(\pi U^\alpha)^{-\gamma}\frac{2^{4-\gamma}}{1-2^{2-\gamma}}\right.$$
$$\left.+\,5N^3(\pi U^\alpha)^{-2\gamma}\left(\frac{2^{4-\gamma}}{1-2^{2-\gamma}}\right)^2\right)$$
$$+\,8\sigma_0^2(f)(V-U)\left(1+\gamma_0+(2\pi)^{-\gamma}\frac{\gamma_0^{1-\gamma}}{\gamma-1}\right)$$
$$\cdot\left(3+2\log_2\left(1+\frac{(2U)^{1+\kappa}+\alpha_0\kappa_0}{U^\alpha}\right)\right)\frac{(2U)^{1+\kappa+\beta}}{U^\alpha}, \tag{32.34}$$

assuming

$$U\ge\max\left\{\alpha_0,\beta_0,\left(\frac{\gamma_0}{2\pi}\right)^{1/\alpha}\right\} \quad\text{and}\quad \gamma>2. \tag{32.35}$$

Now we are ready to complete the proof of Theorem 14.2 by using the usual Cauchy–Schwarz argument (see e.g. the deduction of Theorem 4.2 from Theorem 9.1 at the end of Sec. 9). Let $2^{k-1}U < W \le 2^k U$. It is easy to find a sequence $W_0 = U < W_1 < \cdots < W_k = W$ such that $W_i < W_{i+1} \le 2W_i$ for all $0 \le i < k$. Assume that $\alpha\gamma > 1$. Combining the Cauchy–Schwarz inequality and (32.34) with the choice $U = W_i$ and $V = W_{i+1}$, $0 \le i < k$, and using the simple fact that $x^{1-\alpha\gamma}$ is a monotone decreasing function of x (since $\alpha\gamma > 1$), we have

$$\begin{aligned}
&\Delta_f^2(\mathcal{G};\vec{\mathcal{R}};U,W)\\
&\quad= \int_{\omega\in\Omega_{\mathcal{G}}} |D_f(\vec{\mathcal{R}};\omega;U,W)|^2 \, d\text{ProdMeas}_{\mathcal{G}}(\omega)\\
&\quad= \int_{\omega\in\Omega_{\mathcal{G}}} \left|\sum_{i=0}^{k-1} D_f(\vec{\mathcal{R}};\omega;W_i,W_{i+1})\right|^2 d\text{ProdMeas}_{\mathcal{G}}(\omega)\\
&\quad\le \int_{\omega\in\Omega_{\mathcal{G}}} \left(\sum_{i=0}^{k-1} \left|D_f(\vec{\mathcal{R}};\omega;W_i,W_{i+1})\right|^2\right)\left(\sum_{i=0}^{k-1} 1\right) d\text{ProdMeas}_{\mathcal{G}}(\omega)\\
&\quad= \left(\sum_{i=0}^{k-1} \int_{\omega\in\Omega_{\mathcal{G}}} |D_f(\vec{\mathcal{R}};\omega;W_i,W_{i+1})|^2 \, d\text{ProdMeas}_{\mathcal{G}}(\omega)\right)\left(\sum_{i=0}^{k-1} 1\right)\\
&\quad= \left(\sum_{i=0}^{k-1} \Delta_f^2(\mathcal{G};\vec{\mathcal{R}};W_i,W_{i+1})\right)\left(\sum_{i=0}^{k-1} 1\right) \le k\cdot\sigma_0^2(f)\left(\sum_{i=0}^{k-1}\Bigg((W_{i+1}-W_i)\right.\\
&\qquad \cdot\left(25N^2\pi^{-\gamma}W_i^{1-\alpha\gamma}\frac{2^{4-\gamma}}{1-2^{2-\gamma}} + 5N^3\pi^{-2\gamma}W_i^{1-2\alpha\gamma}\left(\frac{2^{4-\gamma}}{1-2^{2-\gamma}}\right)^2\right)\Bigg)\\
&\qquad + 8(W_{i+1}-W_i)\left(1+\gamma_0+(2\pi)^{-\gamma}\frac{\gamma_0^{1-\gamma}}{\gamma-1}\right)\\
&\qquad \left.\cdot\left(3+2\log_2\left(1+\frac{(2W_i)^{1+\kappa}+\alpha_0\kappa_0}{W_i^{\alpha}}\right)\right)\frac{(2W_i)^{1+\kappa+\beta}}{W_i^{\alpha}}\right)\\
&\quad\le k\cdot\sigma_0^2(f)(W-U)\Bigg(25N^2\pi^{-\gamma}U^{1-\alpha\gamma}\frac{2^{4-\gamma}}{1-2^{2-\gamma}}
\end{aligned}$$

$$+ 5N^3\pi^{-2\gamma}U^{1-2\alpha\gamma}\left(\frac{2^{4-\gamma}}{1-2^{2-\gamma}}\right)^2 + 8\left(1+\gamma_0+(2\pi)^{-\gamma}\frac{\gamma_0^{1-\gamma}}{\gamma-1}\right)$$

$$\times\left(3+2\log_2\left(1+\frac{(2W)^{1+\kappa}+\alpha_0\kappa_0}{W^\alpha}\right)\right)(2W)^{1+\kappa+\beta-\alpha}\Bigg), \tag{32.36}$$

assuming

$$U \geq \max\left\{\alpha_0, \beta_0, \left(\frac{\gamma_0}{2\pi}\right)^{1/\alpha}\right\}, \ \gamma > 2 \quad \text{and} \quad \alpha\gamma > 1. \tag{32.37}$$

Since $k = \lceil \log_2(W/U)\rceil$ (upper integral part), (32.36)–(32.37) complete the proof of Theorem 14.2. □

33. Multiple Mixedupness (II): Proof of Lemma 12.2

By (12.38) and (12.39) we have

$$\int_0^U (|\mathcal{Z}_J(\mathbf{v}_1,\ldots,\mathbf{v}_N;t)\cap B| - |J|\text{vol}(B))^2\,dt$$

$$= \int_0^U \left(\sum_{j\in J}\chi_B(j;t) - |J|\text{vol}(B)\right)^2 dt$$

$$= \sum_{j\in J}\int_0^U \chi_B(j;t)\,dt + \sum_{j_1,j_2\in J:\ j_1\neq j_2}\int_0^U \chi_B(j_1;t)\chi_B(j_2;t)\,dt$$

$$+ |J|^2U(\text{vol}(B))^2 - 2|J|\text{vol}(B)\sum_{j\in J}\int_0^U \chi_B(j;t)\,dt$$

$$= \sum_{j\in J}\left(\int_0^U \chi_B(j;t)\,dt - U\text{vol}(B)\right)$$

$$+ \sum_{\substack{(j_1,j_2)\in J\times J:\\ j_1\neq j_2}}\left(\int_0^U \chi_B(j_1;t)\chi_B(j_2;t)\,dt - U\,(\text{vol}(B))^2\right)$$

$$+|J|U(\mathrm{vol}(B)-(\mathrm{vol}(B))^2)-2|J|\mathrm{vol}(B)$$
$$\times\sum_{j\in J}\left(\int_0^U \chi_B(j;t)\,dt-U\mathrm{vol}(B)\right). \tag{33.1}$$

By using (12.8)–(12.10) in the 3-dimensional case, we have

$$\sum_{j\in J}\left|\int_0^U \chi_B(j;t)\,dt-U\mathrm{vol}(B)\right|$$
$$\leq\sum_{j\in J}\left(\frac{3}{2}\right)^3\left(\frac{2U}{H_j+1}+\sum_{\substack{\mathbf{h}\in\mathbb{Z}^3:\\1\leq\|\mathbf{h}\|_\infty\leq H_j}}\frac{1}{\Lambda(\mathbf{h})}\min\left\{\frac{1}{\pi|\mathbf{h}\cdot\mathbf{v}_j|},U\right\}\right), \tag{33.2}$$

where $H_j\geq 1$ are arbitrary integers, and (see (12.39))

$$\chi_B(j;t)=\begin{cases}1 & \text{if } \mathbf{z}_j+t\mathbf{v}_j\in B \text{ modulo one},\\ 0 & \text{otherwise.}\end{cases}$$

If $j_1\neq j_2$ then

$$\chi_B(j_1;t)\chi_B(j_2;t)=\begin{cases}1 & \text{if both } \mathbf{z}_{j_1}+t\mathbf{v}_{j_1}\in B,\\ & \qquad\mathbf{z}_{j_2}+t\mathbf{v}_{j_2}\in B \text{ mod one},\\ 0 & \text{otherwise.}\end{cases} \tag{33.3}$$

We can clearly rewrite (33.3) in an equivalent form about 6-dimensional torus line

$$\chi_B(j_1;t)\chi_B(j_2;t)=\begin{cases}1 & \text{if } \mathbf{z}_{j_1}\oplus\mathbf{z}_{j_2}+t\mathbf{v}_{j_1}\oplus\mathbf{v}_{j_2}\in B\times B \text{ mod one},\\ 0 & \text{otherwise,}\end{cases} \tag{33.4}$$

where, for given vectors $\mathbf{u}=(u_1,\ldots,u_{d_1})$ and $\mathbf{w}=(w_1,\ldots,w_{d_2})$, we use the notation

$$\mathbf{u}\oplus\mathbf{w}=(u_1,\ldots,u_{d_1},w_1,\ldots,w_{d_2}).$$

Applying (12.8)–(12.10) for the 6-dimensional box $B \times B$ of 6-dimensional volume $(\mathrm{vol}(B))^2$, we have

$$\sum_{\substack{(j_1,j_2)\in J\times J:\\ j_1\neq j_2}} \left| \int_0^U \chi_B(j_1;t)\chi_B(j_2;t)\,dt - U\left(\mathrm{vol}(B)\right)^2 \right|$$

$$\leq \sum_{\substack{(j_1,j_2)\in J\times J:\\ j_1\neq j_2}} \left(\frac{3}{2}\right)^6 \left(\frac{2U}{H_{j_1,j_2}+1} + \sum_{\substack{\mathbf{h}\in\mathbb{Z}^6:\\ 1\leq\|\mathbf{h}\|_\infty\leq H_{j_1,j_2}}} \right.$$

$$\left. \times \frac{1}{\Lambda(\mathbf{h})} \min\left\{ \frac{1}{\pi|\mathbf{h}\cdot\mathbf{v}_{j_1}\oplus\mathbf{v}_{j_2}|}, U \right\} \right), \tag{33.5}$$

where $H_{j_1,j_2} \geq 1$ are arbitrary integers.

We estimate the sums (33.2) and (33.5) by using the Separation Condition (12.26)–(12.27). We start with (33.2); let $H_j = N^{1/3}$. Then by (33.2)

$$\sum_{j\in J} \left| \int_0^U \chi_B(j;t)\,dt - U\mathrm{vol}(B) \right|$$

$$\leq \left(\frac{3}{2}\right)^3 \sum_{j\in J} \left(\frac{2U}{N^{1/3}+1} + \sum_{\substack{\mathbf{h}\in\mathbb{Z}^3:\\ 1\leq\|\mathbf{h}\|_\infty\leq N^{1/3}}} \frac{1}{\Lambda(\mathbf{h})} \min\left\{ \frac{1}{\pi|\mathbf{h}\cdot\mathbf{v}_j|}, U \right\} \right)$$

$$\leq \frac{8|J|U}{N^{1/3}} + 4\sum_{j\in J} \sum_{\substack{\mathbf{h}\in\mathbb{Z}^3:\ |\mathbf{h}\cdot\mathbf{v}_j|<1/(\pi U)\\ 1\leq\|\mathbf{h}\|_\infty\leq N^{1/3}}} \frac{U}{\Lambda(\mathbf{h})}$$

$$+ \frac{4}{\pi}\sum_{j\in J} \sum_{\substack{\mathbf{h}\in\mathbb{Z}^3:\ |\mathbf{h}\cdot\mathbf{v}_j|\geq 1/(\pi U)\\ 1\leq\|\mathbf{h}\|_\infty\leq N^{1/3}}} \frac{1}{\Lambda(\mathbf{h})|\mathbf{h}\cdot\mathbf{v}_j|}. \tag{33.6}$$

For every lattice point $\mathbf{h} \in \mathbb{Z}^3$ with $1 \leq \|\mathbf{h}\|_\infty \leq N^{1/3}$, let $r(\mathbf{h}) = r(\mathbf{h}; U; N)$ denote the smallest integer such that

$$\frac{1}{\pi U} \leq \frac{2^{r(\mathbf{h})}|\mathbf{h}|}{N}. \tag{33.7}$$

To estimate (33.6) we distinguish two cases.

Case 1: $U \leq \frac{N^{1/3}}{\sqrt{3}\pi}$

Then (33.7) and $\sqrt{3}\|\mathbf{h}\|_\infty \geq |\mathbf{h}|$ imply

$$2^{r(\mathbf{h})} \geq \frac{N}{\pi U|\mathbf{h}|} \geq \frac{N}{\pi(N^{1/3}/\sqrt{3}\pi)\sqrt{3}\|\mathbf{h}\|_\infty} \geq \frac{N}{N^{1/3}N^{1/3}} = N^{1/3}. \tag{33.8}$$

Applying (12.27) with $\ell = -1$ and 0, (33.8) gives that the interval

$$\left[-\frac{2^{r(\mathbf{h})}|\mathbf{h}|}{N}, \frac{2^{r(\mathbf{h})}|\mathbf{h}|}{N}\right] \tag{33.9}$$

contains less than $2^r + 2^r = 2^{r+1}$ dot products $\mathbf{h} \cdot \mathbf{v}_k$, $1 \leq k \leq N$ (with $r = r(\mathbf{h})$). Combining this information with (33.7) and (33.9), we obtain the upper bound

$$\begin{aligned}\sum_{j\in J} \sum_{\substack{\mathbf{h}\in\mathbb{Z}^3:\ |\mathbf{h}\cdot\mathbf{v}_j|<1/(\pi U)\\ 1\leq\|\mathbf{h}\|_\infty\leq N^{1/3}}} \frac{U}{\Lambda(\mathbf{h})} &\leq \sum_{\substack{\mathbf{h}\in\mathbb{Z}^3:\ |\mathbf{h}\cdot\mathbf{v}_j|<1/(\pi U)\\ 1\leq\|\mathbf{h}\|_\infty\leq N^{1/3}}} 2^{r(\mathbf{h})+1}\frac{U}{\Lambda(\mathbf{h})}\\ &\leq \sum_{\substack{\mathbf{h}\in\mathbb{Z}^3:\ |\mathbf{h}\cdot\mathbf{v}_j|<1/(\pi U)\\ 1\leq\|\mathbf{h}\|_\infty\leq N^{1/3}}} \frac{4N}{\pi|\mathbf{h}|U}\frac{U}{\Lambda(\mathbf{h})}\\ &= \sum_{\substack{\mathbf{h}\in\mathbb{Z}^3:\ |\mathbf{h}\cdot\mathbf{v}_j|<1/(\pi U)\\ 1\leq\|\mathbf{h}\|_\infty\leq N^{1/3}}} \frac{4N}{\pi|\mathbf{h}|\Lambda(\mathbf{h})},\end{aligned} \tag{33.10}$$

where in the last step we used the following consequence of (33.7):

$$\frac{2^{r(\mathbf{h})}|\mathbf{h}|}{N} < \frac{2}{\pi U} \iff 2^{r(\mathbf{h})} < \frac{2N}{\pi|\mathbf{h}|U}. \tag{33.11}$$

Using the definition (12.6) of $\Lambda(\mathbf{h})$ and the trivial inequality

$$|\mathbf{h}| = \sqrt{h_1^2 + h_2^2 + h_3^2} \geq \frac{1 + |h_i|}{2}, \quad i = 1, 2, 3$$

for all $\mathbf{h} \in \mathbb{Z}^3 \setminus \mathbf{0}$, we can easily estimate the infinite sum

$$\sum_{\mathbf{h}\in\mathbb{Z}^3\setminus\mathbf{0}} \frac{1}{|\mathbf{h}|\Lambda(\mathbf{h})} \leq \sum_{\mathbf{h}\in\mathbb{Z}^3\setminus\mathbf{0}} \prod_{i=1}^{3} \frac{2}{(1+|h_i|)^{4/3}}$$

$$\leq 2^3 \left(\sum_{h\in\mathbb{Z}} \frac{1}{(1+|h|)^{4/3}} \right)^3 \leq 10^4. \tag{33.12}$$

Combining (33.10) and (33.12), in Case 1 we obtain the upper bound

$$\sum_{j\in J} \sum_{\substack{\mathbf{h}\in\mathbb{Z}^3:\ |\mathbf{h}\cdot\mathbf{v}_j|<1/(\pi U)\\ 1\leq\|\mathbf{h}\|_\infty\leq N^{1/3}}} \frac{U}{\Lambda(\mathbf{h})} \leq \sum_{\mathbf{h}\in\mathbb{Z}^3\setminus\mathbf{0}} \frac{4N}{\pi|\mathbf{h}|\Lambda(\mathbf{h})} \leq \frac{4\cdot 10^4 N}{\pi}, \tag{33.13}$$

which will be used later in the estimation of (33.6).

Next we estimate the last sum in (33.6) — as usual, we assume Case 1. Again (33.8) gives that every interval

$$\left[\frac{\ell 2^{r(\mathbf{h})}|\mathbf{h}|}{N}, \frac{(\ell+1)2^{r(\mathbf{h})}|\mathbf{h}|}{N} \right] \subset [-|\mathbf{h}|, |\mathbf{h}|] \tag{33.14}$$

contains less than $2^{r(\mathbf{h})}$ dot products $\mathbf{h}\cdot\mathbf{v}_k$, $1 \leq k \leq N$. Notice that (33.7) implies $|\ell| \leq \pi U|\mathbf{h}|$.

Similarly to (33.10), (33.14) yields the upper bound

$$\sum_{j\in J} \sum_{\substack{\mathbf{h}\in\mathbb{Z}^3:\ |\mathbf{h}\cdot\mathbf{v}_j|\geq 1/(\pi U)\\ 1\leq\|\mathbf{h}\|_\infty\leq N^{1/3}}} \frac{1}{\Lambda(\mathbf{h})\pi|\mathbf{h}\cdot\mathbf{v}_j|}$$

$$\leq \sum_{\substack{\mathbf{h}\in\mathbb{Z}^3:\ |\mathbf{h}\cdot\mathbf{v}_j|\geq 1/(\pi U)\\ 1\leq\|\mathbf{h}\|_\infty\leq N^{1/3}}} \frac{1}{\Lambda(\mathbf{h})} 2^{r(\mathbf{h})+2} \sum_{\ell=1}^{\pi U|\mathbf{h}|} \frac{N}{\pi\ell 2^{r(\mathbf{h})}|\mathbf{h}|}$$

$$= \frac{4N}{\pi} \sum_{\substack{\mathbf{h}\in\mathbb{Z}^3:\ |\mathbf{h}\cdot\mathbf{v}_j|\geq 1/(\pi U)\\ 1\leq\|\mathbf{h}\|_\infty\leq N^{1/3}}} \frac{1}{|\mathbf{h}|\Lambda(\mathbf{h})} \left(\sum_{\ell=1}^{\pi U|\mathbf{h}|} \frac{1}{\ell} \right), \tag{33.15}$$

where again we used (33.11).

Using the fact

$$\sum_{\ell=1}^{n} \frac{1}{\ell} \le 2\log n$$

we have

$$\sum_{\ell=1}^{\pi U|\mathbf{h}|} \frac{1}{\ell} \le 2\left(\log\pi + \log U + \log|\mathbf{h}|\right). \tag{33.16}$$

Applying (33.16) in (33.15), we obtain

$$\sum_{j\in J} \sum_{\substack{\mathbf{h}\in\mathbb{Z}^3:\ |\mathbf{h}\cdot\mathbf{v}_j|\ge 1/(\pi U)\\ 1\le\|\mathbf{h}\|_\infty\le N^{1/3}}} \frac{1}{\Lambda(\mathbf{h})\pi|\mathbf{h}\cdot\mathbf{v}_j|}$$

$$\le 4N \sum_{\mathbf{h}\in\mathbb{Z}^3\setminus\mathbf{0}} \frac{1}{|\mathbf{h}|\Lambda(\mathbf{h})}(1+\log U) + 4N \sum_{\mathbf{h}\in\mathbb{Z}^3\setminus\mathbf{0}} \frac{\log|\mathbf{h}|}{|\mathbf{h}|\Lambda(\mathbf{h})}. \tag{33.17}$$

Repeating the argument of (33.12), and using the simple inequality

$$\frac{1}{2}\prod_{i=1}^{3}(1+\log(1+h_i^2)) \ge \frac{1}{2}\sum_{i=1}^{3}\log(1+h_i^2) = \frac{1}{2}\log\prod_{i=1}^{3}(1+h_i^2)$$

$$\ge \frac{1}{2}\log(h_1^2+h_2^2+h_3^2) = \log|\mathbf{h}|,$$

we have

$$\sum_{\mathbf{h}\in\mathbb{Z}^3\setminus\mathbf{0}} \frac{\log|\mathbf{h}|}{|\mathbf{h}|\Lambda(\mathbf{h})} \le 2^2 \sum_{\mathbf{h}\in\mathbb{Z}^3\setminus\mathbf{0}} \prod_{i=1}^{3} \frac{1+\log(1+h_i^2)}{(1+|h_i|)^{4/3}}$$

$$\le 4\left(\sum_{h\in\mathbb{Z}} \frac{1+\log(1+h^2)}{(1+|h|)^{4/3}}\right)^3 \le 10^6. \tag{33.18}$$

Next we use (33.18) and (33.12) in (33.17):

$$\sum_{j\in J} \sum_{\substack{\mathbf{h}\in\mathbb{Z}^3:\ |\mathbf{h}\cdot\mathbf{v}_j|\ge 1/(\pi U)\\ 1\le\|\mathbf{h}\|_\infty\le N^{1/3}}} \frac{1}{\Lambda(\mathbf{h})|\mathbf{h}\cdot\mathbf{v}_j|} \le 8\cdot 10^4 N(1+\log U) + 8\cdot 10^6 N. \tag{33.19}$$

Applying (33.13) and (33.18) in (33.16), we obtain the upper bound

$$\sum_{j\in J}\left|\int_0^U \chi_B(j;t)\,dt - U\mathrm{vol}(B)\right|$$
$$\le 8|J|UN^{-1/3} + 16\cdot 10^4 N/\pi + 16\cdot 10^4 N(1+\log U) + 16\cdot 10^6 N$$
$$\le 8|J|UN^{-1/3} + 16\cdot 10^4 N\log U + 2\cdot 10^7 N, \tag{33.20}$$

assuming Case 1.

Next we estimate (33.6) under the condition that Case 1 fails.

Case 2: $U > \frac{N^{1/3}}{\sqrt{3}\pi}$

We repeat the argument of (33.7)–(33.20) with the minor change that we use (12.26) instead of (12.27), which gives an extra factor of $6\log N$. Thus we obtain the following analog of (33.20):

$$\sum_{j\in J}\left|\int_0^U \chi_B(j;t)\,dt - U\mathrm{vol}(B)\right|$$
$$\le 8|J|UN^{-1/3} + 6\log N\cdot(16\cdot 10^4 N\log U + 2\cdot 10^7 N), \tag{33.21}$$

assuming Case 2.

Note that in Case 2 we have the upper bound $\log N \le 4\log U$, and using it in (33.21),

$$\sum_{j\in J}\left|\int_0^U \chi_B(j;t)\,dt - U\mathrm{vol}(B)\right|$$
$$\le 8|J|UN^{-1/3} + 24\log U\cdot(16\cdot 10^4 N\log U + 2\cdot 10^7 N)$$
$$< 8|J|UN^{-1/3} + 4\cdot 10^6 N(\log U)^2 + 5\cdot 10^8 N\log U. \tag{33.22}$$

Combining (33.20) and (33.22), we obtain the upper bound

$$\sum_{j\in J}\left|\int_0^U \chi_B(j;t)\,dt - U\mathrm{vol}(B)\right|$$
$$\le 8|J|UN^{-1/3} + 4\cdot 10^6 N(\log U)^2 + 5\cdot 10^8 N\log U, \tag{33.23}$$

which holds in *both* Case 1 and Case 2.

Next we estimate the sum (33.5). Again we use the Separation Condition (12.26)–(12.27), and again let $H_{j_1,j_2} = N^{1/3}$. Then by (33.5) we obtain the following analog of (33.6):

$$\sum_{\substack{(j_1,j_2)\in J\times J:\\ j_1\neq j_2}} \left| \int_0^U \chi_B(j_1;t)\chi_B(j_2;t)\,dt - U(\text{vol}(B))^2 \right|$$

$$\leq \left(\frac{3}{2}\right)^6 \sum_{\substack{(j_1,j_2)\in J\times J:\\ j_1\neq j_2}} \left(\frac{2U}{N^{1/3}+1} + \sum_{\substack{\mathbf{h}\in\mathbb{Z}^6:\\ 1\leq\|\mathbf{h}\|_\infty\leq N^{1/3}}} \right.$$

$$\left. \times \frac{1}{\Lambda(\mathbf{h})} \min\left\{ \frac{1}{\pi|\mathbf{h}\cdot\mathbf{v}_{j_1}\oplus\mathbf{v}_{j_2}|}, U \right\} \right)$$

$$\leq \frac{30|J|^2U}{N^{1/3}} + 15 \sum_{\substack{(j_1,j_2)\in J\times J:\\ j_1\neq j_2}} \sum_{\substack{\mathbf{h}\in\mathbb{Z}^6:\ |\mathbf{h}\cdot\mathbf{v}_{j_1}\oplus\mathbf{v}_{j_2}|<1/(\pi U)\\ 1\leq\|\mathbf{h}\|_\infty\leq N^{1/3}}} \frac{U}{\Lambda(\mathbf{h})}$$

$$+ 15 \sum_{\substack{(j_1,j_2)\in J\times J:\\ j_1\neq j_2}} \sum_{\substack{\mathbf{h}\in\mathbb{Z}^6:\ |\mathbf{h}\cdot\mathbf{v}_{j_1}\oplus\mathbf{v}_{j_2}|\geq 1/(\pi U)\\ 1\leq\|\mathbf{h}\|_\infty\leq N^{1/3}}} \frac{1}{\Lambda(\mathbf{h})\pi|\mathbf{h}\cdot\mathbf{v}_{j_1}\oplus\mathbf{v}_{j_2}|}. \tag{33.24}$$

Note that

$$\mathbf{h}\cdot\mathbf{v}_{j_1}\oplus\mathbf{v}_{j_2} = \mathbf{h}^{(1)}\cdot\mathbf{v}_{j_1} + \mathbf{h}^{(2)}\cdot\mathbf{v}_{j_2}, \tag{33.25}$$

where $\mathbf{h} = (h_1, h_2, \ldots, h_6)$, $\mathbf{h}^{(1)} = (h_1, h_2, h_3)$ and $\mathbf{h}^{(2)} = (h_4, h_5, h_6)$.

By using (33.25) and (12.6), we can rewrite a critical part of (33.24) as follows:

$$\sum_{\substack{(j_1,j_2)\in J\times J:\\ j_1\neq j_2}} \sum_{\substack{\mathbf{h}\in\mathbb{Z}^6:\ |\mathbf{h}\cdot\mathbf{v}_{j_1}\oplus\mathbf{v}_{j_2}|<1/(\pi U)\\ 1\leq\|\mathbf{h}\|_\infty\leq N^{1/3}}} \frac{U}{\Lambda(\mathbf{h})}$$

$$= \sum_{\substack{(j_1,j_2)\in J\times J:\\ j_1\neq j_2}} \sum_{\substack{\mathbf{h}=\mathbf{h}^{(1)}\oplus\mathbf{h}^{(2)}\in\mathbb{Z}^6:\ 1\leq\|\mathbf{h}\|_\infty\leq N^{1/3}\\ |\mathbf{h}^{(1)}\cdot\mathbf{v}_{j_1}+\mathbf{h}^{(2)}\cdot\mathbf{v}_{j_2}|<1/(\pi U)}} \frac{U}{\Lambda(\mathbf{h}^{(1)})\Lambda(\mathbf{h}^{(2)})}. \tag{33.26}$$

To estimate (33.26) we distinguish two cases.

Case 1: $U \le \frac{N^{1/3}}{\sqrt{3}\pi}$

We recall (33.7) and (33.8): for every 3-dimensional lattice point $\mathbf{h}' \in \mathbb{Z}^3$ with $1 \le \|\mathbf{h}'\|_\infty \le N^{1/3}$, let $r(\mathbf{h}') = r(\mathbf{h}'; U; N)$ denote the smallest integer such that

$$\frac{1}{\pi U} \le \frac{2^{r(\mathbf{h}')}|\mathbf{h}'|}{N}, \tag{33.27}$$

and in Case 1 we have the lower bound

$$2^{r(\mathbf{h}')} \ge \frac{N}{\pi U|\mathbf{h}'|} \ge \frac{N}{\pi\left(N^{1/3}/\sqrt{3}\pi\right)\sqrt{3}\|\mathbf{h}'\|_\infty} \ge \frac{N}{N^{1/3}N^{1/3}} = N^{1/3}. \tag{33.28}$$

Moreover, for every unit vector $\mathbf{v}_j \in \mathbf{S}^2$, $1 \le j \le N$ there is an integer $\ell = \ell(\mathbf{v}_j)$ such that

$$\frac{\ell 2^{r(\mathbf{h}')}\,|\mathbf{h}'|}{N} \le \mathbf{h}' \cdot \mathbf{v}_j \le \frac{(\ell+1)2^{r(\mathbf{h}')}\,|\mathbf{h}'|}{N} \tag{33.29}$$

and

$$|\ell| \le \pi U|\mathbf{h}'|, \tag{33.30}$$

where the latter follows from (33.27).

Combining (12.27) and (33.28), we obtain that the interval

$$\left[\frac{\ell 2^{r(\mathbf{h}')}|\mathbf{h}'|}{N}, \frac{(\ell+1)2^{r(\mathbf{h}')}|\mathbf{h}'|}{N}\right] \tag{33.31}$$

contains less than $2^{r(\mathbf{h}')}$ dot products $\mathbf{h}' \cdot \mathbf{v}_k$, $1 \le k \le N$. Using (33.29)–(33.31) we obtain the following analog of (33.10):

$$\sum_{\substack{(j_1,j_2)\in J\times J:\\ j_1\ne j_2}} \sum_{\substack{\mathbf{h}^{(1)}\oplus\mathbf{h}^{(2)}\in\mathbb{Z}^6:\ 1\le\|\mathbf{h}^{(i)}\|_\infty\le N^{1/3}\\ \left|\mathbf{h}^{(1)}\cdot\mathbf{v}_{j_1}+\mathbf{h}^{(2)}\cdot\mathbf{v}_{j_2}\right|<1/(\pi U)}} \frac{U}{\Lambda(\mathbf{h}^{(1)})\Lambda(\mathbf{h}^{(2)})}$$

$$= \sum_{j_1\in J} \sum_{\substack{\mathbf{h}^{(1)}\in\mathbb{Z}^3:\\ 1\le\|\mathbf{h}^{(1)}\|_\infty\le N^{1/3}}} \frac{1}{\Lambda(\mathbf{h}^{(1)})} \sum_{\substack{j_2\in J:\\ j_2\ne j_1}} \sum_{\substack{\mathbf{h}^{(2)}\in\mathbb{Z}^3:\ 1\le\|\mathbf{h}^{(2)}\|_\infty\le N^{1/3}\\ |\mathbf{h}^{(1)}\cdot\mathbf{v}_{j_1}+\mathbf{h}^{(2)}\cdot\mathbf{v}_{j_2}|<1/(\pi U)}} \frac{U}{\Lambda(\mathbf{h}^{(2)})}$$

$$\leq 2\sum_{j_1\in J}\sum_{\substack{\mathbf{h}^{(1)}\in\mathbb{Z}^3:\\ 1\leq\|\mathbf{h}^{(1)}\|_\infty\leq N^{1/3}}}\frac{1}{\Lambda(\mathbf{h}^{(1)})}\sum_{\substack{\mathbf{h}^{(2)}\in\mathbb{Z}^3:\\ \|\mathbf{h}^{(1)}\|_\infty\leq\|\mathbf{h}^{(2)}\|_\infty\leq N^{1/3}}} 2^{r(\mathbf{h}^{(2)})+2}\cdot\frac{U}{\Lambda(\mathbf{h}^{(2)})}$$

$$\leq 2|J|\sum_{\substack{\mathbf{h}^{(1)}\in\mathbb{Z}^3:\\ 1\leq\|\mathbf{h}^{(1)}\|_\infty\leq N^{1/3}}}\frac{1}{\Lambda(\mathbf{h}^{(1)})}\sum_{\substack{\mathbf{h}^{(2)}\in\mathbb{Z}^3:\\ \|\mathbf{h}^{(1)}\|_\infty\leq\|\mathbf{h}^{(2)}\|_\infty\leq N^{1/3}}}\frac{8N}{\pi|\mathbf{h}^{(2)}|U}\cdot\frac{U}{\Lambda(\mathbf{h}^{(2)})}$$

$$\leq 2|J|\sum_{\substack{\mathbf{h}^{(1)}\in\mathbb{Z}^3:\\ 1\leq\|\mathbf{h}^{(1)}\|_\infty\leq N^{1/3}}}\frac{1}{\Lambda(\mathbf{h}^{(1)})}\sum_{\substack{\mathbf{h}^{(2)}\in\mathbb{Z}^3:\\ \|\mathbf{h}^{(1)}\|_\infty\leq\|\mathbf{h}^{(2)}\|_\infty\leq N^{1/3}}}\frac{8N}{\pi\|\mathbf{h}^{(2)}\|_\infty U}$$

$$\cdot\frac{U}{\Lambda(\mathbf{h}^{(2)})}\leq\frac{16N|J|}{\pi}\sum_{\substack{\mathbf{h}^{(1)}\in\mathbb{Z}^3:\\ 1\leq\|\mathbf{h}^{(1)}\|_\infty\leq N^{1/3}}}\sum_{\substack{\mathbf{h}^{(2)}\in\mathbb{Z}^3:\\ 1\leq\|\mathbf{h}^{(2)}\|_\infty\leq N^{1/3}}}$$

$$\times\frac{1}{\sqrt{\|\mathbf{h}^{(1)}\|_\infty}\sqrt{\|\mathbf{h}^{(2)}\|_\infty}\Lambda(\mathbf{h}^{(1)})\Lambda(\mathbf{h}^{(2)})},\tag{33.32}$$

where we used that for every $\mathbf{h}^{(1)}\cdot\mathbf{v}_{j_1}$ there are at most $4\cdot 2^{r(\mathbf{h}^{(2)})}$ dot products $\mathbf{h}^{(2)}\cdot\mathbf{v}_{j_2}$ such that

$$|\mathbf{h}^{(1)}\cdot\mathbf{v}_{j_1}+\mathbf{h}^{(2)}\cdot\mathbf{v}_{j_2}|<1/(\pi U),$$

at the end of (33.32) we used the following consequence of (33.27):

$$\frac{2^{r(\mathbf{h}^{(2)})}|\mathbf{h}^{(2)}|}{N}<\frac{2}{\pi U}\iff 2^{r(\mathbf{h}^{(2)})}<\frac{2N}{\pi|\mathbf{h}^{(2)}|U},$$

and finally, we used

$$\|\mathbf{h}^{(1)}\|_\infty\leq\|\mathbf{h}^{(2)}\|_\infty\leq|\mathbf{h}^{(2)}|.$$

By (33.32) we have the upper bound

$$\sum_{\substack{(j_1,j_2)\in J\times J:\\ j_1\neq j_2}}\sum_{\substack{\mathbf{h}^{(1)}\oplus\mathbf{h}^{(2)}\in\mathbb{Z}^6:\ 1\leq\|\mathbf{h}^{(i)}\|_\infty\leq N^{1/3}\\ |\mathbf{h}^{(1)}\cdot\mathbf{v}_{j_1}+\mathbf{h}^{(2)}\cdot\mathbf{v}_{j_2}|<1/(\pi U)}}\frac{U}{\Lambda(\mathbf{h}^{(1)})\Lambda(\mathbf{h}^{(2)})}$$

$$\leq\frac{16N|J|}{\pi}\left(\sum_{\substack{\mathbf{h}\in\mathbb{Z}^3:\\ 1\leq\|\mathbf{h}\|_\infty\leq N^{1/3}}}\frac{1}{\sqrt{\|\mathbf{h}\|_\infty}\Lambda(\mathbf{h})}\right)^2.\tag{33.33}$$

Using the definition (12.6) of $\Lambda(\mathbf{h})$ and the trivial inequality

$$\|\mathbf{h}\|_\infty \ge \frac{1+|h_i|}{2}, \quad i=1,2,3$$

for $\mathbf{h} \in \mathbb{Z}^3 \setminus \mathbf{0}$, we can easily estimate the infinite sum

$$\sum_{\mathbf{h}\in\mathbb{Z}^3\setminus\mathbf{0}} \frac{1}{\sqrt{\|\mathbf{h}\|_\infty}\Lambda(\mathbf{h})} \le \sum_{\mathbf{h}\in\mathbb{Z}^3\setminus\mathbf{0}} \prod_{i=1}^{3} \frac{\sqrt{2}}{(1+|h_i|)^{7/6}}$$
$$\le 2^{3/2}\left(\sum_{h\in\mathbb{Z}} \frac{1}{(1+|h|)^{7/6}}\right)^3 \le 10^5. \quad (33.34)$$

Combining (33.33) and (33.34), in Case 1 we have

$$\sum_{\substack{(j_1,j_2)\in J\times J:\\ j_1\ne j_2}} \sum_{\substack{\mathbf{h}^{(1)}\oplus\mathbf{h}^{(2)}\in\mathbb{Z}^6:\ 1\le\|\mathbf{h}^{(i)}\|_\infty\le N^{1/3}\\ |\mathbf{h}^{(1)}\cdot\mathbf{v}_{j_1}+\mathbf{h}^{(2)}\cdot\mathbf{v}_{j_2}|<1/(\pi U)}} \frac{U}{\Lambda(\mathbf{h}^{(1)})\Lambda(\mathbf{h}^{(2)})} \le \frac{16N|J|}{\pi}\cdot 10^{10}. \quad (33.35)$$

Let us return to (33.26); we clearly have

$$\sum_{\substack{(j_1,j_2)\in J\times J:\\ j_1\ne j_2}} \sum_{\substack{\mathbf{h}=\mathbf{h}^{(1)}\oplus\mathbf{h}^{(2)}\in\mathbb{Z}^6:\ 1\le\|\mathbf{h}\|_\infty\le N^{1/3}\\ |\mathbf{h}^{(1)}\cdot\mathbf{v}_{j_1}+\mathbf{h}^{(2)}\cdot\mathbf{v}_{j_2}|<1/(\pi U)}} \frac{U}{\Lambda(\mathbf{h}^{(1)})\Lambda(\mathbf{h}^{(2)})}$$
$$= \sum_{\substack{(j_1,j_2)\in J\times J:\\ j_1\ne j_2}} \sum_{\substack{\mathbf{h}^{(1)}\oplus\mathbf{h}^{(2)}\in\mathbb{Z}^6:\ 1\le\|\mathbf{h}^{(i)}\|_\infty\le N^{1/3}\\ |\mathbf{h}^{(1)}\cdot\mathbf{v}_{j_1}+\mathbf{h}^{(2)}\cdot\mathbf{v}_{j_2}|<1/(\pi U)}} \frac{U}{\Lambda(\mathbf{h}^{(1)})\Lambda(\mathbf{h}^{(2)})}$$
$$+ 2(|J|-1)\sum_{j\in J} \sum_{\substack{\mathbf{h}'\in\mathbb{Z}^3:\ |\mathbf{h}'\cdot\mathbf{v}_j|<1/(\pi U)\\ 1\le\|\mathbf{h}'\|_\infty\le N^{1/3}}} \frac{U}{\Lambda(\mathbf{h}')}$$
$$\le \frac{16N|J|}{\pi}\cdot 10^{10} + 2(|J|-1)\frac{4\cdot 10^4 N}{\pi} \le \frac{17\cdot 10^{10}N|J|}{\pi}, \quad (33.36)$$

where in the last step we used (33.35) and (33.13). Note that (33.36) holds under the condition of Case 1.

Next we estimate the last sum in (33.24). The same way as we derived (33.15) from (33.10), we can derive the following inequality from (33.32):

$$\sum_{\substack{(j_1,j_2)\in J\times J:\\ j_1\neq j_2}} \sum_{\substack{\mathbf{h}^{(1)}\oplus\mathbf{h}^{(2)}\in\mathbb{Z}^6:\ 1\le\|\mathbf{h}^{(i)}\|_\infty\le N^{1/3}\\ |\mathbf{h}^{(1)}\cdot\mathbf{v}_{j_1}+\mathbf{h}^{(2)}\cdot\mathbf{v}_{j_2}|\ge 1/(\pi U)}}$$

$$\times \frac{1}{\Lambda(\mathbf{h}^{(1)})\Lambda(\mathbf{h}^{(2)})|\mathbf{h}^{(1)}\cdot\mathbf{v}_{j_1}+\mathbf{h}^{(2)}\cdot\mathbf{v}_{j_2}|} \le \frac{16N|J|}{\pi}$$

$$\times \sum_{\substack{\mathbf{h}^{(1)}\in\mathbb{Z}^3:\\ 1\le\|\mathbf{h}^{(1)}\|_\infty\le N^{1/3}}} \sum_{\substack{\mathbf{h}^{(2)}\in\mathbb{Z}^3:\\ \|\mathbf{h}^{(1)}\|_\infty\le\|\mathbf{h}^{(2)}\|_\infty\le N^{1/3}}}$$

$$\times \frac{1}{|\mathbf{h}^{(2)}|\Lambda(\mathbf{h}^{(1)})\Lambda(\mathbf{h}^{(2)})}\left(\sum_{\ell=1}^{\pi U|\mathbf{h}^{(2)}|}\frac{1}{\ell}\right), \tag{33.37}$$

where the upper end $\pi U|\mathbf{h}^{(2)}|$ in the last sum comes from (33.30) with the choice $\mathbf{h}'=\mathbf{h}^{(2)}$.

Using the fact

$$\sum_{\ell=1}^{n}\frac{1}{\ell}\le 2\log n$$

we have

$$\sum_{\ell=1}^{\pi U|\mathbf{h}^{(2)}|}\frac{1}{\ell}\le 2(\log\pi+\log U+\log|\mathbf{h}^{(2)}|),$$

and applying it for the end part of (33.37), we obtain

$$\sum_{\substack{\mathbf{h}^{(1)}\in\mathbb{Z}^3:\\ 1\le\|\mathbf{h}^{(1)}\|_\infty\le N^{1/3}}} \sum_{\substack{\mathbf{h}^{(2)}\in\mathbb{Z}^3:\\ \|\mathbf{h}^{(1)}\|_\infty\le\|\mathbf{h}^{(2)}\|_\infty\le N^{1/3}}} \frac{1}{|\mathbf{h}^{(2)}|\Lambda(\mathbf{h}^{(1)})\Lambda(\mathbf{h}^{(2)})}$$

$$\times\left(\sum_{\ell=1}^{\pi U|\mathbf{h}^{(2)}|}\frac{1}{\ell}\right)\le 2(\log\pi+\log U)\sum_{\substack{\mathbf{h}^{(1)}\in\mathbb{Z}^3:\\ 1\le\|\mathbf{h}^{(1)}\|_\infty\le N^{1/3}}}$$

$$\times\sum_{\substack{\mathbf{h}^{(2)}\in\mathbb{Z}^3:\\ \|\mathbf{h}^{(1)}\|_\infty\le\|\mathbf{h}^{(2)}\|_\infty\le N^{1/3}}}\frac{1}{|\mathbf{h}^{(2)}|\Lambda(\mathbf{h}^{(1)})\Lambda(\mathbf{h}^{(2)})}$$

$$+2 \sum_{\substack{\mathbf{h}^{(1)}\in\mathbb{Z}^3:\\ 1\le\|\mathbf{h}^{(1)}\|_\infty\le N^{1/3}}} \sum_{\substack{\mathbf{h}^{(2)}\in\mathbb{Z}^3:\\ \|\mathbf{h}^{(1)}\|_\infty\le\|\mathbf{h}^{(2)}\|_\infty\le N^{1/3}}} \frac{\log|\mathbf{h}^{(2)}|}{|\mathbf{h}^{(2)}|\Lambda(\mathbf{h}^{(1)})\Lambda(\mathbf{h}^{(2)})}$$

$$\le 2(\log\pi+\log U) \sum_{\substack{\mathbf{h}^{(1)}\in\mathbb{Z}^3:\\ 1\le\|\mathbf{h}^{(1)}\|_\infty\le N^{1/3}}} \cdot \sum_{\substack{\mathbf{h}^{(2)}\in\mathbb{Z}^3:\\ 1\le\|\mathbf{h}^{(2)}\|_\infty\le N^{1/3}}}$$

$$\times \frac{1}{\sqrt{\|\mathbf{h}^{(1)}\|_\infty}\sqrt{\|\mathbf{h}^{(2)}\|_\infty}\Lambda(\mathbf{h}^{(1)})\Lambda(\mathbf{h}^{(2)})} + 2 \sum_{\substack{\mathbf{h}^{(1)}\in\mathbb{Z}^3:\\ 1\le\|\mathbf{h}^{(1)}\|_\infty\le N^{1/3}}}$$

$$\times \sum_{\substack{\mathbf{h}^{(2)}\in\mathbb{Z}^3:\\ 1\le\|\mathbf{h}^{(2)}\|_\infty\le N^{1/3}}} \frac{\sqrt{1+\log\|\mathbf{h}^{(1)}\|_\infty}\sqrt{1+\log\|\mathbf{h}^{(2)}\|_\infty}}{\sqrt{\|\mathbf{h}^{(1)}\|_\infty}\sqrt{\|\mathbf{h}^{(2)}\|_\infty}\Lambda(\mathbf{h}^{(1)})\Lambda(\mathbf{h}^{(2)})}$$

$$= 2(\log\pi+\log U)\left(\sum_{\substack{\mathbf{h}\in\mathbb{Z}^3:\\ 1\le\|\mathbf{h}\|_\infty\le N^{1/3}}} \frac{1}{\sqrt{\|\mathbf{h}\|_\infty}\Lambda(\mathbf{h})}\right)^2$$

$$+2\left(\sum_{\substack{\mathbf{h}\in\mathbb{Z}^3:\\ 1\le\|\mathbf{h}\|_\infty\le N^{1/3}}} \frac{\sqrt{1+\log\|\mathbf{h}\|_\infty}}{\sqrt{\|\mathbf{h}\|_\infty}\Lambda(\mathbf{h})}\right)^2, \tag{33.38}$$

where we used the fact that the function

$$f(x) = \frac{1+\log x}{x} \quad \text{is monotone decreasing for } x \ge 1.$$

Using the definition (12.6) of $\Lambda(\mathbf{h})$ and the simple inequalities

$$\|\mathbf{h}\|_\infty \ge \frac{1+|h_i|}{2}$$

and

$$\prod_{i=1}^{3}(1+\log(1+|h_i|)) \ge 1+\sum_{i=1}^{3}\log(1+|h_i|) = 1+\log\prod_{i=1}^{3}(1+|h_i|)$$
$$\ge 1+\log(|h_1|+|h_2|+|h_3|) \ge 1+\log\|\mathbf{h}\|_\infty$$

for $\mathbf{h} = (h_1, h_2, h_3) \in \mathbb{Z}^3 \setminus \mathbf{0}$, we can easily estimate the infinite sum

$$\sum_{\mathbf{h}\in\mathbb{Z}^3\setminus\mathbf{0}} \frac{\sqrt{1+\log\|\mathbf{h}\|_\infty}}{\sqrt{\|\mathbf{h}\|_\infty}\Lambda(\mathbf{h})} \le 2^{3/2} \sum_{\mathbf{h}\in\mathbb{Z}^3\setminus\mathbf{0}} \prod_{i=1}^{3} \frac{\sqrt{1+\log(1+|h_i|)}}{(1+|h_i|)^{7/6}}$$
$$\le 2^{3/2}\left(\sum_{h\in\mathbb{Z}} \frac{\sqrt{1+\log(1+|h|)}}{(1+|h|)^{7/6}}\right)^3 \le 10^6. \tag{33.39}$$

Combining (33.37)–(33.39) and (33.34), in Case 1 we have

$$\sum_{\substack{(j_1,j_2)\in J\times J:\\ j_1\neq j_2}} \sum_{\substack{\mathbf{h}^{(1)}\oplus\mathbf{h}^{(2)}\in\mathbb{Z}^6:\ 1\le\|\mathbf{h}^{(i)}\|_\infty\le N^{1/3}\\ |\mathbf{h}^{(1)}\cdot\mathbf{v}_{j_1}+\mathbf{h}^{(2)}\cdot\mathbf{v}_{j_2}|\ge 1/(\pi U)}}$$
$$\times \frac{1}{\Lambda(\mathbf{h}^{(1)})\Lambda(\mathbf{h}^{(2)})|\mathbf{h}^{(1)}\cdot\mathbf{v}_{j_1}+\mathbf{h}^{(2)}\cdot\mathbf{v}_{j_2}|}$$
$$\le \frac{16N|J|}{\pi}(2(\log\pi+\log U)10^{10}+2\cdot 10^{12}). \tag{33.40}$$

Clearly

$$\sum_{\substack{(j_1,j_2)\in J\times J:\\ j_1\neq j_2}} \sum_{\substack{\mathbf{h}\in\mathbb{Z}^6:\ |\mathbf{h}\cdot\mathbf{v}_{j_1}\oplus\mathbf{v}_{j_2}|\ge 1/(\pi U)\\ 1\le\|\mathbf{h}\|_\infty\le N^{1/3}}}$$
$$\times \frac{1}{\Lambda(\mathbf{h})\pi|\mathbf{h}\cdot\mathbf{v}_{j_1}\oplus\mathbf{v}_{j_2}|} = \sum_{\substack{(j_1,j_2)\in J\times J:\\ j_1\neq j_2}}$$
$$\times \sum_{\substack{\mathbf{h}^{(1)}\oplus\mathbf{h}^{(2)}\in\mathbb{Z}^6:\ 1\le\|\mathbf{h}^{(i)}\|_\infty\le N^{1/3}\\ |\mathbf{h}^{(1)}\cdot\mathbf{v}_{j_1}+\mathbf{h}^{(2)}\cdot\mathbf{v}_{j_2}|\ge 1/(\pi U)}} \frac{1}{\Lambda(\mathbf{h}^{(1)})\Lambda(\mathbf{h}^{(2)})|\mathbf{h}^{(1)}\cdot\mathbf{v}_{j_1}+\mathbf{h}^{(2)}\cdot\mathbf{v}_{j_2}|}$$
$$+2(|J|-1)\sum_{j\in J} \sum_{\substack{\mathbf{h}'\in\mathbb{Z}^3:\ 1\le\|\mathbf{h}'\|_\infty\le N^{1/3}\\ |\mathbf{h}'\cdot\mathbf{v}_j|\ge 1/(\pi U)}} \frac{1}{\Lambda(\mathbf{h}')|\mathbf{h}'\cdot\mathbf{v}_j|}$$
$$\le \frac{32N|J|}{\pi}((\log\pi+\log U)10^{10}+10^{12})$$
$$+2(|J|-1)(8\cdot 10^4 N(1+\log U)+8\cdot 10^6 N)$$
$$\le 2\cdot 10^{11}N|J|\log U + 2\cdot 10^{13}N|J|, \tag{33.41}$$

where in the last step we used (33.40) and (33.19). Note that (33.41) holds under the condition of Case 1.

Applying (33.36) and (33.41) in (33.24), we obtain the upper bound

$$\sum_{\substack{(j_1,j_2)\in J\times J:\\ j_1\neq j_2}} \left|\int_0^U \chi_B(j_1;t)\chi_B(j_2;t)\,dt - U(\mathrm{vol}(B))^2\right|$$

$$\leq \frac{30|J|^2U}{N^{1/3}} + 15\cdot\frac{17\cdot 10^{10}N|J|}{\pi} + 15\cdot(2\cdot 10^{11}N|J|\log U + 2\cdot 10^{13}N|J|)$$

$$\leq \frac{30|J|^2U}{N^{1/3}} + 3\cdot 10^{12}N|J|\log U + 31\cdot 10^{13}N|J|, \tag{33.42}$$

assuming Case 1.

Next we estimate (33.6) under the condition that Case 1 fails.

Case 2: $U > \frac{N^{1/3}}{\sqrt{3}\pi}$

Just like before, we can repeat the argument of Case 1 with the minor change that we use (12.26) instead of (12.27), which gives an extra factor of $6\log N$. Thus we obtain the following analog of (33.42):

$$\sum_{\substack{(j_1,j_2)\in J\times J:\\ j_1\neq j_2}} \left|\int_0^U \chi_B(j_1;t)\chi_B(j_2;t)\,dt - U(\mathrm{vol}(B))^2\right|$$

$$\leq \frac{30|J|^2U}{N^{1/3}} + 6\log N(3\cdot 10^{12}N|J|\log U + 31\cdot 10^{13}N|J|), \tag{33.43}$$

assuming Case 2.

Note that in Case 2 we have the upper bound $\log N \leq 4\log U$, and using it in (33.43),

$$\sum_{\substack{(j_1,j_2)\in J\times J:\\ j_1\neq j_2}} \left|\int_0^U \chi_B(j_1;t)\chi_B(j_2;t)\,dt - U(\mathrm{vol}(B))^2\right|$$

$$\leq \frac{30|J|^2U}{N^{1/3}} + 24\log U(3\cdot 10^{12}N|J|\log U + 31\cdot 10^{13}N|J|). \tag{33.44}$$

Combining (33.41) and (33.44), we obtain the upper bound

$$\sum_{\substack{(j_1,j_2)\in J\times J:\\ j_1\neq j_2}} \left| \int_0^U \chi_B(j_1;t)\chi_B(j_2;t)\,dt - U(\mathrm{vol}(B))^2 \right|$$

$$\leq 30|J|^2UN^{-1/3} + 72\cdot 10^{12}N|J|(\log U)^2 + 75\cdot 10^{14}N|J|\log U, \quad (33.45)$$

which holds in *both* Case 1 and Case 2.

Finally, combining (33.1), (33.23) and (33.45), we conclude

$$\int_0^U \left(\sum_{j\in J} \chi_B(j;t) - |J|\mathrm{vol}(B) \right)^2 dt$$

$$\leq \sum_{j\in J} \left| \int_0^U \chi_B(j;t)\,dt - U\mathrm{vol}(B) \right|$$

$$+ \sum_{\substack{(j_1,j_2)\in J\times J:\\ j_1\neq j_2}} \left| \int_0^U \chi_B(j_1;t)\chi_B(j_2;t)\,dt - U\,(\mathrm{vol}(B))^2 \right|$$

$$+ |J|U + 2|J| \sum_{j\in J} \left| \int_0^U \chi_B(j;t)\,dt - U\mathrm{vol}(B) \right|$$

$$\leq 8|J|UN^{-1/3} + 4\cdot 10^6 N(\log U)^2 + 5\cdot 10^8 N\log U$$

$$+ 30|J|^2UN^{-1/3} + 72\cdot 10^{12}N|J|(\log U)^2 + 75\cdot 10^{14}N|J|\log U$$

$$+ |J|U + 2|J|(8|J|UN^{-1/3} + 4\cdot 10^6 N(\log U)^2 + 5\cdot 10^8 N\log U)$$

$$\leq 10^{14}N|J|(\log U)^2 + 10^{16}N|J|\log U + 54|J|^2UN^{-1/3},$$

which completes the proof of Lemma 12.2. □

34. Multiple Mixedupness (III): Proof of Theorem 12.2

Similarly to the proof of Lemma 12.2 (see (12.36)–(12.39)), we *fix* a sequence $\mathbf{v}_1, \mathbf{v}_2, \ldots, \mathbf{v}_N \in \mathbf{S}^2$ of N unit vectors on the unit sphere that satisfies the

Separation Condition (12.26)–(12.27). Let $J \subseteq \{1,2,3,\ldots,N\}$ be a subset, and consider the 3-dimensional torus system

$$\mathcal{Z}_J(t) = \mathcal{Z}_J(\mathbf{v}_j, j \in J; t) = \{\mathbf{z}_j + t\mathbf{v}_j \text{ modulo one}: \ j \in J\},\ 0 \le t \le U, \tag{34.1}$$

where of course modulo one means the fractional part applied coordinate-wise, $0 \le t \le U$ refers to the time evolution, and

$$\mathcal{Z}_J(0) = \{\mathbf{z}_j \text{ modulo one}: \ j \in J\} \tag{34.2}$$

is the initial configuration (i.e., a $|J|$-element point set in the unit torus $I^3 = [0,1)^3$). Let $A \subset [0,1)^3$ be a Lebesgue measurable set; we study the integral

$$\begin{aligned}
&\int_{\mathbf{w}\in[0,1)^3} \int_0^U (|\mathcal{Z}_J(t) \cap (A+\mathbf{w})| - |J|\text{vol}(A))^2 \, dt \, d\mathbf{w} \\
&= \int_{\mathbf{w}\in[0,1)^3} \int_0^U (|\mathcal{Z}_J(\mathbf{v}_j, j \in J; t) \cap (A+\mathbf{w})| - |J|\text{vol}(A))^2 \, dt \, d\mathbf{w} \\
&= \int_{\mathbf{w}\in[0,1)^3} \int_0^U \left(\sum_{j\in J} \chi_{A+\mathbf{w}}(j;t) - |J|\text{vol}(A) \right)^2 dt \, d\mathbf{w},
\end{aligned} \tag{34.3}$$

where

$$\chi_S(j;t) = \begin{cases} 1 & \text{if } \mathbf{z}_j + t\mathbf{v}_j \in S \text{ modulo one} \\ 0 & \text{otherwise.} \end{cases} \tag{34.4}$$

We need the following simple lemma.

Lemma 34.1. *For every Lebesgue measurable test set $A \subset [0,1)^3 = I^3$ with $\text{vol}(A) = p$ and $q = 1-p$, we have*

$$\begin{aligned}
&\int_{\mathbf{w}\in[0,1)^3} \frac{1}{U} \int_0^U (|\mathcal{Z}_J(\mathbf{v}_j, j \in J; t) \cap (A+\mathbf{w})| - |J|\text{vol}(A))^2 \, dt \, d\mathbf{w} \\
&\le pq|J| + \sum_{\substack{\mathbf{n}\in\mathbb{Z}^3: \\ \mathbf{n}\ne\mathbf{0}}} |a_{\mathbf{n}}|^2 \sum_{j_1,j_2\in J:\ j_1\ne j_2} \min\left\{ \frac{1}{\pi U|\mathbf{n}\cdot\mathbf{v}_{j_1} - \mathbf{n}\cdot\mathbf{v}_{j_2}|}, 1 \right\},
\end{aligned}$$

where

$$a_{\mathbf{n}} = \int_A e^{-2\pi i \mathbf{n}\cdot\mathbf{z}} \, d\mathbf{z} \quad \text{and} \quad \sum_{\substack{\mathbf{n}\in\mathbb{Z}^3: \\ \mathbf{n}\ne\mathbf{0}}} |a_{\mathbf{n}}|^2 = pq.$$

Proof. For notational convenience we work with the test function $f = \chi_A$ instead of the test set A; we define f on the torus I^3 (i.e., we extend f periodically), and consider its Fourier expansion

$$f(\mathbf{z}) = \sum_{\mathbf{n}\in\mathbb{Z}^3} a_{\mathbf{n}} e^{2\pi \mathrm{i}\mathbf{n}\cdot\mathbf{z}} \quad \text{with} \quad a_{\mathbf{n}} = a_{\mathbf{n}}(f) = \int_{I^3} f(\mathbf{z}) e^{-2\pi \mathrm{i}\mathbf{n}\cdot\mathbf{z}}\, d\mathbf{z}, \tag{34.5}$$

where of course $\mathbf{n}\cdot\mathbf{z} = n_1z_1 + n_2z_2 + n_3z_3$ and $\mathrm{i} = \sqrt{-1}$. Clearly $a_{\mathbf{0}} = \int_{I^3} f\, d\mathrm{V} = \mathrm{vol}(A)$.

If f is replaced with a translated function $f_{\mathbf{w}}(\mathbf{z}) = f(\mathbf{z}-\mathbf{w})$ (i.e., f is translated modulo one by a fixed vector $\mathbf{w}\in I^3$), then the corresponding Fourier coefficient changes as follows:

$$\begin{aligned} a_{\mathbf{n}}(f_{\mathbf{w}}) &= \int_{I^3} f(\mathbf{z}-\mathbf{w}) e^{-2\pi \mathrm{i}\mathbf{n}\cdot\mathbf{z}}\, d\mathbf{z} = \int_{I^3} f(\mathbf{z}) e^{-2\pi \mathrm{i}\mathbf{n}\cdot(\mathbf{z}+\mathbf{w})}\, d\mathbf{z} \\ &= e^{-2\pi \mathrm{i}\mathbf{n}\cdot\mathbf{w}} a_{\mathbf{n}}(f). \end{aligned} \tag{34.6}$$

By (34.5) and (34.6) we have

$$|\mathcal{Z}_J(t)\cap(A+\mathbf{w})| - |J|\mathrm{vol}(A) = \sum_{j\in J} \sum_{\substack{\mathbf{n}\in\mathbb{Z}^3:\\ \mathbf{n}\neq\mathbf{0}}} e^{-2\pi \mathrm{i}\mathbf{n}\cdot\mathbf{w}} a_{\mathbf{n}} e^{2\pi \mathrm{i}\mathbf{n}\cdot(\mathbf{z}_j + t\mathbf{v}_j)}. \tag{34.7}$$

By squaring (34.7) we have

$$\begin{aligned} &(|\mathcal{Z}_J(t)\cap(A+\mathbf{w})| - |J|\mathrm{vol}(A))^2 \\ &= \sum_{j\in J} \sum_{\substack{\mathbf{n}\in\mathbb{Z}^3:\\ \mathbf{n}\neq\mathbf{0}}} |a_{\mathbf{n}}|^2 + \sum_{\substack{\mathbf{n}\in\mathbb{Z}^3:\\ \mathbf{n}\neq\mathbf{0}}} |a_{\mathbf{n}}|^2 \sum_{j_1,j_2\in J:\ j_1\neq j_2} e^{2\pi \mathrm{i}\mathbf{n}\cdot(\mathbf{z}_{j_1}-\mathbf{z}_{j_2})} e^{2\pi \mathrm{i}t\mathbf{n}\cdot(\mathbf{v}_{j_1}-\mathbf{v}_{j_2})} \\ &\quad + \sum_{\substack{\mathbf{n}_\ell\in\mathbb{Z}^3:\ \ell=1,2\\ \mathbf{0}\neq\mathbf{n}_1\neq\mathbf{n}_2\neq\mathbf{0}}} e^{-2\pi \mathrm{i}(\mathbf{n}_1-\mathbf{n}_2)\cdot\mathbf{w}} a_{\mathbf{n}_1}\overline{a_{\mathbf{n}_2}} \\ &\quad \times \sum_{j_1,j_2\in J} e^{2\pi \mathrm{i}(\mathbf{n}_1\cdot\mathbf{z}_{j_1} - \mathbf{n}_2\cdot\mathbf{z}_{j_2})} e^{2\pi \mathrm{i}t(\mathbf{v}_{j_1}\cdot\mathbf{n}_1 - \mathbf{v}_{j_2}\cdot\mathbf{n}_2)}, \end{aligned} \tag{34.8}$$

where, as usual, $\overline{z} = x - \mathrm{i}y$ denotes the complex conjugate of a complex number $z = x + \mathrm{i}y$.

Next we integrate (34.8) as t runs from 0 to U, and divide the integral by U:

$$\frac{1}{U}\int_0^U (|\mathcal{Z}_J(t)\cap(A+\mathbf{w})| - |J|\mathrm{vol}(A))^2\,dt$$

$$= \sum_{j\in J}\sum_{\substack{\mathbf{n}\in\mathbb{Z}^3:\\ \mathbf{n}\neq\mathbf{0}}} |a_{\mathbf{n}}|^2 + \sum_{\substack{\mathbf{n}\in\mathbb{Z}^3:\\ \mathbf{n}\neq\mathbf{0}}} |a_{\mathbf{n}}|^2$$

$$\times \sum_{j_1,j_2\in J:\ j_1\neq j_2} e^{2\pi\mathrm{i}\mathbf{n}\cdot(\mathbf{z}_{j_1}-\mathbf{z}_{j_2})}\frac{e^{2\pi\mathrm{i}U\mathbf{n}\cdot(\mathbf{v}_{j_1}-\mathbf{v}_{j_2})}-1}{2\pi\mathrm{i}U\mathbf{n}\cdot(\mathbf{v}_{j_1}-\mathbf{v}_{j_2})}$$

$$+ \sum_{\substack{\mathbf{n}_\ell\in\mathbb{Z}^3:\ \ell=1,2\\ \mathbf{0}\neq\mathbf{n}_1\neq\mathbf{n}_2\neq\mathbf{0}}} e^{-2\pi\mathrm{i}(\mathbf{n}_1-\mathbf{n}_2)\cdot\mathbf{w}}a_{\mathbf{n}_1}\overline{a_{\mathbf{n}_2}}$$

$$\times \sum_{j_1,j_2\in J} e^{2\pi\mathrm{i}(\mathbf{n}_1\cdot\mathbf{z}_{j_1}-\mathbf{n}_2\cdot\mathbf{z}_{j_2})}\frac{e^{2\pi\mathrm{i}U(\mathbf{v}_{j_1}\cdot\mathbf{n}_1-\mathbf{v}_{j_2}\cdot\mathbf{n}_2)}-1}{2\pi\mathrm{i}U(\mathbf{v}_{j_1}\cdot\mathbf{n}_1-\mathbf{v}_{j_2}\cdot\mathbf{n}_2)}. \tag{34.9}$$

Since

$$\int_{I^3} e^{-2\pi\mathrm{i}\mathbf{n}\cdot\mathbf{w}}\,d\mathbf{w} = 0 \tag{34.10}$$

for all $\mathbf{n}\in\mathbb{Z}^3$ except the trivial case $\mathbf{n}=\mathbf{0}$, by (34.9) and (34.10) we have

$$\int_{\mathbf{w}\in[0,1)^3}\frac{1}{U}\int_0^U (|\mathcal{Z}_J(\mathbf{v}_j, j\in J;t)\cap(A+\mathbf{w})| - |J|\mathrm{vol}(A))^2\,dt\,d\mathbf{w}$$

$$= \sum_{j\in J}\sum_{\substack{\mathbf{n}\in\mathbb{Z}^3:\\ \mathbf{n}\neq\mathbf{0}}} |a_{\mathbf{n}}|^2 + \sum_{\substack{\mathbf{n}\in\mathbb{Z}^3:\\ \mathbf{n}\neq\mathbf{0}}} |a_{\mathbf{n}}|^2$$

$$\times \sum_{j_1,j_2\in J:\ j_1\neq j_2} e^{2\pi\mathrm{i}\mathbf{n}\cdot(\mathbf{z}_{j_1}-\mathbf{z}_{j_2})}\frac{e^{2\pi\mathrm{i}U\mathbf{n}\cdot(\mathbf{v}_{j_1}-\mathbf{v}_{j_2})}-1}{2\pi\mathrm{i}U\mathbf{n}\cdot(\mathbf{v}_{j_1}-\mathbf{v}_{j_2})}. \tag{34.11}$$

Applying Parseval's formula for the function $f=\chi_A$, we have

$$\sum_{\substack{\mathbf{n}\in\mathbb{Z}^3:\\ \mathbf{n}\neq\mathbf{0}}} |a_{\mathbf{n}}|^2 = \sum_{\mathbf{n}\in\mathbb{Z}^3} |a_{\mathbf{n}}|^2 - |a_{\mathbf{0}}|^2 = \mathrm{vol}(A) - (\mathrm{vol}(A))^2$$

$$= p - p^2 = p(1-p) = pq. \tag{34.12}$$

Finally note that

$$\left|\frac{e^{2\pi iU\mathbf{n}\cdot(\mathbf{v}_{j_1}-\mathbf{v}_{j_2})}-1}{2\pi iU\mathbf{n}\cdot(\mathbf{v}_{j_1}-\mathbf{v}_{j_2})}\right| \le \min\left\{\frac{1}{\pi U\left|\mathbf{n}\cdot(\mathbf{v}_{j_1}-\mathbf{v}_{j_2})\right|},1\right\}. \tag{34.13}$$

Lemma 34.1 follows from (34.11), (34.12) and (34.13). □

We estimate the critical sum in Lemma 34.1 by distinguishing three cases according as

$$1 \le |\mathbf{n}| < \frac{\sqrt{N}}{U} \text{ or } \frac{\sqrt{N}}{U} \le |\mathbf{n}| \le \frac{N}{\pi U} \text{ or } |\mathbf{n}| > \frac{N}{\pi U}. \tag{34.14}$$

Formally,

$$\sum_{\substack{\mathbf{n}\in\mathbb{Z}^3:\\ \mathbf{n}\neq\mathbf{0}}} |a_{\mathbf{n}}|^2 \sum_{j_1,j_2\in J:\ j_1\neq j_2} \min\left\{\frac{1}{\pi U|\mathbf{n}\cdot\mathbf{v}_{j_1}-\mathbf{n}\cdot\mathbf{v}_{j_2}|},1\right\}$$
$$= \text{Part One} + \text{Part Two} + \text{Part Three}, \tag{34.15}$$

where

$$\text{Part One} = \sum_{\substack{\mathbf{n}\in\mathbb{Z}^3:\\ 1\le|\mathbf{n}|<\sqrt{N}/U}} |a_{\mathbf{n}}|^2$$
$$\times \sum_{j_1,j_2\in J:\ j_1\neq j_2} \min\left\{\frac{1}{\pi U|\mathbf{n}\cdot\mathbf{v}_{j_1}-\mathbf{n}\cdot\mathbf{v}_{j_2}|},1\right\}, \tag{34.16}$$

$$\text{Part Two} = \sum_{\substack{\mathbf{n}\in\mathbb{Z}^3:\\ \sqrt{N}/U\le|\mathbf{n}|\le N/(\pi U)}} |a_{\mathbf{n}}|^2$$
$$\times \sum_{j_1,j_2\in J:\ j_1\neq j_2} \min\left\{\frac{1}{\pi U|\mathbf{n}\cdot\mathbf{v}_{j_1}-\mathbf{n}\cdot\mathbf{v}_{j_2}|},1\right\}, \tag{34.17}$$

and

$$\text{Part Three} = \sum_{\substack{\mathbf{n}\in\mathbb{Z}^3:\\ |\mathbf{n}|>N/(\pi U)}} |a_{\mathbf{n}}|^2 \sum_{j_1,j_2\in J:\ j_1\neq j_2} \min\left\{\frac{1}{\pi U|\mathbf{n}\cdot\mathbf{v}_{j_1}-\mathbf{n}\cdot\mathbf{v}_{j_2}|},1\right\}. \tag{34.18}$$

We start with Part Two; we rewrite (34.17) as follows:

$$\text{Part Two} = \sum_{\substack{\mathbf{n}\in\mathbb{Z}^3:\\ \sqrt{N}/U\le|\mathbf{n}|\le N/(\pi U)}} |a_{\mathbf{n}}|^2 \cdot \sum_{j_1\in J}\sum_{\substack{m\in\mathbb{Z}:\\ m\ge 0}} \times \sum_{\substack{j_2\in J\setminus\{j_1\}:\\ m\le\pi U|\mathbf{n}\cdot\mathbf{v}_{j_1}-\mathbf{n}\cdot\mathbf{v}_{j_2}|<m+1}} \min\left\{\frac{1}{\pi U|\mathbf{n}\cdot\mathbf{v}_{j_1}-\mathbf{n}\cdot\mathbf{v}_{j_2}|},1\right\}. \tag{34.19}$$

We have the trivial implication

$$m \le \pi U|\mathbf{n}\cdot\mathbf{v}_{j_1}-\mathbf{n}\cdot\mathbf{v}_{j_2}| < m+1$$
$$\Longrightarrow \mathbf{n}\cdot\mathbf{v}_{j_1} - \frac{1}{\pi U} \le \mathbf{n}\cdot\mathbf{v}_{j_2} \le \mathbf{n}\cdot\mathbf{v}_{j_1} + \frac{1}{\pi U}. \tag{34.20}$$

We recall (12.26): for every nonzero lattice point $\mathbf{n}\in\mathbb{Z}^3\setminus\mathbf{0}$ and every integer $-N\le\ell\le N-1$,

$$\text{the number of dot products } \mathbf{n}\cdot\mathbf{v}_k \in \left[\frac{\ell|\mathbf{n}|}{N}, \frac{(\ell+1)|\mathbf{n}|}{N}\right],$$
$$1\le k\le N \text{ is less than } 6\log N. \tag{34.21}$$

The hypothesis of Part Two

$$\frac{\sqrt{N}}{U} \le |\mathbf{n}| \le \frac{N}{\pi U} \tag{34.22}$$

implies

$$\frac{1}{\pi U} > \frac{|\mathbf{n}|}{N}. \tag{34.23}$$

Using (34.20)–(34.23), we can estimate the end sum of (34.19) as follows:

$$\sum_{j_1\in J}\sum_{\substack{m\in\mathbb{Z}:\\ m\ge 0}} \sum_{\substack{j_2\in J\setminus\{j_1\}:\\ m\le\pi U|\mathbf{n}\cdot\mathbf{v}_{j_1}-\mathbf{n}\cdot\mathbf{v}_{j_2}|<m+1}} \min\left\{\frac{1}{\pi U|\mathbf{n}\cdot\mathbf{v}_{j_1}-\mathbf{n}\cdot\mathbf{v}_{j_2}|},1\right\}$$

$$\le \sum_{j_1\in J}\sum_{\substack{m\in\mathbb{Z}:\\ 0\le m\le 2U|\mathbf{n}|}} 6\log N\left(\frac{N}{\pi U|\mathbf{n}|}+2\right)\frac{1}{\max\{1,m\}}$$

$$\leq |J| 6 \log N \left(\frac{N}{\pi U |\mathbf{n}|} + 2 \right) \cdot 2 \log(2U|\mathbf{n}|)$$

$$\leq |J| 6 \log N \left(\frac{N}{\pi \sqrt{N}} + 2 \right) \cdot 2 \log N \leq 15|J|\sqrt{N}(\log N)^2, \tag{34.24}$$

since by (34.22),

$$\sqrt{N} \leq U|\mathbf{n}| < N/2.$$

Using (34.24) in (34.19), we have

$$\text{Part Two} \leq 15|J|\sqrt{N}(\log N)^2 \sum_{\substack{\mathbf{n}\in\mathbb{Z}^3: \\ \mathbf{n}\neq\mathbf{0}}} |a_{\mathbf{n}}|^2$$

$$= 15pq|J|\sqrt{N}(\log N)^2, \tag{34.25}$$

where we used Parseval's formula (see (34.12))

$$\sum_{\substack{\mathbf{n}\in\mathbb{Z}^3: \\ \mathbf{n}\neq\mathbf{0}}} |a_{\mathbf{n}}|^2 = pq. \tag{34.26}$$

Next we switch to Part Three. Then

$$\rho = \rho(\mathbf{n}) = \frac{\pi U|\mathbf{n}|}{N} > 1, \tag{34.27}$$

and we have the following analog of (34.19)–(34.25):

$$\text{Part Three} = \sum_{\substack{\mathbf{n}\in\mathbb{Z}^3: \\ |\mathbf{n}|>N/(\pi U)}} |a_{\mathbf{n}}|^2 \sum_{j_1\in J} \sum_{\substack{m\in\mathbb{Z}: \\ m\geq 0}} \cdot \sum_{\substack{j_2\in J\setminus\{j_1\}: \\ m\rho(\mathbf{n})\leq \pi U|\mathbf{n}\cdot\mathbf{v}_{j_1}-\mathbf{n}\cdot\mathbf{v}_{j_2}|<(m+1)\rho(\mathbf{n})}}$$

$$\min\left\{ \frac{1}{\pi U|\mathbf{n}\cdot\mathbf{v}_{j_1} - \mathbf{n}\cdot\mathbf{v}_{j_2}|}, 1 \right\}$$

$$\leq \sum_{\substack{\mathbf{n}\in\mathbb{Z}^3: \\ |\mathbf{n}|>N/(\pi U)}} |a_{\mathbf{n}}|^2 |J| \sum_{\substack{m\in\mathbb{Z}: \\ 0\leq m\rho(\mathbf{n})\leq 2U|\mathbf{n}|}} 3\cdot 6\log N \frac{1}{\max\{1, m\rho(\mathbf{n})\}}$$

$$\leq \sum_{\substack{\mathbf{n}\in\mathbb{Z}^3: \\ |\mathbf{n}|>N/(\pi U)}} |a_{\mathbf{n}}|^2 |J| \sum_{\substack{m\in\mathbb{Z}: \\ 0\leq m\leq 2U|\mathbf{n}|/\rho(\mathbf{n})}} 3\cdot 6\log N \frac{1}{\max\{1, m\}}$$

$$= \sum_{\substack{\mathbf{n}\in\mathbb{Z}^3:\\ |\mathbf{n}|>N/(\pi U)}} |a_{\mathbf{n}}|^2 |J| 18\log N \sum_{\substack{m\in\mathbb{Z}:\\ 0\le m\le 2N/\pi}} \frac{1}{\max\{1,m\}}$$

$$\le 18pq|J|\log N \cdot 2\log N = 36pq|J|(\log N)^2, \tag{34.28}$$

where we used (34.26), (34.27) and the equality

$$\frac{2U|\mathbf{n}|}{\rho(\mathbf{n})} = \frac{2U|\mathbf{n}|}{\pi U|\mathbf{n}|/N} = 2N/\pi.$$

Finally, we estimate Part One, i.e., we assume

$$1 \le |\mathbf{n}| < \sqrt{N}/U. \tag{34.29}$$

For every lattice point $\mathbf{n} \in \mathbb{Z}^3$ satisfying (34.29), let $r(\mathbf{n}) = r(\mathbf{n}; U; N)$ denote the smallest integer such that

$$\frac{1}{\pi U} \le \frac{2^{r(\mathbf{n})}|\mathbf{n}|}{N}. \tag{34.30}$$

By (34.29) and (34.30),

$$2^{r(\mathbf{n})} \ge \frac{N}{\pi U|\mathbf{n}|} > \frac{N}{\pi\sqrt{N}} > N^{1/3}. \tag{34.31}$$

(12.27) and (34.31) give that every interval

$$\left[\frac{\ell 2^{r(\mathbf{n})}|\mathbf{n}|}{N}, \frac{(\ell+1)2^{r(\mathbf{n})}|\mathbf{n}|}{N}\right] \tag{34.32}$$

contains less than 2^r dot products $\mathbf{n}\cdot\mathbf{v}_k$, $1 \le k \le N$ (with $r = r(\mathbf{n})$). Combining this information with (34.30) and (34.32), we obtain the upper bound

$$\text{Part One} = \sum_{\substack{\mathbf{n}\in\mathbb{Z}^3:\\ 1\le|\mathbf{n}|<\sqrt{N}/U}} |a_{\mathbf{n}}|^2 \sum_{j_1\in J} \sum_{j_2\in J:\ j_1\ne j_2} \min\left\{\frac{1}{\pi U|\mathbf{n}\cdot\mathbf{v}_{j_1} - \mathbf{n}\cdot\mathbf{v}_{j_2}|}, 1\right\}$$

$$\le |J| \sum_{\substack{\mathbf{n}\in\mathbb{Z}^3:\\ 1\le|\mathbf{n}|<\sqrt{N}/U}} |a_{\mathbf{n}}|^2 2^{r(\mathbf{n})+2} \sum_{\ell=0}^{\pi U|\mathbf{n}|} \frac{1}{\max\{1,\ell\}}$$

$$\leq |J| \sum_{\substack{\mathbf{n}\in\mathbb{Z}^3:\\ 1\leq|\mathbf{n}|<\sqrt{N}/U}} |a_{\mathbf{n}}|^2 \frac{8N}{\pi U|\mathbf{n}|} \cdot 2\log(2\pi U|\mathbf{n}|)$$

$$= \frac{16|J|N\log(2\pi U)}{\pi U} \sum_{\substack{\mathbf{n}\in\mathbb{Z}^3:\\ 1\leq|\mathbf{n}|<\sqrt{N}/U}} \frac{|a_{\mathbf{n}}|^2}{|\mathbf{n}|}$$

$$+ \frac{16|J|N}{\pi U} \sum_{\substack{\mathbf{n}\in\mathbb{Z}^3:\\ 1\leq|\mathbf{n}|<\sqrt{N}/U}} |a_{\mathbf{n}}|^2 \frac{\log N}{|\mathbf{n}|}$$

$$\leq \frac{16pq|J|N(1+\log(2\pi U))}{\pi U}, \tag{34.33}$$

where at the end we used (34.26).

Combining Lemma 34.1, (34.15), (34.25), (34.28) and (34.33), we have

$$\int_{\mathbf{w}\in[0,1)^3} \frac{1}{U}\int_0^U (|\mathcal{Z}_J(\mathbf{v}_j, j\in J; t)\cap(A+\mathbf{w})| - |J|\text{vol}(A))^2\, dt\, d\mathbf{w}$$

$$\leq pq|J| + \text{Part One} + \text{Part Two} + \text{Part Three}$$

$$\leq pq|J| + 15pq|J|\sqrt{N}(\log N)^2 + 36pq|J|(\log N)^2$$

$$+ \frac{16pq|J|N(1+\log(2\pi U))}{\pi U}$$

$$\leq \frac{6pq|J|N(1+\log(2\pi U))}{U} + 16pq|J|\sqrt{N}(\log N)^2,$$

if $U \geq 10$ and $N \geq 10^6$. Thus we obtain the following analog of Lemma 12.2.

Lemma 34.2. *Let $\mathbf{v}_1, \mathbf{v}_2, \ldots, \mathbf{v}_N \in \mathbf{S}^2$ be a sequence of $N \geq 10^6$ unit vectors on the unit sphere that satisfies the Separation Condition* (12.26)–(12.27). *Then for every subset $J \subseteq \{1, 2, 3, \ldots, N\}$, every initial configuration* (12.37), *every Lebesgue measurable subset $A \subset [0,1)^3$ with* vol$(A) = p$ *and $q = 1 - p$, and every real number $U \geq 10$, we have*

$$\int_{\mathbf{w}\in[0,1)^3} \int_0^U (|\mathcal{Z}_J(\mathbf{v}_j, j\in J; t)\cap(A+\mathbf{w})| - |J|\text{vol}(A))^2\, dt\, d\mathbf{w}$$

$$\leq 6pq|J|N(1+\log(2\pi U)) + 16pqU|J|\sqrt{N}(\log N)^2.$$

□

We derive Theorem 12.2 from Lemma 34.2 the same way — by repeated application — as we derived Theorem 12.1 from Lemma 12.2. Again we choose an arbitrary sequence $\mathbf{v}_1, \mathbf{v}_2, \ldots, \mathbf{v}_N \in \mathbf{S}^2$ of N unit vectors on the unit sphere that satisfies the Separation Condition (12.26)–(12.27); in the applications of Lemma 34.2 we use the short notation

$$\mathcal{Z}_J(t) = \mathcal{Z}_J(\mathbf{v}_j, j \in J; t).$$

We choose a sequence $A_1, \ldots, A_k$ of k Lebesgue measurable sets in the unit cube $[0,1)^3$; choose a positive ε such that

$$0 < \varepsilon < p_0 = \prod_{i=1}^{k} p_i \quad \text{with } p_i = \mathrm{vol}(A_i) \tag{34.34}$$

(and so $q_i = 1 - p_i$), choose an initial configuration

$$\mathcal{Y} = \{\mathbf{y}_1, \ldots, \mathbf{y}_N\} \subset [0,1]^3 = I^3,$$

and choose a real number $T_0 > 0$. We apply Lemma 34.2 with $A = A_1$,

$$J = J_0 = \{1, 2, 3, \ldots, N\},\ \mathcal{Z}_{J_0}(0) = \{\mathbf{y}_j + T_0\mathbf{v}_j \text{ modulo one}: \ j \in J_0\} \tag{34.35}$$

representing the initial configuration, and obtain the upper bound for every real $U_1 \geq 10$

$$\int_{\mathbf{w}\in I^3} \int_0^{U_1} (|\mathcal{Z}_{J_0}(t) \cap (A_1 + \mathbf{w})| - Np_1)^2 \, dt \, d\mathbf{w}$$
$$\leq 6p_1q_1N^2(1 + \log(2\pi U_1)) + 16p_1q_1U_1N^{3/2}(\log N)^2. \tag{34.36}$$

For every $\eta_1 > 0$ let

$$S_1(U_1;\eta_1) = \left\{(t, \mathbf{w}) \in [0, U_1] \times I^3 : \ \left|\frac{1}{N}|\mathcal{Z}_{J_0}(t) \cap (A_1 + \mathbf{w})| - p_1\right| \geq \eta_1\right\}. \tag{34.37}$$

Combining (34.36) and (34.37), we have the following upper bound for the 4-dimensional Lebesgue measure

$$\mathrm{vol}_4(S_1(U_1;\eta_1)) \leq \frac{1}{\eta_1^2}(6p_1q_1(1 + \log(2\pi U_1)) + 16p_1q_1U_1N^{-1/2}(\log N)^2). \tag{34.38}$$

Next we choose an arbitrary $(\tau_1, \mathbf{w}_1) \in [0, U_1] \times I^3 \setminus S_1(U_1;\eta_1)$, and define

$$J_1(\tau_1;\mathbf{w}_1) = \{j \in J_0 : \ \mathbf{y}_j + (T_0 + \tau_1)\mathbf{v}_j \in (A_1 + \mathbf{w}_1) \text{ modulo one}\}. \tag{34.39}$$

By definition

$$|J_1(\tau_1;\mathbf{w}_1)| = |\mathcal{Z}_{J_0}(\tau_1)\cap(A_1+\mathbf{w}_1)|. \tag{34.40}$$

By (34.37) and (34.40),

$$||J_1(\tau_1;\mathbf{w}_1)| - p_1N| \le \eta_1 N. \tag{34.41}$$

Next we apply Lemma 34.2 with the initial configuration

$\mathcal{Z}_{J_1(\tau_1;\mathbf{w}_1)}(0) = \{\mathbf{y}_j + (T_0+\tau_1)\mathbf{v}_j \text{ modulo one}:\ j\in J_1(\tau_1;\mathbf{w}_1)\},\ A=A_2,$

and obtain the upper bound for every real $U_2\ge 10$

$$\begin{aligned}\int_{\mathbf{w}\in I^3}\int_0^{U_2} &(|\mathcal{Z}_{J_1(\tau_1;\mathbf{w}_1)}(t)\cap(A_2+\mathbf{w})| - |J_1(\tau_1;\mathbf{w}_1)|p_2)^2\,dt\,d\mathbf{w}\\ &\le 6p_2q_2|J_1(\tau_1;\mathbf{w}_1)|N(1+\log(2\pi U_2))\\ &\quad + 16p_2q_2U_2|J_1(\tau_1;\mathbf{w}_1)|\sqrt{N}(\log N)^2.\end{aligned} \tag{34.42}$$

For every $\eta_2>0$ let

$$\begin{aligned}S_2(\tau_1;\mathbf{w}_1;U_2;\eta_2) = \Bigg\{(t,\mathbf{w})\in[0,U_2]\times I^3:\ &\Bigg|\frac{1}{|J_1(\tau_1;\mathbf{w}_1)|}\\ &|\mathcal{Z}_{J_1(\tau_1;\mathbf{w}_1)}(t)\cap(A_2+\mathbf{w})| - p_2\Bigg| \ge \eta_2\Bigg\}.\end{aligned} \tag{34.43}$$

Combining (34.42) and (34.43), we have

$$\begin{aligned}&\mathrm{vol}_4(S_2(\tau_1;\mathbf{w}_1;U_2;\eta_2))\\ &\le \frac{1}{\eta_2^2}\left(\frac{6p_2q_2N\,(1+\log(2\pi U_2))}{|J_1(\tau_1;\mathbf{w}_1)|} + \frac{16p_2q_2U_2\sqrt{N}(\log N)^2}{|J_1(\tau_1;\mathbf{w}_1)|}\right).\end{aligned} \tag{34.44}$$

Next we choose an arbitrary $(\tau_2,\mathbf{w}_2)\in[0,U_2]\times I^3\setminus S_1(\tau_1;\mathbf{w}_1:U_2;\eta_2)$, and define

$$\begin{aligned}J_2(\tau_1;\mathbf{w}_1;\tau_2;\mathbf{w}_2) &= \{j\in J_1(\tau_1;\mathbf{w}_1):\ \mathbf{y}_j + (T_0+\tau_1+\tau_2)\\ &\qquad\times\mathbf{v}_j\in(A_2+\mathbf{w}_2)\text{ modulo one}\}\\ &= \Bigg\{j\in J_0:\ \mathbf{y}_j + \left(T_0+\sum_{i=1}^h\tau_i\right)\mathbf{v}_j\in(A_h+\mathbf{w}_h)\\ &\qquad\qquad \text{modulo one for } h=1,2\Bigg\},\end{aligned} \tag{34.45}$$

where we used (34.39). By definition

$$|J_2(\tau_1;\mathbf{w}_1;\tau_2;\mathbf{w}_2)| = |\mathcal{Z}_{J_1(\tau_1;\mathbf{w}_1)}(\tau_2)\cap(A_2+\mathbf{w}_2)|. \tag{34.46}$$

By (34.43) and (34.46),

$$||J_2(\tau_1;\mathbf{w}_1;\tau_2;\mathbf{w}_2)| - p_2|J_1(\tau_1;\mathbf{w}_1)|| \le \eta_2|J_1(\tau_1;\mathbf{w}_1)|,$$

and combining it with (34.41), we have

$$\begin{aligned} ||J_2(\tau_1;\mathbf{w}_1;\tau_2;\mathbf{w}_2)| - p_1p_2N| &\le \eta_2(p_1N+\eta_1N)+p_2\eta_1N \\ &\le (\eta_1+\eta_2+\eta_1\eta_2)N \\ &= ((1+\eta_1)(1+\eta_2)-1)N. \end{aligned} \tag{34.47}$$

Next we apply Lemma 34.2 with $A = A_3$ and the initial configuration

$$\begin{aligned} \mathcal{Z}_{J_2(\tau_1;\mathbf{w}_1;\tau_2;\mathbf{w}_2)}(0) = \{&\mathbf{y}_j + (T_0+\tau_1+\tau_2)\mathbf{v}_j \text{ modulo one} : \\ & j \in J_2(\tau_1;\mathbf{w}_1;\tau_2;\mathbf{w}_2)\}, \end{aligned} \tag{34.48}$$

and obtain the upper bound for every real $U_3 \ge 10$

$$\begin{aligned} &\int_{\mathbf{w}\in I^3}\int_0^{U_3} (|\mathcal{Z}_{J_2(\tau_1;\mathbf{w}_1;\tau_2;\mathbf{w}_2)}(t)\cap(A_3+\mathbf{w})| - |J_2(t_1;t_2)|p_3)^2\,dt\,d\mathbf{w} \\ &\quad \le (6p_3q_3N(1+\log(2\pi U_3)) + 16p_3q_3U_3\sqrt{N}(\log N)^2)|J_2(\tau_1;\mathbf{w}_1;\tau_2;\mathbf{w}_2)|. \end{aligned} \tag{34.49}$$

For every $\eta_3 > 0$ let

$$\begin{aligned} S_3(\tau_1;\mathbf{w}_1;\tau_2;\mathbf{w}_2;U_3;\eta_3) = \Bigg\{(t,\mathbf{w}) \in [0,U_3]\times I^3 : \Bigg|&\frac{1}{|J_2(\tau_1;\mathbf{w}_1;\tau_2;\mathbf{w}_2)|} \\ &\times|\mathcal{Z}_{J_2(\tau_1;\mathbf{w}_1;\tau_2;\mathbf{w}_2)}(t)\cap(A_3+\mathbf{w})| - p_3\Bigg| \ge \eta_3\Bigg\}. \end{aligned} \tag{34.50}$$

Combining (34.49) and (34.50), we have

$$\begin{aligned} &\mathrm{vol}_4(S_3(\tau_1;\mathbf{w}_1;\tau_2;\mathbf{w}_2;U_3;\eta_3)) \\ &\quad \le \frac{1}{\eta_3^2}\,\frac{6p_3q_3N(1+\log(2\pi U_3)) + 16p_3q_3U_3\sqrt{N}(\log N)^2}{|J_2(\tau_1;\mathbf{w}_1;\tau_2;\mathbf{w}_2)|}. \end{aligned} \tag{34.51}$$

Next we choose an arbitrary $(\tau_3; \mathbf{w}_3) \in [0, U_3] \times I^3 \setminus S_2(\tau_1; \mathbf{w}_1; \tau_2; \mathbf{w}_2; U_3; \eta_3)$, and define

$$\begin{aligned} J_3(\tau_1; \mathbf{w}_1; \tau_2; \mathbf{w}_2; \tau_3; \mathbf{w}_3) &= \{j \in J_2(\tau_1; \mathbf{w}_1; \tau_2; \mathbf{w}_2) : \mathbf{y}_j + (T_0 + \tau_1 + \tau_2 + \tau_3) \\ &\qquad \times \mathbf{v}_j \in (A_3 + \mathbf{w}_3) \text{ modulo one}\} \\ &= \left\{ j \in J_0 : \mathbf{y}_j + \left(T_0 + \sum_{i=1}^{h} \tau_i \right) \mathbf{v}_j \in (A_h + \mathbf{w}_h) \right. \\ &\qquad \left. \text{modulo one for } h = 1, 2, 3 \right\}, \end{aligned} \tag{34.52}$$

where we used (34.45).

By definition

$$|J_3(\tau_1; \mathbf{w}_1; \tau_2; \mathbf{w}_2; \tau_3; \mathbf{w}_3)| = |\mathcal{Z}_{J_2(\tau_1; \mathbf{w}_1; \tau_2; \mathbf{w}_2)}(\tau_3) \cap (A_3 + \mathbf{w}_3)|. \tag{34.53}$$

By (34.50) and (34.53),

$$||J_3(\tau_1; \mathbf{w}_1; \tau_2; \mathbf{w}_2; \tau_3; \mathbf{w}_3)| - p_3 |J_2(\tau_1; \mathbf{w}_1; \tau_2; \mathbf{w}_2)|| \le \eta_3 |J_2(\tau_1; \mathbf{w}_1; \tau_2; \mathbf{w}_2)|,$$

and combining it with (34.47), we have

$$\begin{aligned} &||J_3(\tau_1; \mathbf{w}_1; \tau_2; \mathbf{w}_2; \tau_3; \mathbf{w}_3)| - p_1 p_2 p_3 N| \\ &\quad \le \eta_3 (p_1 p_2 N + ((1 + \eta_1)(1 + \eta_2) - 1)N) + p_3((1 + \eta_1)(1 + \eta_2) - 1)N \\ &\quad \le \eta_3 N + \eta_3((1 + \eta_1)(1 + \eta_2) - 1)N + ((1 + \eta_1)(1 + \eta_2) - 1)N \\ &\quad = ((1 + \eta_1)(1 + \eta_2)(1 + \eta_3) - 1)N, \end{aligned} \tag{34.54}$$

which is the analog of (34.47).

We keep repeating this argument until we cover all k given measurable test sets $A_1, \ldots, A_k$. At the end of the process we obtain a subset of $J_0 = \{1, 2, 3, \ldots, N\}$

$$\begin{aligned} J_k(\tau_1; \mathbf{w}_1; \ldots; \tau_k; \mathbf{w}_k) &= \left\{ j \in J_0 : \mathbf{y}_j + \left(T_0 + \sum_{i=1}^{h} \tau_i \right) \mathbf{v}_j \in (A_h + \mathbf{w}_h) \right. \\ &\qquad \left. \text{modulo one for } h = 1, 2, \ldots, k \right\} \end{aligned} \tag{34.55}$$

with

$$\left| |J_k(\tau_1; \mathbf{w}_1 : \ldots ; \tau_k; \mathbf{w}_k)| - N \prod_{i=1}^{k} p_i \right| \leq \left(\prod_{i=1}^{k} (1+\eta_i) - 1 \right) N \tag{34.56}$$

(a generalization of (34.54)), where $p_i = \mathrm{vol}(A_i)$ (see (34.34)),

$$(\tau_i, \mathbf{w}_i) \in [0, U_i] \times I^3 \setminus S_i(\tau_1; \mathbf{w}_1; \ldots ; \tau_{i-1}; \mathbf{w}_{i-1}; U_i; \eta_i) \text{ for } i \geq 2 \text{ and}$$
$$(\tau_1, \mathbf{w}_1) \in [0, U_1] \times I^3 \setminus S_1(U_1; \eta_1), \tag{34.57}$$

and (see (34.51) for $i = 3$)

$$\begin{aligned} &\mathrm{vol}_4(S_i(\tau_1; \mathbf{w}_1; \ldots ; \tau_{i-1}; \mathbf{w}_{i-1}; U_i; \eta_i)) \\ &\leq \frac{1}{\eta_i^2} \left(\frac{6p_iq_iN(1+\log(2\pi U_i)) + 16p_iq_iU_i\sqrt{N}(\log N)^2}{|J_{i-1}(\tau_1; \mathbf{w}_1; \ldots ; \tau_{i-1}; \mathbf{w}_{i-1})|} \right) \end{aligned} \tag{34.58}$$

for $i \geq 2$, and (see (34.38))

$$\mathrm{vol}_4(S_1(U_1; \eta_1)) \leq \frac{1}{\eta_1^2}(6p_1q_1(1+\log(2\pi U_1)) + 16p_1q_1U_1N^{-1/2}(\log N)^2). \tag{34.59}$$

Since

$$\left| |J_h(\tau_1; \mathbf{w}_1; \ldots ; \tau_h; \mathbf{w}_h)| - N \prod_{i=1}^{h} p_i \right| \leq \left(\prod_{i=1}^{h} (1+\eta_i) - 1 \right) N$$

holds for every $1 \leq h \leq k$, by (34.58) we have for $i \geq 2$

$$\begin{aligned} &\mathrm{vol}_4(S_i(\tau_1; \mathbf{w}_1; \ldots ; \tau_{i-1}; \mathbf{w}_{i-1}; U_i; \eta_i)) \\ &\leq \frac{1}{\eta_i^2} \left(\frac{6p_iq_i(1+\log(2\pi U_i)) + 16p_iq_iU_iN^{-1/2}(\log N)^2}{\prod_{\ell=1}^{i-1} p_\ell - \left(\prod_{\ell=1}^{i-1}(1+\eta_\ell) - 1 \right)} \right). \end{aligned} \tag{34.60}$$

Next we specify the parameters U_i and η_i: let

$$U_i = kT \quad \text{and} \quad \eta_i = \frac{\varepsilon}{4k}, \; 1 \leq i \leq k, \tag{34.61}$$

which implies (see (12.68))

$$\prod_{i=1}^{k} (1+\eta_i) - 1 < \varepsilon/2, \tag{34.62}$$

and so by (34.56),

$$\left| |J_k(\tau_1; \mathbf{w}_1; \ldots; \tau_k; \mathbf{w}_k)| - N \prod_{i=1}^{k} p_i \right| \leq \varepsilon N/2. \tag{34.63}$$

Next we use (34.61) and (34.62) in (34.60), and also use

$$0 < \varepsilon < p_0 = \prod_{i=1}^{k} p_i$$

(see (34.34)): thus we obtain

$$\begin{aligned} &\mathrm{vol}_4(S_i(\tau_1; \mathbf{w}_1; \ldots; \tau_{i-1}; \mathbf{w}_{i-1}; U_i; \eta_i)) \\ &\quad \leq \frac{16k^2}{\varepsilon^2} \left(\frac{6p_i q_i (1 + \log(2\pi kT)) + 16 p_i q_i k T N^{-1/2} (\log N)^2}{\varepsilon - \frac{\varepsilon}{2}} \right) \\ &\quad = \frac{32 p_i q_i k^2}{\varepsilon^3} (6(1 + \log(2\pi kT)) + 16kTN^{-1/2}(\log N)^2). \end{aligned} \tag{34.64}$$

Finally, simple calculation gives that the inequality

$$\frac{32 p_i q_i k^2}{\varepsilon^3} (6(1 + \log(2\pi kT)) + 16kTN^{-1/2}(\log N)^2) \leq \frac{\varepsilon}{k} T \tag{34.65}$$

holds if

$$T \geq C_3(\varepsilon; k) = \frac{10^4 k^5}{\varepsilon^5} \tag{34.66}$$

and

$$N \geq C_4(\varepsilon; k) = \frac{10^6 k^6}{\varepsilon^3}. \tag{34.67}$$

Thus by (34.54)–(34.67), we have

$$\mathrm{vol}_4(S_i(\tau_1; \mathbf{w}_1; \ldots; \tau_{i-1}; \mathbf{w}_{i-1}; U_i; \eta_i)) \leq \frac{\varepsilon}{k} T$$

for $1 \leq i \leq k$, which implies

$$\frac{1}{T} \sum_{i=1}^{k} \mathrm{vol}_4(S_i(\tau_1; \mathbf{w}_1; \ldots; \tau_{i-1}; \mathbf{w}_{i-1}; U_i; \eta_i)) \leq \varepsilon. \tag{34.68}$$

Combining the Separation Condition (see (12.26)–(12.27)) with (34.55), (34.57) and (34.68), Theorem 12.2 follows with the choice

$$T_0 + (h-1)T < t_h = \sum_{i=1}^{h} \tau_i < t_0 + hT \quad \text{and} \quad \mathbf{w}_h \in I^3, \ 1 \le h \le k.$$

Thus the proof of Theorem 12.2 is complete. □

References

[Be2015] J. Beck, From Khinchin's conjecture on strong uniformity to superuniform motions, *Mathematika* **61** (2015) 591–707.

[Ch-Ma2006] N. Chernov and R. Makarian, *Chaotic Billiards*, Mathematical Surveys and Monographs, Vol. 127 (Amer. Math. Soc., 2006).

[C-F-S82] I. P. Cornfeld, S. V. Fomin and Ya. G. Sinai, *Ergodic Theory* (Springer Verlag, 1982).

[Dr-Ti97] M. Drmota and R. F. Tichy, *Sequences, Discrepancies and Applications*, Lecture Notes in Math. Vol. 1651 (Springer Verlag, 1997).

[F71] W. Feller, *An Introduction to Probability Theory and its Applications*, Vols. 1–2, 2nd edition (Wiley, 1971).

[Gr-St92] G. R. Grimmett and D. R. Stirzaker, *Probability and Random Processes*, 2nd edition (Clarendon Press, 1992).

[Ke-Ma-Sm86] S. Kerckhoff, H. Masur and J. Smillie, Ergodicity of billiard flows and quadratic differentials, *Annals of Math.* **124** (1986) 293–311.

[Kh23] A. Khinchin, Eins Satz über Kettenbrüche mit arithmetischen Adwendungen, *Math. Z.* **18** (1923) 289–306.

[Kh49] A. Khinchin, *Mathematical Foundations of Statistical Mechanics* (Dover Publ., 1949).

[Kö-Sz13] D. König and A. Szücs, Mouvement d'un point abandonne a l'interier d'un cube, *Rend. Circ. Mat. Palermo* **36** (1913) 79–90.

[Ku-Ni74] L. Kuipers and H. Niederreiter, *Uniform Distribution of Sequences* (Wiley-Interscience, 1974).

[Mac51] A. M. Macbeath, An extremal property of the hypersphere, *Math. Proc. Cambridge Phil. Soc.* **47** (1951) 245–247.

[Mar70] J. M. Marstrand, On Khinchin's conjecture about strong uniform distribution, *Proc. London Math. Soc.* **21** (1970) 540–556.

[Ra36] D. A. Raikov, On some arithmetical properties of summable functions, *Mat. Sbornik* **1** (1936) 377–384.

[Re62] A. Renyi, *Wahrscheinlichkeitsrechnung*, (Probability Theory in German) (Deutscher Verlag der Wissenschaften, 1962).

[Ri45] F. Riesz, Sur la théorie ergodique, *Comment. Math. Helv.* **17** (1945) 221–239.

[Sa39] E. Sas, Über eine Extremumeigenshaft der Ellipsen, *Compositio Math.* **6** (1939) 468–470.

[Sch75] W. M. Schmidt, Irregukarities of distribution. IX, *Acta Arithmetica* **27** (1975) 385–396.

[Wal82] P. Walters, *An Introduction to Ergodic Theory*, Graduate Texts in Math. Vol. 79, (Springer, 1982).

[Wat58] G. N. Watson, *A Treatise of Bessel Functions* (Cambridge Univ. Press, 1958).

[We16] H. Weyl, Über die Gleichverteilung von Zahlen mod Eins, *Math. Ann.* **77** (1916) 313–352.

Index